Handbook of Experimental Pharmacology

Volume 151/II

Springer-Verlag Berlin Heidelberg GmbH

Purinergic and Pyrimidinergic Signalling II

Cardiovascular, Respiratory, Immune, Metabolic and Gastrointestinal Tract Function

Contributors

I. Biaggioni, R.M. Broad, G. Burnstock, J. Chapal, B.N. Cronstein, F. Di Virgilio, G.R. Dubyak, I. Feoktistov, D. Ferrari, J. Ferrier, D. Hillaire-Buys, S.M.O. Hourani, E.K. Jackson, J. Linden, M.M. Loubatières-Mariani, M.C. Montesinos, A. Pelleg, P. Petit, V. Ralevic, G. Vassort, V. Vishwanath, C.D. Wegner, M. Williams

Editors

M.P. Abbracchio and M. Williams

Springer

Professor
MARIA PIA ABBRACCHIO
Institute of Pharmacological Sciences
University of Milan
Via Balzaretti, 9
20133 Milan
ITALY
e-mail: mariapia.abbracchio@unimi.it

Dr. MICHAEL WILLIAMS
Department of Molecular Pharmacology and Biological Chemistry
Northwestern University Medical School
Searle 8-477 (S-215)
303 East Chicago Avenue
Chicago, IL 60611
e-mail: ruerivoli@aol.com

With 59 Figures and 7 Tables

ISBN 978-3-642-63223-5

Library of Congress Cataloging-in-Publication Data applied for
Purinergic and pyrimidinergic signalling /
contributors M.P. Abbracchio . . . Ed.: M.P. Abbracchio and M. Williams. –
Berlin ; Heidelberg ; New York ; Barcelona ; Hong Kong ; London ; Milan ; Paris ; Singapore ;
Tokyo : Springer (Handbook of experimental pharmacology ;vol. 151)
2. Cardiovascular, respiratory, immune, metabolic and gastrointestinal tract function. – 2001
ISBN 978-3-642-63223-5 ISBN 978-3-642-56921-0 (eBook)
DOI 10.1007/978-3-642-56921-0

Originally published by Springer-Verlag Berlin Heidelberg New York in 2001
Softcover reprint of the hardcover 1st edition 2001

Cover design: design & production GmbH, Heidelberg
Typesetting: Best-set Typesetter Ltd., Hong Kong

SPIN: 10776823 27/3020xv – 5 4 3 2 1 0 – printed on acid-free paper

Preface

The field of purinergic receptor research has gradually emerged from that of a controversial "left-field" intellectual curiosity of the 1970s to a mainstream biomedical research activity for the twenty-first century that promises – with the appropriate degree of rigor in the selection of disease targets in which purines play a key role in affecting tissue pathophysiology – to deliver novel medications within the next decade or two.

While many gifted individuals, both biologists and chemists, have contributed to the body of knowledge related to purinergic receptor function generated over the past 30 years, the genesis of the field is irrevocably associated with the vision, tenacity, passion, and proselytizing of Professor Geoffrey Burnstock. In two seminal reviews, Burnstock (1972) and Burnstock (1978), Geoff described the conceptual nucleus that provided the framework for the creation of the present body of knowledge that is contained in this volume. The spectacular advances in the molecular biology and functional pharmacology/physiology of P1 (adenosine) and P2 (ATP, ADP, UTP, UDP) receptors over the past decade are a direct result of: (a) Geoff's instance, in the face of considerable and highly vocal skepticism, that ATP and adenosine were key mediators of extracellular communication, acting via discrete cell surface receptors; and (b) the many collaborators, now numbering more than eight hundred (King and North 2000), who have worked with Geoff over the past three decades.

The present two-volume *Handbook of Experimental Pharmacology*, on *Purine and Pyrimidinergic Signalling* (HEP 151/I and HEP 151/II), presents a "state of the art" overview of the many aspects of P1 and P2 receptor function. Volume I covers *Molecular, Nervous and Urogenitary System Function* and Vol. II *Cardiovascular, Respiratory, Immune, Metabolic and Gastrointestinal System Function.*

Volume I starts with a brief historical overview of the field provided by the volume editors, Maria P. Abbracchio and Michael Williams (Chap. 1). Among other things, the events and discoveries that lead to adopting the currently used P1 and P2 receptor nomenclature are reviewed.

Molecular Aspects of Purinergic Receptors

The task of reviewing P1 (adenosine) receptors, their structure and signal transduction properties is covered by ANNA LORENZEN and ULRICH SCHWABE in Chap. 2. In Chap. 3, IAN CHESSELL, ANTON D. MICHEL and PAT HUMPHREY review the molecular properties of the ionotropic P2X receptor family and in Chap. 4, MIKE BOARDER and TANIA WEBB review the structure and function of the P2Y G-protein coupled, metabotropic receptor family.

In the first of three chapters contributed to this monograph, GEOFF BURNSTOCK discusses developmental aspects of P2 receptor function in Chap. 5.

Medicinal Chemistry

In a key chapter that provides much of the structural information for other chapters in both parts of the monograph, KEN JACOBSON and LARS KNUTSEN review progress in the characterization and optimization of P1 and P2 purine and pyrimidine receptor ligands (Chap. 6).

Purine Release

In Chap. 7, BEATA SPERLAGH and SYLVESTER VIZI discuss the various physiological and pathophysiological stimuli involved in the release of purines into the extracellular space, while the expanding knowledge of the diversity of the ectonucleoside (E-NTPase) family of enzymes involved in the metabolism of purines and pyrimidines is covered in detail by HERBERT ZIMMERMAN (Chap. 8), highlighting the crucial role of these enzymes in regulating the functional effects of nucleotides and nucleosides.

Purines in the Nervous System

SUSAN MASINO and TOM DUNWIDDIE review the role of P1 and P2 receptors and their interactions in central nervous system function (Chap. 9), while CHARLES KENNEDY extends this topic to consider the role of purinergic receptors in the peripheral nervous system (Chap. 10).

In Chapter 11, JOE NEARY and MARIA ABBRACCHIO discuss the involvement of P1 and P2 receptor mechanisms in maintaining nervous system viability via modulation of the production, release and effects of trophic factors (both neurotrophins and pleiotrophins). GARY HOUSLEY (Chap. 12) reviews data on the distribution and function of P1 and P2 receptors (the latter in the form of unique splice variants) in auditory and ocular function, while MIKE SALTER and ALF SOLLEVI discuss the role(s) of adenosine and ATP in nociception (Chap. 13), covering in detail recent advances on the role(s) of these purines in modulating pain perception at the preclinical level and in the clinical setting. In Chap. 14, GERALD CONNOLLY focuses on the emerging functional evidence for distinct pyrimidine receptors sensitive to UTP and UDP, complimenting

Boarder and Webb's chapter on the molecular biology of P2Y receptors (Chap. 3).

In the final chapter in Vol. I of this monograph, Geoff Burnstock discusses the role of purines in urogenital function (Chap. 15), building on the "tube and sac" hypothesis (Burnstock 1999). The latter focuses on increases in organ lumen volume, e.g., gut and gall bladder distension as a trigger for ATP release from endothelial sources which can then mediate peristalsis (physiological stimulus) and pain (pathophysiological stimulus).

Cardiovascular Function

In the first chapter (Chap. 16) of the second volume of this *Handbook*, Michael Broad and Joel Linden provide an update on the well documented role of adenosine in modulating cardiovascular system function, a data base that has been instrumental in helping develop the purinergic nerve hypothesis. Edwin Jackson (Chap. 17) extends this update to a detailed overview of the role of purines in renal system physiology and its contribution to cardiovascular function.

Amir Pelleg and Guy Vassort (Chap. 18) review the role of ATP and P2 receptor activation in cardiovascular system function and Vera Ralevic (Chap. 19) discusses the roles of purines and pyrimidines in endothelial function.

Susanna Hourani (Chap. 20) provides information on the role of purines in the modulation of platelet function via ADP-sensitive receptor mechanisms, a highly topical subject given (a) the advancement of novel ADP antagonists to the clinic as improved antithrombotic agents and (b) the recent cloning of the highly elusive P_{2T}/$P2Y_T$ receptor as the $P2Y_{12}$ receptor (Hollopeter et al. 2000).

Gastrointestinal Tract Function

In his final contribution to this monograph, Geoff Burnstock provides a comprehensive discourse on the role of purines in gut function (Chap. 21), highlighting potential opportunities for drug development.

Respiratory Tract Function

Italo Biaggioni and Igor Feoktistov (Chap. 22) review the involvement of adenosine mechanisms in pulmonary function with Craig Wegner (Chap. 23), complimenting this chapter with an overview of the effects of ATP in respiratory tract function.

Immune Function

Carmen Montesinos and Bruce Cronstein discuss the role of P1 receptors in inflammation and in the action of known antiinflammatory agents

(Chap. 24) and GEORGE DUBYAK (Chap. 25) discusses the dynamics of P2 receptors in immune system function. This is complimented by a chapter from FRANCESCO DI VIRGILIO, VENKATERAMAN VISHAWANATH and DAVIDE FERRARI (Chap. 26) on the emerging role of the cytolytic, pore forming $P2X_7$ receptor in immune cell apoptotic and necrotic processes.

Metabolic Function

PIERRE PETIT and his colleagues, D. HILLAIRE-BUYS, M.M. LOUBATIÈRES-MARIANI and J. CHAPAL, discuss the role of purinergic receptors in the pathophysiology of type 2 diabetes.(Chap. 27), and JACK FERRIER (Chap. 28) discusses similar roles for P1 and P2 receptors in the regulation of bone formation and function.

In the final chapter (Chap. 29), MICHAEL WILLIAMS provides an update on advances in purinergic based therapeutics in a variety of disease areas and also highlights the challenges in taking the vast body of basic research in the area of purinergic receptor function into the clinic arena.

The Editors are indebted to all of the authors for their hard work and patience in providing the excellent contributions in this monograph, most of which arrived in the time frame requested, and would also like to thank our principal Editor, Doris Walker at Springer-Verlag in Heidelberg, not only for her support and advice, but also for her gentle but constant prodding via e-mail to get this volume completed, one of the last projects she worked on before retiring. We would also like to thank Anja McKellar, who seamlessly took over the production of this two-volume set from Doris and brought it to fruition.

The editors are also deeply indebted to Angelo, Alessandro, Holly and Heather for their patience and continuous support as their spouse/parent spent additional evenings and weekends beyond the usual science-based activities that do not fit into the regular workday, editing the contributions of the authors.

Finally, they would like to thank Geoff Burnstock for his friendship, counsel and very active support in helping bring this project to completion. Without Geoff's vision, it is highly unlikely that the field would have reached a stage that the *Handbook of Experimental Pharmacology* would need to devote more than 1000 pages to cover the topic of *Purine and Pyrimidinergic Signalling*.

MARIA P. ABBRACCHIO and MICHAEL WILLIAMS
MILAN and LAKE FOREST, August 2000

References and Additional Reading

Burnstock G (1972) Purinergic nerves. Pharmacol Rev 24:509–581
Burnstock G (1978) A basis for distinguishing two types of purinergic receptor. In: Straub RW, Bolis L (eds) Cell membrane receptors for drugs and hormones: a multidisciplinary approach. Raven, New York, pp 107–118

Burnstock G (1999) Release of vasoactive substances from endothelial cells by shear stress and purinergic mechano-sensory transduction. J Anat 194:335–342
Hollopeter G, Jantzen HM, Vincent D, Li G, England L, Ramakrishnan V, Yang RB, Nurden P, Nurden A, Julius D, Couley PB (2000) Identification of the Platelet ADP Receptor Targeted by Antithrombotic Drugs. Nature, in press
Jacobson KA, Jarvis MF (1997) Purinergic approaches in experimental therapeutics. Wiley-Liss, New York, pp 55–84
King BF, North RA (2000) Purines and the autonomic nervous system: from controversy to the clinic. J Auton Nerv Syst 81:1–300

NB: Because of the various changes in P1 receptor nomenclature over the years, many early papers described A_2 receptors. Since this receptor actually exists in two subforms, A_{2A} and A_{2B}, the terminology A_2-like is used in the present volumes to describe receptors that have been designated as A_2 receptor.

In addition, to achieve the best use of space and avoid duplication, the structures for many of the compounds referred to throughout the text are collated in Chap. 6.

List of Contributors

BIAGGIONI, I., Vanderbilt University, 3500 TVAV, Nashville, TN 37212, USA
e-mail: italo.biaggioni@mcmail.vanderbilt.edu

BROAD, R.M., Box 801395, Health Sciences Center, University of Virginia, Charlottesville, VA 22908, USA
e-mail: mb6x@virginia.edu

BURNSTOCK, G., Autonomic Neuroscience Institute, Royal Free and University College Medical School, Royal Free Campus, Rowland Hill Street, London NW3 2PF, UK
e-mail: g.burnstock@ucl.ac.uk

CHAPAL, J., Laboratory of Pharmacology (Research Unit UPRES EA 1677), Faculty of Medicine, Montpellier I University, France

CRONSTEIN, B.N., New York University Medical Center, Department of Medicine/Division of Rheumatology, 550 First Ave, New York, NY 10016, USA
e-mail: cronsb01@mcrcr6.med.nyu.edu

DI VIRGILIO, F., Department of Experimental and Diagnostic Medicine, Section of General Pathology, University of Ferrara, Via Borsari, 46, I-44100 Ferrara, Italy
e-mail: FDV@dns.unife.it

DUBYAK, G.R., Department of Physiology and Biophysics, Case Western Reserve University, School of Medicine, Cleveland, Ohio 44106, USA
e-mail: gxd3@po.cwru.edu

FEOKTISTOV, I., Vanderbilt University, AA-3228 MCN, Nashville, TN 37232–2195, USA

Ferrari, D., Department of Experimental and Diagnostic Medicine, Section of General Pathology, University of Ferrara, Via Borsari, 46, I-44100 Ferrara, Italy

Ferrier, J., Medical Research Council Group in Periodontal Physiology, University of Toronto, Toronto M5S 3E2, Canada
e-mail: j.ferrier@utoronto.ca

Hillaire-Buys, D., Laboratory of Pharmacology (Research Unit UPRES EA 1677), Faculty of Medicine, Montpellier I University, France

Hourani, S.M.O., School of Biological Sciences, University of Surrey, Guildford, Surrey GU2 7XH, UK
e-mail: s.hourani@surrey.ac.uk

Jackson, E.K., Center for Clinical Pharmacology, University of Pittsburgh Medical Center, 623 Scaife Hall, 200 Lothrop Street, Pittsburgh, PA 15213-2582, USA
e-mail: edj@pitt.edu

Linden, J., Box 801395, Health Sciences Center, University of Virginia, Charlottesville, VA 22908, USA
e-mail: jlinden@virginia.edu

Loubatières-Mariani, M.M., Laboratory of Pharmacology (Research Unit UPRES EA 1677), Faculty of Medicine, Montpellier I University, France

Montesinos, M.C., New York University Medical Center, Department of Medicine/Division of Rheumatology, 550 First Ave, New York, NY 10016, USA

Pelleg, A., MCP Hahnemann University, Departments of Medicine and Pharmacology, Philadelphia, PA 19102–1192, USA
e-mail: ap33@drexel.edu

Petit, P., Laboratory of Pharmacology (Research Unit UPRES EA 1677), Faculty of Medicine, Montpellier I University, France
e-mail: pharmaco@sc.univ-montp1.fr

Ralevic, V., School of Biomedical Sciences, Queen's Medical Centre, The University of Nottingham, Nottingham NG7 2UH, UK

Vassort, G., Physiopathologie Cardiovasculaire, INSERM U-390, Montpellier, France

VISHWANATH, V., Department of Experimental and Diagnostic Medicine, Section of General Pathology, University of Ferrara, Via Borsari, 46, I-44100 Ferrara, Italy

WEGNER, C.D., Skokie Discovery Biology, Pharmacia Discovery Research, 4901 Searle Parkway, Room J-322A, Skokie, Illinois 60077, USA
e-mail: craig.d.wegner@monsanto.com

WILLIAMS, M., Department of Molecular Pharmacology and Biological Chemistry, Northwestern University Medical School, Searle 8-477 (S-215), 303 East Chicago Avenue, Chicago, IL 60611, USA
e-mail: ruerivoli@aol.com

[illegible] ne Medicine, [illegible] Dorset, [illegible]

[illegible] Research [illegible] USA [illegible]

[illegible] School, [illegible] IL 60611, U[illegible]

Contents

Section III: Cardiovascular and Renal Systems

CHAPTER 16

P1 Receptors in the Cardiovascular System

The first 15 chapters of this monograph are found in the companion volume (HEP 151/I). Its Contents are reprinted immediately after those of the present volume.

CHAPTER 17

P1 and P2 Receptors in the Renal System

CHAPTER 18

P2 Receptors in the Cardiovascular System

CHAPTER 19

Roles of Purines and Pyrimidines in Endothelium

CHAPTER 28

Purinergic and Pyrimidinergic Receptor Signaling in Bone Cells

CHAPTER 29

Clinical Opportunities in Purinergic Neuromodulation

Contents of Companion Volume 151/I

Purinergic and Pyrimidinergic Signalling I:
Molecular, Nervous and Urogenitary System Function

Section III
Cardiovascular and Renal Systems

CHAPTER 16

P1 Receptors in the Cardiovascular System

R.M. BROAD and JOEL LINDEN

A. Adenosine Receptor Subtypes in the Cardiovascular System

The presence of adenosine receptors on cardiovascular tissues has previously been established on the basis of functional responses to adenosine and adenosine analogs. Specific functions attributed to each of the four adenosine receptor subtypes (A_1, A_{2A}, A_{2B}, and A_3) are discussed in detail below following a summary of the direct evidence for adenosine receptors in cardiovascular tissues based on the detection of receptor transcripts or protein.

I. Transcripts in Heart and Vascular Tissues

Using polymerase chain reaction (PCR), transcripts for all four receptor subtypes have been detected in heart homogenates, with A_1 and A_{2A} transcripts being more prevalent than A_{2B} and A_3 (DIXON et al. 1996; cf. MENG et al. 1994; PETERFREUND et al. 1996; LINDEN and JACOBSON 1998). When adenosine receptors were first cloned in the early 1990s, the tissue distribution was surveyed initially by Northern blotting. A_1 transcript was found in heart, but at substantially lower levels than in brain, spinal cord, testis, thyroid, and adipose tissue (LIBERT et al. 1991; REPPERT et al. 1991). Rat myocardial A_1 mRNA levels and receptor protein increase during development up to birth and then decline with age (MATHERNE et al. 1996). A_{2A} transcript was easily detected in Northern blots of heart (STEHLE et al. 1992), although it is not known how much of this signal is due to non-myocardial cells. Transcript for A_{2B} and A_3 receptors cannot be detected in the adult rat heart by Northern blotting (MEYERHOF et al. 1991; STEHLE et al. 1992). A_3 transcript has been identified in Northern blots of rat aorta (ZHAO et al. 1997) and low levels were detected in rat heart by PCR (ZHOU et al. 1992).

II. Radioligand Binding Studies

Low levels of A_1 receptors have also been detected by radioligand binding to heart membranes. Specific binding to GPCRs was initially detected on rat ventricle membranes with the agonist, [^{125}I]-ABA, at a density 40 times lower than in rat brain (Linden et al. 1985). A somewhat higher density of [^{125}I]-ABA binding sites was found in atria (30 fmol/mg protein) than in ventricles (23 fmol/mg) (Linden et al. 1990). [^{125}I]-ABA also binds to A_1 receptors on atrial myocardium from guinea pig, rat and rabbit (Froldi and Belardinelli 1990). Studies using the A_1-selective antagonist, [^{3}H]CPX, in pig, rat, rabbit, and guinea pig confirmed that there is a somewhat higher density of A_1 receptors in atria, 76 fmol/mg, than ventricles, 34 fmol/mg (Musser et al. 1993). Autoradiography using [^{3}H]CPX demonstrated that the conductive cells of the atrioventricular node are enriched in A_1 receptor binding sites relative to ventricular myocardial cells in guinea pig hearts (Parkinson and Clanachan 1991). Long-term treatment of rats with the A_1-selective agonist, *R*-PIA, leads to a decrease in [^{125}I]-ABA binding in rat heart, indicative of receptor down-regulation (Lee et al. 1993a). Conversely, feeding theophylline to guinea pigs (Wu et al. 1989) or rats (Lee et al. 1993b) significantly increases the number of A_1 receptor binding sites within two days. These data suggest that endogenous levels of adenosine in the heart are sufficient to maintain the receptor in a partially down-regulated state.

Radioligand binding studies have demonstrated the presence of A_{2A} receptors on coronary arteries. A very low density (<10 fmol/mg protein) of high affinity binding sites for the agonist radioligand, [^{125}I]-APE, was detected on porcine coronary artery membranes, and a much high density (ca. 900 fmol/mg protein) of binding sites for the antagonist, [^{3}H]SCH 58261, (Belardinelli et al. 1996). Since agonists bind with high affinity only to G protein coupled receptors, this discrepancy may be due in part to the fact that A_{2A} receptors in membrane homogenates appear to be poorly coupled to GTP binding proteins (Luthin et al. 1995).

B. Control of Adenosine Levels in the Heart

In this section the regulation, during normoxia and hypoxia, of the key enzymes involved in adenosine metabolism is discussed. Extracellular levels of adenosine can be increased by the dephosphorylation of extracellular AMP through the action of ecto 5′-nucleotidase or by export of intracellular adenosine via nucleoside transporters (Zimmerman, Chap. 9, first volume). Intracellular adenosine is also formed from the hydrolysis of *S*-adenosylhomocysteine to adenosine and homocysteine or by AMP dephosphorylation through cytosolic 5′-nucleotidase. Due to the activity of nucleoside transporters, raising intracellular adenosine will lead to an increased efflux of the purine. Since the nucleoside transporters that are most abundant in the heart are bidirectional, they also participate in the uptake of adenosine. Once taken up into the cells, adenosine can be rephosphorylated to AMP by adenosine kinase. When high concentrations of adenosine accumulate, adenosine kinase becomes saturated and most adenosine is converted to inosine by adenosine deaminase, which has a higher capacity but lower affinity than adenosine kinase. Adenosine deaminase is primarily an intracellular soluble enzyme, but also exists as a membrane-associated ecto-enzyme.

I. Normoxia

Under normal circumstances, most of the adenosine formed intracellularly is rephosphorylated to AMP by adenosine kinase. Adenosine formation is 30-fold greater than adenosine release (DECKING and SCHRADER 1998). Infusion of the adenosine kinase inhibitors, 5′-iodotubercidin or 5′-amino-5′-deoxyadenosine, into the coronary vasculature of a normally oxygenated guinea pig heart causes a dose-dependent increase in adenosine release (KROLL et al. 1993; LLOYD and SCHRADER 1993). In contrast, blockade of SAH hydrolase with adenosine dialdehyde does not substantially alter adenosine release, suggesting that the formation of adenosine from SAH hydrolysis does not contribute significantly to the basal production of adenosine (KROLL et al. 1993). Adenine nucleotide metabolism also appears to contribute to the formation of adenosine in the extracellular space since inhibiting ecto 5′-nucleotidase activity with AOPCP in isolated guinea pig hearts decreases venous adenosine levels (HEADRICK et al. 1992). Adenosine deaminase also contributes to the metabolism of adenosine since inhibition of adenosine deaminase activity with EHNA increases interstitial adenosine (DECKING et al. 1997)

Although interstitial adenosine usually is maintained at low concentrations during normoxia, the concentrations may be high enough to exert a tonic influence on adenosine receptors. Thus, the A_1-selective antagonist, CPX, alone gives rise to a significant increase in the heart rate of embryonic mouse hearts (HOFMAN et al. 1997), while the A_{2A}-selective antagonist, SCH 58261, alone increases systolic and diastolic blood pressure in freely moving rats (MONOPOLI et al. 1998). Central A_1 and A_{2A} adenosine receptors and inhibitory A_1 receptors on sympathetic nerve terminals (SMITS et al. 1991; KAROON et al. 1995) may contribute to these responses.

II. Hypoxia

The release of adenosine, inosine, and hypoxanthine is dramatically elevated under conditions of hypoxia (MIURA et al. 1996), beginning as soon as 30 s after the initiation of ischemia (FELIX et al. 1997). Using microdialysis, interstitial adenosine levels during 5 min of ischemia were increased threefold to between 2 μmol and 7 μmol/l, depending on the species examined, and returned to baseline levels after 5 min of reperfusion (HARRISON et al. 1998). Perfusion of guinea pig hearts with a 30% O_2 solution caused a tenfold increase in adenosine release above basal levels (SCHÜTZ et al. 1981). DEUSSEN et al. (1988) measured the formation of SAH in the presence of excess *l*-homocysteine as an index of interstitial adenosine concentration and showed that adenosine levels increase concomitantly with decreased coronary blood flow, being accompanied by a decrease in myocardial ATP and an increase in ADP levels. Similarly in guinea pig hearts, increased myocardial O_2 consumption (BARDENHEUER and SCHRADER 1986) or decreased oxygenation of the perfusion buffer (CLEMO and BELARDINELLI 1986) can be correlated with increased adenosine release. DECKING et al. (1997) showed that adenosine release from guinea pig hearts is enhanced with decreased oxygenation, and provided evidence that the amount of adenosine formed exceeds the amount released, indicating the presence of a strong salvaging process even during hypoxia.

Perfusion of hearts with [^{3}H]adenosine leads to selective labeling of endothelial adenosine stores, suggesting that the endothelium is a barrier to adenosine movement across the vascular wall (NEES et al. 1985). Therefore, changes in the specific activity of released [^{3}H]adenosine can be used as an indicator of the relative contribution of endothelium to overall adenosine release. Relative to basal conditions, hypoxia leads to a dramatic increase in adenosine release, yet causes a significant decrease in the specific activity of purine release from hearts in which endothelial cells were selectively preloaded with [^{3}H]adenosine (DEUSSEN et al. 1986). Hence, coronary endothelial cells are not the sole source of the adenosine released during hypoxia. Since isolated ventricular myocytes have been shown to release adenosine and inosine during hypoxia (KITAKAZE et al. 1996) and zero-glucose hypoxia (SMOLENSKI et al. 1992), cardiomyocytes are also a likely source of the purine. Although the hypoxia-mediated release of adenosine has been firmly established, the exact mechanisms involved in this release remains unclear. Conflicting data implicate stimulation of cytosolic 5′-nucleotidase (DARVISH and METTING 1993; PRZYKLENK et al. 1997), stimulation of ecto 5′-nucleotidase, and inhibition of adenosine kinase (KROLL et al. 1993; DECKING et al. 1997).

C. Functional Cardiovascular Responses to Adenosine

Adenosine is known to evoke a variety of responses in the heart and vascular tissues. In this section the adenosine receptor subtypes that mediate the individual cardiac and vascular responses are discussed.

I. Negative Dromotropic

Perfusion of guinea pig hearts with an N_2-saturated solution causes release of adenosine that is correlated positively with prolongation of the atria-to-His (A–H) interval, a measurement of conduction velocity or dromotropy. This effect is blocked by the adenosine receptor antagonist, 8-SPT, and by the addition of exogenous ADA, providing evidence that such changes in conduction are mediated by adenosine receptors. The EC_{50} value for adenosine-mediated negative dromotropy is 260 nmol/l (CLEMO and BELARDINELLI 1986). A reversible, negative dromotropic effect is observed with the A_1-selective agonist, CVT-510 (SNOWDY et al. 1999) and an irreversible effect is observed with an irreversible A_1 agonist, *m*-DITC-ADAC (ZHANG et al. 1997). Adenosine-mediated negative dromotropy is blocked by the A_1-selective antagonist, CPX (MEESTER et al. 1998), but not by the A_{2A}-selective antagonist SCH 58261 (BELARDINELLI et al. 1998). In isolated rabbit AV nodal cells, adenosine has been shown to cause a decrease in action potential amplitude and duration and to increase the refractory period in nodal cells, resulting in an increase in rate-dependent activation failure and an increased threshold for action potential activation, likely via activation of IK, Ado, the inwardly-rectified current mediated by the adenosine-activated K^+ channel (WANG et al. 1996a) This may explain how adenosine maintains rhythmicity in the ventricles during periods of atrial arrhythmia.

II. Negative Chronotropic

The ability of adenosine to reduce heart rate was among its first recognized effects (DRURY and SZENT-GYORGYI 1929). These findings have been confirmed by many investigators and the predominant receptor subtype responsible has been identified as A_1, with a characteristic potency order of CPA > NECA > 2-chloroadenosine (WEBB et al. 1990). CPA, but not CGS 21680 or A_3-selective concentrations of APNEA, were shown to reduce heart rate in anesthetized rats (HANNON et al. 1995). A_3 receptors appear to contribute to a predominant A_1 receptor mediated chronotropic response to adenosine in fetal mice (HOFMAN et al. 1997). The bradycardiac actions of CHA given ICV to rats are much more pronounced than those of APNEA (STELLA et al. 1998). CPX alone was shown to cause a 20% increase in heart rate in fetal mice (HOFMAN et al. 1997), suggesting that endogenous adenosine is sufficiently high to activate A_1 receptors partially. Dose-response curves for adenosine and NECA to cause bradycardia in guinea pig hearts are right-shifted by CPX, resulting in a linear Schild plot with a slope near unity, suggesting that A_1 is the only adenosine receptor subtype that mediates the chronotropic response in this species (MEESTER et al. 1998). Further evidence for the role of A_1 receptors in the bradycardic actions of adenosine is provided by an observed significant decrease in the basal heart rate of mice with transgenically overexpressed A_1 receptors relative to wild-type mice (GAUTHIER et al. 1998). The negative chronotropic response to CPA is blocked by pertussis toxin, indicating that bradycardic responses mediated via the A_1 receptor is coupled through Gi/Go proteins. Based on the effect of various ion channel modulators on the A_1 response, it was proposed that A_1 receptor-mediated bradycardia is the result of a combination of effects on I_{Ca}, I_{Cl}, I_f (the pacemaker current) and I_{Na} currents (HOFMAN et al. 1997). In the SA node, A_1 receptor-mediated increases in IK,Ado are thought to be responsible for hyperpolarization of membrane potential, which decreases pacemaker firing, leading to negative chronotropy. In addition, adenosine reverses isoproterenol-stimulated increases in I_{Ca} and I_f (BELARDINELLI et al. 1988).

The effects of adenosine and ACh on heart rate have been investigated in mice deficient in GIRK4, one of the subunits of IK_{ACh} required for its function (WICKMAN et al. 1998). IK_{ACh} mediates about half of the chronotropic effects of adenosine and vagal stimulation. Some of the residual direct effects of adenosine and ACh may be mediated by direct inhibition of adenylate cyclase activity in atria (LINDEN et al. 1990).

III. Inotropic

Adenosine receptor stimulation has been shown to have only a modest direct effect (in the absence of catecholamines) on contractility in mammalian ventricles. A substantial direct negative inotropic action is produced by A_1 adenosine receptor activation in atria. Adenosine exerts a strong anti-adrenergic effect on catecholamine-induced positive inotropy in both the atria and ventricles (SHRYOCK and BELARDINELLI 1997). Accordingly, the following examination of adenosinergic effects on myocardial contractility has been divided into direct and indirect components.

1. Direct Atrial

In the canine left atrium, a negative inotropic response to NECA is antagonized by 8-PT, CPX, and the A_1-selective antagonist N-0861, indicating an action at the A_1 subtype (MARTIN 1992). Although the agonist potency order for negative inotropy in the right and left atria was reported to be atypical of an A_1 response, it is blocked by CPX but not DMPX, suggesting an A_1-mediated effect (OGUCHI et al. 1995). Similarly, the inotropic actions of NECA in guinea pig are attenuated by the A_1-selective antagonist, WRC-0571 (MARTIN et al. 1996). A CPA-mediated negative inotropic response in rats is significantly enhanced by the A_1-specific allosteric enhancer, PD 81,723 (MUDUMBI et al. 1993). A series of alkylxanthines were shown to elicit a positive inotropic effect on the guinea pig left atrium in the absence of exogenous agonist, providing evidence that basal levels of adenosine are sufficient to exert an inhibitory tone on myocardial contractility (SANAE et al. 1995).

Adenosine was shown to reduce action potential duration in myocytes isolated from atria but not ventricles (WANG et al. 1996a). This contributes to an attenuation of twitch amplitude with an EC_{50} value for exogenous adenosine in the region of 600 nmol/l and a maximum response at approximately 10 μmol/l (WANG and BELARDINELLI 1994). The ability to shorten atrial, but not ventricular, action potential duration (APD) has also been investigated using the A_1-selective agonist, CVT-510 (SNOWDY et al. 1999). In the conscious pig model, A_1 receptor activation causes a decrease in left atrial pressure and a reduced rate of left ventricular contraction, +dP/dt (HUANG et al. 1999).

Adenosine activates an outward K^+ current in a dose-dependent manner that is blocked by theophylline (KURACHI et al. 1986). The ion current modulated by the purine has been called IK_{Ado} (WANG and BELARDINELLI 1994) which probably is identical to IK_{ACh}. 4-Aminopyridine (4-AP), a blocker of K^+ channels, causes a dose-dependent attenuation of the effects of NECA on rabbit atrial contractility (MCKINLEY et al. 1990).

2. Indirect Ventricular

Adenosine and ACh produce an "indirect" effect to counteract the positive inotropic effects of catecholamines, which is more pronounced than their "direct" effects produced in the absence of added catecholamines or adrenergic tone. This is particularly true in ventricular myocardium. Thus, stimulation of β-adrenergic receptors on rat ventricular myocytes leads to an increase in contractile strength and a decrease in the duration of contraction that is attenuated by R-PIA. This response is blocked by co-administration of CPX, providing evidence for actions at an A_1 receptor (BELLARDINELLI et al. 1989; DOBSON and FENTON 1998). In isolated rabbit cardiac myocytes, CCPA inhibits isoproterenol-mediated contractions and this effect is reversed by CPX (WANG et al. 1997). Furthermore, the overexpression of A_1 receptors in mouse heart causes an attenuation of isopreterenol-stimulated increases in +dP/dt. Although without effect on their own, adenosine and R-PIA were shown to reduce the $[Ca^{2+}]i$ transients induced by isopreterenol in ventricular myocytes (TANG et al. 1998; NORTON et al. 1999). Adenosine counteracts the effects of β-adrenergic receptor activation in the heart, but it also acts to inhibit norepinephrine release from sympathetic nerve terminals (MUNCH et al. 1996).

The anti-adrenergic effects of adenosine are likely mediated in large part by the inhibition of norepinephrine-stimulated adenylate cyclase activity. In the heart, protein kinase A (PKA) phosphorylates Ca^{2+} channels and increases their conductance (TRAUTWEIN and HESCHELER 1990). The anti-adrenergic effects of adenosine on $[Ca^{2+}]i$ transients are reversed by CPX, but are also enhanced by the A_{2A} antagonist, CSC, suggesting that A_{2A} receptor activation partially counteracts the effects of A_1 receptor activation. In addition, the A_{2A} agonist, DPMA, was shown to augment $[Ca^{2+}]i$ current, an effect that was blocked by the A_{2A} antagonist, DMPX (TANG et al. 1998). CSC and ZM 241285 enhanced, while CGS 21680 attenuated, the anti-adrenergic effects of adenosine on +dP/dt (NORTON et al. 1999). CGS 21680 alone was shown to enhance the strength of contraction in rat ventricular myocytes and to prolong the duration of contraction (DOBSON and FENTON 1998). This effect on action potential duration by CGS 21680 is opposite to the shortening effect of isopreterenol, indicating that the effects of A_{2A} receptor activation are not simply due to stimulation of norepinephrine release. Since isoproterenol increases contractility by increasing intracellular cAMP through activation of Gs, these data may also imply that the inotropic actions mediated by A_{2A} receptors are caused in part by some action that is independent of changes in cAMP.

IV. Systemic Blood Pressure

Adenosine modulates systemic blood pressure through a variety of mechanisms, including direct effects on peripheral blood vessels, indirect effects via modulation of the specific centers in the CNS that control blood pressure, and by altering the peripheral release of other vasoactive mediators such as renin. It was initially thought that the A_2 subtypes were responsible for these effects, although further investigation has provided evidence for actions mediated by all four adenosine receptor subtypes.

Activation of A_1, A_{2A}, and A_3 receptors was found to reduce systemic blood pressure (HANNON et al. 1995), while the A_{2B} receptor is primarily responsible for relaxing the aorta (MARTIN et al. 1993). The non-selective agonist CV-1808 decreases peripheral vascular resistance in hind limb, kidney, and mesenteric blood vessels of conscious rats, leading to increased blood flow in these regions and a reduction in mean arterial blood pressure. The effect on MABP is reduced by administration of 8-SPT (WEBB et al. 1990). Adenosine evokes the formation and release of the vasodilator, nitric oxide (NO), from vascular tissues (LI et al. 1995). In endothelial cells, NO synthase is activated by calcium. It is notable in this regard that A_{2B} adenosine receptors were recently shown to be dually coupled to both Gs and Gq (LINDEN et al. 1999). Coupling via Gq may be responsible for activation of NO synthase. In free moving rats, an A_{2A}-selective agonist, 2-HE-NECA, elicits a hypotensive response that is blocked by the A_{2A} antagonist SCH 58261. The antagonist alone slightly elevates blood pressure and catecholamine release, suggesting an inhibitory input by A_{2A} receptors on sympathetic tone (MONOPOLI et al. 1998). Intravenous (IV) injection of CGS 21680 leads to a dose-dependent reduction in mean arterial pressure with an EC_{50} value of 11 nmol/l (MATHOT et al. 1995). Microinjection of adenosine into the nucleus tractus solitarii (NTS) and area postrema reduces systolic and diastolic blood pressure. Reversal of this effect by 8-SPT confirms that it is adenosine receptor-mediated (TSENG et al. 1988). Intracerebroventricular (ICV) administration of the A_1-selective agonist, CPA, causes a decrease in arterial blood pressure in anesthetized rats. The effects of CPA are blocked by co-administration of CPX. Employing A_1- and A_2-selective agonists and comparing the effects of IV vs ICV injections of these compounds indicates that central A_2 receptors are responsible for control of systemic blood pressure (STELLA et al. 1993). Adenosine receptor antagonists that easily cross the blood-brain barrier are relatively more effective at attenuating the hypotensive actions of NECA than is XAC, an antagonist that does not have ready access to the CNS, supporting a central locus for some of the hypotensive effects of adenosine (FREDHOLM et al. 1987). Caffeine and the polar antagonist 8-SPT that does not cross the blood brain barrier were comparable in their ability to antagonize 2-CADO-mediated hypotension (EVONIUK et al. 1987), implying that adenosine also acts to lower blood pressure at peripheral adenosine receptors.

Blood pressure can also be controlled by increases in plasma renin, which acts to increase angiotensin II (Jackson, Chap. 17, this volume). Caffeine and DPSPX both attenuated adenosine-mediated hypotension and increased baseline levels of plasma renin to a similar extent, indicating not only that physiological concentrations of adenosine are sufficient to exert tonic inhibitory control on renin release, but that this control occurs at peripheral adenosine receptors (Tofovic et al. 1991).

V. Coronary Vasodilation

The coronary vasculature is very sensitive to adenosine. The EC_{50} value for adenosine-mediated coronary vasodilation was calculated to be 3 μmol, with a potency order of NECA > R-PIA > CHA, indicating an action at A_2 receptors (Kusachi et al. 1983). The net EC_{50} value for coronary vasodilation was subsequently divided into a high (10 nmol/l) and low affinity components (10 μmol/l), with only the high affinity effects being blocked by 8-PT (Harden et al. 1996). This nanomolar affinity binding site provides for the possibility that basal levels of adenosine, released even under normoxic conditions, can influence coronary vascular tone. The potency order for coronary vasodilation has been repeated using receptor subtype-selective agonists: CGS 21680 > NECA > adenosine > CCPA. Coronary vascular responses to these compounds are blocked by the A_{2A}-selective antagonists, ZM 241385 and SCH 58261 (Shryock et al. 1998). A rightward shift in the dose-response curve of CGS 21680-mediated vasodilation by SCH 58261 results in a linear Schild plot with a slope of 1, implicating the A_{2A} receptor as the only subtype contributing to dilation in the coronaries. Increased coronary flow evoked by hypercapnic or metabolic acidosis is eliminated by SCH 58261 but not CPX (Phillis et al. 1998), supporting a role for adenosine acting at A_{2A} receptors under pathological conditions.

The early phase of hypoxia-induced coronary hyperemia is associated with NO release and is blocked by NO-synthase inhibition whereas only the late phase is associated with adenosine release and is blocked by BW-A1433. This would imply a temporal dissociation of the NO and adenosine pathways in contributing to hyperemia (Park et al. 1992).

Since A_{2A}-selective antagonists attenuate adenosine-mediated coronary relaxation, it follows that the A_{2A} receptor is primarily responsible for such effects. Nevertheless, CPA has been noted to produce a dose-dependent relaxation of intact canine coronary arteries. The effects of CPA are unaltered by ischemia and reperfusion, whereas the slope of the dose-response curve to DPMA is significantly altered (Cox et al. 1994), suggesting that dilatations mediated by CPA and DPMA occur through distinct mechanisms. Similarly, relaxation of porcine coronary arteries can be elicited by both CGS 21680 and CPA, but only the actions of the former are sensitive to blockade by NO-synthase inhibitors. Only CGS 21680 elicits NO production in this tissue (Abebe et al. 1995).

VI. Mast Cell Degranulation

Adenosine evokes the release of various mediators from tissue-resident mast cells, which can alter various cardiovascular responses. Intravenous infusion of CGS 21680 and APNEA caused hypotension in pithed rats with the actions of APNEA being attenuated in a dose-dependent manner by the mast cell stabilizing agents sodium cromoglycate or lodoxamide or by prior mast cell depletion with compound 48/80. These data suggest that the hypotensive actions of A_3 receptors occur indirectly via release of mast cell mediators such as histamine which produces an endothelial dependent dilation of large blood vessels (HANNON et al. 1995). In the microcirculation, histamine is a vasoconstrictor. Thus, in the hamster cheek pouch, I-ABA induces an arteriolar vasoconstriction that is attenuated only by high concentrations of BW-A1433, implicating A_3 adenosine receptors. This effect is concomitant with mast cell degranulation as measured by ruthenium red uptake (SHEPHERD et al. 1996). The activation of A_3 adenosine receptors on rodent perivascular mast cells provokes vasoconstriction that is mediated by histamine and thromboxane (SHEPHERD and DULING 1996). In rodents, inosine at concentrations above 1 μmol/l triggers mast cell degranulation via A_3 receptors (JIN et al. 1997). Using receptor subtype-selective agonists, the potency order for enhancement of antigen-stimulated serotonin release from rat mast cells was found to be IB-MECA > CPA > CGS 21680, also implicating the A_3 receptor. Intradermal injection of adenosine agonists into rodents causes plasma protein extravasation. This response is blocked by the antihistamine, cyproheptidine, or by mast cell depletion using compound 48/80 (REEVES et al. 1997). In contrast to rodent mast cells, β-hexosaminidase release from canine BR mast cells occurs with the potency order NECA > R-PIA >> CGS 21680 = IB-MECA indicating that, under these circumstances, degranulation of mast cells is mediated primarily by actions at A_{2B} receptors. In this study, NECA, but not IB-MECA, also increased intracellular levels of cAMP, IP_3, and $[Ca^{2+}]i$ (AUCHAMPACH et al. 1997).

In contrast to the effects of A_3 or A_{2B} receptors in stimulating degranulation of various mast cells, A_{2A} receptors have been shown in some instances to inhibit stimulated mediator release (JIN et al. 1997). Thus, adenosine inhibits antigen-induced tryptase release from human mast cells with an EC_{50} value of approximately 1 μmol/l. The order of potency for inhibition is CGS 21680 > adenosine > CPA and the effects of adenosine are blocked by the A_{2A} antagonist, ZM 241385.

VII. Cardiovascular Apoptosis

Stimulation of adenosine receptors, especially the A_{2A} and A_3 subtypes, can be both protective and detrimental to cell survival depending on the concentration of agonist used and the receptor subtype involved. In several tissues, A_3 agonists are protective in the nanomolar range yet promote apoptotic cell death at concentrations of 10 μmol/l or higher (JACOBSON et al. 1998). Thus, a 12h exposure of rat microglial cells to micromolar levels of the A_1 agonist, 2-CADO caused DNA fragmentation, a measure of apoptosis (OGATA and SCHUBERT 1996), whereas a 16-h exposure of cultured human neutrophils to adenosine agonists was protective in the nanomolar range with a potency order of CGS 21680 > NECA > CPA > IB-MECA (WALKER et al. 1997). In the latter study, CGS 21680 was shown to be protective through its ability to double the half-life for the appearance of apoptosis. Micromolar concentrations of IB-MECA and Cl-IB-MECA have been shown to cause a reduction in the number of human astrocytoma cells while nanomolar levels led to a reduction in spontaneous apoptosis and modifications in the intracellular distribution of an anti-apoptotic protein (ABBRACCHIO et al. 1998). The pro-apoptotic effects of adenosine and 2-CADO are attenuated by nucleoside transport inhibitors, suggesting that, in addition to actions at extracellular receptors, a component of purine-mediated cell apoptosis is due to interactions with an intracellular site (DAWICKI et al. 1997; BARBIERI et al. 1998). It was found that adenosine and IB-MECA caused chromatin condensation and DNA fragmentation in rat cardiomyocytes. Dose-response analysis showed an inverse relationship between the concentration of IB-MECA and nuclear size or cell survival. Since these effects are not blocked by CPX or mimicked by R-PIA, it would appear that apoptosis is mediated through A_3 receptors in cardiomyocytes (SHNEYVAYS et al. 1998).

Anoxia or ischemia and subsequent reperfusion have been shown to cause DNA fragmentation and the production of Fas, an apoptosis-specific protein, in rat and rabbit cardiomyocytes, respectively (TANAKA et al. 1994; GOTTLIEB et al. 1994). Since adenosine and inosine concentrations can reach micromolar levels in hypoxic cardiac tissue, it is possible that apoptosis may be mediated by adenosine or inosine under these conditions, although the connection between these events has not yet been established.

The proteolytic formation of a proapoptotic protein, caspase-3, and receptor-induced apoptosis were recently shown to be inhibited by increased intracellular levels of cAMP (NIWA et al. 1999). Since A_2-like receptors are generally thought to act through an enhancement of adenylate cyclase activity, it is possible that A_2 receptor-mediated protective effects ultimately act via the caspase pathway.

VIII. Angiogenesis

The process of new blood vessel growth, or angiogenesis, is sometimes modeled experimentally as the proliferation of cultured endothelial cells. Proliferation of bovine microvascular endothelial cells can be accelerated by adenosine, with these effects becoming apparent within the first 24h of exposure. Hypoxia is a powerful stimulus for neovascularization. The proliferative effect of hypoxia on endothelial cells is blocked by 8-PT, implicating actions at adenosine receptors. Endothelial cells also display a positive chemotaxis response along an adenosine concentration gradient, providing further evidence for a role by adenosine in angiogenesis (MEININGER et al. 1988). In human endothelial cells (HUVECs), adenosine also causes proliferation, although this effect is not seen for at least 72h (ETHIER et al. 1993). The dose-response curve for this effect is shallow, but occurs with an EC_{50} value in the low-, or even sub-nanomolar range. Curiously, in light of apoptosis studies, adenosine continues to stimulate cell proliferation well into the high micromolar range.

Another indicator of cell proliferation is an increase in transcription and translation of VEGF, the vascular endothelial growth factor. Hypoxia has been shown to evoke VEGF release from bovine retinal capillary endothelial cells, an effect that is mimicked by CGS 21680, DPMA, and NECA, but not by CPA. Furthermore, adenosine deaminase and CSC block hypoxia-induced VEGF, mRNA, or protein production, whereas CPX does not, supporting a role for the A_{2A} receptor as a mediator of this process (TAKAGI et al. 1996a). Transcript and protein expression of KDR, a VEGF receptor, is diminished during hypoxia, and the specific binding of VEGF to KDR is reduced. Since this effect is mimicked by DPMA and blocked by CSC, it is also thought to be mediated via A_{2A} receptors (TAKAGI et al. 1996b).

IX. Cardioprotection

NEWBY (1984) referred to adenosine as a "retaliatory metabolite" since it acts to maintain a metabolic balance in the ischemic heart by increasing oxygen supply and reducing oxygen demand. Myocardial ischemia causes deleterious effects including the induction of arrhythmias, depletion of ATP and other high energy phosphates, a decrease in myocardial contractility, and irreversible necrosis of cardiac muscle. Adenosine released during hypoxia or ischemia protects the heart from injury. For example, in perfused rat hearts, prolonged ischemia leads to depleted tissue [ATP], induction of contracture, and depression of left ventricular developed pressure (LVDP) during the reperfusion phase. Adenosine treatment reduces the deleterious effects of ischemia on LVDP, extends the time to onset of ischemic contracture, and maintains tissue levels of ATP (ELY et al. 1985). The following sections will provide an overview of the proposed mechanisms for adenosine-mediated protection of the heart from ischemic damage, followed by a more detailed examination of the individual processes that contribute to protection.

The myocardial damage and functional deficits brought about by an extended period of ischemia in the heart can be reduced by one or more shorter, non-deleterious periods of ischemia, a phenomenon called "ischemic preconditioning" (MURRY et al. 1986). The protective actions of preconditioning (PC) on relative infarct size in rabbit hearts could be blocked by the adenosine receptor antagonists, 8-SPT, PD 115199, CPX, and BW-A1433 (ARMSTRONG and GANOTE 1995). Administration of exogenous adenosine or R-PIA prior to ischemia mimics the actions of PC, implicating adenosine and adenosine receptors as mediators of ischemic PC (LIU et al. 1991). Moreover, if 8-SPT is given between the preconditioning ischemic episode and the prolonged ischemic period, the ability of PC to reduce relative infarct area is eliminated, indicating that the adenosine-mediated component of PC persists into the longer ischemic period (MIURA and TSUCHIDA 1999). As with ischemic preconditioning, several shorter "pulses" of adenosine are more protective against ischemic damage than is a single, extended exposure to the purine. The beneficial effects of a 5-min period of adenosinergic preconditioning was shown to last up to 30 min in the isolated rabbit heart (ARMSTRONG et al. 1997). A second or late phase of protection against myocardial stunning and infarction occurs after 24 h. In the rabbit, the second phase appears to result from activation of NFκB in response to NO, reactive oxygen species, PKC, and tyrosine kinase, and is blocked by diethyldithiocarbamate, an inhibitor of NFκB (XUAN et al. 1999).

Preconditioning is evoked by interactions with a variety of receptors including A_1 adenosine, and in some cases, A_3 adenosine, α1 adrenergic, B_2 bradykinin, M_2 muscarinic, AT_1 angiotensin, and ET_1 endothelin receptors (WANG et al. 1996b). The common denominator of these is activation of protein kinase C (PKC). Adenosine activates PKC activity during ischemia and this activation is blocked by calphostin C (ARMSTRONG and GANOTE 1995; IKONOMIDIS et al. 1997). The concentration of adenosine released during ischemic preconditioning is sufficient to increase PKC activity in human myocytes (IKONOMIDIS et al. 1997). Furthermore, the phorbol ester, PMA, mimics preconditioning. The actions of PMA are not affected by 8-SPT, providing evidence that adenosine receptor interactions occur prior to PKC activation (LIANG 1997). In support of this, it was shown that the reduction in infarct size afforded by PMA in the ischemic rabbit heart is not altered by adding 8-SPT, whereas the protective effects of R-PIA are blocked by chelerythrine (ILIODROMITIS et al. 1998). In rat and human ventricular strips, adenosine increases myocyte viability at the end of ischemia and during reperfusion in a manner that is sensitive to both 8-SPT and chelerythrine (CLEVELAND et al. 1996). Preconditioning-mediated inhibition of LDH release during ischemia is concomitant with a translocation of PKC and PKC from the cytosol to the plasma membrane (MIYAWAKI and ASHRAF 1997). The 8-SPT- and chelerythrine-sensitive preconditioning effects derived from several short ischemic episodes in the anesthetized rabbit are associated with the translocation of PKC from the cytosol to the plasma membrane (MIURA et al. 1998). It is notable that A_1 receptor activation, which occurs via a Gi/o G protein acts synergistically with Gq-coupled receptors to activate PLC (LINDEN 1991). This suggests that A_1 receptor activation may precondition the heart synergistically with receptors that signal via Gq, e.g., α1-adrenergic receptors.

In contrast to the above contention that adenosine receptor interactions temporally precede stimulation of membrane-associated PKC activity is the finding that adenosine released from endothelium during ischemia is mostly blocked by calphostin C but not by 8-SPT (ZHOU et al. 1996). This suggests that some adenosine release occurs subsequent to PKC activation. Similarly, in isolated rat cardiomyocytes, the ability of PMA to reduce LDH release during late ischemia is blocked by 8-SPT and the ecto 5′-nucleotidase inhibitor, AOPCP (Kitakaze et al. 1996). These data do not preclude the possibility that adenosine receptor activation stimulates PKC activity during preconditioning, but it does support the contention that PKC activation during PC leads to adenosine release, or adenosine formation from adenine nucleotides, and subsequent adenosine receptor stimulation. Such findings could be suggestive of a reverberating pathway between adenosine formation and PKC activation, possibly contributing to the temporal persistence of preconditioning.

It is generally accepted that the downstream effector of PKC during preconditioning is the inwardly rectified, ATP-sensitive K^+ channel. K_{ATP} channels are heteromultimers with a 4:4 stoichiometry of an inwardly rectifying K^+ channel (K_{IR}) subunit plus a sulfonylurea receptor (SUR) (Dorschner et al. 1999). SUR2A/KIR6.2 and SUR2B/(KIR6.1 or 6.2) constitute the cardiac and vascular smooth muscle type K_{ATP} channels, respectively. The sulfonylurea, glibenclamide, can abolish preconditioning and prevent the increase of ischemic threshold observed during the second of two sequential exercise tests (Tomai et al. 1999). Glibenclamide has also been shown to prevent adenosine induced preconditioning in all species tested, including human right trabeculae (Cleveland et al. 1997). K_{ATP} channels were first identified on the sarcolemma. A distinct channel also has been detected on the inner membranes of mitochondria (Hu et al. 1999), although the molecular identity of the mitochondrial K_{ATP} channel is yet to be elucidated. Activation of the mitochondrial channel triggers flavoprotein oxidation. Cell surface K_{ATP} channels are activated by pinacidil but only weakly by 100 mol/l diazoxide, and these channels are blocked by 10 μmol/l glibenclamide but are insensitive to 500 μmol/l 5-hydroxydecanoate (5-HD). Mitochondrial K_{ATP} channels are activated by pinacidil and diazoxide and are blocked by 5-HD (Hu et al. 1999; Liu et al. 1998; Sato et al. 1998). Glibenclamide and 5-HD reduce the protection of chick myocytes afforded by PMA during ischemia (Liang 1997) and decrease the protection from infarction afforded by the A_3 agonist, Cl-IB-MECA in rabbit hearts (Tracey et al. 1998). The effects of glibenclamide on mitochondrial K_{ATP} channels are difficult to assess because the sulfonylurea independently activates flavoprotein fluorescence, possibly by directly uncoupling mitochondria (Hu et al. 1999). The K_{ATP} channel purified from inner mitochondrial membranes contains thio groups that may respond to redox potential and free radicals, and is activated by low concentrations of ADP, up to 200 μmol/l (Grigoriev et al. 1999). It has been proposed that the role of the mitochondrial K_{ATP} channel is to maintain mitochondrial volume, which is coupled to metabolic control in the organelle (Szewczyk and Marban 1999).

Diazoxide causes cardioprotection and reversible oxidation of flavoproteins with an EC_{50} value of 27 μmol/l. Both redox changes and myocardial protection are inhibited by 5-HD. It has been proposed that diazoxide acts via PKC to activate mitochondrial K_{ATP} channels since the effect of diazoxide to protect the rat heart from injury due to the Ca^{2+} paradox is completely abolished by chelerythrine, a PKC inhibitor, and diazoxide causes translocation of PKC to mitochondria (WANG and ASHRAF 1999). Diazoxide produces a dose-dependent attenuation of the reduction in LVDP that is normally observed during reperfusion, in addition to limiting ischemic contracture and preserving tissue viability in isolated rat heart (GARLID et al. 1997).

Vascular K_{ATP} channels may contribute to PC. Thus, PC, CHA, and the K_{ATP} channel opener pinacidil all protect the coronary vasculature from the deleterious effects of ischemia and reperfusion. Protection is reversed by glibenclamide (MACZEWSKI and BERESEWICZ 1998).

Adenosine has been shown to activate the sarcolemmal K_{ATP} channel (IK_{ATP}) in cardiac myocytes and to decrease the sensitivity of the channel to ATP (HU et al. 1999). It was originally thought that intracellular ATP could not regulate K_{ATP} channels since the IC_{50} of ATP for inhibition of IK_{ATP} is in the low micromolar range whereas the concentration of ATP present in most cells is >1 mmol/l. However, phosphatidylinositol phosphates can decrease the sensitivity of the K_{ATP} channel for ATP. Thus, a 2-min exposure of recombinant K_{ATP} channels to PIP_2 and PIP increases the IC_{50} of ATP by approximately 1000-fold (BAUKROWITZ et al. 1998). PIP, PIP_2, and PIP_3, but not IP_3, can reduce the inhibitory effects of ATP on IK_{ATP} in ventricular myocytes such that 1 mmol/l ATP does not exert a substantial inhibition of K_{ATP} channel activity (SHYNG and NICHOLS 1998).

1. Myocardial Survival and Function

The involvement of adenosine release in ischemic preconditioning was first demonstrated for cardiomyocyte survival, where it was shown that PC reduces infarct size (i.e., the area of infarct relative to the area at risk) and this protection is blocked by adenosine receptor antagonists. In addition, application of adenosine or R-PIA prior to prolonged ischemia mimics the protection afforded by PC (LIU et al. 1991). The protective actions of adenosine are known to be dose-dependent, with an EC_{50} value of approximately 1 μmol/l (RICE et al. 1996). If the supernatant from ischemic cultured human myocytes is added to non-ischemic myocytes, it protects these cells from ischemic damage, indicating that the preconditioning process occurs through the actions of a released mediator. Blockade by 8-SPT provided evidence that the mediator involved is adenosine (IKONOMIDIS et al. 1997). This contention is further supported by evidence that, during ischemia, the nucleoside transport inhibitor draflazine increases the concentration of adenosine and concomitantly reduces the release of creatine kinase, an indicator of myocardial cell death (SOMMERSCHILD et al. 1997). Similarly, dipyridamole enhances the ability of exogenous adenosine to improve ischemia-induced myocyte survival (AUCHAMPACH and GROSS 1993).

Protection of pig myocardial survival by ischemic preconditioning can be mimicked by treatment with the A_1-selective agonist, GR79236, and blocked by the A_1-selective antagonist, CPX (LOUTTIT et al. 1999). The A_1 receptor has also been implicated as the locus mediating myocardial survival in the rabbit (LASLEY et al. 1995; RICE et al. 1996; HILL et al. 1998; ILIODROMITIS et al. 1998), chick (STRICKLER et al. 1996; LIANG and JACOBSON 1998; LIANG 1998; DOUGHERTY et al. 1998), and human (IKONOMIDIS et al. 1997). When the A_1 agonist, R-PIA was chronically infused into rabbits, it caused a down-regulation of A_1 receptor-mediated events, such as the bradycardic response to A_1 agonists. In these desensitized animals, ischemic preconditioning failed to protect the myocardium, even though adrenergic and muscarinic agonists still conferred protection (HASHIMI et al. 1998). These data suggest that, although stimulation of other receptors that activate PKC can mimic the protection afforded by preconditioning, the A_1 adenosine receptors is primarily responsible for preconditioning due to ischemia in the rabbit. However, although CPX completely blocked the inotropic aspects of CCPA in ischemic rabbit cardiac myocytes, it had limited effects on the prevention of contracture by PC. A combination of CPX and BW-A1433 did eliminate PC-mediated prevention of contracture (WANG et al. 1997). In contrast, the ability of PC to improve the recovery of developed pressure in rat heart during reperfusion was blunted by both CPX or 8-SPT (WINTER et al. 1997).

Over-expression of the A_1 receptor in mouse hearts leads to an increase in the time to ischemic contracture, and renders the hearts more resistant to ischemia-induced increases in diastolic tension and decreases in LVDP and +dP/dt (MATHERNE et al. 1997; GAUTHIER et al. 1998). Similarly, the A_1 agonist CHA maintains EDP and LVDP during ischemia and reperfusion in rat and rabbits (HEADRICK 1996), while R-PIA improves the recovery of LVDP from ischemic insult.

Selective agonists of A_3 receptors have been noted to be protective in some species. The A_3-selective agonist IB-MECA was shown to be equipotent with R-PIA for the prevention of ischemia-induced damage in the ischemic rabbit heart. The protective effects of IB-MECA or CB-MECA were not influenced by the same concentration of BW-A1433 that blocked the actions of R-PIA (TRACEY et al. 1997, 1998). One interesting possibility is that A_3 receptors act in part by triggering the release of mediators from resident mast cells in the heart with these mediators activating PLC-coupled receptors that contribute to myocardial preconditioning. Mammalian hearts have been shown to release large amounts of histamine in response to reperfusion following ischemia (LINDEN 1994) while treatment of perfused rat hearts with lodoxamide eliminates ischemia/reperfusion-induced myocyte mortality and preserves cardiac output (PARENTEAU and CLARK 1991), supporting this possibility.

There appear to be A_1 and A_3 receptors functionally coupled to myocyte protection on embryonic chicken ventricular myocytes. Thus, CCPA and Cl-IB-MECA reduced the creatine kinase release and cell mortality that result from hypoxia in these cells. The protection afforded by CCPA or Cl-IB-MECA was selectively blocked by the A_1 antagonist 8-CPT or the A_3 selective antagonist MRS 1191, respectively (LIANG and JACOBSON 1998). In the embryonic chicken, neither CPX nor MRS 1191 attenuated the protective effects of PC as effectively as a combination of the two antagonists (LEE et al. 1998). Inosine, at high concentrations (>1 μmol/l), is known to activate A_3 receptors (JIN et al. 1997). As with adenosine, inosine is released in large amounts during ischemia (HARRISON et al. 1998), and may contribute to cardioprotection via A_3 receptors.

In the chick ventricular myocyte model, blockade of A_{2A} receptors with CSC causes a slight increase in the protective actions of PC and R-PIA, while CGS 21680 attenuates the ability of PC to reduce creatine kinase release and cell death during ischemia and reperfusion (STRICKLER et al. 1996). These data suggest that A_{2A} receptors are not responsible for PC in the heart. However, A_{2A} receptors may protect tissues by a different mechanism. In bovine coronary endothelial cells, R-PIA does not mimic the ability of PC to protect cultured endothelial cells from ischemia/reperfusion injury, whereas CGS 21680 does, suggesting an action at A_{2A} receptors (ZHOU et al. 1996). In the open chest dog model, CGS 21680 administered for 60 min, beginning in the last 5 min of coronary occlusion, significantly reduced infarct size (SCHLACK et al. 1993). It is possible that these protective effects of A_{2A} agonists are mediated by inhibitory effects on neutrophils and other inflammatory cells that are activated by ischemia/reperfusion (BARNES et al. 1995; SULLIVAN and LINDEN 1998). It was also found that 2HE-NECA and CGS 21680 were more potent than CCPA for the maintenance of EDP and recovery of LVDP during reperfusion (LOZZA et al. 1997). CGS 21680, administered during the end of ischemia and the beginning of reperfusion in anesthetized dogs, improves the restoration of systolic wall thickening, an indicator of recovery from myocardial stunning (SCHLACK et al. 1993).

During prolonged ischemia and reperfusion, a loss in contractile function is associated with approximately a sixfold increase in myocardial hydrogen peroxide (H_2O_2) formation in rat hearts (SLEZAK et al. 1995). Moreover, perfusion of isolated rat hearts with H_2O_2 leads to a greater than 90% loss of contractile function and an increase in resting tension. Peroxide-induced contractile deficits are reduced by CPA, but not CGS 21680, with CPA-mediated protection being reversed by addition of CPX providing evidence that A_1 receptor interactions can also improve the contractile function of ischemic hearts by reducing the deleterious effects of hydrogen peroxide (KARMAZYN and COOK 1992). It was subsequently shown that CPA, but not CGS 21680 or IB-MECA, restore the imbalance in I_{Ca},L brought about by H_2O_2 in isolated cardiac myocytes (THOMAS et al. 1998), providing a potential mechanism to explain these actions.

2. ATP/Cellular Energetics

The concept of ischemic preconditioning, established by MURRY et al. (1986), was supported by the demonstration that, although a 10-min period of ischemia caused a decline in the tissue ATP content of canine hearts, there were no further decreases in ATP levels over three subsequent periods of ischemia (REIMER et al. 1986). These authors also showed that a 40-min period of ischemia caused a much more dramatic depletion of ATP compared to four 10-min periods. Subsequent studies have shown, in several animal models, that PC leads to a conservation of ATP (MIYAWAKI and ASHRAF 1997), PCr and GATP (HEADRICK 1996; HARRISON et al. 1998), and [ATP]/[ADP][Pi] (HEADRICK 1996). The protective effects of PC are blocked by 8-SPT.

Prior to the discovery that preconditioning protects cardiac ATP levels during ischemia, it was shown that adenosine prevents ATP depletion from cardiac tissue during an ischemic period of up to 12 min. It was also shown that adenosine enhances ATP recovery following ischemia for up to 60 min (ELY et al. 1985), indicating that both ischemic preconditioning and adenosine receptor activation can preserve ATP levels in the heart. Ischemic preconditioning and R-PIA both mitigate a decrease in ATP and PCr, and attenuate the rise in lactate levels that results from ischemia in canine myocardium (PISARENKO et al. 1997). In chick myocytes, adenosine, CCPA, and IB-MECA all preserve ATP levels during ischemia and reperfusion. ATP levels also are maintained by endogenous levels of adenosine in ischemic chick myocytes with overexpressed A_1 and A_3 receptors. Blockade of these receptors with CPX or MRS 1191, respectively, eliminates these protective effects (DOUGHERTY et al. 1998). It was demonstrated that the ATP-preserving actions of ischemic preconditioning in chick myocytes are attenuated by CPX or MRS 1191, with combined blockade by the two antagonists leading to a complete elimination of the beneficial effects of PC. This suggests that the overall preconditioning response may be the result of a contribution by both A_1 and A_3 receptors (LEE et al. 1998).

3. Rhythmicity

Effects of PC on electrophysiological parameters provides evidence for preconditioning in humans, although similar observations have been made in other species. It was first demonstrated during percutaneous transluminal coronary angioplasty (PTCA; also known as "balloon angioplasty") that a decrease in the ST segment of cardiac action potentials (ST shift) was associated with the ischemia produced during balloon inflation, but that this shift was substantially diminished during a subsequent inflation (DEUTSCH et al. 1990; TAGGART et al. 1993). Theophylline has been shown to attenuate the ability of preconditioning to reduce ST shift during the second balloon inflation (HASHIMURA et al. 1997), while 8-SPT blocks preconditioning of ST shift during the second and third inflations (COHEN et al. 1997). Similar effects have been demonstrated for the adenosine receptor antagonists aminophylline and bamophylline (CLAEYS et al. 1996; TOMAI et al. 1996). Taken together, these data provide strong evidence that the reduction in ST shift afforded by preconditioning is mediated via an adenosine receptor. The receptor subtype involved in man awaits further study.

Similar to its action on ST shift is the finding that ischemic preconditioning protects against the incidence of ventricular tachycardia, ventricular fibrillation, and premature ventricular complexes. This effect lasts for up to 30 min of reperfusion, with the strongest protection occurring between 10 min and 20 min (SHIKI and HEARSE 1987). In this study, the incidence of ventricular tachycardia and ventricular fibrillation during a second ischemic episode was inversely related to the length of the first ischemic period for times between 30 s and 5 min. The mixed A_1/A_2 agonist AMP 579 can be used in lieu of preconditioning to prevent ventricular fibrillation during ischemia (SMITS et al. 1998), suggesting that adenosine receptors are responsible for preconditioning-mediated prevention of arrhythmias. Blockade of adenosine receptors with theophylline, XAC, 8-SPT, or CSC, but not CPX, increased the occurrence of ischemia-induced ventricular fibrillation, while inhibition of adenosine deaminase activity with EHNA reduced it (SCHREIECK and RICHARDT 1999). These data suggest that not only are the levels of adenosine released during ischemia sufficient to prevent the formation of arrhythmias but also that these actions may be mediated by an A_2, and not the A_1, adenosine receptor.

4. Coronary Flow

It is well known that adenosine is released from cardiac tissue in response to a decrease in the O_2 supply/demand ratio and that a purine elicits a compensatory dose-dependent increase in coronary flow (BARDENHEUER and SCHRADER 1986). Similarly, it has been shown that increasing myocardial O_2 consumption causes a decrease in coronary vascular resistance. This response was attenuated by 8-SPT, indicating the involvement of endogenous adenosine (DUNCKER et al. 1998). The dose-response curve for coronary vasodilation by adenosine is very steep; raising adenosine levels by only 62% increases coronary flow from 5% of maximum to 50% of maximum. The EC_{50} for coronary vasodilation by endogenous interstitial adenosine was estimated to be in the region of 150 nmol/l, with this value increasing substantially in the added presence of the antagonist, 8-PT (STEPP et al. 1996). In isolated guinea-pig hearts, the dilatory effect of PC is blocked by the A_1-selective antagonist, N-0861 and mimicked by CHA and IB-MECA, implicating actions at A_1 and A_3 receptors (GIANNELLA et al. 1997). Also, 8-SPT blocks, and CHA mimics, PC-mediated maintenance of coronary flow during reperfusion of rat and rabbit hearts (HEADRICK 1996). Although it was mimicked by CHA, PC cannot be blocked by either 8-SPT or CPX, indicating that PC-mediated preservation of vasodilation during reperfusion may be distinct from interactions with adenosine receptors in the endothelium. Furthermore, preconditioning is eliminated by glibenclamide and mimicked by pinacidil, suggesting the involvement of K_{ATP} channels in this effect (MACZEWSKI and BERESEWICZ 1998).

X. Summary

Stimulation of adenosine receptors leads to a variety of responses in the cardiovascular system. A_1 receptors are largely responsible for mediating adenosine effects in the heart including negative inotropic, dromotropic, and chromotropic responses. Myocardial preconditioning is mediated by A_1 receptors with some direct or possibly indirect contribution by A_3 receptors (see Fig. 1). Functional effects of myocardial A_{2A} receptors are slight, but these may act to counteract partially the effects of A_1 receptor activation. All adenosine

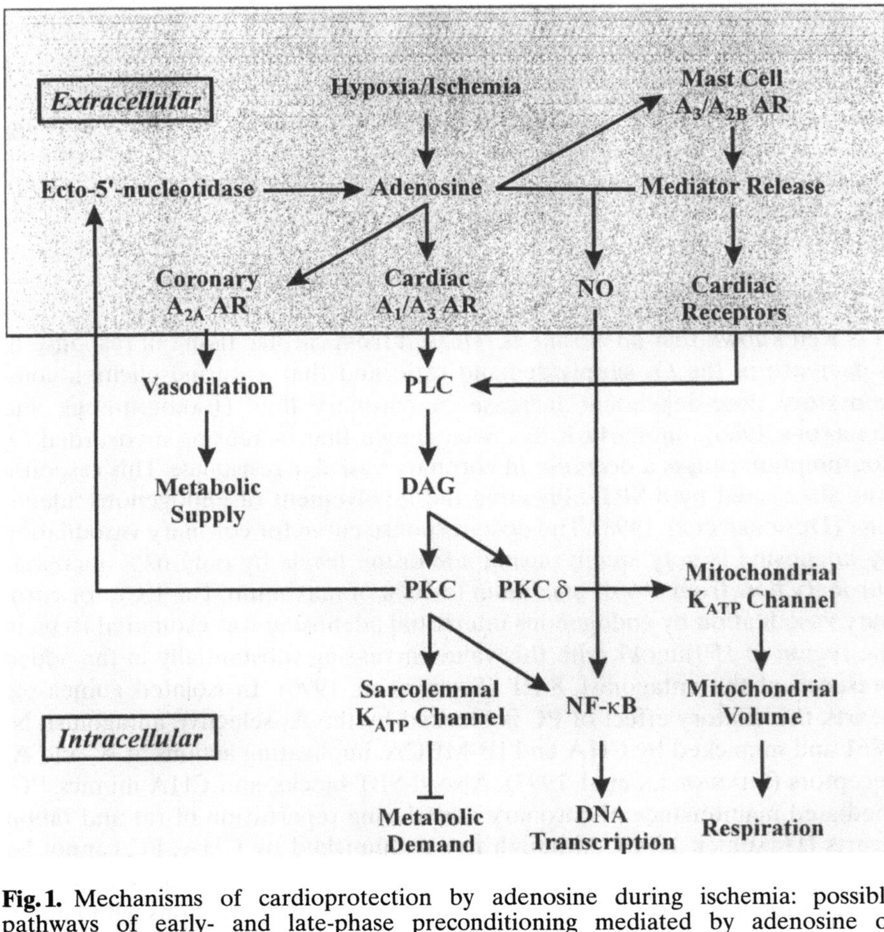

Fig. 1. Mechanisms of cardioprotection by adenosine during ischemia: possible pathways of early- and late-phase preconditioning mediated by adenosine or hypoxia/ischemia. For details of individual pathways, see text

receptors are involved in the drop in blood pressure produced by adenosine. Activation of central A_{2A} receptors decreases sympathetic tone. Coronary vascular smooth muscle dilates in response to A_{2A} receptor activation and both A_{2A} and A_{2B} receptors contribute to dilation of other vascular beds. NO produced in response to adenosine may be derived from activation of Gq-coupled A_{2B} adenosine receptors on endothelial cells. Activation of A_1 receptors inhibits the release of norepinephrine from sympathetic nerve terminals. A_3 and/or A_{2B} receptors are responsible for perivascular mast cell degranulation which can produce microvascular vasoconstriction and dilation of large blood vessels. Angiogenesis appears to involve A_2, possibly A_{2A}, receptors.

References

Abbracchio MP, Ceruti S, Brambilla R, Barbieri D, Camurri A, Franceschi C, Giammarioli AM, Jacobson KA, Cattabeni F, Malorni W (1998) Adenosine A(3) receptors and viability of astrocytes. Drug Dev Res 45:379–386

Abebe W, Hussain T, Olanrewaju H, Mustafa SJ (1995) Role of nitric oxide in adenosine receptor-mediated relaxation of porcine coronary artery. Am J Physiol 269:H1672–H1678

Armstrong S, Ganote CE (1995) In vitro ischaemic preconditioning of isolated rabbit cardiomyocytes: effects of selective adenosine receptor blockade and calphostin C. Cardiovascular Research 29:647–652

Armstrong SC, Hoover DB, Shivell LC, Ganote CE (1997) Preconditioning of isolated rabbit cardiomyocytes – no evident separation of induction, memory and protection. Journal of Molecular & Cellular Cardiology 29:2285–2298

Auchampach JA, Gross GJ (1993) Adenosine A_1 receptors, KATP channels, and ischemic preconditioning in dogs. Am J Physiol Heart Circ Physiol 264: H1327–H1336

Auchampach JA, Jin J, Wan TC, Caughey GH, Linden J (1997) Canine mast cell adenosine receptors: cloning and expression of the A_3 receptors and evidence that degranulation is mediated by the A_{2B} receptor. Mol Pharmacol 52:846–860

Barbieri D, Abbracchio MP, Salvioli S, Monti D, Cossarizza A, Ceruti S, Brambilla R, Cattabeni F, Jacobson KA, Franceschi C (1998) Apoptosis by 2-chloro-2′-deoxy-adenosine and 2-chloro-adenosine in human peripheral blood mononuclear cells. Neurochem Int 32:493–504

Bardenheuer H, Schrader J (1986) Supply-to-demand ratio for oxygen determines formation of adenosine by the heart. Am J Physiol 250:H173–80

Barnes CR, Mandell GL, Carper HT, Luong S, Sullivan GW (1995) Adenosine modulation of tumor necrosis factor-a-induced neutrophil activation. Biochem Pharmacol 50:1851–1857

Baukrowitz T, Schulte U, Oliver D, Herlitze S, Krauter T, Tucker SJ, Ruppersberg JP, Fakler B (1998) PIP2 and PIP as determinants for ATP inhibition of KATP channels [see comments]. Science 282:1141–1144

Belardinelli L, Giles WR, West A (1988) Ionic mechanisms of adenosine actions in pacemaker cells from rabbit heart. J Physiol 405:615–633

Belardinelli L, Shryock JC, Ruble J, Monopoli A, Dionisotti S, Ongini E, Dennis DM, Baker SP (1996) Binding of the novel nonxanthine A_{2A} adenosine receptor antagonist [^{3}H]SCH58261 to coronary artery membranes. Circ Res 79:1153–1160

Belardinelli L, Shryock JC, Snowdy S, Zhang Y, Monopoli A, Lozza G, Ongini E, Olsson RA, Dennis DM (1998) The A_{2A} adenosine receptor mediates coronary vasodilation. J Pharmacol Exp Ther 284:1066–1073

Bellardinelli L, Linden J, Berne RM (1989) The cardiac effects of adenosine. Prog Cardiovasc Dis 32:73–97
Claeys MJ, Vrints CJ, Bosmans JM, Conraads VM, Snoeck JP (1996) Aminophylline inhibits adaptation to ischaemia during angioplasty. Role of adenosine in ischaemic preconditioning [see comments]. Eur Heart J 17:539–544
Clemo HF, Belardinelli L (1986) Effect of adenosine on atrioventricular conduction. 2. Modulation of atrioventricular node transmission by adenosine in hypoxic isolated guinea pig hearts. Circ Res 59:437–446
Cleveland JCJ, Meldrum DR, Rowland RT, Banerjee A, Harken AH (1997) Adenosine preconditioning of human myocardium is dependent upon the ATP-sensitive K+ channel. J Mol Cel Cardiol 29:175–182
Cleveland JCJ, Meldrum DR, Rowland RT, Sheridan BC, Banerjee A, Harken AH (1996) The obligate role of protein kinase C in mediating clinically accessible cardiac preconditioning. Surgery 120:345–352
Cohen MV, Yang XM, Downey JM (1997) Attenuation of S-T segment elevation during repetitive coronary occlusions truly reflects the protection of ischemic preconditioning and is not an epiphenomenon. Basic Res Cardiol 92:426–434
Cordeiro JM, Ferrier GR, Howlett SE (1995) Effects of adenosine in simulated ischemia and reperfusion in guinea pig ventricular myocytes. Am J Physiol 269: H121–H129
Cox BF, Greenland BD, Perrone MH, Merkel LA (1994) Ischaemia/reperfusion selectively attenuates coronary vasodilatation to an adenosine A_2- but not to an A_1-agonist in the dog. Br J Pharmacol 111:1233–1239
Darvish A, Metting PJ (1993) Purification and regulation of an AMP-specific cytosolic 5′-nucleotidase from dog heart. Am J Physiol 264:H1528–H1534
Dawicki DD, Chatterjee D, Wyche J, Rounds S (1997) Extracellular ATP and adenosine cause apoptosis of pulmonary artery endothelial cells. Am J Physiol 273: L485–L494
Decking UK, Schlieper G, Kroll K, Schrader J (1997) Hypoxia-induced inhibition of adenosine kinase potentiates cardiac adenosine release. Circ Res 81:154–164
Decking UKM, Schrader J (1998) Role of adenosine kinase and AMP deminase in the regulation of cardaic purine release. Drug Dev Res 45:295–303
Deussen A, Borst M, Kroll K, Schrader J (1988) Formation of *S*-adenosylhomocysteine in the heart. 2. A sensitive index for regional myocardial underperfusion. Circ Res 63:250–261
Deussen A, Moser G, Schrader J (1986) Contribution of coronary endothelial cells to cardiac adenosine production. Pflugers Arch 406:608–614
Deutsch E, Berger M, Kussmaul WG, Hirshfeld JWJ, Herrmann HC, Laskey WK (1990) Adaptation to ischemia during percutaneous transluminal coronary angioplasty. Clinical, hemodynamic, and metabolic features [see comments]. Circ 82:2044–2051
Dixon AK, Gubitz AK, Sirinathsinghji DJ, Richardson PJ, Freeman TC (1996) Tissue distribution of adenosine receptor mRNAs in the rat. Br J Pharmacol 118: 1461–1468
Dobson JG, Fenton RA (1998) Cardiac physiology of adenosine. In: Burnstock G, Dobson JGJ, Liang BT, Linden J (eds) Cardovascular biology of purines. Kluwer, Dordrecht, p 21
Dorschner H, Brekardin E, Uhde I, Schwanstecher C, Schwanstecher M (1999) Stoichiometry of sulfonylurea-induced ATP-sensitive potassium channel closure. Mol Pharmacol 55:1060–1066
Dougherty C, Barucha J, Schofield PR, Jacobson KA, Liang BT (1998) Cardiac myocytes rendered ischemia resistant by expressing the human adenosine A_1 or A_3 receptor. FASEB J 12:1785–1792
Drury AN, Szent-Gyorgyi A (1929) The physiological activity of adenine compounds with especial reference to their action upon the mammalian heart. J Physiol (London) 68:213–237

Duncker DJ, Stubenitsky R, Verdouw PD (1998) Role of adenosine in the regulation of coronary blood flow in swine at rest and during treadmill exercise. Am J Physiol 275:H1663–H1772
Duncker DJ, Van Zon NS, Ishibashi Y, Bache RJ (1996) Role of K+ ATP channels and adenosine in the regulation of coronary blood flow during exercise with normal and restricted coronary blood flow. J Clin Invest 97:996–1009
Ely SW, Mentzer RMJ, Lasley RD, Lee BK, Berne RM (1985) Functional and metabolic evidence of enhanced myocardial tolerance to ischemia and reperfusion with adenosine. J Thorac Cardiovasc Surg 90:549–556
Ethier MF, Chander V, Dobson JGJ (1993) Adenosine stimulates proliferation of human endothelial cells in culture. Am J Physiol 265:H131–H138
Evoniuk G, von Borstel RW, Wurtman RJ (1987) Antagonism of the cardiovascular effects of adenosine by caffeine or 8-(*p*-sulfophenyl)theophylline. J Pharmacol Exp Ther 240:428–432
Felix SB, Stangl V, Frank TM, Harms C, Berndt T, Kastner R, Baumann G (1997) Release of a stable cardiodepressant mediator after myocardial ischaemia during reperfusion. Cardiovasc Res 35:68–79
Fredholm BB, Jacobson KA, Jonzon B, Kirk KL, Li YO, Daly JW (1987) Evidence that a novel 8-phenyl-substituted xanthine derivative is a cardioselective adenosine receptor antagonist in vivo. J Cardiovasc Pharmacol 9:396–400
Froldi G, Belardinelli L (1990) Species-dependent effects of adenosine on heart rate and atrioventricular nodal conduction. Mechanism and physiological implications. Circ Res 67:960–978
Garlid KD, Paucek P, Yarov-Yarovoy V, Murray HN, Darbenzio RB, D'Alonzo AJ, Lodge NJ, Smith MA, Grover GJ (1997) Cardioprotective effect of diazoxide and its interaction with mitochondrial ATP-sensitive K+ channels. Possible mechanism of cardioprotection. Circ Res 81:1072–1082
Gauthier NS, Morrison RR, Byford AM, Jones R, Headrick JP, Matherne GP (1998) Functional genomics of transgenic overexpression of A(1) adenosine receptors in the heart. Drug Dev Res 45:402–409
Giannella E, Mochmann HC, Levi R (1997) Ischemic preconditioning prevents the impairment of hypoxic coronary vasodilatation caused by ischemia/reperfusion: role of adenosine A_1/A_3 and bradykinin B_2 receptor activation. Circ Res 81: 415–422
Gottlieb RA, Burleson KO, Kloner RA, Babior BM, Engler RL (1994) Reperfusion injury induces apoptosis in rabbit cardiomyocytes. J Clin Invest 94:1621–1628
Grigoriev SM, Skarga YY, Mironova GD, Marinov BS (1999) Regulation of mitochondrial K-ATP channel by redox agents. Biochim Biophys Acta 1410:91–96
Hannon JP, Pfannkuche HJ, Fozard JR (1995) A role for mast cells in adenosine A_3 receptor-mediated hypotension in the rat. Br J Pharmacol 115:945–952
Harden FA, Harrison GJ, Headrick J, Jordan LR, Willis RJ (1996) A biphasic response to adenosine in the coronary vasculature of the K(+)-arrested perfused rat heart. Eur J Pharmacol 307:49–53
Harrison GJ, Willis RJ, Headrick JP (1998) Extracellular adenosine levels and cellular energy metabolism in isochemically preconditioned rat heart. Cardiovasc Res 40:74–87
Hashimi MW, Thornton JD, Downey JM, Cohen MV (1998) Loss of myocardial protection from ischemic preconditioning following chronic exposure to *R*(–)-N6-(2-phenylisopropyl)adenosine is related to defect at the adenosine A_1 receptor. Mol Cell Biochem 186:19–25
Hashimura K, Kijima Y, Matsu-ura Y, Ueda T, Kato Y, Mori I, Minamino T, Kitakaze M, Hori M (1997) Effect of theophylline on adaptation of the heart to myocardial ischemia during percutaneous transluminal coronary angioplasty in patients with stable angina pectoris. Am J Cardiol 79:475–477
Headrick JP (1996) Ischemic preconditioning: bioenergetic and metabolic changes and the role of endogenous adenosine. J Mol Cell Cardiol 28:1227–1240

Headrick JP, Matherne GP, Berne RM (1992) Myocardial adenosine formation during hypoxia: effects of ecto-5′-nucleotidase inhibition. J Mol Cell Cardiol 24:295–303
Hill RJ, Oleynek JJ, Magee W, Knight DR, Tracey WR (1998) Relative importance of adenosine A_1 and A_3 receptors in mediating physiological or pharmacological protection from ischemic myocardial injury in the rabbit heart. J Mol Cell Cardiol 30:579–585
Hofman PL, Hiatt K, Yoder MC, Rivkees SA (1997) A_1 adenosine receptors potently regulate heart rate in mammalian embryos. Am J Physiol 273:R1374–R1380
Hu H, Sato T, Seharaseyon J, Liu YG, Johns DC, O'Rourke B, Marban E (1999) Pharmacological and histochemical distinctions between molecularly defined sarcolemmal K-ATP channels and native cardiac mitochondrial K-ATP channels. Mol Pharmacol 55:1000–1005
Hu K, Li GR, Nattel S (1999) Adenosine-induced activation of ATP-sensitive K^+ channels in excised membrane patches is mediated by PKC. Am J Physiol 276: H488–H495
Huang CH, Kim SJ, Ghaleh B, Kudej RK, Shen YT, Bishop SP, Vatner SF (1999) An adenosine agonist and preconditioning shift the distribution of myocardial blood flow in conscious pigs. Am J Physiol 276:H368–H375
Ikonomidis JS, Shirai T, Weisel RD, Derylo B, Rao V, Whiteside CI, Mickle DA, Li RK (1997) Preconditioning cultured human pediatric myocytes requires adenosine and protein kinase C. Am J Physiol 272:H1220–H1230
Iliodromitis EK, Miki T, Liu Gs, Downey JM, Cohen MV, Kremastinos DT (1998) The PKC activator PMA preconditions rabbit heart in the presence of adenosine receptor blockade: is 5′-nucleotidase important? J Mol Cell Cardiol 30:2201–2211
Jacobson KA, Moro S, Kim YC, Li AH (1998) A(3) adenosine receptors: protective vs. damaging effects identified using novel agonists and antagonists. Drug Dev Res 45:113–124
Jin X, Shepherd RK, Duling BR, Linden J (1997) Inosine binds to A_3 adenosine receptors and stimulates mast cell degranulation. J Clin Invest 100:2849–2857
Karczewski P, Hendrischke T, Wolf WP, Morano I, Bartel S, Schrader J (1998) Phosphorylation of phospholamban correlates with relaxation of coronary artery induced by nitric oxide, adenosine, and prostacyclin in the pig. J Cell Biochem 70:49–59
Karmazyn M, Cook MA (1992) Adenosine A_1 receptor activation attenuates cardiac injury produced by hydrogen peroxide. Circ Res 71:1101–1110
Karoon P, Rubino A, Burnstock G (1995) Enhanced sympathetic neurotransmission in the tail artery of 1,3-dipropyl-8-sulphophenylxanthine (DPSPX)-treated rats. Br J Pharmacol 116:1918–1922
Kitakaze M, Minamino T, Node K, Komamura K, Inoue M, Hori M, Kamada T (1996) Activation of ecto-5′-nucleotidase by protein kinase C attenuates irreversible cellular injury due to hypoxia and reoxygenation in rat cardiomyocytes. J Mol Cell Cardiol 28:1945–1955
Kroll K, Decking UK, Dreikorn K, Schrader J (1993) Rapid turnover of the AMP-adenosine metabolic cycle in the guinea pig heart. Circ Res 73:846–856
Kurachi Y, Nakajima T, Sugimoto T (1986) On the mechanism of activation of muscarinic potassium channels by adenosine in isolated atrial cells: involvement of GTP-binding proteins. Pflugers Arch 407:264–274
Kusachi S, Thompson RD, Olsson RA (1983) Ligand selectivity of dog coronary adenosine receptor resembles that of adenylate cyclase stimulatory (Ra) receptors. J Pharmacol Exp Ther 227:316–321
Lasley RD, Noble MA, Konyn PJ, Mentzer RM, Jr. (1995) Different effects of an adenosine A_1 analogue and ischemic preconditioning in isolated rabbit hearts. Ann Thorac Surg 60:1698–1703
Lee AE, Wilcox K, Jacobson KA, Dichter M, Liang BT (1998) Adenosine receptor subtypes and cardioprotection. Drug Dev Res 45:394–401

Lee HT, Thompson CI, Hernandez A, Lewy JL, Belloni FL (1993a) Cardiac desensitization to adenosine analogues after prolonged R-PIA infusion in vivo. Am J Physiol 265:H1916–H1927
Lee HT, Thompson CI, Linden J, Belloni FL (1993b) Differential sensitization of cardiac actions of adenosine in rats after chronic theophylline treatment. Am J Physiol 264:H1634–H1643
Li JM, Fenton RA, Cutler BS, Dobson JGJ (1995) Adenosine enhances nitric oxide production by vascular endothelial cells. Am J Physiol 269:C519–C523
Liang BT (1997) Protein kinase C-mediated preconditioning of cardiac myocytes: role of adenosine receptor and KATP channel. Am J Physiol 273:H847–H853
Liang BT (1998) Protein kinase C-dependent activation of KATP channel enhances adenosine-induced cardioprotection. Biochem J 336:337–343
Liang BT, Jacobson KA (1998) A physiological role of the adenosine A_3 receptor: sustained cardioprotection. Proc Natl Acad Sci USA 95:6995–6999
Libert F, Schiffmann SN, Lefort A, Parmentier M, Gérard C, Dumont JE, Vanderhaeghen J-J, Vassart G (1991) The orphan receptor cDNA RDC7 encodes an A_1 adenosine receptor. EMBO J 10:1677–1682
Linden J (1991) Structure and function of the A_1 adenosine receptor. FASEB J 5:2668–2676
Linden J (1994) Cloned adenosine A_3 receptors: pharmacological properties, species differences and receptor functions. Trends Pharmacol Sci 15:298–306
Linden J, Hollen CE, Patel A (1990) The mechanism by which adenosine and cholinergic agents reduce contractility in rat myocardium: correlation with cyclic adenosine monophosphate and receptor densities. Circ Res 56:728–735
Linden J, Jacobson KA (1998) Molecular biology and pharmacology of recombinant human adenosine receptors. In: Burnstock G, Dobson JG Jr, Liang BT, Linden J (eds) Cardiovascular biology of purines. Kluwer, Dordrecht, p 1
Linden J, Patel A, Sadek S (1985) [125I]Aminobenzyladenosine, a new radioligand with improved specific binding to adenosine receptors in heart. Circ Res 56:279–284
Linden J, Thai T, Figler H, Robeva AS (1999) Characterization of human A_{2B} adenosine receptors: radioligand binding, Western blotting and coupling to Gq in HEK-293 and HMC-1 mast cells. Mol Pharmacol 56:705–713
Liu GS, Thornton J, Van Winkle DM, Stanley AW, Olsson RA, Downey JM (1991) Protection against infarction afforded by preconditioning is mediated by A_1 adenosine receptors in rabbit heart. Circ 84:350–356
Liu Y, Sato T, O'Rourke B, Marban E (1998) Mitochondrial ATP-dependent potassium channels: novel effectors of cardioprotection? Circ 97:2463–2469
Lloyd HG, Schrader J (1993) Adenosine metabolism in the guinea pig heart: the role of cytosolic *S*-adenosyl-L-homocysteine hydrolase, 5′-nucleotidase and adenosine kinase. Eur Heart J 14 [Suppl I]:27–33
Louttit JB, Hunt AAE, Maxwell MP, Drew GM (1999) The time course of cardioprotection induced by GR79236, a selective adenosine A(1)-receptor agonist, in myocardial ischaemia-reperfusion injury in the pig. J Cardiovasc Pharmacol 33: 285–291
Lozza G, Conti A, Ongini E, Monopoli A (1997) Cardioprotective effects of adenosine A_1 and A_{2A} receptor agonists in the isolated rat heart. Pharmacol Res 35:57–64
Luthin DR, Olsson RA, Thompson RD, Sawmiller DR, Linden J (1995) Characterization of two affinity states of adenosine A_{2A} receptors with a new radioligand, 2-[2-(4-amino-3-[125I]iodophenyl)ethylamino]adenosine. Mol Pharmacol 47:307–313
Maczewski M, Beresewicz A (1998) The role of adenosine and ATP-sensitive potassium channels in the protection afforded by ischemic preconditioning against the post-ischemic endothelial dysfunction in guinea-pig hearts. J Mol Cell Cardiol 30:1735–1747
Martin PL (1992) Evidence that adenosine receptors in the dog left atrium are not of the typical A_1 or A_2 adenosine receptor subtypes. Eur J Pharmacol 214:199–205

Martin PL, Ueeda M, Olsson RA (1993) 2-Phenylethoxy-9-methyladenine: an adenosine receptor antagonist that discriminates between A_2 adenosine receptors in the aorta and the coronary vessels from the guinea pig. J Pharmacol Exp Ther 265:248–253
Martin PL, Wysocki RJ, Barrett RJ, May JM, Linden J (1996) Characterization of 8-(*N*-methylisopropyl)amino-*N*6-(5′-endohydroxy-endonorbornyl)-9-methyladenine (WRC-0571), a highly potent and selective, non-xanthine antagonist of A_1 adenosine receptors. J Pharmacol Exp Ther 276:490–499
Matherne GP, Byford AM, Gilrain JT, Dalkin AC (1996) Changes in myocardial A_1 adenosine receptor and message levels during fetal development and postnatal maturation. Biol Neonate 70:199–205
Matherne GP, Linden J, Byford AM, Gauthier NS, Headrick JP (1997) Transgenic A(1) adenosine receptor overexpression increases myocardial resistance to ischemia. Proc Natl Acad Sci USA 94:6541–6546
Mathot RA, Cleton A, Soudijn W, Ijzerman AP, Danhof M (1995) Pharmacokinetic modelling of the haemodynamic effects of the A_{2A} adenosine receptor agonist CGS 21680C in conscious normotensive rats. Br J Pharmacol 114:761–768
McKinley JB, Dahlman D, MacLeod KM (1990) The interaction of adenosine analogues with cAMP-generating and cAMP-independent positive inotropic agents in rabbit left atrium. Naunyn-Schmiedebergs Arch Pharmacol 342:605–612
Meester BJ, Shankley NP, Welsh NJ, Wood J, Meijler FL, Black JW (1998) Pharmacological classification of adenosine receptors in the sinoatrial and atrioventricular nodes of the guinea-pig. Br J Pharmacol 124:685–692
Meininger CJ, Schelling ME, Granger HJ (1988) Adenosine and hypoxia stimulate proliferation and migration of endothelial cells, Am J Physiol 255:H554–H562
Meng F, Xie G, Chalmers D, Morgan C, Watson SJ Jr, Akil H (1994) Cloning and expression of the A_{2A} adenosine receptor from guinea pig brain. Neurochem Res 19:613–621
Meyerhof W, Müller-Brechlin R, Richter D (1991) Molecular cloning of a novel putative G-protein coupled receptor expressed during rat spermiogenesis. Febs Lett 284:155–160
Minamino T, Kitakaze M, Komamura K, Node K, Takeda H, Inoue M, Hori M, Kamada T (1995) Activation of protein kinase C increases adenosine production in the hypoxic canine coronary artery through the extracellular pathway. Arterioscler Thromb Vasc Biol 15:2298–2304
Miura T, Kawamura S, Goto M, Sakamoto J, Tsuchida A, Matsuzaki M, Shimamoto K (1998) Effect of protein kinase C inhibitors on cardioprotection by ischemic preconditioning depends on the number of preconditioning episodes. Cardiovasc Res 37:700–709
Miura T, Suzuki K, Shimamoto K, Iimura O (1996) Suppression of the degradation of adenine nucleotides during ischemia may not be a sufficient mechanism for infarct size limitation by preconditioning. Basic Res Cardiol 91:425–432
Miura T, Tsuchida A (1999) Adenosine and preconditioning revisited [Review]. Clin Exp Pharmacol Physiol 26:92–99
Miyawaki H, Ashraf M (1997) Ca^{2+} as a mediator of ischemic preconditioning. Circ Res 80:790–799
Mizumura T, Auchampach JA, Linden J, Bruns RF, Gross GJ (1996) PD 81,723, an allosteric enhancer of the A_1 adenosine receptor, lowers the threshold for ischemic preconditioning in dogs. Circ Res 79:415–423
Monopoli A, Casati C, Lozza G, Forlani A, Ongini E (1998) Cardiovascular pharmacology of the A_{2A} adenosine receptor antagonist, SCH 58261, in the rat. J Pharmacol Exp Ther 285:9–15
Mudumbi RV, Montamat SC, Bruns RF, Vestal RE (1993) Cardiac functional responses to adenosine by PD 81723, an allosteric enhancer of the adenosine A_1 receptor. Am J Physiol 264:H1017–H1022

Munch G, Kurz T, Urlbauer T, Seyfarth M, Richardt G (1996) Differential presynaptic modulation of noradrenaline release in human atrial tissue in normoxia and anoxia. Br J Pharmacol 118:1855–1861

Murry CE, Jennings RB, Reimer KA (1986) Preconditioning with ischemia: a delay of lethal cell injury in ischemic myocardium. Circ 74:1124–1136

Musser B, Morgan ME, Leid M, Murray TF, Linden J, Vestal RE (1993) Species comparison of adenosine and β-adrenoceptors in mammalian atrial and ventricular myocardium. Eur J Pharmacol 246:105–111

Nees S, Herzog V, Becker BF, Bock M, Des RC, Gerlach E (1985) The coronary endothelium: a highly active metabolic barrier for adenosine. Basic Res Cardiol 80:515–529

Newby AC (1984) Adenosine and the concept of "retaliatory metabolites." Trends Biol Sci 9:42–44

Niwa M, Hara A, Kanamori Y, Matsuno H, Kozawa O, Yoshimi N, Mori H, Uematsu T (1999) Inhibition of tumor necrosis factor-alpha induced neutrophil apoptosis by cyclic AMP: involvement of caspase cascade. Eur J Pharmacol 371:59–67

Norton GR, Woodiwiss AJ, McGinn RJ, Lorbar M, Chung ES, Honeyman TW, Fenton RA, Dobson JGJ, Meyer TE (1999) Adenosine A_1 receptor-mediated antiadrenergic effects are modulated by A_{2A} receptor activation in rat heart. Am J Physiol 276:H341–H349

Ogata T, Schubert P (1996) Programmed cell death in rat microglia is controlled by extracellular adenosine. Neurosci Lett 218:91–94

Oguchi T, Furukawa Y, Sawaki S, Kasama M, Chiba S (1995) Are negative chronotropic and inotropic responses to adenosine differentiated at the receptor or postreceptor levels in isolated dog hearts? J Pharmacol Exp Ther 272:838–844

Parenteau GL, Clark RE (1991) Prevention of ischemia-reperfusion injury by the allergy drug lodoxamide tromethamine. Ann Thorac Surg 52:832–838

Park KH, Rubin LE, Gross SS, Levi R (1992) Nitric oxide is a mediator of hypoxic coronary vasodilatation. Relation to adenosine and cyclooxygenase-derived metabolites. Circ Res 71:992–1001

Parkinson FE, Clanachan AS (1991) Adenosine receptors and nucleoside transport sites in cardiac cells. Br J Pharmacol 104:399–405

Peterfreund RA, MacCollin M, Gusella J, Fink JS (1996) Characterization and expression of the human A_{2A} adenosine receptor gene. J Neurochem 66:362–368

Phillis JW, Song D, O'Regan MH (1998) The role of adenosine in rat coronary flow regulation during respiratory and metabolic acidosis. Eur J Pharmacol 356:199–206

Pisarenko OI, Tskitishvily OV, Studneva IM, Serebryakova LI, Timoshin AA, Ruuge EK (1997) Metabolic and antioxidant effects of *R*(+/-)-N6-(2-phenylisopropyl)-adenosine following regional ischemia and reperfusion in canine myocardium. Biochim Biophys Acta 1361:295–303

Przyklenk K, Hata K, Zhao L, Kloner RA, Elliott GT (1997) Disparate effects of preconditioning and MLA on 5′-NT and adenosine levels during coronary occlusion. Am J Physiol 273:H945–H951

Raatikainen MJ, Peuhkurinen KJ, Hassinen IE (1994) Contribution of endothelium and cardiomyocytes to hypoxia-induced adenosine release. J Mol Cell Cardiol 26:1069–1080

Reeves JJ, Jones CA, Sheehan MJ, Vardey CJ, Whelan CJ (1997) Adenosine A_3 receptors promote degranulation of rat mast cells both in vitro and in vivo. Inflamm Research 46:180–184

Reimer KA, Murry CE, Yamasawa I, Hill ML, Jennings RB (1986) Four brief periods of myocardial ischemia cause no cumulative ATP loss or necrosis. Am J Physiol 251:H1306–H1315

Reppert SM, Weaver DR, Stehle JH, Rivkees SA (1991) Molecular cloning and characterization of a rat A_1-adenosine receptor that is widely expressed in brain and spinal cord. Mol Endocrinol 5:1037–1048

Rice PJ, Armstrong SC, Ganote CE (1996) Concentration-response relationships for adenosine agonists during preconditioning of rabbit cardiomyocytes. J Mol Cell Cardiol 28:1355–1365
Sanae F, Ohmae S, Kurita M, Sawanishi H, Takagi K, Miyamoto K (1995) Structure-activity relationships of alkylxanthines: alkyl chain elongation at the N1- or N7-position decreases cardiotonic activity in the isolated guinea pig heart. Jpn J Pharmacol 69:75–82
Sato T, O'Rourke B, Marban E (1998) Modulation of mitochondrial ATP-dependent K^+ channels by protein kinase C. Circ Res 83:110–114
Schlack W, Schafer M, Uebing A, Schafer S, Borchard U, Thamer V (1993) Adenosine A_2-receptor activation at reperfusion reduces infarct size and improves myocardial wall function in dog heart. J Cardiovasc Pharmacol 22:89–96
Schreieck J, Richardt G (1999) Endogenous adenosine reduces the occurrence of ischemia-induced ventricular fibrillation in rat heart. J Mol Cell Cardiol 31: 123–134
Schütz W, Schrader J, Gerlach E (1981) Different sites of adenosine formation in the heart, Am J Physiol 240:H963–H970
Shepherd RK, Duling BR (1996) Inosine-induced vasoconstriction is mediated by histamine and thromboxane derived from mast cells. Am J Physiol 270:H560–H566
Shepherd RK, Linden J, Duling BR (1996) Adenosine-induced vasoconstriction in vivo – role of the mast cell and A(3) adenosine receptor. Circ Res 78:627–634
Shiki K, Hearse DJ (1987) Preconditioning of ischemic myocardium: reperfusion-induced arrhythmias. Am J Physiol 253:H1470–H1476
Shneyvays V, Nawrath H, Jacobson KA, Shainberg A (1998) Induction of apoptosis in cardiac myocytes by an A_3 adenosine receptor agonist. Exp Cell Res 243:383–397
Shryock JC, Belardinelli L (1997) Adenosine and adenosine receptors in the cardiovascular system: biochemistry, physiology, and pharmacology [Review]. Am J Cardiol 79:2–10
Shryock JC, Snowdy S, Baraldi PG, Cacciari B, Spalluto G, Monopoli A, Ongini E, Baker SP, Belardinelli L (1998) A(2A)-Adenosine receptor reserve for coronary vasodilation. Circ 98:711–718
Shyng SL, Nichols CG (1998) Membrane phospholipid control of nucleotide sensitivity of KATP channels [see comments]. Science 282:1138–1141
Slezak J, Tribulova N, Pristacova J, Uhrik B, Thomas T, Khaper N, Kaul N, Singal PK (1995) Hydrogen peroxide changes in ischemic and reperfused heart. Cytochemistry and biochemical and X-ray microanalysis. Am J Pathol 147:772–781
Smits GJ, McVey M, Cox BF, Perrone MH, Clark KL (1998) Cardioprotective effects of the novel adenosine A_1/A_2 receptor agonist AMP 579 in a porcine model of myocardial infarction. J Pharmacol Exp Ther 286:611–618
Smits P, Lenders JWM, Willemsen JJ, Thien T (1991) Adenosine attenuates the response to sympathetic stimuli in humans. Hypertension 18:216–223
Smolenski RT, Suitters A, Yacoub MH (1992) Adenine nucleotide catabolism and adenosine formation in isolated human cardiomyocytes [errata in J Mol Cell Cardiol 1993, 25(7):887 and 1993, 25:1381]. J Mol Cell Cardiol 24:91–96
Snowdy S, Liang HX, Blackburn B, Lum R, Nelson M, Wang L, Pfister J, Sharma BP, Wolff A, Belardinelli L (1999) A comparison of an A(1) adenosine receptor agonist (CVT-510) with diltiazem for slowing of AV nodal conduction in guinea-pig. Br J Pharmacol 126:137–146
Sommerschild HT, Grund F, Offstad J, Jynge P, Ilebekk A, Kirkeboen KA (1997) Importance of endogenous adenosine during ischemia and reperfusion in neonatal porcine hearts. Circ 96:3094–3103
Stehle JH, Rivkees SA, Lee JJ, Weaver DR, Deeds JD, Reppert SM (1992) Molecular cloning and expression of the cDNA for a novel A_2-adenosine receptor subtype. Mol Endo 6(3):384–393
Stella L, Berrino L, Maione S, de Novellis V, Rossi F (1993) Cardiovascular effects of adenosine and its analogs in anaesthetized rats. Life Sci 53:755–763

Stella L, de Novellis V, Marabese I, Berrino L, Maione S, Filippelli A, Rossi F (1998) The role of A_3 adenosine receptors in central regulation of arterial blood pressure. Br J Pharmacol 125:437–440

Stepp DW, Van Bibber R, Kroll K, Feigl EO (1996) Quantitative relation between interstitial adenosine concentration and coronary blood flow. Circ Res 79:601–610

Strickler J, Jacobson KA, Liang BT (1996) Direct preconditioning of cultured chick ventricular myocytes novel functions of cardiac adenosine a(2a) and a(3) receptors. J Clin Invest 98:1773–1779

Sullivan GW, Linden J (1998) Role of A_{2A} adenosine receptors in inflammation. Drug Dev Res 45:103–112

Szewczyk A, Marban E (1999) Mitochondria: a new target for K+ channel openers? Trends Pharmacol Sci 20:157–161

Taggart PI, Sutton PMI, Oliver RM, Swanton RH (1993) Ischemic preconditioning may activate potassium channels in humans? Circ 88 [Suppl I]:69

Takagi H, King GL, Ferrara N, Aiello LP (1996b) Hypoxia regulates vascular endothelial growth factor receptor KDR/Flk gene expression through adenosine A_2 receptors in retinal capillary endothelial cells. Invest Ophthalmol Vis Sci 37:1311–1321

Takagi H, King GL, Robinson GS, Ferrara N, Aiello LP (1996a) Adenosine mediates hypoxic induction of vascular endothelial growth factor in retinal pericytes and endothelial cells. Invest Ophthalmol Vis Sci 37:2165–2176

Tanaka M, Ito H, Adachi S, Akimoto H, Nishikawa T, Kasajima T, Marumo F, Hiroe M (1994) Hypoxia induces apoptosis with enhanced expression of Fas antigen messenger RNA in cultured neonatal rat cardiomyocytes. Circ Res 75:426–433

Tang XL, Wang HX, Cho CH, Wong TM (1998) Reduced responsiveness of $[Ca^{2+}]i$ to adenosine A_1- and A_2-receptor stimulation in the isoproterenol-stimulated ventricular myocytes of spontaneously hypertensive rats. J Cardiovasc Pharmacol 31:493–498

Thomas GP, Sims SM, Cook MA, Karmazyn M (1998) Hydrogen peroxide-induced stimulation of L-type calcium current in guinea pig ventricular myocytes and its inhibition by adenosine A_1 receptor activation. J Pharmacol Exp Ther 286:1208–1214

Tofovic SP, Branch KR, Oliver RD, Magee WD, Jackson EK (1991) Caffeine potentiates vasodilator-induced renin release. J Pharmacol Exp Ther 256:850–860

Tomai F, Crea F, Gaspardone A, Versaci F, De Paulis R, Polisca P, Chiariello L, Gioffre PA (1996) Effects of A_1 adenosine receptor blockade by bamiphylline on ischaemic preconditioning during coronary angioplasty [see comments]. Eur Heart J 17:846–853

Tomai F, Danesi A, Ghini AS, Crea F, Perino M, Gaspardone A, Ruggeri G, Chiariello L, Gioffre PA (1999) Effects of K-ATP channel blockade by glibenclamide on the warm-up phenomenon. Eur Heart J 20:196–202

Tracey WR, Magee W, Masamune H, Kennedy SP, Knight DR, Buchholz RA, Hill RJ (1997) Selective adenosine a(3) receptor stimulation reduces ischemic myocardial injury in the rabbit heart. Cardiovas Res 33:410–415

Tracey WR, Magee W, Masamune H, Oleynek JJ, Hill RJ (1998) Selective activation of adenosine A_3 receptors with *N*6-(3-chlorobenzyl)-5′-*N*-methylcarboxamidoadenosine (CB-MECA) provides cardioprotection via KATP channel activation. Cardiovas Res 40:138–145

Trautwein W, Hescheler J (1990) Regulation of cardiac L-type calcium current by phosphorylation and G proteins. Annu Rev Physiol 52:257–274

Tseng CJ, Biaggioni I, Appalsamy M, Robertson D (1988) Purinergic receptors in the brainstem mediate hypotension and bradycardia. Hypertension 11:191–197

Walker BA, Rocchini C, Boone RH, Ip S, Jacobson MA (1997) Adenosine A_{2A} receptor activation delays apoptosis in human neutrophils. J Immunol 158:2926–2931

Wang D, Belardinelli L (1994) Mechanism of the negative inotropic effect of adenosine in guinea pig atrial myocytes. Am J Physiol 267:H2420–H2429

Wang D, Shryock JC, Belardinelli L (1996a) Cellular basis for the negative dromotropic effect of adenosine on rabbit single atrioventricular nodal cells. Circ Res 78:697–706
Wang J, Drake L, Sajjadi F, Firestein GS, Mullane KM, Bullough DA (1997) Dual activation of adenosine A_1 and A_3 receptors mediates preconditioning of isolated cardiac myocytes. Eur J Pharmacol 320:241–248
Wang P, Downey JM, Cohen MV (1996b) Mast cell degranulation does not contribute to ischemic preconditioning in isolated rabbit hearts. Basic Res Cardiol 91:458–467
Wang YG, Ashraf M (1999) Role of protein kinase C in mitochondrial K-ATP channel-mediated protection against Ca^{2+} overload injury in rat myocardium. Circ Res 84:1156–1165
Webb RL, McNeal RBJ, Barclay BW, Yasay GD (1990) Hemodynamic effects of adenosine agonists in the conscious spontaneously hypertensive rat. J Pharmacol Exp Ther 254:1090–1099
Wickman K, Nemec J, Gendler SJ, Clapham DE (1998) Abnormal heart rate regulation in GIRK4 knockout mice. Neuron 20:103–114
Winter CB, Cleveland JC, Butler KL, Bensard DB, Mitchell MB, Harken AH, Banerjee A (1997) Facilitative interactions between noradrenergic and purinergic signaling during preconditioning of the rat heart. J Mol Cell Cardiol 29:163–173
Wu S-N, Linden J, Visentin S, Boykin M, Belardinelli L (1989) Enhanced sensitivity of heart cells to adenosine and up-regulation of receptor number after treatment of guinea pigs with theophylline. Circ Res 65:1066–1077
Xuan YT, Tang XL, Banerjee S, Takano H, Li RCX, Han H, Qiu YM, Li JJ, Bolli R (1999) Nuclear factor-kappa B plays an essential role in the late phase of ischemic preconditioning in conscious rabbits. Circ Res 84:1095–1109
Zhang J, Belardinelli L, Jacobson KA, Otero DH, Baker SP (1997) Persistent activation by and receptor reserve for an irreversible A_1-adenosine receptor agonist in DDT1 MF-2 cells and in guinea pig heart. Mol Pharmacol 52:491–498
Zhao Z, Francis CE, Ravid K (1997) An A_3-subtype adenosine receptor is highly expressed in rat vascular smooth muscle cells: its role in attenuating adenosine-induced increase in cAMP. Microvasc Res 54:243–252
Zhou QY, Li C, Olah ME, Stiles GL, Civelli O (1992) Molecular cloning and characterization of an adenosine receptor: the A_3 adenosine receptor. Proc Natl Acad Sci USA 89:7432–7436
Zhou XB, Zhai XL, Ashraf M (1996) Preconditioning of bovine endothelial cells – the protective effect is mediated by an adenosine A_2 receptor through a protein kinase C signaling pathway. Circ Res 78:73–81

CHAPTER 17

P1 and P2 Receptors in the Renal System

E.K. JACKSON

A. Introduction

Modulation of the function of excretory organs by purine receptors is an ancient regulatory strategy adopted early in the course of vertebrate evolution. For instance, in the shark rectal gland, an excretory organ that evolved more than 400 million years ago, primordial adenosine receptors strongly influence the rate of sodium chloride transport and elimination (FORREST 1996). It is not surprising therefore that P1 and P2 receptors are important regulators of renal function in mammals. The purpose of this chapter is to summarize the available information regarding the physiology and pharmacology of renal P1 and P2 receptors and to discuss where this field will likely yield important advances in pharmacotherapeutics.

Because little to nothing is known regarding the role of A_3 receptors in renal physiology, these P1 receptor subtypes are not discussed. On the other hand, even though the renal P2 receptor field is relatively new and currently available information is narrow in scope, important recent findings in this regard enable a useful, if preliminary, analysis of the role of P2 receptors in renal physiology. The available information regarding renal A_1 and A_2 receptors is broad and deep, thus allowing a detailed discussion of receptor distribution, signal transduction mechanisms, proximate physiological responses, and opportunities for novel pharmacotherapeutic strategies based on interventions that affect A_1 and A_2 receptors.

B. Renal A_1 Receptors

I. Renal A_1 Receptor Distribution and Signal Transduction Mechanisms

A_1 receptors exist on a number of important renal structures, both vascular and tubular. Ligand binding experiments identify A_1 receptors in renal microvessels (FREISSMUTH et al. 1987), glomeruli (PALACIOS et al. 1987; FREISSMUTH et al. 1987; TOYA et al. 1993), and collecting tubules (PALACIOS et al. 1987). Ligand binding studies also detect A_1 receptors in several renal epithelial cell model systems including cells derived from cortical collecting tubular cells (SPIELMAN et al. 1992) and proximal epithelium-like OK and LLC-PK_1 cells (COULSON et al. 1996). Inasmuch as the proximal tubule appears to be the major diuretic site of action of A_1 receptor antagonists (see below), it is surprising that A_1 receptors are not detectable using ligand binding or autoradiographic approaches in whole kidney membranes, renal cortex, cortical tubules, or proximal tubular brush border membranes (WU and CHURCHILL 1985; FREISSMUTH et al. 1987; PALACIOS et al. 1987; WEBER et al. 1988; BLANCO et al. 1992). However, in situ hybridization studies reveal A_1 receptor mRNA expression in collecting tubules and in the juxtaglomerular apparatus (WEAVER and REPPERT 1992). Moreover, the more sensitive RT-PCR technique detects A_1 receptor mRNA expression in glomeruli, medullary collecting ducts, cortical and medullary thick ascending limbs, proximal convoluted and straight tubules (YAMAGUCHI et al. 1995) and in the outer medullary descending vasa recta (KREISBERG et al. 1997).

In renal epithelial cells, A_1 receptors appear to engage at least two different signal transduction pathways:

(1) $G_i \rightarrow$ decreased adenylate cyclase activity $\rightarrow$ decreased cAMP; and
(2) $G_{q/11} \rightarrow$ stimulation of PLC $\rightarrow$ increased IP_3, intracellular calcium and DAG $\rightarrow$ PKC activation.

The evidence supporting the two aforementioned signal transduction systems in the context of renal epithelial A_1 receptors is currently incomplete.

Evidence for the G_i signal transduction pathway is based mainly on cyclic AMP measurements. A_1 receptor activation decreases cyclic AMP in cultured proximal epithelium-like LLC-PK_1 cells (LEVIER et al. 1992) and OK cells (COULSON et al. 1991), in the perfused rectal gland of the dogfish shark, a model of system for the thick ascending limb of Henle's loop (KELLEY et al. 1990), in cortical collecting tubule cells (AREND et al. 1987b, 1988), in inner medullary collecting duct cells (YAGIL 1990, 1992), and in perfused inner medullary collecting ducts (EDWARDS and SPIELMAN 1994). Presumably, the reduction in cyclic AMP levels by A_1 receptor simulation is due to activation of G_i, since G_i is present in renal epithelial cells and is negatively coupled to adenylate cyclase. Moreover, inhibition of G_i with pertussis toxin inhibits the diuretic/natriuretic response to A_1 receptor blockade (KOST et al. 1998) and pertussis toxin blocks the effects of an A_1 receptor agonist on sodium transport in cultured amphibian epithelial (A6) cells (HAYSLETT et al. 1995); however, negative results with pertussis toxin are reported (MIZUMOTO et al. 1993) and more studies are required to clarify the role of G_i in the tubular effects of A_1 receptors.

Several studies support the involvement of the $G_{q/11}$ signal transduction pathway in cells with proximal and distal epithelial phenotypes. In cultured LLC-PK_1 cells, A_1 receptor activation increases intracellular levels of inositol phosphates (LEVIER et al. 1992), and in cultured OK cells, A_1 receptor activation stimulates protein kinase C activity (COULSON et al. 1991). In cultured cortical collecting tubule cells, A_1 receptor activation augments intracellular levels of calcium (AREND et al. 1988) and diacylglycerol (SCHWIEBERT et al. 1992) and increases protein kinase C activity (SCHWIEBERT et al. 1992), and in A_6 cells (distal phenotype) A_1 receptor activation increases intracellular calcium (CASAVOLA et al. 1996).

In addition to the classical signal transduction processes described above, A_1 receptors may also activate unknown pathways in renal epithelial cells that do not involve adenylate cyclase or protein kinase C. In epithelial cells from rabbit connecting tubules and cortical collecting ducts, A_1 receptor activation stimulates transepithelial calcium transport without significantly elevating intracellular levels of cyclic AMP (HOENDEROP et al. 1998). Also, neither chelation of intracellular calcium with BAPTA/AM nor down-regulation of protein kinase C with prolonged phorbol ester treatment affects A_1 receptor-induced transepithelial calcium transport (HOENDEROP et al. 1998).

Little is known regarding the signal transduction pathways by which A_1 receptor activation vasoconstricts the preglomerular microvessels, mediates/modulates tubuloglomerular feedback (TGF), and inhibits renin release. However, the effect of A_1 receptor activation on renin release is pertussis toxin sensitive (ROSSI et al. 1987a), suggesting that G_i-mediated inhibition of adenylate cyclase importantly contributes to this response. This conclusion is consistent with the well-known role of cyclic AMP as an intracellular mediator of renin secretion in the juxtaglomerular cell (JACKSON 1991).

II. Renal Systems Regulated by A_1 Receptors

1. Regulation of Preglomerular Vascular Resistance

Infusion of a selective A_1 receptor agonist into the renal interstitium decreases both cortical and medullary blood flow indicating that A_1 receptor stimulation causes vasoconstriction in both superficial and deep nephrons (Fig. 1) (AGMON et al. 1993). The reduction in blood flow induced by A_1 receptor stimulation is due primarily to an increase in the vascular resistance of preglomerular microvessels (OSSWALD et al. 1978b; MURRAY and CHURCHILL 1984, 1985), although vasoconstriction of the outer medullary descending vasa recta may contribute to the decrease in blood flow of deeper nephrons (SILLDORFF et al. 1996). A_1 receptor-mediated preglomerular vasoconstriction is strongly modulated by endogenous autacoids, particularly nitric oxide and angiotensin II. In this regard, nitric oxide (NO) attenuates A_1 receptor-mediated preglomerular vasoconstriction (BARRETT and DROPPLEMAN 1993), while angiotensin II potentiates A_1 receptor-mediated preglomerular vasoconstriction (OSSWALD et al. 1975; SPIELMAN and OSSWALD 1979; HALL and GRANGER 1986a; DERAY et al. 1990b; WEIHPRECHT et al. 1994; MUNGER and JACKSON 1994; TRAYNOR et al. 1998). The fact that not all studies report a positive synergy when angiotensin II receptors and A_1 receptors are co-activated (ROSSI et al. 1987b; BARRETT and DROPPLEMAN 1993; CARMINES and INSCHO 1994) suggests

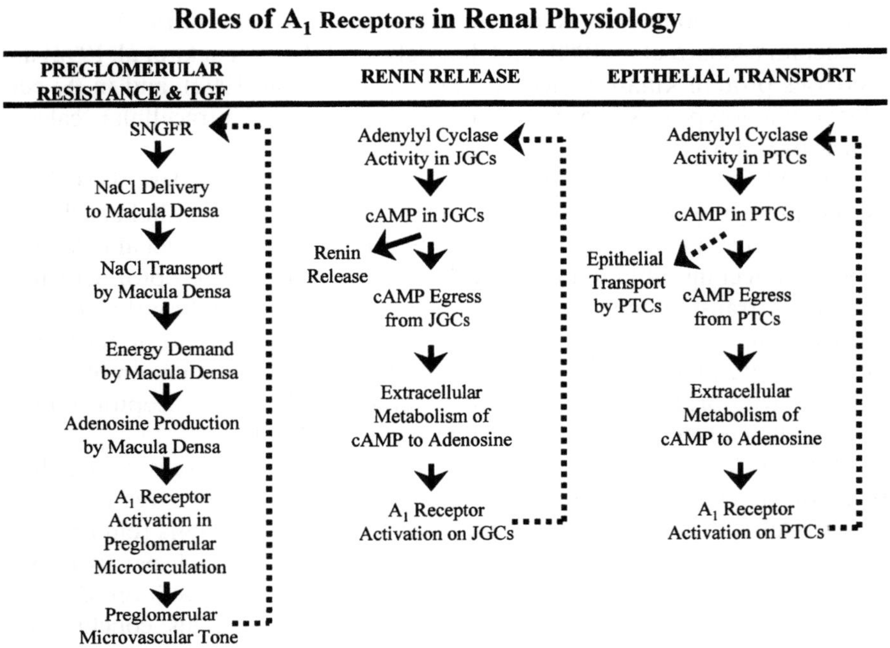

Fig. 1. Possible mechanisms by which A_1 receptors participate in the regulation of preglomerular resistance, tubuloglomerular feedback (TGF), renin release, and renal epithelial transport. *Solid and broken arrows* indicate positive and negative causal connections, respectively. JGCs, juxtaglomerular cells; PTCs, proximal tubular cells

that this interaction is itself influenced by additional factors that vary among model systems.

2. Regulation of Tubuloglomerular Feedback

Tubuloglomerular feedback (TGF) is a mechanism by which variations in loop of Henle flow rate elicit inverse changes in single nephron glomerular filtration rate (SNGFR), thus assuring that the reabsorptive capacity of the nephron is not exceeded by the load of ultrafiltrate. Whenever SNGFR exceeds the reabsorptive capacity of the proximal tubule of a given nephron, the concentration of NaCl reaching the macula densa of that nephron increases. This in turn causes the macula densa to dispatch a signal to the afferent arteriole of that nephron, which causes the afferent arteriole to constrict, thereby reducing capillary pressure, single nephron blood flow and SNGFR, thus completing the feedback loop.

As illustrated in Fig. 1 (left panel), a proposed mechanism (OSSWALD et al. 1980, 1996) for TGF is: increased NaCl concentration at the macula densa → enhanced NaCl reabsorption by the macula densa → increased energy demand and augmentation of ATP breakdown in the macula densa →

increased rate of adenosine biosynthesis from AMP in the macula densa → diffusion of adenosine from the macula densa to the adjacent afferent arteriole → afferent arteriolar vasoconstriction due to A_1 receptor activation. This aforementioned mechanism is supported by the following experimental evidence:

1. Increasing the energy utilization by the kidney markedly diminishes renal tissue ATP levels and concomitantly augments renal adenosine levels (OSSWALD et al. 1980).
2. Hypertonic saline increases adenosine release from mouse thick ascending limbs (BAUDOUIN-LEGROS et al. 1995).
3. In animals, a high sodium diet increases renal interstitial levels of adenosine by as much as 18-fold (SIRAGY and LINDEN 1996) and increases total tissue levels of adenosine in the renal cortex and medulla by approximately 2-fold (ZOU et al. 1999a).
4. Adenosine receptor antagonists such as theophylline (OSSWALD et al. 1980), DPSPX (FRANCO et al. 1989), and DPCPX (SCHNERMANN et al. 1990) inhibit TGF.
5. Exogenous adenosine deaminase, the enzyme that metabolizes adenosine to inosine, attenuates TGF (OSSWALD et al. 1980).
6. Inhibition of adenosine transport with dipyridamole or inhibition of adenosine deaminase with erythro-9-(2-hydroxy-3-nonyl)adenine augments TGF (OSSWALD et al. 1982).
7. Intraluminal (FRANCO et al. 1989) or peritubular (SCHNERMANN et al. 1990) infusions of selective A_1 agonists reduce glomerular hydrostatic pressure.
8. The reduction in renal blood flow induced by intrarenal infusions of hypertonic saline, a model system for TGF, is blocked by adenosine receptor antagonists (GERKENS et al. 1983a; CALLIS et al. 1989; DERAY et al. 1990b).

The TGF model by Osswald et al. (1980) posits that adenosine mediates TGF, and the aforementioned evidence is consistent with this hypothesis. However, the available evidence is also consistent with adenosine subserving a modulatory, rather than mediator, role in TGF. In this regard, recent experiments by THOMSON et al. (1998) support a modulator role for adenosine generated by ecto-5′-nucleotidase.

3. Regulation of Renin Release

Inasmuch as A_1 receptors are coupled to adenylate cyclase via inhibitory G-proteins and since stimulation of renin release from juxtaglomerular cells by many physiological and pharmacological stimuli involves activation of adenylate cyclase, juxtaglomerular A_1 receptors may function to restrain renin release responses (Fig. 1, middle panel). This theory is called the *adenosine-brake hypothesis* (JACKSON 1991).

The adenosine-brake hypothesis is supported by considerable evidence. For instance, both in vitro and in vivo, exogenous adenosine and selective A_1

receptor agonists decrease renin release (Tagawa and Vander 1970; Osswald et al. 1978a; Spielman 1984; Arend et al. 1984; Murray and Churchill 1985; Skøtt and Baumbach 1985; Churchill and Churchill 1985; Deray et al. 1987, 1989b; Rossi et al. 1987a; Barchowsky et al. 1987; Churchill and Bidani 1987; Lorenz et al. 1993). Moreover, pharmacological agents that increase levels of endogenous adenosine reduce renin release (Arend et al. 1986; Kuan et al. 1990b).

Another line of evidence supporting the adenosine-brake hypothesis is that pharmacological blockade of A_1 adenosine receptors with either non-selective [theophylline, caffeine and 1,3-dipropyl-8-sulfophenylxanthine (DPSPX)] or A_1 selective [1,3-dipropyl-8-cyclopentylxanthine (DPCPX) and FK453] adenosine receptor antagonists potentiates renin release in many different model systems. For example, theophylline enhances renin release in dogs (Reid et al. 1972; Langård et al. 1983), isolated rabbit afferent arterioles (Cannon et al. 1989) and in isolated perfused rabbit (Viskoper et al. 1977) and rat (Peart et al. 1975) kidneys. Caffeine stimulates renin release induced by renal artery hypotension in dogs (Deray et al. 1989a), by furosemide (Paul et al. 1989), hydralazine (Tofovic et al. 1991), and salt-depletion (Tseng et al. 1993) in rats, and by diazoxide in humans (Brown et al. 1991). DPSPX, an adenosine receptor antagonist that is restricted to the extracellular space, increases renin release in rats induced by sodium restriction (Kuan et al. 1989), hydralazine (Kuan et al. 1990b; Tofovic et al. 1991), and renal artery clipping (Kuan et al. 1990b). Moreover, the A_1 receptor selective antagonist DPCPX stimulates renin release induced by isoproterenol in rats (Pfeifer et al. 1995) and stimulates basal renin release in superfused juxtaglomerular cells (Albinus et al. 1998), and the A_1 receptor selective antagonist FK453 augments renin release in rats (Pfeifer et al. 1995) and humans (van Buren et al. 1993; Balakrishnan et al. 1993).

Most likely, a biochemical mechanism called the *cyclic AMP-adenosine pathway* provides the source of endogenous adenosine that mediates the adenosine-brake on renin release (Fig. 1, middle panel). Activation of adenylate cyclase, as occurs whenever renin release is stimulated via β_1-adrenoceptors or prostaglandin receptors, causes egress of intracellular cyclic AMP onto the surface of the cell membrane (Barber and Butcher 1981, 1983). Current evidence supports the conclusion that extracellular cyclic AMP is converted by ecto-phosphodiesterase to AMP and hence to adenosine by ecto-5′-nucleotidase (Mi et al. 1994; Mi and Jackson 1995, 1998; Dubey et al. 1996, 1998; Jackson et al. 1997). Since these reactions occur at the cell surface, stimulation of adenylyl cyclase results in adenosine biosynthesis in the biophase in which the extracellular adenosine receptors reside, thereby generating pharmacologically active concentrations of adenosine locally without increasing adenosine levels at more distal sites.

In addition to inhibiting renin release at the level of the juxtaglomerular cell, adenosine may also inhibit renin secretion via a CNS mechanism. In this regard, injections of adenosine or adenosine agonists into the CNS cause sympathoinhibition (BARRACO et al. 1986; MOSQUEDA-GARCIA 1989; SCISLO et al. 1998), and blockade of CNS adenosine receptors with caffeine enhances central sympathetic tone (SMITS et al. 1983, 1986; MOSQUEDA-GARCIA et al. 1990). Thus, adenosine formed within the CNS may brake renin release responses by inhibiting sympathetic tone and thereby reducing activation of juxtaglomerular β_1-adrenoceptors. This hypothesis is supported by the findings that adenosine receptor antagonists that gain entry into the CNS increase renin release responses more so than do antagonists that are excluded from the CNS (TOFOVIC et al. 1991) and by the observations that selective blockade of CNS adenosine receptors augments renal renin and norepinephrine secretion induced by systemic administration of vasodilators (TOFOVIC et al. 1996).

4. Regulation of Tubular Transport

Several lines of evidence strongly support the conclusion that A_1 receptors importantly regulate transport by renal epithelial cells. Stimulation of A_1 receptors in cultured OK cells (proximal phenotype) increases Na^+-glucose symport as well as Na^+-phosphate symport (COULSON et al. 1991), and in microperfused proximal convoluted tubules activation of A_1 receptors enhances basolateral Na^+–$3HCO_3^-$ symport (TAKEDA et al. 1993). Moreover, KW-3902, an A_1 receptor selective antagonist, inhibits sodium dependent phosphate transport in renal proximal tubular cells by increasing cyclic AMP (CAI et al. 1994, 1995). Thus, A_1 receptors appear to mediate enhancement of tubular transport in proximal tubular epithelial cells.

In contrast, A_1 receptors may cause inhibition of tubular transport in the thick ascending limb and collecting ducts. For instance, in the perfused rectal gland of the dogfish shark (KELLEY et al. 1990; FORREST 1996), in the isolated perfused medullary thick ascending limb (BEACH and GOOD 1992), and in cultured inner medullary collecting duct cells (MOYER et al. 1995), stimulation of A_1 receptors decreases Cl^- transport. Moreover, stimulation of A_1 receptors decreases vasopressin-induced water permeability in the perfused inner medullary collecting duct (EDWARDS and SPIELMAN 1994), and in cultured inner medullary collecting duct cells, A_1 receptors cause an increase in transepithelial resistance as well as a decrease in sodium uptake (YAGIL et al. 1994). Finally, in A6 cells (distal phenotype), A_1 receptor agonists decrease Na^+/H^+ exchange (CASAVOLA et al. 1997). However, another study using A6 cells reports A_1 receptor-mediated increases in the number times open probability of amiloride-sensitive sodium channels (MA and LING 1996), and HOENDEROP et al. (1998) report A_1 receptor-mediated increases in calcium transport in cultured collecting duct cells. Thus, while the bulk of the evidence supports inhibition of transport in the thick ascending limb and collecting duct by A_1 receptors, evidence to the contrary exists and further studies are warranted.

A number of in vivo studies describe the effects of adenosine and adenosine receptor agonists on renal excretory function. In general, intravenous infusions of non-selective and selective A_1 agonists reduce urine volume and sodium excretion (CHURCHILL 1982; COOK and CHURCHILL 1984; CHURCHILL et al. 1984; CHURCHILL and BIDANI 1987; LEVENS et al. 1991a; EDLUND and SOLLEVI 1993). In humans, intravenous infusions of adenosine decrease the renal excretion of sodium, lithium, phosphate, uric acid, chloride, and urea (BALAKRISHNAN et al. 1996b). Likewise, infusions of non-selective A_1 receptor agonists into the renal artery or in the suprarenal aorta usually (TAGAWA and VANDER 1970; OSSWALD et al. 1975, 1978b; OSSWALD 1975; SPIELMAN 1984; DERAY et al. 1987, 1989b), but not always (HALL and GRANGER 1986b; MIYAMOTO et al. 1988; EDLUND et al. 1994), reduce urine volume and sodium excretion. The rat seems to be an exception since intrarenol artery infusions of adenosine or selective A_1 receptor agonists increase, rather than decrease, sodium excretion and urine volume (MIYAMOTO et al. 1988; YAGIL 1994; FRANSEN and KOOMANS 1995). However, interstitial adenosine infusions are antiphosphaturic in diabetic rats via an A_1 receptor-mediated mechanism (PFLUEGER et al. 1998).

It is challenging to infer from the in vivo studies whether adenosine receptors have direct effects on tubular transport and whether the direct tubular effects at physiological levels of adenosine are natriuretic/diuretic or antinatriuretic/antidiuretic. In this regard, changes in blood pressure, renal blood flow (RBF), the intrarenal distribution of RBF, glomerular filtration rate (GFR), and renin release confound interpretation of the renal excretory responses to adenosine receptor agonists. Also, endogenous adenosine levels in the kidney may be sufficient to cause near saturation of A_1 receptors on tubular epithelial cells, thus further confounding the interpretation of the effects of exogenous adenosine receptor agonists on renal function. Thus, investigations with adenosine receptor antagonists are more informative than studies with adenosine receptor agonists with regard to understanding the physiological roles of endogenous adenosine/A_1 receptor interactions on tubular transport.

Numerous studies report that selective blockade of renal A_1 receptors causes diuresis and natriuresis with minimum or no effects on potassium excretion. For example, systemic administration of DPCPX (KNIGHT et al. 1993b) and 8-(dicyclopropylmethyl)-1,3-dipropylxanthine (SHIMADA et al. 1991), two highly selective A_1 receptor antagonists, increase urine volume and sodium excretion in rats. Moreover, systemic administration of equipotent doses of FK453 and DPCPX, two structurally dissimilar, highly potent and selective A_1 antagonists, induces similar increases in sodium excretion and urine volume in rats, while equal doses of the less active enantiomer of FK453, i.e., FR113452, do not alter urine volume or sodium excretion (KUAN et al. 1993). Most recently, GELLAI et al. (1998) report that the S-enantiomer of 1,3-dipropyl-8-[2-(5,6-epoxynorbornyl)] xanthine (CVT-124), one of the most potent and selective A_1 receptor antagonists yet identified, causes a marked diuresis and natriuresis in conscious rats. The maximum diuretic/natriuretic effect of CVT-124 is twice that of hydrochlorothiazide, and CVT-124 enhances the diuretic/natriuretic effect of furosemide without potentiating furosemide-induced potassium excretion. CVT-124 also induces diuresis/natriuresis in sodium-loaded animals (PFISTER et al. 1997). Racemic 1,3-dipropyl-8-[2-(5,6-epoxynorbornyl)] xanthine (CVT-117), like CVT-124, is also diuretic/natriuretic (OBERBAUER et al. 1998). Finally, direct infusions of 8-(noradamantan-3-yl)-1,3-dipropylxanthine (KW-3902), yet another potent and selective A_1 receptor antagonist, into the renal artery of dogs greatly increases urine flow, sodium excretion, and osmolar clearance (AKI et al. 1997). Because in normal animals selective A_1 antagonists have little effect on renal blood flow, glomerular filtration rate, or arterial blood pressure, the natriuretic/diuretic effects of this class of drugs cannot be ascribed to non-tubular effects. In summary, all the studies with A_1 receptor antagonists are consistent with the hypothesis that endogenous adenosine/A_1 receptor interactions enhance tubular reabsorption of sodium.

The available in vitro evidence (see above) suggests that the proximal tubule is the main site at which endogenous adenosine/A_1 receptor interactions enhance tubular sodium reabsorption. Work by VAN BUREN et al. (1993) and BALAKRISHNAN et al. (1993) suggests that blockade of A_1 receptors in humans with FK453 causes diuresis/natriuresis mostly by interfering with proximal tubular transport, although evidence for more distal effects are reported in this study.

What is the source of adenosine in vivo that activates tubular A_1 receptors? Since adenosine is filtered by the glomerulus, and one-half of filtered adenosine escapes tubular uptake (THOMPSON et al. 1985), filtered adenosine may contribute to adenosine levels in the tubular lumen. Also, both filtered nucleotides and nucleotides released by epithelial cells are converted to adenosine by proximal epithelial cells which are well-endowed with ecto-ATPase, ecto-ADPase, and ecto-5′-nucleotidase (CULIC et al. 1990; PAWELCZYK et al. 1992; LE HIR and KAISSLING 1993).

By analogy with juxtaglomerular cells and renin release, it is conceivable that the cyclic AMP-adenosine pathway contributes importantly to local adenosine formation in the tubular lumen (Fig. 1, right panel). In this regard, stimulation of adenylate cyclase increases cyclic AMP in proximal tubular epithelial cells, and cyclic AMP would inhibit epithelial transport by reducing the activity of the Na^+–H^+ antiporter in the luminal membrane as well as the Na^+–$3HCO_3^-$ symport in the basolateral membrane. However, since stimulation of adenylate cyclase would entail egress of cyclic AMP with local conversion of cyclic AMP to adenosine, activation of A_1 receptors would limit the initial stimulus to adenylate cyclase. Given the multiple mechanisms available for adenosine production in the tubular lumen, it is not surprising that the concentration of adenosine in both the urine (SVENSSON and JONZON 1990) and renal interstitium (BARANOWSKI and WESTENFELDER 1994) is several hundred nanomoles per liter, a level that would fully activate epithelial A_1 receptors.

III. Renal Diseases Involving A_1 Receptors

Activation of renal A_1 receptors engages signal transduction mechanisms which trigger several important proximate physiological responses in the kidneys (see above). These proximate responses in turn elicit renal physiological pathways that ultimately are either beneficial or detrimental to the kidneys. Thus, *adenomimetics* and *adenolytics* (herein defined as drugs that directly or indirectly increase or decrease, respectively, adenosine receptor activation) may represent important new pharmacotherapeutic strategies for the treatment of renal disease. In this regard, at least four major renal pathophysiological mechanisms may be amenable to adenomimetics or adenolytics.

1. Chronic Renal Failure

The pathophysiology of chronic renal failure (CRF), regardless of the initial insult, most likely involves a vicious cycle that results in a gradual, but inexorable, loss of renal function. The loss of nephrons due to an initial renal insult results in dilation of the afferent arterioles of the remaining nephrons. Because preglomerular vasodilation increases glomerular pressure, hyperfiltration in the remaining nephrons ensues, thus maintaining a relatively normal GFR. Although this response is adaptive in the short-term, in the long-term the increased glomerular pressure causes glomerulosclerosis with further loss of nephrons and acceleration of the vicious cycle (Brenner et al. 1996). Thus, although drugs that decrease glomerular pressure would result in an acute reduction in GFR, such drugs may delay or prevent the gradual progression to end stage renal disease. Conversely, although drugs that increase glomerular pressure may increase GFR acutely, such drugs may accelerate the rate of GFR decline.

The above considerations suggest that adenomimetics may be useful for the treatment of CRF. Stimulation of renal A_1 adenosine receptors with selective or non-selective A_1 receptor agonists reproducibly decreases GFR (Osswald et al. 1975, 1978b; Churchill 1982; Churchill et al. 1984; Pawlowska et al. 1987; Hall and Granger 1986b; Churchill and Bidani 1987; Nies et al. 1991; Levens et al. 1991a; Barrett and Wright 1994; Yagil 1994; Edlund et al. 1994). This effect is most likely mediated by preglomerular microvascular vasoconstriction induced by activation of A_1 receptors (see above), thus decreasing hydrostatic pressure in the glomerular capillaries and diminishing the driving force for glomerular filtration. In addition, inhibition of renin release, and therefore Ang II formation, by A_1 receptors (see above) may also participate. Because Ang II normally constricts postglomerular vessels, reducing the amount of Ang II would decrease Ang II-mediated postglomerular vasoconstriction which also would reduce the glomerular hydrostatic pressure. At any rate, activation of preglomerular, microvascular, and juxtaglomerular A_1 receptors with adenomimetics should reduce glomerular hydrostatic pressure and preserve renal function in CRF. Indeed, administration of selective A_1 receptor agonists (Weihprecht et al. 1994) decreases stop-flow pressure (an index of glomerular pressure) in rats, and chronic administration of an adenosine uptake inhibitor (VALLON and OSSWALD 1994) to rats with streptozotocin-induced CRF diminishes urinary protein excretion (a marker of glomerular damage). Finally, inhibition of adenosine deaminase prevents puromycin aminonucleoside nephrotoxicity (NOSAKA et al. 1997), although in this case the mechanism may involve inhibition of reactive oxygen metabolites by decreasing the substrate for xanthine oxidase.

The above considerations also imply that adenolytics may be harmful in CRF, and available data support this conclusion. Selective blockade of renal A_1 receptors causes an increase in glomerular capillary pressure (MUNGER and JACKSON 1994), and chronic administration of caffeine enhances the ability of Ang II to increase filtration fraction suggesting that blockade of renal adenosine receptors augments Ang II-induced glomerular hypertension (HOLYCROSS and JACKSON 1992). Also, chronic blockade of adenosine receptors by caffeine administration increases arterial blood pressure and reduces renal function in rats with renovascular hypertension (OHNISHI et al. 1986; KOHNO et al. 1991; CHOI et al. 1993; KOST et al. 1994) as well as in a genetic model of essential hypertension with compromised renal function (TOFOVIC and JACKSON 2000). Finally, since Ang II constricts postglomerular microvessels and increases glomerular pressure, stimulation of the renin-angiotensin system is to be avoided in patients with CRF. Indeed, several large clinical trails clearly show that blocking the renin-angiotensin system with angiotensin converting enzyme inhibitors attenuates the decline in renal function in patients with CRF (LEWIS et al. 1993; RUGGENENTI and REMUZZI 1997). Since blockade of A_1 receptors enhances renin release (see above), this is another reason to avoid adenolytic drugs in CRF. Indeed, studies by BALAKRISHNAN et al. (1996a) demonstrate that FK453 augments renin release in patients with chronic renal failure without improving glomerular filtration.

2. Acute Renal Failure Induced by Radiocontrast Agents and Renal Revascularization

Infusions of hypertonic radiocontrast media cause a rapid reduction in GFR and RBF that occasionally leads to acute renal failure (ARF) in susceptible patients (CHOU et al. 1971; ANSARI and BALDWIN 1976; KATZBERG et al. 1983). Since hypertonic radiocontrast agents cause an osmotic diuresis, the delivery of NaCl to the macula densa is increased. Therefore, it is possible that the pathophysiological mechanism of radiocontrast-induced nephropathy includes activation of TGF. Inasmuch as adenosine either mediates or modulates TGF responses, adenolytic drugs may be useful in the prevention and treatment of radiocontrast-induced ARF.

Several studies support the concept that adenolytic drugs may attenuate radiocontrast-induced nephropathy. AREND et al. (1987a) report that intrarenal injections of hypertonic radiocontrast media in dogs increases the renal excretion rate of adenosine, that theophylline, an adenosine receptor antagonist, attenuates radiocontrast-induced decreases in RBF and GFR, and that dipyridamole, an inhibitor of adenosine uptake, accentuates RBF and GFR responses to hypertonic radiocontrast media. Studies by DERAY et al. (1990a) in dogs confirm that theophylline reduces the renovascular response to radiocontrast media. In humans, although both low- and high-osmolality radiocontrast media increase urinary excretion rate of adenosine, the increase with high-osmolality radiocontrast media is greater than that caused by low-osmolality radiocontrast media, and the negative effects of high-osmolality radiocontrast media on creatinine clearance are greater than those caused by low-osmolality radiocontrast media (KATHOLI et al. 1995). As in dogs, in humans theophylline attenuates the radiocontrast-induced changes in renal function (ERLEY et al. 1994; KATHOLI et al. 1995).

Although the results with theophylline support the hypothesis that endogenous adenosine/A_1 receptor interactions are involved in the pathophysiology of radiocontrast-induced nephropathy, theophylline is a nonselective adenosine receptor antagonist and exerts other effects as well. Importantly, recent studies by ERLEY et al. (1997, 1998) demonstrate that selective blockade of A_1 receptors with DPCPX prevents radiocontrast-induced reductions in RBF and GFR in a rat model of radiocontrast-induced nephropathy. Likewise, in dogs the selective A_1 receptor antagonist, KW-3902 prevents radiocontrast-induced reductions in RBF and GFR (ARAKAWA et al. 1996; NAGASHIMA et al. 1998).

In some cases, successful revascularization of a kidney with a renal artery stenosis is not accompanied by an early improvement in renal function (DEAN 1985). One hypothesis to explain this phenomenon is that the ischemic kidney adapts by diminishing its metabolic and transport activity, so that when sudden reperfusion occurs a large quantity of ultrafiltrate escapes proximal reabsorption and activates TGF. In support of this hypothesis, ABELS et al. (1992) report that in dogs CGS 15943A, an adenosine receptor antagonist, attenuates the increase in renal vascular resistance (RVR) induced by correction of a renal artery stenosis.

In summary, both radiocontrast-induced and renal revascularization-induced ARF may be mediated by TGF, and therefore endogenous adenosine/A_1 receptor interactions may participate in the pathophysiology of these renal disorders. There is sufficient evidence to warrant clinical trials of adenolytic drugs, in particular selective A_1 receptor antagonists, for the prevention and treatment of radiocontrast-induced nephropathy as well as renal revascularization-induced ARF.

3. Acute Tubular Necrosis-Induced Acute Renal Failure

ARF is a precipitous reduction in GFR leading to accumulation of nitrogenous wastes that affects 5% of hospitalized patients and is associated with a high mortality rate. Approximately two-thirds of all intrinsic ARF is secondary to acute tubular necrosis (ATN), and the mortality rate in patients with ATN-induced ARF is approximately 50%. Irrespective of the nephrotoxic insult, the pathophysiology of ATN-induced ARF involves physical obstruction of tubules with epithelial cells, backleak of ultrafiltrate through a disrupted epithelial barrier, and functional dysregulation of the renal microcirculation.

It is theoretically possible that endogenous adenosine/A_1 receptors participate in ATN-induced ARF. Because damage to epithelial cells impairs energy generation, depletion of ATP and ADP and accumulation of AMP and adenosine may occur in injured epithelial cells, and adenosine-mediated constriction of the preglomerular microcirculation due to activation of A_1 receptors could result in a functional reduction in GFR. Moreover, because epithelial cell damage also decreases reabsorptive capacity of the proximal tubule, increased distal delivery of NaCl could diminish GFR via activation of TGF, a process mediated or modulated by endogenous adenosine/A_1 receptor interactions (see above).

The hypothesis that endogenous adenosine/A_1 receptor interactions participate in ATN-induced ARF is supported by numerous studies involving several different nephrotoxins. Administration of adenosine receptor antagonists before or soon after injection of glycerol attenuates glycerol-induced ARF. In this regard, the non-selective antagonists aminophylline (BIDANI and CHURCHILL 1983) and theophylline (BIDANI et al. 1987), the relatively non-selective antagonist 8-phenyltheophylline (BOWMER et al. 1986), and several highly A_1 receptor selective antagonists including DPCPX (KELLETT et al. 1989; PANJEHSHAHIN et al. 1991b), 8-(dicyclopropylmethyl)-1,3-dipropylxanthine (SHIMADA et al. 1991), FK453 (ISHIKAWA et al. 1993a; ANDOH et al. 1991), and KW-3902 (SUZUKI 1992; SUZUKI et al. 1992) attenuate glycerol-induced ARF. The observations in the rat that glycerol increases circulating levels of adenosine (ISHIKAWA et al. 1993b), increases the renovascular response to adenosine (GOULD et al. 1995), and increases A_1 receptor density and mRNA levels (GOULD et al. 1997) add credibility to the hypothesis that adenosine is involved in glycerol-induced ARF.

In addition to glycerol-induced renal failure, pretreatment or co-treatment with adenosine receptor antagonists attenuates ARF caused by several other nephrotoxic drugs that directly damage renal epithelial cells. For instance, the non-selective antagonist aminophylline (HEIDEMANN et al. 1989) and the highly A_1-selective antagonists DPCPX (KNIGHT et al. 1991), FK453 (ANDOH et al. 1991), and KW-3902 (SUZUKI 1992; NAGASHIMA et al. 1995) attenuate cisplatin-induced ARF. Also, maneuvers that block TGF, such as a high salt diet and furosemide, attenuate cisplatin-induced ARF (HEIDEMANN et al. 1985), suggesting that TGF, and by implication adenosine, participates in this form of ARF. The A_1 selective antagonists FK453 (ANDOH et al. 1991) and KW-3902 (YAO et al. 1994) reduce gentamicin-induced ARF, and KW-3902 (NAGASHIMA et al. 1994) attenuates cephaloridine-induced nephrotoxicity. Amphotericin B nephrotoxicity is reduced by aminophylline (GERKENS et al. 1983b; HEIDEMANN et al. 1983a), theophylline (HEIDEMANN et al. 1991), and inhibition of TGF with salt repletion (HEIDEMANN et al. 1983b; BRANCH et al.

1987; Ohnishi et al. 1989; Llanos et al. 1991), but not by DPSPX, an adenosine receptor antagonist that is restricted to the extracellular space (Kuan et al. 1990a). Also, amphotericin B does not increase urinary adenosine excretion in patients (Carlson et al. 1997). $HgCl_2$-induced ARF (Rossi et al. 1990) and cyclosporin-induced (Gerkens and Smith 1985; Panjehshahin et al. 1991a) ARF are resistant to adenosine receptor blockade. However, caffeine and theophylline attenuate cyclosporin-induced glomerular contraction in vitro (Potier et al. 1997) and the selective A_1 receptor antagonist FK453 modestly attenuates the decline in GFR and RBF induced by cyclosporin (Balakrishnan et al. 1996c). Moreover, a recent study demonstrates that cyclosporin in kidney transplantation recipients decreases adenosine uptake into red blood cells and increases circulating levels of adenosine (Guieu et al. 1998). Additional studies are required to clarify the role of adenosine in both amphotericin B as well as cyclosporin-induced nephrotoxicity.

Renal ischemia/hypoxia is a powerful stimulus for the renal production of adenosine (Osswald et al. 1977; Miller et al. 1978; Ramos-Salazar and Baines 1986), and adenosine may participate in ARF caused by renal ischemia/hypoxia. Theophylline (Osswald et al. 1977; Sakai et al. 1979; Lin et al. 1986, 1988; Gouyon and Buignard 1988) and adenosine α,β-methylene diphosphate (van Waarde et al. 1989), an inhibitor of ecto-5′-nucleotidase, attenuate, whereas dipyridamole (Lin et al. 1987), an inhibitor of adenosine uptake, potentiates ischemia/hypoxia-induced ARF.

Although hypothyroidism-induced renal dysfunction is not due to ATN, hypothyroidism is associated with renal dysfunction characterized by renal vasoconstriction and alterations in tubular transport. Blockade of adenosine receptors with DPSPX increases single nephron GFR, single nephron blood flow, and the ultrafiltration coefficient, while decreasing afferent and efferent resistances in thyroidectomized rats (Franco et al. 1996). In contrast, in normal rats, DPSPX does not modify glomerular hemodynamics. In the same study, in normal rats exogenous adenosine increases afferent and efferent resistances, while decreasing single nephron GFR, single nephron blood flow, and the ultrafiltration coefficient. Also, adenosine transport in brush border membrane vesicles from hypothyroid rat kidneys appears to be diminished (Martínez et al. 1997). Thus, adenosine may contribute to hypothyroidism-induced renal dysfunction.

Although the aforementioned studies clearly indicate that A_1 receptor antagonists prevent many forms of experimental ATN-induced ARF, they do not address the clinically important issue of whether this class of drugs reverses established ATN-induced ARF. In fact the available data suggest that adenosine is not involved in established nephrotoxin-induced ARF. For instance, intrarenal infusion of an A_1 agonist (N^6-cyclopentyladenosine; CPA) in normal animals cannot mimic the reductions in GFR and the increases in RVR caused by cisplatin as would be anticipated if endogenous adenosine plays an important role with regard to altering renal hemodynamics during established ARF (Jackson et al. 1995). Moreover, in animals with established cisplatin-induced ARF, the GFR and RVR responses to CPA are not altered compared with control animals, suggesting that endogenous adenosine is not occupying A_1 receptors in established cisplatin-induced ARF (JACKSON et al. 1995). Also, in established cisplatin-induced ARF, renal cortical levels of adenosine, inosine, xanthine, and hypoxanthine are not elevated, and DPCPX, a highly selective A_1 receptor antagonist, does not improve renal function in established cisplatin-induced ARF (JACKSON et al. 1995).

The finding that adenosine is not involved in established nephrotoxin-induced ARF indicates a need to re-evaluate the hypothesis that adenosine participates in nephrotoxin-induced ARF by functionally constricting the preglomerular microcirculation. In this regard, a likely explanation for the effectiveness of A_1 receptor antagonists to prevent, but not reverse, ATN-induced ARF is that endogenous adenosine facilitates reabsorption of ultrafiltrate by the proximal tubule. This would increase the concentrations of nephrotoxins in the tubular lumen, result in a greater diffusion of nephrotoxins into proximal epithelial cells, and thus enhance epithelial cell injury. Conversely, antagonism of A_1 receptors in the proximal tubule would decrease reabsorption of ultrafiltrate, would reduce the concentration of nephrotoxins, and thus would protect proximal tubular epithelial cells. Several observations support this hypothesis. First, as already mentioned, A_1 receptor antagonists prevent, but do not reverse, nephrotoxin-induced ARF. Second, the diuretic site-of-action of adenosine receptor antagonists is primarily the proximal tubule, and adenosine receptor antagonists are much more effective against nephrotoxins (e.g., glycerol, cisplatin, gentamicin, and cephaloridine) that selectively injure this region of the nephron compared with nephrotoxins that cause widespread tubular injury (e.g., amphotericin B and cyclosporin). Third, acetazolamide, a diuretic that inhibits transport in the proximal tubule, prevents, but does not reverse, cisplatin-induced ARF (HEIDEMANN et al. 1985).

Other forms of ATN-induced ARF may also be amenable to treatment with adenolytics. Gram-negative sepsis/endotoxemia can induce ARF, and mortality in such patients is more than 60%. The observation that endotoxin can release adenosine from endothelial cells (ROUNDS et al. 1994) suggests a possible role for adenosine, and studies by KNIGHT et al. (1993a) demonstrate that administration of the selective A_1 receptor antagonist DPCPX attenuates somewhat endotoxin-induced ARF in rats.

4. Edema

Adenolytics, particularly A_1 receptor antagonists, may be clinically useful diuretics. As described above, A_1 receptor antagonists inhibit tubular transport, predominantly in the proximal tubule but also in the collecting duct, and induce a brisk natriuresis/diuresis with little or no effect on potassium excretion (TERAI et al. 1990; SHIMADA et al. 1991; SUZUKI et al. 1992; KUAN et al. 1993; KNIGHT et al. 1993b; MIZUMOTO et al. 1993; VAN BUREN et al. 1993; BALAKRISHNAN et al. 1993). Moreover, CVT-124, a highly selective A_1 receptor antagonist, uncouples proximal tubular reabsorption from single nephron GFR, thus preventing TGF-mediated reductions in GFR and RBF from being triggered by increases in distal delivery of NaCl (WILCOX et al. 1999). Therefore, adenolytic-based diuretic strategies may be useful for the treatment of edematous conditions such as heart failure, liver disease, and renal disease, particularly in patients with diuretic resistance or in patients who cannot tolerate other diuretics because of reductions in GFR or RBF. Also, antagonism of renal A_1 receptors may be antihypertensive in salt-sensitive hypertension (NOMURA et al. 1995). Clinical trials are warranted to establish whether chronic administration of A_1 receptor antagonists induces a sustained decrease in total body sodium and whether A_1 receptor antagonists augment the diuretic efficacy of loop diuretics and/or thiazide diuretics by sequential blockade of nephron segments.

C. Renal A_2 Receptors

It is difficult to distinguish A_{2A} and A_{2B} receptors using currently available A_2 receptor agonists and antagonists, and consequently the majority of published pharmacological studies do not unequivocally discriminate between these A_2 receptor subtypes. In the material that follows, except where the data is clearly stated to address the issue, the A_{2A} and A_{2B} receptors are not differentiated. Therefore, the term "A_2 receptor", meaning either subtype, is mostly employed in the subsequent discussion.

I. Renal A_2 Receptor Distribution and Signal Transduction Mechanisms

Radioligand binding studies reveal A_2 receptors in whole kidney membranes (WU and CHURCHILL 1985) and A_{2A} receptor mRNA, but not A_{2B} receptor mRNA, is expressed in the renal papilla and inner medulla at a location outside the collecting duct (WEAVER and REPPERT 1992). Using RT-PCR, KREISBERG et al. (1997) report expression of both A_{2A} and A_{2B} receptor mRNA in the outer medulla and outer medullary descending vasa recta. In cultured cortical collecting duct cells (AREND et al. 1987b; HOENDEROP et al. 1998), connecting tubule cells (HOENDEROP et al. 1998), LLC-PK$_1$ cells (LEVIER et al. 1992), and A6 cells (CASAVOLA et al. 1997), and in isolated renal cortical tubules (LANG et al. 1985), glomeruli (FREISSMUTH et al. 1987), and renal microvessels (FREISSMUTH et al. 1987), renal A_2 receptors stimulate cyclic AMP production. Most likely stimulation of adenylate cyclase represents the predominant signal transduction pathway employed by renal A_2 receptors.

II. Renal Systems Regulated by A_2 Receptors

Currently there is little evidence that A_2 receptors *directly* regulate tubular epithelial transport. For instance, the diuretic/natriuretic effects of adenosine

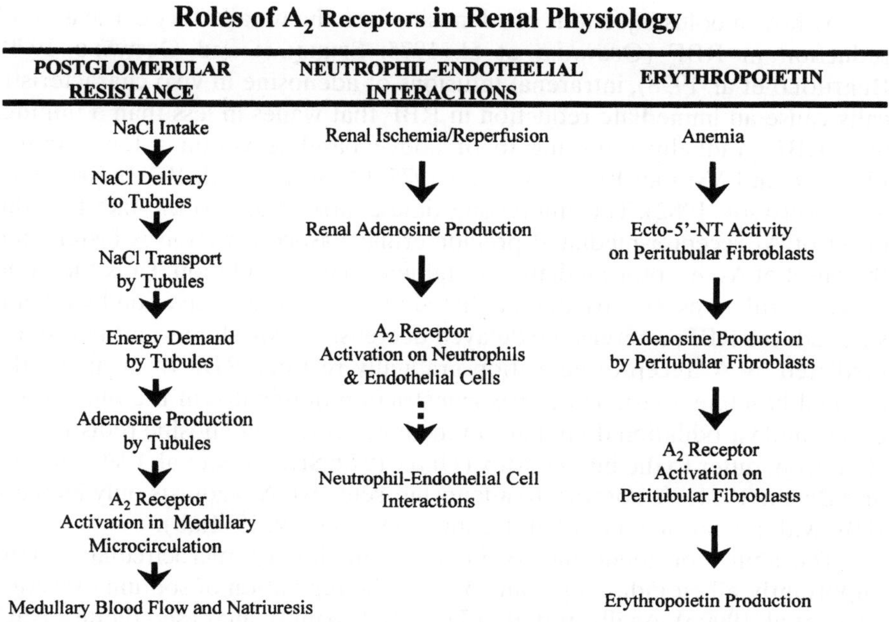

Fig. 2. Possible mechanisms by which A_2 receptors participate in the regulation of postglomerular resistance, sodium excretion, neutrophil-endothelial cell interactions, and erythropoietin production. *Solid and broken arrows* indicate positive and negative causal connections, respectively. Ecto-5′-NT, ecto-5′-nucleotidase

receptor antagonists correlate with A_1, rather than A_2, receptor affinity (Suzuki et al. 1992). Although in A6 cells A_2 receptors inhibit amiloride-sensitive sodium channels (Ma and Ling 1996), A_2 receptors stimulate Na^+/H^+ exchange activity (Casavola et al. 1997), and the net effect on sodium transport appears to be positive (Lang et al. 1985). A_2 receptors also stimulate renin release (Churchill and Churchill 1985); however, the physiological relevance of this finding is unclear.

Three important renal systems may be regulated by A_2 receptors (Fig. 2):

1. Postglomerular vascular resistance
2. Neutrophil-endothelial cell interactions in the kidney
3. Erythropoietin (EPO) production by peritubular fibroblasts

1. Regulation of Postglomerular Vascular Resistance

In contrast to A_1 receptor activation, which causes predominantly vasoconstriction of superficial and deep preglomerular microvessels, A_2 receptor activation in the renal microcirculation leads to vasodilation of postglomerular vessels in the inner cortex and medulla, but not in the outer cortex (Spielman et al. 1980; Murray and Churchill 1984, 1985; Agmon et al. 1993).

Although bolus injections of adenosine into the renal artery cause a rapid reduction in RBF (Osswald et al. 1975; Spielman and Osswald 1979; Berthold et al. 1998), intrarenal infusions of adenosine in vivo characteristically cause an immediate reduction in RBF that wanes in less than a minute, with RBF gradually returning to or above baseline within a few minutes (Tagawa and Vander 1970; Osswald 1975; Osswald et al. 1978b; Spielman and Thompson 1982). This interesting time-course is due to the fact that the onset of A_1 receptor-mediated preglomerular vasoconstriction is faster than the onset of A_2 receptor-mediated postglomerular vasodilation. Thus, the rapid preglomerular vasoconstriction mediated by A_1 receptor activation by adenosine reduces RBF; whereas the delayed decrease in postglomerular resistance mediated by A_2 receptor activation gradually restores RBF to or above the original baseline. Inasmuch as vasoconstriction dominates in the outer renal cortex and vasodilation dominates in the inner cortex, adenosine redistributes RBF from outer to the inner cortex (Ueda 1972; Spielman et al. 1980; Dinour and Brezis 1991). In contrast to adenosine, selective A_2 agonists only increase RBF without causing an initial decline (Levens et al. 1991a,b).

The ability of adenosine to dilate the medullary microcirculation may importantly, albeit indirectly, contribute to the regulation of sodium excretion (Zou et al. 1999b). As illustrated in Fig. 2 (left panel), increased dietary NaCl intake results in greater NaCl delivery to the renal tubules, which in turn increases tubular transport and energy demand. This sequence of events increases adenosine production, particularly in the medulla, thus activating A_2 receptors in the medullary microcirculation. Stimulation of A_2 receptors in the medullary microcirculation increases medullary blood flow, which alters peritubular forces governing sodium reabsorption. The net result is an enhanced excretion of NaCl. Therefore, whereas A_1 receptors directly enhance NaCl reabsorption by stimulating transport by renal epithelial cells, A_2 receptors indirectly decrease NaCl reabsorption by augmenting renal medullary blood flow.

2. Regulation of Neutrophil-Endothelial Cell Interactions in the Kidney

A_2 receptor activation is well known to inhibit neutrophil-endothelial cell interactions in vitro (Cronstein 1994). More importantly, A_2 receptor activation in vivo attenuates neutrophil-endothelial interactions following ischemia/reperfusion injury (Nolte et al. 1992). In this regard, Okusa et al. (1998) demonstrate that selective activation of A_{2A} receptors markedly diminishes the infiltration of neutrophils into kidneys subjected to 45 min of renal ischemia followed by 48 h of reperfusion. As illustrated in Fig. 2 (middle panel), it is possible that renal adenosine production in response to renal ischemia functions to protect the kidney from neutrophil attack.

3. Regulation of Erythropoietin Production by Peritubular Fibroblasts

Several reports indicate that adenosine receptors importantly regulate erythropoietin production by peritubular fibroblasts. In this regard, Paul et al.

(1988) report that in the isolated, perfused rat kidney exogenous adenosine increases the secretion of erythropoietin and that in hypoxic rats adenosine deaminase inhibition increases and adenosine deaminase diminishes erythropoietin levels. UENO et al. (1988) found that A_2 receptor activation increases erythropoietin levels in exhypoxic polycythemic mice. Additional reports confirm that stimulation of A_2 receptors enhances erythropoietin synthesis and release by a mechanism involving stimulation of adenylate cyclase (NAKASHIMA et al. 1991, 1993a,b, 1994; OHIGASHI et al. 1993). Consistent with a role for adenosine in the regulation of erythropoietin production are the results by KAISSLING et al. (1993) demonstrating an increase in the expression of ecto-5′-nucleotidase on peritubular fibroblasts in response to anemia. Also, renal peritubular type I fibroblasts co-express erythropoietin, 5′-nucleotidase, and an oxygen sensing mechanism (BACHMANN and RAMASUBBU 1997). It is possible, therefore, that anemia may stimulate erythropoietin production by peritubular fibroblasts via a mechanism involving activation of ecto-5′-nucleotidase, adenosine formation, and A_2 receptor activation (Fig. 2, right panel).

However, not all investigators record positive results. TAN and RATCLIFF (1992) describe no change in renal erythropoietin mRNA levels in perfused rat kidneys exposed to an adenosine receptor agonist. GLEITER et al. (1996, 1997a) report that neither theophylline nor dipyridamole modify the increase in plasma erythropoietin levels induced by hemorrhage or hypobaric hypoxia in humans. GLEITER et al. (1997b) also find no evidence in rats for modulation of carbon monoxide-induced erythropoietin production by endogenous or exogenous adenosine. Additional studies are required to clarify the current controversy regarding the role of adenosine in the regulation of erythropoietin production.

III. Renal Diseases Involving A_2 Receptors

1. Renal Ischemia/Reperfusion Injury

In animals subjected to 45 min of ischemia followed by 48 h of reperfusion, serum creatinine and blood urea nitrogen levels are markedly elevated, extensive tubular epithelial necrosis is present, vascular congestion in the outer medulla occurs, and neutrophils infiltrate the injured kidney. These changes are mostly abrogated by the activation of A_{2A} receptors (OKUSA et al. 1998). Thus, A_{2A} receptor agonists may have clinical utility to protect the kidneys against ischemia/reperfusion injury.

2. Transplantation-Induced Erythrocytosis

Approximately 15% of renal-transplant recipients develop erythrocytosis and A_2 receptors may be involved since adenosine stimulates erythropoietin production via A_2 receptors (see above). Administration of theophylline, a nonselective A_2 receptor antagonist, for eight weeks to renal-transplant recipients with erythrocytosis reduces erythropoietin levels, hematocrit and red cell mass, and eliminates the need for weekly phlebotomy (BAKRIS et al. 1990). More than 50% of patients may benefit from theophylline therapy (ILAN et al. 1994). Additional clinical trials examining the effects of selective A_2 receptor antagonists on red cell mass in renal-transplant recipients with erythrocytosis are warranted.

D. Renal P2 Receptors

I. Renal P2 Receptor Distribution and Signal Transduction

Autoradiographic studies in the rat kidney demonstrate $[^3H]\alpha,\beta$-methylene ATP, a P2X receptor agonist, binding to interlobular arteries and afferent arterioles, but not to efferent arterioles (CHAN et al. 1998a). Of the seven known P2X subtypes, only $P2X_1$ and $P2X_3$ bind α,β-methylene ATP with high affinity (RALEVIC and BURNSTOCK 1998), and $P2X_3$ receptors are found predominantly on sensory neurons (CHEN et al. 1995). Therefore, as a first approximation, $[^3H]\alpha,\beta$-methylene ATP binding in the renal microcirculation can be equated with $P2X_1$ receptor expression. Immunohistochemical studies with antibodies to $P2X_1$, $P2X_2$, $P2X_3$, $P2X_4$, $P2X_5$, $P2X_6$, and $P2X_7$ receptors detect expression of only $P2X_1$, $P2X_2$, and $P2X_6$ in the rat renal vasculature with localization to the preglomerular microcirculation (CHAN et al. 1998b,c). In this regard, $P2X_1$ receptors are dominant in the afferent arterioles, whereas $P2X_2$ receptors are predominant in the interlobar, arcuate, and interlobular arteries. Immunohistochemical experiments do not detect $P2X_1$ receptors in efferent arterioles or tubules (CHAN et al. 1998b,c); however, such studies do detect $P2X_5$ and $P2X_6$ receptor expression in the tubules with $P2X_5$ receptors localized to the thick ascending limb, distal tubule and collecting ducts, whereas $P2X_6$ receptors are present all along the tubule (CHAN et al. 1998d). Interestingly, in rats made diabetic with streptozotocin, $P2X_7$ receptors can be detected with immunohistochemistry in the renal microcirculation and glomeruli (CHAN et al. 1998a). Also, Northern blot analysis detects $P2X_7$ receptor mRNA in cultured rat mesangial cells (SCHULZE-LOHOFF et al. 1998). Since $P2X_7$ receptors may function to induce cell lysis (RALEVIC and BURNSTOCK 1998), $P2X_7$ receptors may be involved in the pathophysiology of diabetic nephropathy.

P2X receptors are ligand-gated ion channels (RALEVIC and BURNSTOCK 1998), and this mechanism most likely accounts for the renovascular effects of activation of P2X receptors in the kidney. INSCHO et al. (1995) find that the initial and sustained renovascular effects of P2X receptor activation are diminished and abolished, respectively, by blockade of L-type (voltage-dependent) calcium channels. Moreover, removal of extracellular calcium abolishes both the initial and sustained renovascular effects of P2X receptor activation, whereas depletion of intracellular calcium stores with thapsigargin does not modify the renovascular response to P2X receptor activation. These data are consistent with the conclusion that P2X receptor activation causes renovascular effects via influx of cations via P2X receptor-gated ion channels followed by depolarization with activation of voltage-dependent calcium channels. This inference is strongly supported by the observation that the increase in intracellular calcium level induced by ATP in preglomerular vascular smooth muscle cells is markedly attenuated by either blockade of voltage-dependent calcium channels or removal of calcium from the extracellular compartment (INSCHO et al. 1999).

Scant information regarding renal P2Y receptor distribution is available in the public domain. KISHORE et al. (1998) report that $P2Y_2$ receptor mRNA is abundantly expressed in renal cortex, inner medulla, outer medulla, and microdissected collecting ducts. Moreover, these same investigators find that $P2Y_2$ protein (immunoblots and immunocytochemistry) is expressed in the inner medulla, principal cells of the collecting duct (both basolateral and apical aspects), thin ascending limb of Henle's loop, and vascular structures of the inner medulla (KISHORE et al. 1998). In many vascular beds, endothelial cells express P2Y receptors linked to stimulation of constitutive nitric oxide synthase (RALEVIC and BURNSTOCK 1998), and it appears that the renal vascular bed is no exception (see below).

II. Renal Systems Regulated by P2 Receptors

1. Regulation of Renovascular Resistance

P2 receptors may importantly contribute to the regulation of the preglomerular microcirculation (INSCHO et al. 1994; NAVAR et al. 1996). The pharmacological evidence indicates that P2X and P2Y receptors co-exist in the renal vasculature, with P2X receptors located on vascular smooth muscle cells causing vasoconstriction and P2Y receptors residing on endothelial cells causing indirect vasodilation by stimulating the release of NO. For instance, in the dog, ATP infusions into the renal artery decrease renovascular resistance (TAGAWA and VANDER 1970; MAJID and NAVAR 1992). In contrast, in dog kidneys pretreated with an NO synthase inhibitor, ATP induces renal vasoconstriction (MAJID and NAVAR 1992). Although in perfused rat and rabbit kidneys ATP is a renal vasoconstrictor (NEEDLEMAN et al. 1970; CHURCHILL and ELLIS 1993a; VARGAS et al. 1996), 2-methylthio-ATP, a non-selective P2Y receptor agonist, vasodilates the perfused rat kidney, whereas α,β-methylene-ATP, a selective P2X receptor agonist, vasoconstricts this preparation (CHURCHILL and ELLIS 1993a). Importantly, NO synthase inhibition ablates the renal vasodilation induced by 2-methylthio-ATP in the perfused rat kidney (CHURCHILL and ELLIS 1993a). In an extensive series of experiments in the perfused rat kidney using P2 agonists, P2 antagonists, NO synthase inhibitors and removal of endothelium, ELTZE and ULLRICH (1996) confirm the hypothesis that smooth muscle cell P2X receptors mediate renal vasoconstriction and endothelial cell P2Y receptors induce renal vasodilation. The role of endothelial P2Y receptors in mediating renal vasodilation is extended to rabbits and humans by the studies of RUMP et al. (1996, 1998) in isolated renal artery segments. However, COX and SMITS (1996) report that, whereas marked renal vasoconstriction follows intravenous infusions of the P2X agonist β,γ-methylene-ATP, renal vasodilation is not induced by intravenous infusions of the P2Y agonist 2-methylthio-ATP. Moreover, using the in vitro blood-perfused juxtamedullary nephron technique, INSHCO et al. (1998) find evidence for vasoconstrictor, rather than vasodilator, $P2Y_2$ receptors in rat afferent arterioles. Therefore, the functional significance of vasodilatory P2Y receptors in the renal microcirculation depends strongly on the experimental model.

The renovascular effects of P2X receptor activation are restricted to the preglomerular circulation. Using the in vitro blood-perfused juxtamedullary nephron technique, INSCHO et al. (1991, 1992) demonstrate that P2X receptor agonists vasoconstrict arcuate and interlobular arteries and afferent, but not efferent, arterioles. WEIHPRECHT et al. (1992) also report P2X receptor-induced vasoconstriction of the isolated rabbit afferent arteriole preparation. Interestingly, vasoconstriction of arcuate and interlobular arteries in response to P2X receptor agonists is transient, whereas the vasoconstriction of afferent arterioles is sustained (INSCHO et al. 1992). It is likely that the different time course of the response of preglomerular vascular elements to P2X agonists is due to the fact that $P2X_1$ receptors predominate in afferent arterioles, whereas $P2X_2$ receptors prevail in arcuate and interlobular arteries (CHAN et al. 1998b,c).

Studies by Inscho and coworkers (INSCHO 1996; INSCHO et al. 1996) strongly indicate that P2X receptors in the afferent arteriole make an important contribution to autoregulatory modulation of afferent arteriolar caliber. In this regard, either desensitization of, saturation of, or blockade of P2 receptors attenuates pressure-induced vasoconstriction of the afferent arteriole.

2. Regulation of Tubuloglomerular Feedback

MITCHELL and NAVAR (1993) found that saturating doses of either ATP or β,γ-methylene-ATP attenuate TGF changes in stop-flow pressure induced by late proximal perfusion. These findings, along with the findings of Inscho and coworkers that sustained P2X receptor-mediated renovascular vasoconstriction is confined to the afferent arteriole, suggest that P2X receptors importantly mediate or modulate TGF.

3. Regulation of Renal Sympathetic Neurotransmission

In the rat kidney, renal vasoconstriction induced by low frequency nerve stimulation is resistant to α-adrenoceptor blockade, yet is inhibited by α,β-methylene ATP (SCHWARTZ and MALIK 1989). These data suggest that ATP serves as a co-transmitter in renal sympathetic nerves and contributes to sympathetically-induced renal vasoconstriction by activating P2X receptors. However, subsequent studies in the rat kidney demonstrate that the blockade of renovascular responses to sympathetic nerve stimulation by α,β-methylene ATP is not due to desensitization of P2 receptors (SEHIC et al. 1994). Nonetheless, studies by BOHMANN et al. (1997) demonstrate using a variety of pharmacological approaches that co-released ATP acts on prejunctional P2 receptors to inhibit neurotransmitter release. Parenthetically, P2X receptors in the nucleus tractus solitarius appear to mediate inhibition of renal sympathetic nerve activity (SCISLO et al. 1997). Additional studies are required to clarify the role of ATP and P2 receptors in renal sympathetic neurotransmission.

4. Regulation of Tubular Transport

There is growing evidence that P2 receptors modulate tubular transport (FRIEDLANDER and AMIEL 1995). In MDCK cells, a cell line with distal phenotype, ATP increases short circuit current, the transepithelial potential difference, ionic conductance, potassium conductance, transepithelial chloride transport, phosphatidylcholine hydrolysis via phospholipase C activation, arachidonic acid release via phospholipase A_2 activation, cellular cyclic AMP levels via increased prostaglandin production, and inositol phosphate accumulation (SIMMONS 1981; LANG et al. 1988; PAULMICHL and LANG 1988; FRIEDRICH et al. 1989; ZEGARRA-MORAN et al. 1995; FIRESTEIN et al. 1996; POST et al. 1996; YANG et al. 1997). In A6 cells, another cell line with distal phenotype, P2 receptor agonists modulate Na^+–K^+–$2Cl^-$ symport and cation conductance (MIDDLETON et al. 1993; MORI et al. 1996). Evidence in cultured LLC-PK_1, a cell line with mixed proximal/distal phenotype, also suggests that P2 receptors cause release of intracellular calcium and activation of protein kinase C activity (WEINBERG et al. 1989; ANDERSON et al. 1991). In microdissected rat terminal medullary collecting ducts, ATP causes a rapid and sustained increase in intracellular calcium (ECELBARGER et al. 1994). In addition, in perfused rat terminal medullary collecting ducts, ATP inhibits vasopressin-induced osmotic water permeability via a mechanism involving activation of protein kinase C (KISHORE et al. 1995). Similarly, in perfused rabbit cortical collecting tubules, ATP inhibits the hydrosmotic effects of vasopressin (ROUSE et al. 1994). Also, ATP increases intracellular calcium in microdissected rat thin limbs of Henle (BAILEY et al. 1998). Finally, NANOFF et al. (1990) report P2 receptor-mediated enhancement of inositol phosphate formation in rat renal cortical slices.

Given the large array of P2 receptor subtypes and lack of adequately selective agonists and antagonists, it is difficult to deduce precisely which P2 receptor subtypes are involved in a given cellular response. However, in most of the studies cited in the previous paragraph, ATP and UTP elicit similar effects with similar potency. Thus, it appears that P2Y receptors, most likely $P2Y_2$ receptors, importantly modulate tubular transport, particularly in the more distal reaches of the nephron. However, a P2X receptor in the proximal tubule may inhibit sodium-dependent phosphate transport (LEDERER et al. 1995).

With regard to P2 receptors and renal excretory function, no doubt the reader notes the lack of in vivo studies. Well designed studies in intact animals are needed to determine the net effects of P2 receptor agonists and antagonists on renal excretory function in the intact organism.

5. Regulation of Other Renal Systems

P2Y receptors stimulate renin release via increased production of NO (CHURCHILL and ELLIS 1993b) and activate the hydrolysis of polyphosphoinositides in mesangial cells (PFEILSCHIFTER 1990a,b). P2 receptors stimulate proliferation of mesangial cells (SCHULZE-LOHOFF et al. 1992) and enhance DNA synthesis in and proliferation of proximal tubular cells (HUMES and CIESLINSKI 1991; PALLER et al. 1998). The ability of ATP infusions to enhance recovery from renal ischemia may be due to P2 receptor stimulation of cellular proliferation (PALLER et al. 1998).

III. Renal Diseases Involving P2 Receptors

1. Renal Ischemia/Reperfusion Injury

Adenine nucleotides protect kidneys and isolated tubular segments from ischemic/reperfusion injury (SIEGEL et al. 1980, 1983; GAUDIO et al. 1982; WEINBERG and HUMES 1986; TAKANO et al. 1985; STROMSKI et al. 1986; MANDEL et al. 1988). This effect may in part involve P2X receptor-mediated stimulation of the proliferative repair process (PALLER et al. 1998). It is conceivable that A_{2A} receptor agonists combined with P2X receptor agonists would be useful for treating renal ischemia since A_{2A} receptor agonists may limit the initial cellular injury (see above) whereas P2 agonists may facilitate repair of injury nephrons.

2. Glomerulopathies

Proliferation of mesangial cells and expansion of the mesangial cell matrix makes an important contribution to the development of glomerular diseases of various etiologies. In this regard, $P2Y_2$ receptors stimulate proliferation of mesangial cells (SCHULZE-LOHOFF et al. 1992; HUWILER and PFEILSCHIFTER 1994) and $P2X_7$ receptors cause apoptosis and necrosis of mesangial cells (SCHULZE-LOHOFF et al. 1998). Therefore, antagonism of mesangial $P2Y_2$ receptors and stimulation of mesangial $P2X_7$ receptors could prove beneficial in the treatment of glomerulopathies.

E. Conclusion

Since the benchmark discovery by Drury and Szent-Gyorgyi in 1929 that adenosine reduces renal blood flow and urine flow (DRURY and SZENT-GYORGYI 1929), a vast amount of information has accumulated regarding the physiological and pathophysiological roles of renal P1 receptors. There is little doubt that in the near future this knowledge will finally pay off in the form of important new pharmacotherapeutic agents for renal diseases. Several highly selective A_1 receptor antagonists are currently being tested in patients as eukalemic diuretics in edematous states such as heart failure. Once these agents are readily available, physicians doubtless will employ these drugs to prevent radiocontrast media-induced ARF, as well as nephrotoxin-induced ARF, and as combination therapy with loop diuretics and thiazide diuretics to combat refractory edema and diuretic resistance. It is also possible that adenomimetics, such as adenosine receptor agonists, adenosine deaminase inhibitors, and/or adenosine kinase inhibitors, will be explored in clinical studies to determine efficacy in such renal diseases as chronic renal failure and renal ischemia/reperfusion injury.

The scope of discovery in the field of renal P2 receptor physiology and pharmacology is much narrower than the span of discovery in the renal P1 receptor area. Nonetheless, several talented investigative groups are now working in the renal P2 receptor area, and progress in the molecular biology and medicinal chemistry of P2 receptors is accelerating. Thus, our current state of ignorance regarding the physiology and pharmacology of renal P2 receptors will soon end. Given the likely role of P2 receptors in the regulation of preglomerular resistance, TGF, renal sympathetic neurotransmission, renin release, and tubular transport, it is almost certain that drugs acting on renal P2 receptors will find applications in renal diseases.

Acknowledgements. This work was supported in part by NIH grants HL40319 and HL35909.

References

Abels BC, Callis JT, Branch RA, Sabra R (1992) Influence of the adenosine receptor antagonist, CGS 15943A, on renal function after reconstruction of chronic renal artery stenosis in the dog. J Pharmacol Exp Ther 262:166–172

Agmon Y, Dinour D, Brezis M (1993) Disparate effects of adenosine A_1- and A_2-receptor agonists on intrarenal blood flow. Am J Physiol 265:F802–F806

Aki Y, Tomohiro A, Nishiyama A, Kiyomoto K, Kimura S, Abe Y (1997) Effects of KW-3902, a selective and potent adenosine A_1 receptor antagonists, on renal hemodynamics and urine formation in anesthetized dogs. Pharmacology 55:193–201

Albinus M, Finkbeiner E, Sosath B, Osswald H (1998) Isolated superfused juxtaglomerular cells from rat kidney: a model for study of renin secretion. Am J Physiol 275:F991–F997

Anderson RJ, Breckon R, Dixon BS (1991) ATP receptor regulation of adenylate cyclase and protein kinase C activity in cultured renal LLC-PK_1 cells. J Clin Invest 87:1732–1738

Andoh TF, Terai T, Nakano K, Horiai H, Mori J, Kohsaka M (1991) Protective effects of FR113453, an adenosine A_1-receptor antagonist, on experimental acute renal failure in rats. In: Proceedings of International Symposium on Acute Renal Failure; October 2–4, 1991; University of North Carolina at Chapel Hill; page 92

Ansari Z, Baldwin DS (1976) Acute renal failure due to radio-contrast agents. Nephron 17:28–40

Arakawa K, Suzuki H, Naitoh M, Matsumoto A, Hayashi K, Matsuda H, Ichihara A, Kubota E, Saruta T (1996) Role of adenosine in the renal responses to contrast medium. Kidney Int 49:1199–1206
Arend LJ, Bakris GL, Burnett JC Jr, Megerian C, Spielman WS (1987a) Role for intrarenal adenosine in the renal hemodynamic response to contrast media. J Lab Clin Med 110:406–411
Arend LJ, Burnatowska-Hledin MA, Spielman WS (1988) Adenosine receptor-mediated calcium mobilization in cortical collecting tubule cells. Am J Physiol 255:C581–C588
Arend LJ, Haramati A, Thompson CI, Spielman WS (1984) Adenosine-induced decrease in renin release: dissociation from hemodynamic effects. Am J Physiol 247:F447–F452
Arend LJ, Sonnenburgh WK, Smith WL, Spielman WS (1987b) A_1 and A_2 adenosine receptors in rabbit cortical collecting tubule cells. J Clin Invest 79:710–714
Arend LJ, Thompson CI, Brandt MA, Spielman WS (1986) Elevation of intrarenal adenosine by maleic acid decreases GFR and renin release. Kid Int 30:656–661
Bachmann S, Ramasubbu K (1997) Immunohistochemical colocalization of the α-subunit of neutrophil NADPH oxidase and ecto-5′-nucleotidase in kidney and liver. Kidney Int 51:479–482
Bailey MA, Imbert-Teboul M, Burnstock GT, Unwin RJ (1998) P2-purinoceptors in the rat thin limbs of Henle. Journal of the American Society of Nephrology 9:424 A
Bakris GL, Sauter ER, Hussey JL, Fisher JW, Gaber AO, Winsett R (1990) Effects of theophylline on erythropoietin production in normal subjects and in patients with erythrocytosis after renal transplantation. N Engl J Med 323:86–90
Balakrishnan VS, Coles GA, Williams JD (1993) A potential role for endogenous adenosine in control of human glomerular and tubular function. Am J Physiol 265:F504–F510
Balakrishnan VS, Coles GA, Williams JD (1996a) Functional role of endogenous adenosine in human chronic renal disease. Exp Nephrol 4:26–36
Balakrishnan VS, Coles GA, Williams JD (1996b) Effects of intravenous adenosine on renal function in healthy human subjects. Am J Physiol 271:F374–F381
Balakrishnan VS, von Ruhland CJ, Griffiths DFR, Coles GA, Williams JD (1996c) Effects of a selective adenosine A_1 receptor antagonist on the development of cyclosporin nephrotoxicity. Br J Pharmacol 117:879–884
Baranowski RL, Westenfelder C (1994) Estimation of renal interstitial adenosine and purine metabolites by microdialysis. Am J Physiol 267:F174–F182
Barber R, Butcher RW (1981) The quantitative relationship between intracellular concentration and egress of cyclic AMP from cultured cells. Mol Pharmacol 19:38–43
Barber R, Butcher RW (1983) The egress of cyclic AMP from metazoan cells. Adv Cyclic Nucleotide Res 15:119–138
Barchowsky A, Data JL, Whorton RA (1987) Inhibition of renin release by analogues of adenosine in rabbit renal cortical slices. Hypertension 9:619–623
Barraco RA, Phillis JW, Campbell WR, Marcantonio PR, Salah RS (1986) The effects of central injections of adenosine analogs on blood pressure and heart rate in the rat. Neuropharmacology 25:675–680
Barrett RJ, Droppleman DA (1993) Interactions of adenosine A_1 receptor-mediated renal vasoconstriction with endogenous nitric oxide and ANG II. Am J Physiol 265:F651–F659
Barrett RJ, Wright KF (1994) A selective adenosine A_1 receptor antagonist attenuates renal dysfunction during controlled hypotension with adenosine in rats. Anesth Analg 79:460–465
Baudouin-Legros M, Badou A, Paulais M, Hammet M, Teulon J (1995); Hypertonic NaCl enhances adenosine release and hormonal cAMP production in mouse thick ascending limb. Am J Physiol 269:F103–F109
Beach RE, Good DW (1992) Effects of adenosine on ion transport in rat medullary thick ascending limb. Am J Physiol 263:F482–F487

Berthold H, Just A, Kirchheim HR, Osswald H, Ehmke H (1998) Renal haemodynamic responses to exogenous and endogenous adenosine in conscious dogs. J Physio 510:321–330
Bidani AK, Churchill PC (1983) Aminophylline ameliorates glycerol-induced acute renal failure in rats. Can J Physiol Pharmacol 61:567–571
Bidani AK, Churchill PC, Packer W (1987) Theophylline-induced protection in myoglobinuric acute renal failure: further characterization. Can J Physiol Pharmacol 65:42–45
Blanco J, Canela EI, Mallol J, Lluis C, Franco R (1992) Characterization of adenosine receptors in brush-border membranes from pig kidney. Brit J Pharmacol 107:671–678
Bohmann C, von Kügelgen I, Rump LC (1997) P_2-receptor modulation of noradrenergic neurotransmission in rat kidney. Brit J Pharmacol 121:1255–1262
Bowmer CJ, Collis MG, Yates MS (1986) Effect of the adenosine antagonist 8-phenyltheophylline on glycerol-induced acute renal failure in the rat. Br J Pharmac 88:205–212
Branch RA, Jackson EK, Jacqz E, Stein R, Ray WA, Ohnhaus EE, Meusers P, Heidemann H (1987) Amphotericin-B nephrotoxicity in humans decreased by sodium supplements with coadministration of ticarcillin or intravenous saline. Klin Wochenschr 65:500–506
Brenner BM, Lawler EV, Mackenzie HS (1996) The hyperfiltration theory: a paradigm shift in nephrology. Kidney Int 49:1774–1777
Brown NJ, Porter J, Ryder D, Branch RA (1991) Caffeine potentiates the renin response to diazoxide in man. Evidence for a regulatory role of endogenous adenosine. J Pharmacol Exp Ther 256:56–61
Cai H, Batuman V, Puschett DB, Puschett JB (1994) Effect of KW-3902, a novel adenosine A_1 receptor antagonist, on sodium-dependent phosphate and glucose transport by the rat renal proximal tubular cell. Life Sci 55:839–845
Cai H, Puschett DB, Guan S, Batuman V, Puschett JB (1995) Phosphate transport inhibition by KW-3902, an adenosine A_1 receptor antagonist, is mediated by cyclic adenosine monophosphate. Am J Kidney Dis 26:825–830
Callis JT, Kuan CJ, Branch KR, Abels, BC, Sabra R, Jackson EK, Branch RA (1989) Inhibition of renal vasoconstriction induced by intrarenal hypertonic saline by the nonxanthine adenosine antagonist CGS 15843A. J Pharmacol Exp Ther 248:1123–1129
Cannon ME, Twu BM, Yang CS, Hsu CH (1989) The effect of theophylline and cyclic adenosine 3′,5′-monophosphate on renin release by afferent arterioles. J Hyper 7:569–576
Carlson MA, Ferraz AAB, Condon RE (1997) Urinary adenosine excretion in patients receiving amphotericin B. Surgery 121:190–193
Carmines PK, Inscho EW (1994) Renal arteriolar angiotensin responses during varied adenosine receptor activation. Hypertension 23:I114–I119
Casavola V, Guerra L, Reshkin SJ, Jacobson KA, Murer H (1997) Polarization of adenosine effects on intracellular pH in A_6 renal epithelial cells. Molecular Pharmacology 51:516–523
Casavola V, Guerra L, Reshkin SJ, Jacobson KA, Verrey F, Murer H (1996) Effect of adenosine on Na^+ and Cl^- currents in A6 monolayers. Receptor localization and messenger involvement. J Membr Biol 151:237–245
Chan CM, Hillman K, Moss S, Unwin RJ, Burnstock G (1998a) Putative cytolytic $P2X_7$ receptor in diabetic rat kidney and its stable transfection in lymphocytes. Journal of the American Society of Nephrology 9:420A
Chan CM, Unwin RJ, Bardini M, Oglesby IB, Ford APDW, Townsend-Nicholson A, Burnstock G (1998b) Localization of $P2X_1$ purinoceptors by autoradiography and immunohistochemistry in rat kidneys. Am J Physiol 274: F799–F804

Chan CM, Unwin RJ, Burnstock G (1998c) Variable co-localization of $P2X_1$ and $P2X_2$ receptors along the rat renal vasculature. Journal of the American Society of Nephrology 9:335A
Chan CM, Unwin RJ, Burnstock G (1998d) Immunohistochemical localization of $P2X_5$ and $P2X_6$ receptors along the normal rat nephron. Journal of the American Society of Nephrology 9:420A
Chen CC, Akopian AN, Sivilotti L, Colquhoun D, Burnstock G, Wood JN (1995) A P2X purinoceptor expressed by a subset of sensory neurons. Nature 377:428–431
Choi KC, Lee J, Moon KH, Park KK, Kim SW, Kim NH (1993) Chronic caffeine ingestion exacerbates 2-kidney, 1-clip hypertension and ameliorates deoxycorticosterone acetate-salt hypertension in rats. Nephron 65:619–622
Chou CC, Hook JB, Hsieh CP, Burns TD, Dabney JM (1971) Effects of radiopaque dyes on renal vascular resistance. J Lab Clin Med 78:705–712
Churchill PC (1982) Renal effects of 2-chloroadenosine and their antagonism by aminophylline in anesthetized rats. J Pharmacol Exp Ther 222:319–323
Churchill PC, Bidani A (1987) Renal effects of selective adenosine receptor agonists in anesthetized rats. Am J Physiol 252:F299–F303
Churchill PC, Churchill MC (1985) A_1 and A_2 adenosine receptor activation inhibits and stimulates renin secretion of rat renal cortical slices. J Pharmacol Exp Ther 232:589–594
Churchill PC, Ellis VR (1993a) Pharmacological characterization of the renovascular P_2 purinergic receptors. J Pharmacol Exp Ther 265:334–338
Churchill PC, Ellis VR (1993b) Purinergic P_{2Y} receptors stimulate renin secretion by rat renal cortical slices. J Pharmacol Exp Ther 266:160–163
Churchill PC, Bidani AK, Churchill MC, Prada J (1984) Renal effects of 2-chloroadenosine in the two-kidney goldblatt rat. J Pharmacol Exp Ther 230:302–306
Cook CB, Churchill PC (1984) Effects of renal denervation on the renal responses of anesthetized rats to cyclohexyladenosine. Can J Physiol Pharmacol 62:934–938
Coulson R, Johnson RA, Olsson RA, Cooper DR, Scheinman SJ (1991) Adenosine stimulates phosphate and glucose transport in opossum kidney epithelial cells. Am J Physiol 260:F921–F928
Coulson R, Proch PS, Olsson RA, Chalfant CE, Cooper DR (1996) Upregulated renal adenosine A_1 receptors augment PKC and glucose transport but inhibit proliferation. Am J Physiol 270:F263–F274
Cox BF, Smits GJ (1996) Regional hemodynamic effects of purinergic P2 receptor subtype agonists in rats. J Pharmacol Exp Ther 277:1492–1500
Cronstein BN (1994) Adenosine, an endogenous anti-inflammatory agent. Journal of Applied Physiology 76:5–13
Culic O, Sabolic I, Zanic-Grubisic T (1990) The stepwise hydrolysis of adenine nucleotides by ectoenzymes of rat renal brush-border membranes. Biochimica et Biophysica Acta 1030:143–151
Dean RH (1985) Complications of renal revascularization. In Bernhard VM, Towne JB (eds) Complications of Vascular Surgery, Orlando: Grune & Stratton, Inc., pp 239–241
Deray G, Branch RA, Herzer WA, Ohnishi A, Jackson EK (1987) Adenosine inhibits β-adrenoceptor but not DBcAMP-induced renin release. Am J Physiol 252:F46–F52
Deray G, Branch RA, Jackson EK (1989a) Methylxanthines augment the renin response to suprarenal-aortic constriction. Naunyn-Schmiedeberg's Arch Pharmacol 339:690–696
Deray G, Branch RA, Ohnishi A, Jackson EK (1989b) Adenosine inhibits renin release induced by suprarenal-aortic constriction and prostacyclin. Naunyn-Schmiedeberg's Arch Pharmacol 339:590–595
Deray G, Martinez F, Cacoub P, Baumelou B, Baumelou A, Jacobs C (1990a) A role for adenosine, calcium and ischemia in radiocontrast-induced intrarenal vasoconstriction. Am J Nephrol 10:316–322

Deray G, Sabra R, Herzer WA, Jackson EK, Branch RA (1990b) Interaction between angiotensin II and adenosine in mediating the vasoconstrictor response to intrarenal hypertonic saline infusions in the dog. J Pharmacol Exp Ther 252:631–635
Dinour D, Brezis M (1991) Effects of adenosine on intrarenal oxygenation. Am J Physiol 261:F787–F791
Drury AN, Szent-Gyorgyi A (1929) The physiological activity of adenine compounds with special reference to their action upon the mammalian heart. J Physiol Lond 68:213–237
Dubey RK, Gillespie DG, Jackson EK (1998) Cyclic AMP-adenosine pathway induces nitric oxide synthesis in aortic smooth muscle cells. Hypertension 31:296–302
Dubey RK, Mi Z, Gillespie DG, Jackson EK (1996) Cyclic AMP-adenosine pathway inhibits vascular smooth cell growth. Hypertension 28:765–771
Ecelbarger CA, Maeda Y, Gibson CC, Knepper MA (1994) Extracellular ATP increases intracellular calcium in rat terminal collecting duct via a nucleotide receptor. Am J Physiol 267:F998–1006
Edlund A, Ohlsén H, Sollevi A (1994) Renal effects of local infusion of adenosine in man. Clin Sci 87:143–149
Edwards RM, Spielman WS (1994) Adenosine A_1 receptor-mediated inhibition of vasopressin action in inner medullary collecting duct. Am J Physiol 266:F791–F796
Eltze M, Ullrich B (1996) Characterization of vascular P_2 purinoceptors in the rat isolated perfused kidney. Eur J Pharmacol 306:139–152
Erley CM, Duda SH, Schlepckow S, Koehler J, Huppert PE, Strohmaier WL, Bohle A, Risler T, Osswald H (1994) Adenosine antagonist theophylline prevents the reduction of glomerular filtration rate after contrast media application. Kidney Int 45:1425–1431
Erley CM, Heyne N, Burgert K, Langanke J, Risler T, Osswald H (1997) Prevention of radiocontrast-induced nephropathy by adenosine antagonists in rats with chronic nitric oxide deficiency. Journal of the American Society of Nephrology 8:1125–1132
Erley CM, Heyne EN, Rossmeier S, Vogel T, Risler T, Osswald H (1998) Adenosine and extracellular volume in radiocontrast media-induced nephropathy. Kidney Int 54:S-192–S-194
Firestein BL, Xing M, Hughes RJ, Corvera CU, Insel PA (1996) Heterogeneity of P_{2u}- and P_{2y}-purinergic receptor regulation of phospholipases in MDCK cells. Am J Physiol 271:F610–F618
Forrest JN Jr (1996) Cellular and molecular biology of chloride secretion in the shark rectal gland: Regulation by adenosine receptors. Kidney Int 4:1557–1562
Franco M, Bell PD, Navar LG (1989) Effect of adenosine A_1 analogue on tubuloglomerular feedback mechanism. Am J Physiol 257:F231–F236
Franco M, Bobadilla NA, Suárez J, Tapia E, Sánchez L, Herrera-Acosta J (1996) Participation of adenosine in the renal hemodynamic abnormalities of hypothyroidism. Am J Physiol 270:F254–F262
Fransen R, Koomans HA (1995) Adenosine and renal sodium handling: Direct natriuresis and renal nerve-mediated antinatriuresis. J Am Soc Nephrol 6:1491–1497
Freissmuth M, Hausleithner V, Tuisl E, Nanoff C, Schütz W (1987) Glomeruli and microvessels of the rabbit kidney contain both A_1- and A_2-adenosine receptors. Naunyn-Schmiedeberg's Arch Pharmacol 335:438–444
Friedlander G, Amiel C (1995) Extracellular nucleotides as modulators of renal tubular transport. Kid Int 47:1500–1506
Friedrich F, Weiss H, Paulimichel M, Lang F (1989) Activation of potassium channels in renal epithelioid cells (MDCK) by extracellular ATP. Am J Physiol 256:C1016–C1021
Gaudio KM, Taylor MR, Chaudry IH, Kashgarian M, Siegel NJ (1982) Accelerated recovery of single nephron function by the postischemic infusion of ATP–MgCl2. Kidney International 22:13–20

Gellai M, Schreiner GF, Ruffolo RR Jr, Fletcher T, DeWolf R, Brooks DP (1998) CVT-124, a novel adenosine A_1 receptor antagonist with unique diuretic activity. J Pharmacol Exp Ther 286:1191–1196
Gerkens, JF, Smith AJ (1985) Captopril and theophylline treatment on cyclosporine-induced nephrotoxicity in rats. Transplantation 40:213–214
Gerkens JF, Heidemann HT, Jackson EK, Branch RA (1983a) Aminophylline inhibits renal vasoconstriction produced by intrarenal hypertonic saline. J Pharmacol Exp Ther 225:611–615
Gerkens JF, Heidemann HT, Jackson EK, Branch RA (1983b) Effect of aminophylline on amphotericin B nephrotoxicity in the dog. J Pharmacol Exp Ther 224:609–613
Gleiter CH, Becker T, Wenzel J (1997a) Erythropoietin production in healthy volunteers subjected to controlled hypobaric hypoxia: further evidence against a role for adenosine. Br J Clin Pharmacol 44:203–205
Gleiter CH, Brause M, Delabar U, Zebski H, Eckardt K-U (1997b) Evidence against a major role of adenosine in oxygen-dependent regulation of erythropoietin in rats. Kidney Int 52:338–344
Gleiter CH, Freudenthaler S, Delabar U, Eckardt KU, Mühlbauer B, Gundert-Remy U, Osswald H (1996) Erythropoietin production in healthy volunteers subjected to controlled haemorrhage: evidence against a major role for adenosine. Br J Clin Pharmacol 42:729–735
Gould J, Bowmer CJ, Yates MS (1995) Renal haemodynamic responses to adenosine in acute renal failure. Nephron 71:184–189
Gould J, Morton MJ, Sivaprasadarao A, Bowmer CJ, Yates MS (1997) Renal adenosine A_1 receptor binding characteristics and mRNA levels during the development of acute renal failure in the rat. Br J Pharmacol 120:947–953
Gouyon JB, Guignard JP (1988) Theophylline prevents the hypoxemia-induced renal hemodynamic changes in rabbits. Kid Int 33:1078–1083
Guieu R, Dussol B, Devaux C, Sampol J, Brunet P, Rochat H, Bechis G, Berland YF (1998) Interactions between cyclosporine A and adenosine in kidney transplant recipients. Kidney Int 53:200–204
Hall JE, Granger JP (1986a) Adenosine alters glomerular filtration control by angiotensin II. Am J Physiol 250:F917–F923
Hall JE, Granger JP (1986b) Renal hemodynamics and arterial pressure during chronic intrarenal adenosine infusion in conscious dogs. Am J Physiol 250:F32–F39
Hayslett JP, Macala LJ, Smallwood JI, Kalghatgi L, Gasalla-Herraiz J, Isales C (1995) Adenosine stimulation of Na^+ transport is mediated by an A_1 receptor and a $[Ca^{2+}]_i$-dependent mechanism. Kid Int 47:1576–1584
Heidemann HT, Bolten M, Inselmann G (1991) Effect of chronic theophylline administration on amphotericin B nephrotoxicity in rats. Nephron 59:294–298
Heidemann HT, Gerkens JF, Jackson EK, Branch RA (1983a) Effect of aminophylline on renal vasoconstriction produced by amphotericin B in the rat. Naunyn-Schmiedeberg's Arch Pharmacol 324:148–152
Heidemann HT, Gerkens JF, Jackson EK, Branch RA (1985) Attenuation of cisplatinum-induced nephrotoxicity in the rat by high salt diet, furosemide and acetazolamide. Naunyn-Schmiedeberg's Arch Pharmacol 329:201–205
Heidemann HT, Gerkens JF, Spickard WA, Jackson EK, Branch RA (1983b) Amphotericin B nephrotoxicity in humans decreased by salt depletion. Am J Med 75:476–481
Heidemann HT, Müller S, Mertins L, Stepan G, Hoffmann K, Ohnhaus E (1989) Effect of aminophylline on cisplatin nephrotoxicity in the rat. Br J Pharmacol 97:313–318
Hoenderop JGJ, Hartog A, Willems PHGM, Bindels RJM (1998) Adenosine-stimulated Ca^{2+} reabsorption is mediated by apical A_1 receptors in rabbit cortical collecting system. Am J Physiol 274:F736–F743
Holycross BJ, Jackson EK (1992) Effects of chronic treatment with caffeine on kidneys responses to angiotensin II. Eur J Pharmacol 219:361–367

Humes HD, Cieslinski DA (1991) Adenosine triphosphate stimulates thymidine incorporation but does not promote cell growth in primary cultures or renal proximal tubular cells. Renal Physiol Biochem 14:253–258
Huwiler A, Pfeilschifter J (1994) Stimulation by extracellular ATP and UTP of the mitogen-activated protein kinase cascade and proliferation of rat renal mesangial cells. Br J Pharmacol 113:1455–1463
Ilan Y, Dranitzki-Elhallel M, Rubinger D, Silver J, Popovitzer MM (1994) Erythrocytosis after renal transplantation. Transplantation 57:661–664
Inscho EW (1996) Purinoceptor-mediated regulation of the renal microvasculature 16:385–388
Inscho EW, Carmines PK, Navar LG (1991) Juxtamedullary afferent arteriolar responses to P_1 and P_2 purinergic stimulation. Hypertension 17:1033–1037
Inscho EW, Cook AK, Mui V, Miller J (1998) Direct assessment of renal microvascular responses to P_2-purinoceptor agonists. Am J Physiol 274:F718–F727
Inscho EW, Cook AK, Navar LG (1996) Pressure-mediated vasoconstriction of juxtamedullary afferent arterioles involves P_2-purinoceptor activation. Am J Physiol 271:F1077–F1085
Inscho EW, LeBlanc EA, Pham BT, White SM, Imig JD (1999) Purinoceptor-mediated calcium signaling in preglomerular smooth muscle cells. Hypertension 33:195–200
Inscho EW, Mitchell KD, Navar LG (1994) Extracellular ATP in the regulation of renal microvascular function. FASEB J 8:319–328
Inscho EW, Ohishi K, Cook AK, Belott TP, Navar LG (1995) Calcium activation mechanisms in the renal microvascular response to extracellular ATP. Am J Physiol 268:F867–F884
Inscho EW, Ohishi K, Navar LG (1992) Effect of ATP on pre- and postglomerular juxtamedullary microvasculature. Am J Physiol 263:F886–F893
Ishikawa I, Shikura N, Takada K (1993a) Amelioration of glycerol-induced acute renal failure in rats by an adenosine A_1 receptor antagonist (FR-113453). Renal Failure 15:1–5
Ishikawa I, Shikura N, Takada K, Sato Y (1993b) Changes of adenosine levels in the carotid artery, renal vein and inferior vena cava after glycerol or mercury injection in the rat. Nephron 64:605–608
Jackson EK (1991) Adenosine: A physiological brake on renin release. Annu Rev Pharmacol Toxicol 31:1–35
Jackson EK, Mi Z, Gillespie DG, Dubey RK (1997) Metabolism of cAMP to adenosine in the renal vasculature. J Pharmacol Exp Ther 283:177–182
Jackson, EK, Mi Z, Herzer WA (1995) Studies on the mechanism by which adenosine receptor antagonists attenuate acute renal failure. In Belardinelli L, Pelleg A (eds) "Adenosine and Adenine Nucleotides: From Molecular Biology to Integrative Physiology," Boston: Kluwer Acad Pub., pp 415–423
Kaissling B, Spiess S, Rinne B, LeHir M (1993) Effects of anemia on morphology of rat renal cortex. Am J Physiol 264:F608–F617
Katholi RE, Taylor GJ, McCann WP, Woods WT Jr, Womach KA, McCoy CD, Katholi CR, Moses HW, Mishkel GJ, Lucore CL, Holloway RM, Miller BD, Woodruff RC, Dove JT, Mikell FL, Schneider JA (1995) Nephrotoxicity from contrast media: attenuation with theophylline. Radiology 195:17–22
Katzberg RW, Schulman G, Meggs LG, Caldicott WJH, Damieno MM, Hollenberg NK (1983) Mechanism of renal response to contrast media in dogs: decrease in renal function due to hypertonicity. Invest Radiol 18:74–80
Kellett R, Bowmer CJ, Collis MG, Yates MS (1989) Amelioration of glycerol-induced acute renal failure in the rat with 8-cyclopentyl-1,3-dipropylxanthine. Br J Pharmacol 98:1066–1074
Kelley GG, Poeschla EM, Barron HV, Forrest JN Jr. (1990) A_1 adenosine receptors inhibit chloride transport in the shark rectal gland: Dissociation of inhibition and cyclic AMP. J Clin Invest 85:1629–1636

Kishore BK, Chou C-L, Knepper MA (1995) Extracellular nucleotide receptor inhibits AVP-stimulated water permeability in inner medullary collecting duct. Am J Physiol 269:F863–F869

Kishore BK, Ginns SN, Nielsen S, Knepper MA (1998) Localization of P2Y2-purinoceptor in rat inner medulla. Journal of the American Society of Nephrology 9:425A

Knight RJ, Bowmer CJ, Yates MS (1993a) Effect of the selective A_1 adenosine antagonist 8-cyclopentyl-1,3-dipropylxanthine on acute renal dysfunction induced by *Escherichia Coli* endotoxin in rats. J Pharm Pharmacol 45:979–984

Knight RJ, Bowmer CJ, Yates MS (1993b) The diuretic action of 8-cyclopentyl-1,3-dipropylxanthine, a selective A_1 adenosine receptor antagonist. Br J Pharmacol 109:271–277

Knight RJ, Collis MG, Yates MS, Bowmer CJ (1991) Amelioration of cisplatin-induced acute renal failure with 8-cyclopentyl-1,3-dipropylxanthine. Br J Pharmacol 104:1062–1068

Kohno M, Murakawa K, Horio T, Yokokawa K, Yasunari K, Fukui T, Takeda T (1991) Plasma immunoreactive endothelin-1 in experimental malignant hypertension. Hypertension 18:92–100

Kost CK Jr, Herzer WA, Mi Z, Jackson EK (1998) Diuretic/natriuretic response to A1-receptor blockade in normotensive and hypertensive rats: Role of Gi-proteins. J Am Soc of Nephrol 9:310A

Kost CK Jr, Li P, Pfeifer CA, Jackson (1994) Telemetric blood pressure monitoring in benign 2-kidney, 1-clip renovascular hypertension: Effect of chronic caffeine ingestion. J Pharmacol Exp Ther 270:1063–1070

Kreisberg MS, Silldorff EP, Pallone TL (1997) Localization of adenosine-receptor subtype mRNA in rat outer medullary descending vasa recta by RT-PCR. Am J Physiol 272:H1231–H1238

Kuan CJ, Branch RA, Jackson EK (1990a) Effect of an adenosine receptor antagonist on acute amphotericin B nephrotoxicity. Eur J Pharmacol 178:285–291

Kuan, CJ, Herzer WA, Jackson EK (1993) Cardiovascular and renal effects of blocking A_1 adenosine receptors. J Cardiovasc Pharm 21:822–828

Kuan CJ, Wells JN, Jackson EK (1989) Endogenous adenosine restrains renin release during sodium restriction. J Pharmacol Exp Ther 249:110–116

Kuan, CJ, Wells, JN, Jackson EK (1990b) Endogenous adenosine restrains renin release in conscious rats. Circ Res 66:637–646

Lang F, Plöckinger B, Häussinger D, Paulmichl, M (1988) Effects of extracellular nucleotides on electrical properties of subconfluent Madin Darby canine kidney cells. Biochim Biophys Acta 943:471–476

Lang MA, Preston AS, Handler JS, Forrest JN Jr. (1985) Adenosine stimulates sodium transport in kidney A6 epithelia in culture. Am J Physiol 249:C330–C336

Langård O, Holdaas H, Eide I, Kiil F (1983) Conditions for augmentation of renin release by theophylline. Scand J Clin Lab Invest 43:9–14

Lederer ED, McLeish KR (1995) P2 purinoceptor stimulation attenuates PTH inhibition of phosphate uptake by a G protein-dependent mechanism. Am J Physiol 269:F309–F316

Le Hir M, Kaissling B (1993) Distribution and regulation of renal ecto-5′-nucleotidase: implications for physiological functions of adenosine. Am J Physiol 264:F377–F387

Levens N, Beil M, Jarvis M (1991a) Renal actions of a new adenosine agonist, CGS 21680A selective for the A_2 receptor. J Pharmacol Exp Ther 257:1005–1012

Levens N, Beil M, Schulz R (1991b) Intrarenal actions of the new adenosine agonist CGS 21680A, selective for the A_2 receptor. J Pharmacol Exp Ther 257:1013–1019

LeVier DG, McCoy DE, Spielman WS (1992) Functional localization of adenosine receptor-mediated pathways in the LLC-PK_1 renal cell line. Am J Physiol 263:C729–C735

Lewis EJ, Hunsicker LG, Bain RP, Rohde RD (1993) The effect of angiotensin-converting-enzyme inhibition on diabetic nephropathy. The collaborative study group. N Eng J Med 329:1456–1462
Lin JJ, Churchill PC, Bidani AK (1986) Effect of theophylline on the initiation phase of postischemic acute renal failure in rats. J Lab Clin Med 108:150–154
Lin JJ, Churchill PC, Bidani AK (1987) The effect of dipyridamole on the initiation phase of postischemic acute renal failure in rats. Can J Physiol Pharmacol 65:1491–1495
Lin JJ, Churchill PC, Bidani AK (1988) Theophylline in rats during maintenance phase of post-ischemic acute renal failure. Kid Int 33:24–28
Llanos A, Cieza J, Bernardo J, Echevarria J, Biaggioni I, Sabra R, Branch RA (1991) Effect of salt supplementation on amphotericin B nephrotoxicity. Kid Int 40:302–308
Lorez JN, Weihprecht H, He XR, Skøtt O, Briggs JP, Schnermann J (1993) Effects of adenosine and angiotensin on macula densa-stimulated renin secretion. Am J Physiol 265:F187–F194
Ma H, Ling BN (1996) Luminal adenosine receptors regulate amiloride-sensitive Na^+ channels in A6 distal nephron cells. Am J Physiol 270:F798–F805
Majid DSA, Navar LG (1992) Suppression of blood flow autoregulation plateau during nitric oxide blockade in canine kidney. Am J Physiol 262:F40–F46
Mandel LJ, Takano T, Soltoff SP, Murdaugh S (1988) Mechanisms whereby exogenous adenine nucleotides improve rabbit renal proximal function during and after anoxia. J Clin Invest 81:1255–1264
Martínez F, Franco M, Quintana A, Herrera-Acosta J (1997) Sodium-dependent adenosine transport is diminished in brush border membrane vesicles from hypothyroid rat kidney. Pflügers Arch – Eur J Physiol 433:269–275
Mi Z, Jackson EK (1998) Evidence for an endogenous cAMP-adenosine pathway in the rat kidney. J Pharmacol Exp Ther 287:926–930
Mi Z, Jackson EK (1995) Metabolism of exogenous cyclic AMP to adenosine in the rat kidney. J Pharmacol Exp Ther 273:728–733
Mi Z, Herzer WA, Zhang Y, Jackson EK (1994) 3-isobutyl-1-methylxanthine decreases renal cortical interstitial levels of adenosine and inosine. Life Sci 54:PL277–PL282
Middleton JP, Mangel AW, Basavappa S, Fitz JG (1993) Nucleotide receptors regulate membrane ion transport in renal epithelial cells. Am J Physiol 264:F867–F873
Miller WL, Thomas RA, Berne RM, Rubio R (1978) Adenosine production in the ischemic kidney. Circ Res 43:390–397
Mitchell KD, Navar LG (1993) Modulation of tubuloglomerular feedback responses by extracellular ATP. Am J Physiol 264:F458–F466
Miyamoto M, Yagil Y, Larson T, Robertson C, Jamison RL (1988) Effects of intrarenal adenosine on renal function and medullary blood flow in the rat. Am J Physiol 255:F1230–F1234
Mizumoto H, Karasawa A, Kubo K (1993) Diuretic and renal protective effects of 8-(noradamantan-3-yl)-1,3-dipropylxanthine (KW-3902), a novel adenosine A_1-receptor antagonist, via pertussis toxin insensitive mechanism. J Pharmacol Exp Ther 266:200–266
Mori M, Nishizaki T, Kawahara K, Okada Y (1996) ATP-activated cation conductance in a *Xenopus* renal epithelial cell line. J Physiol 491:281–290
Mosqueda-Garcia R, Killian TJ, Haile V, Tseng CJ, Robertson RM, Robertson D (1990) Effects of caffeine on plasma catecholamines and muscle sympathetic nerve activity in man. Circulation 82:III–335
Mosqueda-Garcia R, Tseng CJ, Appalsamy M, Robertson D (1989) Modulatory effects of adenosine on baroreflex activation in the brainstem of normotensive rats. Eur J Pharmacol 174:119–122
Moyer BD, McCoy DE, Lee B, Kizer N, Stanton BA (1995) Adenosine inhibits arginine vasopressin-stimulated chloride secretion in a mouse IMCD cell line (mIMCD-K2). Am J Physiol 269:F884–F891

Munger KA, Jackson EK (1994) Effects of selective A_1 receptor blockade on glomerular hemodynamics: involvement of renin-angiotensin system. Am J Physiol 267:F783–F790
Murray RD, Churchill PC (1984) Effects of adenosine receptor agonists in the isolated, perfused rat kidney. Am J Physiol 247:H343–H348
Murray RD, Churchill PC (1985) Concentration dependency of the renal vascular and renin secretory responses to adenosine receptor agonists. J Pharmacol Exp Ther 232:189–193
Nagashima K, Kusaka H, Karasawa A (1995) Protective effects of KW-3902, an adenosine A_1-receptor antagonist, against cisplatin-induced acute renal failure in rats. Jpn J Pharmacol 67:349–357
Nagashima K, Kusaka H, Sato K, Karasawa A (1994) Effects of KW-3902, a novel adenosine A_1-receptor antagonist, on cephaloridine-induced acute renal failure in rats. Jpn J Pharmacol 64:9–17
Nagashima K, Yamagata T, Ushiki J, Yamaguchi K, Nishiyama A, Abe Y, Karasawa A (1998) Role for renal interstitial adenosine in the contrast medium-induced renal dysfunction. J Am Soc Nephropath 9:428A
Nakashima J, Brookins J, Beckman B, Fisher JW (1991) Increased erythropoietin secretion in human hepatoma cells by N^6-cyclohexyladenosine. Am J Physiol 261:C455–C460
Nakashima J, Brookins J, Ohigashi T, Fisher JW (1994) Adenosine A_2 receptor modulation of erythropoietin secretion in hepatocellular carcinoma cells. Life Sci 54:109–117
Nakashima J, Ohigashi T, Brookins JW, Bechman BS, Agrawal KC, Fisher JW (1993a) Effects of 5′-N-ethylcarboxamideadenosine (NECA) on erythropoietin production. Kid Int 44:734–740
Nakashima J, Tazak H, Fisher JW (1993b) Increase in erythropoietin secretion mediated by adenosine A_2 receptors. Hum Cell 6:36–46
Nanoff C, Freissmuth M, Tuisl E, Schutz W (1990) P_2-, but not P_1-purinoreceptors mediate formation of 1,4,5-inositol triphosphate and its metabolites via a pertussis toxin-insensitive pathway in the rat renal cortex. Br J Pharmacol 100:62–68
Navar LG, Inscho EW, Majid DSA, Imig JD, Harrison-Bernard LM, Mitchell KD (1996) Paracrine regulation of the renal microcirculation. Physiological Reviews 76:425–536
Needleman P, Minkes MS, Douglas JR (1970) Stimulation of prostaglandin biosynthesis by adenine nucleotides. Profile of prostaglandin release by perfused organs. Circ Res 34:455–460
Nies AS, Beckmann ML, Gerber JG (1991) Contrasting effects of changes in salt balance on the renovascular response to A_1-adenosine receptor stimulation in vivo and in vitro in the rat. J Pharmacol Exp Ther 256:542–546
Nolte D, Lorenzen A, Lehr HA, Zimmer FJ, Klotz KN, Messmer K (1992) Reduction of postischemic leukocyte-endothelium interaction by adenosine via A_2 receptor. Naunyn-Schmiedebergs Arch Pharmacol 346:234–237
Nomura H, Nagashima K, Kusaka H, Karasawa A (1995) Antihypertensive effects of KW-3902, an adenosine A_1-receptor antagonist, in Dahl salt-sensitive rats. Jpn J Pharmacol 68:389–396
Nosaka K, Takahashi T, Hishi T, Imaki H, Suzuki T, Suzuki K, Kurokawa K, Endou H (1997) An adenosine deaminase inhibitor prevents puromycin aminonucleoside nephrotoxicity. Free Radic Biol Med 22:597–605
Oberbauer R, Schreiner GF, Meyer TW (1998) Natriuretic effect of adenosine A_1-receptor blockade in rats. Nephrol Dial Transplant 13:900–903
Ohigashi T, Brookins J, Fisher JW (1993) Adenosine A_1 receptors and erythropoietin production. Am J Physiol 265:C934–C938
Ohnishi A, Branch RA, Jackson K, Hamilton R, Biaggioni I, Deray G, Jackson EK (1986) Chronic caffeine administration exacerbates renovascular, but not genetic, hypertension in rats. J Clin Invest 73:1045–1050

Ohnishi A, Ohnishi T, Stevenhead W, Robinson RD, Glick A, O'Day DM, Sabra R, Jackson EK, Branch RA (1989) Sodium status influences chronic amphotericin B nephrotoxicity in rats. Antimicrob Agents Chemother 33:1222–1227
Okusa MD, Linden J, Macdonald TL, Huang L, Mangrum AJ (1998) Activation of A_{2A}-adenosine receptors (A_{2A}-ARs) reduces polymorphoneutrophil (PMN) infiltration and protects kidneys from ischemia-reperfusion (I/R) injury in the rat. J Am Soc Nephrol 9:583A
Osswald H (1975) Renal effects of adenosine and their inhibition by theophylline in dogs. Naunyn-Schmiedeberg's Arch Pharmacol 288:79–86
Osswald H, Hermes H, Nabakowski G (1982) Role of adenosine in signal transmission of tubuloglomerular feedback. Kidney Int 22:S136–S142
Osswald H, Nabakowski G, Hermes H (1980) Adenosine as a possible mediator of metabolic control of glomerular filtration rate. Int J Biochem 12:263–267
Osswald H, Schmitz HJ, Heidenreich O (1975) Adenosine response of the rat kidney after saline loading, sodium restriction and hemorrhage. Pflügers Arch 357: 323–333
Osswald H, Schmitz HJ, Kemper R (1977) Tissue content of adenosine, inosine and hypoxanthine in the rat kidney after ischemia and postischemic recirculation. Pflügers Arch 371:45–49
Osswald H, Schmitz HJ, Kemper R (1978a) Renal action of adenosine: Effect on renin secretion in the rat. Naunyn-Schmiedeberg's Arch Pharmacol 303:95–99
Osswald H, Spielman WS, Knox FG (1978b) Mechanism of adenosine-mediated decreases in glomerular filtration rate in dogs. Circ Res 43:465–469
Osswald H, Vallon V, Mühlbauer B (1996) Role of adenosine in tubuloglomerular feedback and acute renal failure. Journal of Autonomic Pharmacology 16:377–380
Palacios JM, Fastbom J, Wiederhold KH, Probst A (1987) Visualization of adenosine A_1 receptors in the human and the guinea-pig kidney. Eur J Pharmacol 138:273–276
Paller MS, Schnaith EJ, Rosenberg ME (1998) Purinergic receptors mediate cell proliferation and enhanced recovery from renal ischemia by adenosine triphosphate. J Lab Clin Med 131:174–183
Panjehshahin MR, Chahil RS, Collis MG, Bowmer CJ, Yates MS (1991a) The effect of 8-cyclopentyl-1,3-dipropylxanthine on the development of cyclosporin-induced acute renal failure. J Pharm Pharmacol 43:525–528
Panjehshahin MR, Munsey TS, Collis MG, Bowmer CJ, Yates MS (1991b) Further characterization of the protective effect of 8-cyclopentyl-1,3-dipropylxanthine on glycerol-induced acute renal failure in the rat. J Pharm Pharmacol 44: 109–113
Paul S, Jackson EK, Robertson D, Branch RA, Biaggioni I (1989) Caffeine potentiates the renin response to furosemide in rats. Evidence for a regulatory role of endogenous adenosine. J Pharmacol Exp Ther 251:183–187
Paul P, Rothmann SA, Meagher RC (1988) Modulation of erythropoietin production by adenosine. J Lab Clin Med 112:168–173
Paulmichl M, Lang F (1988) Enhancement of intracellular calcium concentration by extracellular ATP and UTP in Madin Darby canine kidney cells. Biochem Biophys Res Commun 156:1139–1143
Pawelczyk T, Bizon D, Angielski S (1992) The distribution of enzymes involved in purine metabolism in rat kidney. Biochim Biophys Acta 1116:309–314
Pawlowska D, Granger JP, Knox FG (1987) Effects of adenosine infusion into renal interstitium on renal hemodynamics. Am J Physiol 252:F678–F682
Peart WS, Quesada T, Tenyl I (1975) The effects of cyclic adenosine 3′,5′-monophosphate and guanosine 3′,5′-monophosphate and theophylline on renin secretion in the isolated perfused kidney of the rat. Br J Pharmacol 54:55–60
Pfeifer CA, Suzuki F, Jackson EK (1995) Selective A_1 receptor antagonism augments β-adrenergic-induced renin release in vivo. Am J Physiol 269 (4 Pt 2):F469–F479
Pfeilschifter J (1990a) Comparison of extracellular ATP and UTP signalling in rat renal mesangial cells. Biochem J 272:469–472

Pfeilschifter J (1990b) Extracellular ATP stimulates polyphosphoinositide hydrolysis and prostaglandin synthesis in the renal mesangial cells. Cell Signalling 2:129–138

Pfister JR, Belardinelli L, Lee G, Lum RT, Milner P, Stanley WC, Linden J, Baker SP, Schreiner G (1997) Synthesis and biological evaluation of the enantiomers of the potent and selective A_1-adenosine antagonist 1,3-dipropyl-8-[2-(5,6-epoxynorbonyl)]-xanthine. J Med Chem 40:1773–1778

Pflueger AC, Berndt TJ, Knox FG (1998) Effect of renal interstitial adenosine infusion on phosphate excretion in diabetes mellitus rats. Am J Physiol 274:R1228–R1235

Post SR, Jacobson JP, Insel PA (1996) P_2 purinergic receptor agonists enhance cAMP production in Madin-Darby canine kidney epithelial cells via an autocrine/paracrine mechanism. J Biol Chem 271: 2029–2032

Potier M, Aparicio M, Cambar J (1997) Protective effect of three xanthine derivatives (theophylline, caffeine and pentoxifylline) against the cyclosporin A-induced glomerular contraction in isolated glomeruli and cultured mesangial cells. Nephron 77:427–434

Ralevic V, Burnstock G (1998) Receptors for purines and pyrimidines. Pharmacol Rev 50:413–492

Ramos-Salazar A, Baines AD (1986) Role of 5′-nucleotidase in adenosine-mediated renal vasoconstriction during hypoxia. J Pharmacol Exp Ther 236:494–499

Reid IA, Stockigt JR, Goldfien A, Ganong WF (1972) Stimulation of renin secretion in dogs by theophylline. Eur J Pharmacol 17:325–332

Rossi NF, Churchill PC, Churchill MC (1987a) Pertussis toxin reverses adenosine receptor-mediated inhibition of renin secretion in rat renal cortical slices. Life Sci 40:481–487

Rossi NF, Churchill PC, Jacobson KA, Leahy AE (1987b) Further characterization of the renovascular effects of N^6-cyclohexyladenosine in the isolated perfused rat kidney. J Pharmacol Exp Ther 240:911–915

Rossi N, Ellis V, Kontry T, Gunther S, Churchill P, Bidani A (1990) The role of adenosine in $HgCl_2$-induced acute renal failure in rats. Am J Physiol 258:F1554–F1560

Rounds S, Hsieh L, Agarwal KC (1994) Effects of endotoxin injury on endothelial cell adenosine metabolism. J Lab Clin Med 123:309–317

Rouse D, Leite M, Suki WN (1994) ATP inhibits the hydrosmotic effect of AVP in rabbit CCT: evidence for a nucleotide P_{2u} receptor. Am J Physiol 267:F289–F295

Ruggenenti P, Remuzzi G (1997) Angiotensin-converting enzyme inhibitor therapy for non-diabetic progressive renal disease. Curr Opin Nephrol Hypertens 6:489–495

Rump LC, Bohmann C, Schwertfeger E, Krumme B, von Kügelgen I, Schollmeyer P (1996) Extracellular ATP in the human kidney: mode of release and vascular effects. J Auton Pharmacol 16:371–375

Rump LC, Oberhauser V, von Kügelgen I (1998) Purinoceptors mediate renal vasodilation by nitric oxide dependent and independent mechanisms. Kidney Int 54:473–481

Sakai K, Akima M, Nabata H (1979) A possible purinergic mechanism for reactive ischemia in isolated, cross-circulated rat kidney. Jpn J Pharmacol 29:235–242

Schnermann J, Weihprecht H, Briggs JP (1990) Inhibition of tubuloglomerular feedback during $adenosine_1$ receptor blockade. Am J Physiol 258:F553–F561

Schulze-Lohoff E, Zanner S, Ogilvie A, Sterzel RB (1992) Extracellular ATP stimulates proliferation of cultured mesangial cells via P_2-purinergic receptors. Am J Physiol 263:F374–F383

Schulze-Lohoff E, Hugo C, Rost S, Arnold S, Gruber A, Brüne B, Sterzel RB (1998) Extracellular ATP causes apoptosis and necrosis of cultured mesangial cells via $P2Z/P2X_7$ receptors. Am J Physiol 275:F962–F971

Schwartz DD, Malik KU (1989) Renal periarterial nerve-stimulation-induced vasoconstriction at low frequencies is primarily due to release of a purinergic transmitter in the rat. J Pharmacol Exp Ther 250:764–771

Schwiebert EM, Karlson KH, Friedman PA, Dietl P, Spielman WS, Stanton BA (1992) Adenosine regulates a chloride channel via protein kinase C and a G protein in a rabbit cortical collecting duct cell line. J Clin Invest 89:834–841

Scislo TJ, Augustyniak RA, Barraco RA, Woodbury DJ, O'Leary DS (1997) Activation of P_{2x}-purinoceptors in the nucleus tractus solitarius elicits differential inhibition of lumbar and renal sympathetic nerve activity. J Auton Nerv Sys 62:103–110

Scislo TJ, O'Leary DS (1998) Activation of A_{2a} adenosine receptors in the nucleus tractus solitarius inhibits renal but not lumbar sympathetic nerve activity. J Auton Nerv Sys 68:145–152

Sehic E, Ruan Y, Malik KU (1994) Attenuation by α,β-methylenadenosine-5′-triphosphate of periarterial nerve stimulation-induced renal vasoconstriction is not due to desensitization of purinergic receptors. J Pharmacol Exp Ther 271: 983–992

Shimada J, Suzuki F, Nonaka H, Karasawa A, Mizumoto H, Ohno T, Kubo K, Ishii A (1991) 8-(Dicyclopropylmethyl)-1,3-dipropylxanthine: a potent and selective adenosine A_1 antagonist with renal protective and diuretic activities. J Med Chem 34:466–469

Siegel NJ, Avison MJ, Reilly HF, Alger JR, Shulman RG (1983) Enhanced recovery of renal ATP with postischemic infusion of ATP–$MgCl_2$ determined by 31P–NMR. Am J Physiol 245:F530–534

Siegel NJ, Glazier WB, Chaudry IH, Gaudio KM, Lytton B, Baue AE, Kashgarian M (1980) Enhanced recovery from acute renal failure by the postischemic infusion of adenine nucleotides and magnesium chloride in rats. Kidney Int 17:338–349

Silldorff EP, Kreisberg MS, Pallone TL (1996) Adenosine modulates vasomotor tone in outer medullary descending vasa recta of the rat. J Clin Invest 98:18–23

Simmons NL (1981) Identification of a purine (P2) receptor linked to ion transport in cultured renal (MDCK) epithelium. Br J Pharmacol 73:379–384

Siragy HM, Linden J (1996) Sodium intake markedly alters renal interstitial fluid adenosine. Hypertension 27:404–407

Skøtt O, Baumbach L (1985) Effects of adenosine on renin release from isolated rat glomeruli and kidney slices. Pflügers Arch 404:232–237

Smits P, Horman H, Thien T, Houben H, and Van't Laar A (1983) Hemodynamic and humoral effects of coffee after beta-selective and nonselective beta-blockade. Clin Pharmacol Ther 34:153–158

Smits P, Pieters G, Thien T (1986) The role of epinephrine in the circulatory effect of coffee. Clin Pharmacol Ther 40:431–437

Spielman WS (1984) Antagonistic effect of theophylline on the adenosine-induced decrease in renin release. Am J Physiol 247:F246–F251

Spielman WS, Osswald H (1979) Blockade of postocclusive renal vasoconstriction by angiotensin II antagonist: evidence for an angiotensin-adenosine interaction. Am J Physiol 237:F463–F467

Spielman WS, Thompson CI (1982) A proposed role for adenosine in the regulation of renal hemodynamics and renin release. Am J Physiol 242:F423–F435

Spielman WS, Britton SL, Fiken-Olsen MJ (1980) Effect of adenosine on the distribution of renal blood flow in dogs. Circ Res 46:449–456

Spielman WS, Klotz KN, Arend LJ, Olson BA, LeVier DG, Schwabe U (1992) Characterization of adenosine A_1 receptor in a cell line (28A) derived from rabbit collecting tubule. Am J Physiol 263:C502–C508

Stromski ME, Cooper K, Thulin G, Gaudio KM, Siegel NJ, Shulman RG (1986) Chemical and functional correlates of postischemic renal ATP levels. Proc Natl Acad Sci USA 83:6142–6145

Suzuki F (1992) KW-3902. Drugs Future 17:876–878

Suzuki F, Shimada J, Mizumoto H, Karasawa A, Kubo K, Nonaka H, Ishii A, Kawakita T (1992) Adenosine A_1 antagonists. 2. Structure-activity relationships on diuretic activities and protective effects against acute renal failure. J Med Chem 35: 3066–3075

Svensson JO, Jonzon B (1990) Determination of adenosine and cyclic adenosine monophosphate in urine using solid-phase extraction and high-performance liquid chromatography with fluorimetric detection. J Chromatography 529:437–441
Tagawa H, Vander AJ (1970) Effects of adenosine compounds on renal function and renin secretion in dogs. Circ Res 26:327–338
Takano T, Soltoff SP, Murdaugh S, Mandel LJ (1985) Intracellular respiratory dysfunction and cell injury in short-term anoxia of rabbit renal proximal tubules. J Clin Invest 76:2377–2384
Takeda M, Yoshitomi K, Imai M (1993) Regulation of Na^+–$3HCO^-_3$ cotransport in rabbit proximal convoluted tubule via adenosine A_1 receptor. Am J Physiol 265:F511–F519
Tan CC, Ratcliff PJ (1992) Rapid oxygen-dependent changes in erythropoietin mRNA in perfused rat kidneys: Evidence against mediation by cAMP. Kidney Int 41:1581–1587
Terai T, Kita Y, Kusunoki T, Ando T, Shimazaki T, Deguchi Y, Akahane A, Shiokawa Y, Yoshida K (1990) The renal effects of FR-113453, a potent non-xanthine adenosine antagonist. Eur J Pharmacol 183:1057–1058
Thompson CI, Sparks HV, Spielman WS (1985) Renal handling and production of plasma and urinary adenosine. Am J Physiol 248:F545–F551
Thomson SC, Blantz RC, Vallon V (1998) Adenosine (ADO) generated by ecto-5′ nucleotidase (ecto-5NT) modulates macula densa control of glomerular filtration. J Am Soc Nephrol 9:348A
Tofovic SP, Branch KR, Oliver RD, Magee WD, Jackson EK (1991) Caffeine potentiates vasodilator-induced renin release. J Pharmacol Exp Ther 256:850–860
Tofovic SP, Jackson EK (2000) Effects of long-term caffeine consumption on renal function in spontaneously hypertensive heart failure prone rats. J Cardiovasc Pharmacol 33:360–366
Tofovic SP, Kusaka H, Pfeifer CA, Jackson EK (1996) Central effects of caffeine on renal renin secretion and norepinephrine spillover. J Cardiovasc Pharmacol 28: 302–313
Toya Y, Umemura S, Iwamoto T, Hirawa N, Kihara M, Takagi N, Ishii M (1993) Identification and characterization of adenosine A_1 receptor-cAMP system in human glomeruli. Kidney Int 43:928–932
Traynor T, Yang T, Huang YG, Arend L, Oliverio MI, Coffman T, Briggs JP, Schnermann J (1998) Inhibition of adenosine-1 receptor-mediated preglomerular vasoconstriction in AT_{1A} receptor-deficient mice. Am J Physiol 275:F922–F927
Tseng CJ, Kuan CJ, Chu H, Tung CS (1993) Effect of caffeine treatment on plasma renin activity and angiotensin I concentrations in rats on a low sodium diet. Life Sci 52:883–890
Ueda J (1972) Adenine nucleotides and renal function: Special reference with intrarenal distribution of blood flow. Japan J Pharmacol 22:5
Ueno M, Brookins J, Beckman B, Fisher JW (1988) A_1 and A_2 adenosine receptor regulation of erythropoietin production. Life Sci 43:229–237
Vallon V, Osswald H (1994) Dipyridamole prevents diabetes-induced alterations of kidney function in rats. Naunyn-Schmiedeberg's Arch Pharmacol 349:217–222
van Buren M, Bijlsma JA, Boer P, van Rijn HJM, Koomans HA (1993) Natriuretic and hypotensive effect of adenosine-1 blockade in essential hypertension. Hypertension 22:728–734
van Waarde A, Stromski ME, Thulin G, Gaudio KM, Kashgarian M, Shulman RG, Siegel NJ (1989) Protection of the kidney against ischemic injury by inhibition of 5′-nucleotidase. Am J Physiol 256:F298–F305
Vargas F, Osuna A, Fernández-Rivas A (1996) Renal vascular reactivity to ATP in hyper- and hypothyroid rats. Experientia 52:225–229
Viskoper RJ, Maxwell MH, Lupu AN, Rosenfeld S (1977) Renin stimulation by isoproterenol and theophylline in the isolated perfused kidney. Am J Physiol 232:F248–F253

Weaver DR, Reppert SM (1992) Adenosine receptor gene expression in rat kidney. Am J Physiol 263:F991–F995
Weber RG, Jones CR, Palacios JM, Lohse MJ (1988) Autoradiographic visualization of A_1 adenosine receptors in brain and peripheral tissues from rat and guinea-pig using ^{125}I-HPIA. Neurosci Lett 87:215–220
Weihprecht H, Lorenz JN, Briggs JP, Schnermann J (1992) Vasomotor effects of purinergic agonists in isolated rabbit afferent arterioles. Am J Physiol 263:F1026–F1033
Weihprecht H, Lorenz JN, Briggs JP, Schnermann J (1994) Synergistic effects of angiotensin and adenosine in the renal microvasculature. Am J Physiol 266:F227–F239
Weinberg JM, Davis JA, Shayman JA, Knight PR (1989) Alterations of cytosolic calcium in LLC-PK_1 cells induced by vasopressin and exogenous purines. Am J Physiol 256:C967–C976
Weinberg JM, Humes HD (1986) Increases of cell ATP produced by exogenous adenine nucleotides in isolated rabbit kidney tubules. Am J Physiol 250:F720–F733
Wilcox CS, Welch WJ, Schreiner GF, Belardinelli L (1999) Natriuretic and diuretic actions of a highly selective adenosine A_1 receptor antagonist. J Am Soc Nephrol 10:714–720
Wu PH, Churchill PC (1985) 2-chloro-[^{3}H]-adenosine binding in isolated rat kidney membranes. Arch Int Pharmacodyn 273:83–87
Yagil Y (1990) Interaction of adenosine with vasopressin in the inner medullary collecting duct. Am J Physiol 259:F679–F687
Yagil Y (1992) Differential effect of basolateral and apical adenosine on AVP-stimulated cAMP formation in primary culture of IMCD. Am J Physiol 263:F268–F276
Yagil Y (1994) The effects of adenosine on water and sodium excretion. J Pharmacol Exp Ther 268:826–835
Yagil C, Katni G, Yagil Y (1994) The effects of adenosine on transepithelial resistance and sodium uptake in the inner medullary collecting duct. Pflügers Arch 427:225–232
Yamaguchi S, Umemura S, Tamura K, Iwamoto T, Nyui N, Ishigami T, Ishii M (1995) Adenosine A_1 receptor mRNA in microdissected rat nephron segments. Hypertension 26:1181–1185
Yang C-M, Tsai Y-J, Pan S-L, Tsai C-T, Wu W-B, Chiu C-T, Luo S-F, Ou JT (1997) Purinoceptor-stimulated phosphoinositide hydrolysis in Madin-Darby canine kidney (MDK) cells. Naunyn-Schmiedeberg's Arch Pharmacol 356:1–7
Yao K, Kusaka H, Sato K, Karasawa A (1994) Protective effects of KW-3902, a novel adenosine A_1-receptor antagonist, against gentamicin-induced acute renal failure in rats. Jpn J Pharmacol 65:276–170
Zegarra-Moran O, Romeo G, Galietta LJ (1995) Regulation of transepithelial ion transport by two different purinoceptors in the apical membrane of canine kidney (MDCK) cells. Br J Pharmacol 114:1052–1056
Zou AP, Nithipatikom K, Li PL, Cowley AW Jr (1999a) Role of renal medullary adenosine in the control of blood flow and sodium excretion. Am J Physiol 276:R790–R798
Zou A-P, Wu F, Li P-L and Cowley AW Jr (1999b) Effect of chronic salt loading on adenosine metabolism and receptor expression in renal cortex and medulla in rats. Hypertension 33:511–516

CHAPTER 18

P2 Receptors in the Cardiovascular System

A. PELLEG and G. VASSORT

A. Introduction

It is now well established that purine nucleotides play a complex regulatory role in the cardiovascular system (BOARDER and HOURANI 1998). ATP exerts multiple effects on endothelial cells, platelets, perivascular nerves, vascular smooth muscle cells and cardiac myocytes (PEARSON and GORDON 1989; BOARDER and HOURANI 1998; PELLEG and BELARDINELLI 1998; ABBRACCHIO and BURNSTOCK 1998). Other chapters in this volume (JACKSON, Chap. 17, this volume; WEGNER, Chap. 23, this volume) discuss P2 receptors in the renal and respiratory systems as well as the endothelium and platelets. This chapter focuses mainly on P2 receptors in cardiac and vascular myocytes.

DRURY and SZENT-GYORGYI (1929) were first to show that extracellular adenine nucleotides exert pronounced effects on the mammalian heart independent of their pivotal role in cellular metabolism and energetics. Three major effects were noted in that study: a negative chronotropic effect on sinus node automaticity, a negative dromotropic effect on atrio-ventricular (AV) nodal conduction, and lusitropic effect on the coronary vasculature (DRURY and SZENT-GYORGYI 1929). This important study had opened a new field of research in which surprisingly little progress was made for several decades until the observation of BERNE (1963) of the release of ATP into the extracellular space of the hypoxic heart. BURNSTOCK (1978) suggested that the effects of adenosine and adenine nucleotides were mediated by cell surface receptors, purinergic receptor, which he subdivided into P1 and P2, respectively. P2 receptors were subsequently divided into two families: P2X and P2Y; the former are ligand binding cationic channels whose signal transduction pathways are cell membrane delimited and the latter are G protein coupled receptors whose signal transduction pathways involve intracellular second messengers (FREDHOLM et al. 1994). In recent years several different P2X and P2Y receptor proteins have been cloned (NORTH and BARNARD 1997), thereby transforming the original hypotheses of Burnstock from the realm of speculative theory into that of physiological mechanisms.

B. Sources of Extracellular ATP

ATP is released from: (i) ischemic cardiac myocytes (BERNE 1963; PADDLE and BURNSTOCK 1974; FORRESTER and WILLIAMS 1977; WILLIAMS and FORRESTER 1983; KUZMIN et al. 1998), (ii) activated platelets (MILLS et al. 1968; DAY and HOLMSEN 1971; HOLMSEN 1985), (iii) nerve terminals as a co-transmitter (HOLTON 1959; RICHARDSON and BROWN 1987; BURNSTOCK 1990), (iv) inflammatory cells (DIVIRGILIO et al. 1996), (v) erythrocytes (BERGFELD and FORRESTER 1992; ELLSWORTH et al. 1995; SPRAGUE et al. 1996), (vi) endothelial cells (RALEVIC et al. 1992; YANG et al. 1994; BODIN and BURNSTOCK 1996), (vii) smooth muscle cells (PEARSON and GORDON 1979; KATSURAGI et al. 1991), and (viii) exercising muscle cells (FORRESTER 1972; PARKINSON 1973), as well as electrically driven atrial cells challenged by the cardiotonic agents isoproterenol and forskolin (KATSURAGI et al. 1993).

There is strong evidence supporting the notion that ATP is a cotransmitter in perivascular sympathetic nerves and that it plays a role in both local regulation of vascular tone as well as neurotransmitter release via a feedback mechanism (RALEVIC and BURNSTOCK 1991; VON KÜGLGEN and STARKE 1991; SNEDDON et al. 1996). The latter is mediated by adenosine, the product of enzymatic degradation of ATP by ectonucleotidases (PEARSON et al. 1985; PELLEG and BURNSTOCK 1990; KIRKPATRICK and BURNSTOCK 1992; ZIMMERMANN 1996), some of which can be co-released with ATP from nerve terminals (TODOROV et al. 1997). Thus, extracellular ATP can be released into the lumen of blood vessels as well as the interstitial space where it activates P2 receptors located on the surface of endothelial and smooth muscle cells and perivascular nerve terminals. The net effect of extracellular ATP on vascular tone is dependent on the relative contribution of ATP's action at each of these three sites.

The mechanism by which ATP is transported across the cell membrane is not fully understood. In recent years, the existence of an adenine nucleotide binding cassette (ABC) family of proteins has been proposed (AL-AWQATI 1995). These proteins were suggested to be a regulatory component of an ion-channel-regulator complex, such as the cystic fibrosis transmembrane conductance regulator (CFTR) which acts as an ATP channel and enables intracellular ATP to cross the cell membrane and stimulate cell surface receptors (AL-AWQATI 1995; SCHWIEBERT et al. 1995; ABRAHAM et al. 1993, 1997; PASYK and FOSKETT 1997). However, whether the CFTR channel is permeable to ATP has been the subject of some controversy (GRYGORCZYK et al. 1996; GRYGORCZYK and HANRAHAN 1997).

C. Effects of ATP on Vascular Tone

Numerous studies have indicated that ATP and related purine nucleotides can either contract or relax blood vessels. There is evidence that the relaxation is mediated by an endothelium-dependent mechanism (DEMEY and VANHOUTTE 1981; DEMEY et al. 1982), while the contraction is mediated by a direct action on smooth muscle cells (KENNEDY et al. 1985; WHITE et al. 1985). However, at least in the case of the rabbit portal vein, ATP can cause vasodilation by a direct action on smooth muscle cells (RALEVIC and BURNSTOCK 1991). Extracellular ATP is rapidly degraded to adenosine (PEARSON and GORDON 1989) which is a potent vasodilator in different vascular beds (BERNE et al. 1983). Thus, if the endothelial function is intact, luminal ATP would cause vasodilation due to either its own action on endothelial cells or adenosine's action on endothelial and smooth muscle cells. In contrast, interstitial ATP can cause either contraction or relaxation of vascular smooth muscle cells (RALEVIC and BURNSTOCK 1991).

It is now agreed that the vasodilatory action of ATP is mediated by P2Y receptors on endothelial cells and the ATP-induced contraction of smooth muscle cells is mediated by P2X receptors (RALEVIC and BURNSTOCK 1991). However, it has recently been shown that ATP-induced contraction could be mediated by both P2X and P2Y receptors (see below). The mechanism of endothelium-dependent vasodilatory action of ATP is discussed elsewhere in this volume as well as previous reviews (BURNSTOCK 1990; RALEVIC and BURNSTOCK 1991; BOARDER and HOURANI 1998). The direct effects of ATP on smooth muscle cells are discussed below.

D. P2 Receptor-Mediated Signal Transduction in Smooth Muscle Cells

Extracellular ATP evoked a depolarizing transient inward current in smooth muscle cells of the rabbit ear artery (SUZUKI 1985). Using isolated single smooth muscle cells from this artery, it was found that this current resulted from an ATP-activated channel that is cation selective, but one which allows both monovalent and divalent cations to pass across the cell membrane (BENHAM et al. 1987). The channel manifested 3:1 selectivity for Ca^{2+} over Na^{+} at near physiological concentrations and a unitary conductance of ~5pS in 110 mmol/l Ca^{2+} or Ba^{2+} (BENHAM and TSIEN 1987). Its biophysical and pharmacological properties are different from voltage-gated L- or T-type Ca^{2+} channels (BENHAM and TSIEN 1987). The Ca^{2+} influx through this activated channel depolarizes the cell, thereby activating the voltage dependent channels which enhances Ca^{2+} influx (BENHAM 1989). Similar observations were made in cultured rat (TAWADA et al. 1987) and pig (DROOGMANS et al. 1991) aortic smooth muscle cells. In the latter preparation, the release of calcium from intracellular stores induced by ATP, activated a Cl^{-} current which contributed to the depolarization of the cell membrane (DROOGMANS et al. 1991) as well as a K^{+} current (VON DER WEID et al. 1993). Similar activation of a Cl^{-} current by ATP was observed in smooth muscle cells of the rat portal vein (KARASHIMA and TAKATA 1979; XIONG et al. 1991).

The mechanism of the ATP-induced Ca^{2+} influx, inward current, and elevation of intracellular Ca^{2+} could be complex. Ca^{2+} influx could result from either activation of ligand binding channels, subsequent activation of voltage-dependent channels, release of Ca^{2+} from internal stores (sarcoplasmic reticulum (SR)) via either inositol 1,4,5-trisphosphate (IP_3)- or Ca^{2+}-sensitive channels, and activation of Ca^{2+}-dependent channels and combination of all of these factors. Thus, both P2X and P2Y receptors could be involved in the modulation of intracellular Ca^{2+} level ($[Ca^{2+}]_i$) by ATP with P2X as the ligand binding Ca^{2+} channels and P2Y as mediators of ATP-induced formation of IP_3 and subsequent release of Ca^{2+} from internal stores (GERWINS and FREDHOLM 1992). In rat aortic smooth muscle strips evidence was obtained for the P2X- and P2Y-mediated Ca^{2+} influx and P2Y-mediated Ca^{2+} release from internal stores (KITAJIMA et al. 1994). Similarly, in rat pulmonary artery myocytes, extracellular ATP induced Ca^{2+} influx and $[Ca^{2+}]_i$ oscillations by activating P2X and P2Y receptors, respectively (GUIBERT et al. 1996). In contrast, the P2X agonist, α,β-methylene ATP, was ineffective in cultured rat aortic smooth muscle cells (VON DER WEID et al. 1993). Also, in single smooth muscle cells isolated from rat portal vein, ATP released Ca^{2+} from intracellular stores without involvement of (PACAUD et al. 1994) or with minor involvement of IP3 (PACAUD and LOIRAND 1995). More recently, MCLAREN et al., (1998) have shown that the ATP-induced contraction of the rat isolated tail artery is mediated by both $P2X_1$-receptors as well as G protein-coupled P2Y receptors.

In cultured human coronary artery smooth muscle cells, extracellular ATP activated inward current carried by Cl^- and an outward current carried by K^+ (STRØBÆK et al. 1996). Both currents were independent of external Ca^{2+} but dependent on $[Ca^{2+}]_i$, and α,β-methylene ATP, the P2X agonist, failed to activate these currents (STRØBÆK et al. 1996). It seems that there is a phenotypic modulation of vascular myocytes in culture associated with changes in the expression and/or function of P2 receptor subtypes (PACAUD et al. 1995).

The lack of selective and specific P2 receptor antagonists has resulted in an over reliance on the rank order of agonist potency in determining the subtype of P2 receptor responsible for a given action of ATP. However, it has been shown that the rank order of agonist potency at the P2X receptor which mediates contraction of rat isolated tail artery is very different from the potency order for evoking the inward current initiating this contraction (EVANS and KENNEDY 1994). This discrepancy was explained by the relative absence of breakdown of some of the agonists in the single cell preparation used for the electrophysiological study vs the multicellular arterial ring preparation used for the contraction study (EVANS and KENNEDY 1994).

A recent study has indicated that both $P2X_1$ and $P2X_7$ receptors expressed in myocytes of human saphenous vein mediate the contractile effect of extracellular ATP (CARIO-TOUMANIANTZ et al. 1998). Since $P2X_7$ mediates the formation of cell membrane pores permeable to large molecules and thereby the cytotoxic effects of ATP, it has been hypothesized that these receptors in human saphenous vein could be mechanistically involved in smooth cell lysis in the media of varicose veins (CARIO-TOUMANIANTZ et al. 1998). However, further studies are required to determine the physiological role of $P2X_7$ signal transduction in vascular smooth muscle cells.

E. ATP and the Vasculature: Concluding Remark

ATP, an important local regulator of vascular tone, is released into the lumen of blood vessels as well as in the adventitia from endothelial and red blood cells and platelets and perivascular nerves, respectively. It can activate different subtypes of P2-purinergic receptors located on the surface of different cell types, including endothelial and smooth muscle cells as well as sensory nerve endings. In addition, extracellular ATP is degraded by ecto-enzymes to adenosine that can activate P1-purinergic receptor which are also located on these cells. Thus the net effect of extracellular ATP on vascular tone depends on the outcome of several signal transduction pathways triggered by ATP as well as adenosine which could be simultaneously operative. Figure 1 is a schematic diagram illustrating the main sites of actions of extracellular ATP and adenosine which are known to mediate the modulatory effects of these compounds on vascular tone.

F. ATP Modulates Transmembrane Ionic Currents in Cardiac Myocytes

Numerous studies using different experimental models have established the modulatory role of extracellular ATP on ionic currents in cardiac myocytes. These effects of ATP, mediated by P2-purinergic receptor, are summarized in Fig. 2 and discussed in detail below.

I. Ca^{2+} Currents

Two decades ago GOTO et al. (1977) reported that extracellular ATP and ADP enhance calcium inward current (I_{Ca}) and I_{Ca}-dependent phasic tension in muscle bundles isolated from the right atrium of the bullfrog. In a subsequent study, the same group determined that the action of ATP on I_{Ca} and tension does not require the hydrolysis of ATP and is probably mediated by a receptor located at the outer surface of the affected cell membrane (YATANI et al. 1978). This followed the primary observation that ATP is able, like isoprenaline, to induce slow action potentials in K^+-depolarized guinea pig hearts (SCHNEIDER and SPERELAKIS 1975).

A more detailed study under whole-cell patch-clamp demonstrated that the extracellular application of micromolar ATP increases the L-type Ca^{2+} current (I_{CaL}) in mammalian cells isolated from rat ventricular myocardium (SCAMPS et al. 1990, 1992; SCAMPS and VASSORT 1990, 1994a; CHRISTIE et al. 1992) (Fig. 3). ATPγS exerts a similar effect, but adenosine is much less effective

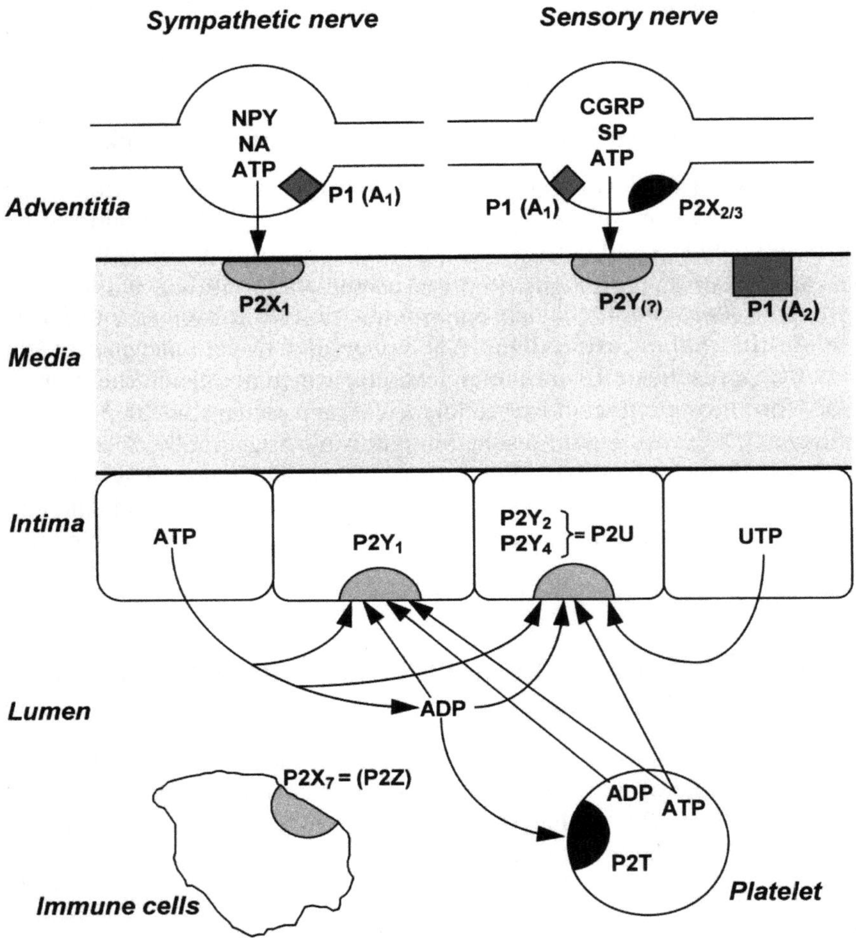

Fig. 1. Schematic diagram illustrating the main receptor subtypes for purine and pyrimidines present in most blood vessels. Perivascular nerves in the adventitia release ATP as a cotransmitter: ATP, released with noradrenaline (*NA*) and neuropeptide Y (*NPY*) from sympathetic nerves, activates $P2X_1$-purinoceptor of smooth muscle resulting in vasoconstriction; ATP, released with calcitonin gene-related peptide (*CGRP*) and substance P (*SP*) from sensory nerves whenever the axon reflex is elicited, activates P2Y-purinoceptors of smooth muscle cells resulting in vasodilation. A_1-adenosine receptors (i.e., P1-purinoceptors) on nerve terminals of sympathetic and sensory nerves mediate the modulation of transmitter release by adenosine, the product of ATP enzymatic degradation. $P2X_{2/3}$-purinoceptors on a subpopulation of sensory nerve terminals mediate nociception. A_2-adenosine receptors on vascular smooth muscle cells mediate vasodilation. ATP and UTP released from endothelial cells during shear stress and hypoxia activate $P2Y_1$ and $P2Y_2$ and $P2Y_4$ receptors, respectively, resulting in the production of nitric oxide and vasodilation. ATP is also released from activated platelets; the latter express ADP-selective purinoceptors ("P2T") while immune cells of various types express $P2X_7$ receptors (P_{2Z}-purinoceptors). (Taken with permission from ABBRACCHIO and BURNSTOCK 1998)

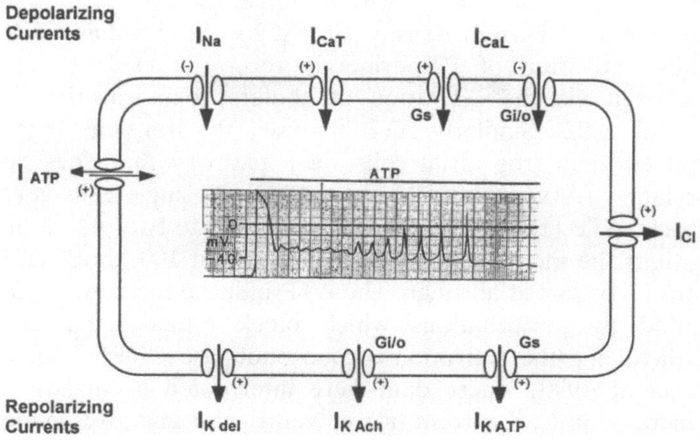

Fig. 2. Schematic presentation of transmembrane ionic currents regulated by extracellular ATP. G_s and $G_{i/o}$ are G proteins which mediate stimulation and inhibition of adenylyl cyclase, respectively. I_{ATP}, non-selective cationic current; I_{Na}, sodium current; I_{CaT} and I_{CaL} are T- and L-type calcium currents, respectively; I_{Cl}, chloride current; I_{Kdel}, delayed outwardly rectifying potassium current; I_{Kach}, inward rectifying K^+ current activated by acetylcholine and adenosine; and I_{KATP}, ATP-sensitive potassium current. I_{Kach} and I_{KATP} stimulation is mediated by a $G_{i/o}$ and a G_s protein, respectively. + and – denote stimulatory or inhibitory effects

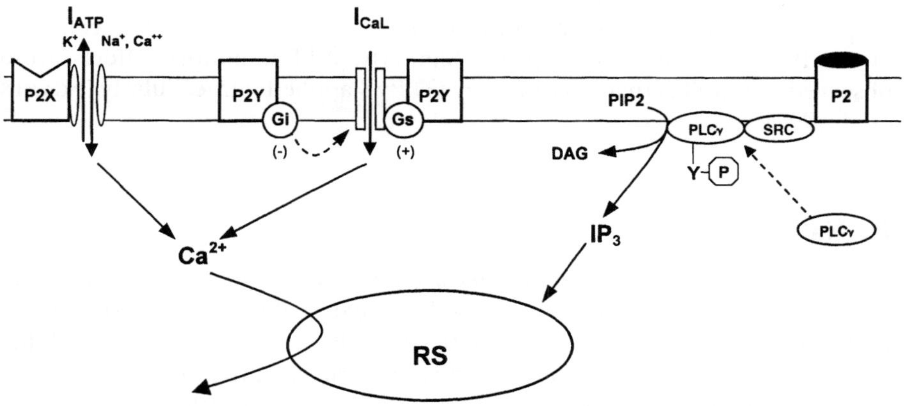

Fig. 3. Schematic outline of signal transduction pathways which mediate the action of extracellular ATP on intracellular Ca^{2+} level (Ca_i^{2+}). On the *left*, activation of the P2X-purinoceptors induces opening of a non-selective cationic channel and thus influx of Ca^{2+} ions, among other cations, that constitute I_{ATP}. The L-type Ca^{2+} current, I_{CaL} is controlled positively by a G_s protein probably directly coupled to the channel protein(s) and negatively by a pertussis toxin-sensitive G_i protein. On the *right*, the activation of another P2-purinergic receptor leads to the phosphorylation and translocation to the sarcolemma of PLCγ which mediates the production of inositol trisphophate (IP_3). IP_3 facilitates or even induces Ca^{2+} release by the sarcoplasmic reticulum (SR) in addition to the well established Ca^{2+}-induced Ca^{2+}-release in cardiac myocytes

and GTP, UTP, CTP, and ITP are without effect (SCAMPS et al. 1990). ATP is unable to alter Ca^{2+} current density after it has been enhanced by cholera toxin. Thus, activation of P2-purinergic receptors leads to an increase of Ca^{2+} current via the activation of cholera toxin-sensitive G_s protein (SCAMPS et al. 1992). Similarly, ATP increases the transient, high-threshold Ca current (I_{CaT}) in frog atrial cells via a pathway that does not involve phosphorylation (ALVAREZ and VASSORT 1992). In single cells isolated from frog ventricle, ATP (1 μmol/l) increases I_{CaL} by up to twofold; at higher ATP concentrations the increase in I_{Ca} is smaller and at 100 μmol/l, ATP reduces this current (ALVAREZ et al. 1990). The ATP-induced increase in Ca^{2+} current is prevented by perturbations which block either signal transduction pathways involving the activation of phospholipase C (PLC) or its activity (ALVAREZ et al. 1990). These data were interpreted to suggest that ATP-induced increase in Ca^{2+} current in frog ventricular myocytes is mediated by P2-purinergic receptors and phosphoinositide turnover (ALVAREZ et al. 1990).

In contrast to its positive effect in rat ventricular myocytes and in guinea pig and rabbit atrial myocytes (HIRANO et al. 1991), extracellular ATP inhibits I_{CaL} in a time- and concentration-manner dependent in isolated ferret ventricular myocytes (QU et al. 1993). This effect of ATP is independent of adenosine receptors (P1) but involves ATP binding to P2Y receptors and the activation of pertussis toxin (PTX) insensitive G protein (QU et al. 1993). Similar observations were obtained in hamster heart cells (VON ZUR MUHLEN et al. 1997). An ATP-induced inhibition, involving neither the P1- nor the P2-purinoceptors, was also reported in guinea pig sinoatrial node cells (QI et al. 1996; QI and KWAN 1996). Similarly, ATP-inhibitory effects were observed after full activation of I_{CaL} by GTPγS applied intracellularly (SCAMPS et al. 1992).

II. Na^+ Current

In addition to its effects on I_{Ca}, extracellular ATP affects the inward Na^+ current (I_{Na}). Under whole-cell patch-clamp, extracellular ATP in the micromolar range causes a leftward shift in both activation and availability characteristics of I_{Na} in rat single cardiac ventricular myocytes (SCAMPS and VASSORT 1994a), similar to its effect on I_{Ca} (SCAMPS et al. 1992). At hyperpolarized potentials, I_{Na} could be slightly increased due to the shift in activation, whereas at cell resting and depolarized potentials, I_{Na} is decreased because of reduced availability (SCAMPS and VASSORT 1994a). ATPγS and α,β-methylene ATP exert similar effects but UTP, β,γ-methylene ATP, ADP, and adenosine are without effect. The shifts observed upon application of extracellular ATP are not affected by cholera toxin treatment suggesting that a G_s protein and cyclic AMP are not involved in this phenomenon (SCAMPS and VASSORT 1994).

III. Cl^- Current

Extracellular ATP (5–50 μmol/l) activates an outwardly rectifying, time-dependent Cl^-current, I_{Cl}, in single guinea-pig atrial myocytes; ADP, AMP, and adenosine also activates I_{Cl} (Matsuura and Ehara 1992). A Cl^- current is activated by extracellular ATP (0.5–100 μmol/l) also in single rat ventricular myocytes (Matsuura and Ehara 1992). This current is blocked by the chloride channel blocker, 4,4*N*-diisothiocyanatostilbene-2,2*N*-disulfonic acid (DIDS), and is not activated by either AMP or adenosine. The differential action of adenosine on I_{Cl}, activated by extracellular ATP in atrial and ventricular myocytes could reflect different Cl^- channels in these cells or species variability. Neither the purinergic receptor subtype, nor the signal transduction pathway mediating the action of ATP on I_{Cl}, is known. However, it has been reported that genistein at 100 μmol/l, but not daidzein, activates the cardiac chloride conductance (Shuba et al. 1996). This effect is antagonized by Na_3VO_4, an inhibitor of phosphotyrosine phosphatase. Comparison of I_{Cl} activated by genistein and by forskolin led the authors to suggest that genistein activates the cAMP-dependent CFTR channel. It has been proposed that activation of the Cl^- current at the resting potential could lead to intracellular Cl^- depletion which can lead to the activation of the HCO_3^-/Cl^- exchanger and subsequent intracellular acidification (Kaneda et al. 1994). Indeed, similar potency and efficacy of ATP was noted for intracellular acidification (Pucéat et al. 1991) and activation of Cl^- current (Kaneda et al. 1994). The physiological importance of ATP enhancement of Cl^- current has not been determined. However, the activation of Cl^- current, whose reversal potential is around –35 mV, is potentially arrhythmogenic because such current could depolarize the membrane potential and shortens action potential plateau. It should also be mentioned that attenuation of isoprenaline-induced Cl^- current in guinea-pig ventricular myocytes after ATP application has been reported (Rankin et al. 1990; Pucéat and Vassort 1996). However, a direct activation of the Cl^-/HCO_3^- exchanger by ATP and the involvement of tyrosine kinase pathways have recently been demonstrated (Pucéat et al. 1998; also see below).

IV. K^+ Currents

A number of K^+ channels are present in cardiac myocytes that determine the shape of the action potential and frequency of contractions. Several observations have indicated that extracellular ATP regulates most of these channels. Friel and Bean (1988) reported that extracellular ATP activates two different ionic conductances in bullfrog atrial cells; one transient, non specific for cations and depolarizing, and the other sustained, probably carrying K^+ ions, and manifesting an inwardly current-voltage relation. Adenosine, AMP, ADP, ITP, and UTP were completely ineffective in activating either conductance. In a subsequent study it was found that extracellular ATP but not adenosine, activates an inwardly rectifying K^+ channel in calf atrial myocytes (FRIEL and BEAN 1990). This channel is nearly identical to the one activated by acetylcholine in these cells; the conductance of the channel activated by ATP and acetylcholine is 30 pS and 31 pS, respectively (FRIEL and BEAN 1990). Also, extracellular ATP and adenosine activate kinetically similar K^+ channels in one- and two-day-old rat atrial myocytes (i.e., single channel conductance and mean open time of 32.0 ± 0.2 pS and 0.5 ± 0.1 ms, respectively, vs 31.3 ± 0.3 pS and 0.9 ± 0.1 ms, respectively) (FU et al. 1995). The muscarinic cholinergic receptor and the A_1-adenosine receptor are known to be directly coupled to a K^+ channel ($K^+_{Ach,Ado}$ channel) via a PTX-sensitive G_K protein. In guinea-pig atrial cells extracellular ATP shortens the action potential (MATSUURA et al. 1996a). This effect is mediated by a P_2-purinergic receptor directly coupled to $K^+_{Ach,Ado}$ channel through a PTX-sensitive G_K protein, analogous to the activation of the channel by either acetylcholine or adenosine (MATSUURA et al. 1996a). In another recent study on isolated guinea-pig atrial myocytes, extracellular ATP (10 μmol/l) transiently activates $I_{K,Ach,Ado}$; however, when this current had been preactivated with either carbochol or adenosine, ATP produced a transient increase followed by a sustained decrease of the current (HARA and NAKAYA 1997). These data were interpreted (HARA and NAKAYA 1997) as a possible explanation for the biphasic inotropic effect (i.e., rapid negative followed by slow positive inotropic effect) of extracellular ATP in rat atrial preparation (FROLDI et al. 1994).

It was also reported that ATP activates the delayed rectifier K^+ current (I_{Ks}), which is slowly activated during the action potential plateau and facilitates repolarization, and whose deactivation contributes to depolarization of pacemaker cells (MATSUURA et al. 1996; MATSUURA and EHARA 1997). ATP-activated K^+ currents are also seen in rat ventricular myocytes. First, ATP in the micromolar range increases the inward rectifying current $I_{K,Ach,Ado}$. In addition, ATP activates a delayed outward K^+ current that requires the presence of 100 nmol/l intracellular Ca^{2+}. The latter effect of ATP is mimicked by the application of arachidonic acid and blocked by $AACOCF_3$, a phospholipase A_2 inhibitor as well as by inhibition of the cyclic AMP pathway (AIMOND et al. 1998).

Another type of K^+ channel, the intracellular ATP-sensitive channel (K^+_{ATP} channel), which is inhibited by intracellular ATP (NOMA 1983), is also regulated by extracellular ATP (Fig. 4). The K^+_{ATP} channel consists of a weak inward rectifier subunit Kir 6.2, plus a member of the adenine nucleotide binding cassette (ABC) superfamily, SUR2 (INAGAKI et al. 1996). K^+_{ATP} channel activation during acute ischemia/hypoxia has been shown to exert a protective effect on the heart (GROVER et al. 1990; GROSS and AUCHAMPACH 1992). Studies in cardiac myocytes in vitro have suggested that the activation of A_1-adenosine receptors could result in the activation of K^+_{ATP} channels (KIRSCH et al. 1990; ITO et al. 1992). However, at least in the hypoxic guinea-pig heart in vivo, endogenous adenosine failed to activate K^+_{ATP} channels (XU et al. 1994). Recent studies have shown that extracellular ATP enhances the current flow

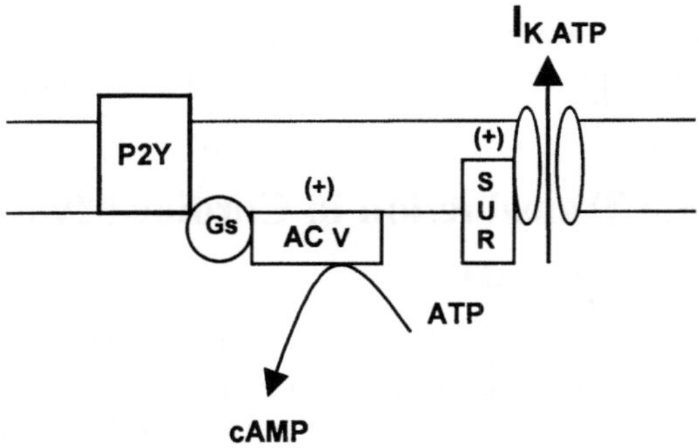

Fig. 4. Schematic description of signal transduction pathways which have been suggested to mediate the activation of ATP-sensitive K^+ current by extracellular ATP in cardiac myocytes. The P2Y-purinoceptor is linked to adenylyl cyclase (*AC*, isoform *V*) via G_s protein. Its activation can reduce the local intracellular level of ATP and thereby lead to the activation of the ATP-sensitive K^+ channel which is blocked by sulfonylurea (*SUR*) derivatives. This effect, like that of β-adrenergic stimulation, can contribute to the shortening of the action potential and energy saving during local myocardial hypoxia

through this channel ($I_{K,ATP}$) once it has been partially activated under conditions of metabolic stress (i.e., 100 μmol/l of intracellular ATP) (Babenko and Vassort 1997a); the enhancement of $I_{K,ATP}$ by extracellular ATP is inhibited by cholera toxin as well as by inhibition of adenylate cyclase (Babenko and Vassort 1997a). Thus, it has been suggested that the mechanism of this effect is the G_s-dependent activation of adenylate cyclase which causes increased cyclic AMP production and thereby reduced levels of intracellular ATP (Babenko and Vassort 1997a). Several analogs of ATP, i.e., α,β-mATP, 2 methyl-S ATP and ATPγS, exert a similar effect to that of ATP, while UTP and ADP have a relatively small effect and AMP and adenosine have no effect (Babenko and Vassort 1997b).

V. Non-Specific Cationic Current

A rapid, desensitizing inward current activated by ATP in the micromolar range was initially reported in frog atrial cells (Friel and Bean 1988). A similar current was found in ventricular rat (Scamps and Vassort 1990) and guinea-pig cardiac myocytes (Parker and Scarpa 1995). It could not be attributed to a Cl^- channel but rather to a non-selective cation channel with a reversal potential near 0 mV, an inwardly rectifying current/voltage relation, and a low unitary conductance. The external application of ATP at 0.5 mmol/l or above consistently activates a weak time-independent weak inward rectifying current in rabbit sino-atrial node cells (Shoda et al. 1997). The channel is nearly equally permeable to K^+, Na^+, or Cs^+, and only five times less to tris and *N*-methyl-D-glucamine. Adenosine, ADP and non-hydrolyzable ATP analogs failed to activate this current.

These currents should be attributed to the activation of P2X-purinoceptors whose precise subtypes remain to be defined.

G. ATP Signal Transduction in Cardiac Myocytes

I. Intracellular Ca^{2+} and pH

In quiescent or stimulated cells, the extracellular application of ATP causes an increase in $[Ca^{2+}]_i$. The increase in $[Ca^{2+}]_i$ results both from the stimulation of the Ca^{2+} currents and from a larger Ca^{2+} release from the SR since caffeine and ryanodine markedly reduce it (DANZIGER et al. 1988; HIRANO et al. 1991; DE YOUNG and SCARPA 1989; PUCÉAT et al. 1991a; CHRISTIE et al. 1992). Moreover, the ATP-induced formation of IP_3 might have some influence on the SR-Ca^{2+} release (see below). Furthermore, ATP directly gates a nonselective cationic channel (FRIEL and BEAN 1988; BJORNSSON et al. 1989; ZHENG et al. 1993) through which a significant Ca^{2+} influx could also occur. Other hypothesized mechanisms include phosphorylation of an extracellular protein leading to activation of a novel ion channel (CHRISTIE et al. 1992), as well as ATP-induced acidosis (PUCÉAT et al. 1991) leading to both the depolarizing effect and the increase in $[Ca^{2+}]_i$ (SCAMPS and VASSORT 1990).

Purinergic receptor stimulation activates the three major pH-regulating systems, the Na^+/H^+ antiporter, the Na^+/HCO_3^- symporter and the Cl^-/HCO_3^- exchanger. Intracellular alkalinization is prevented by amiloride derivatives and is attributable in most part to an activation of the Na^+/H^+ antiport (PUCEAT et al. 1993a). Under experimental acid load, ATP also activates an amiloride-insensitive HCO_3^--dependent alkalinizing mechanism (TERZIC et al. 1992). The signal transduction pathways have not been clearly established in both cases.

The most remarkable pH effect of a sudden application of ATP is a large (0.4 pH unit) and transient (1 min) acidosis that requires Cl^- ions in the extracellular milieu attributed to the activation of the anionic Cl^-/HCO_3^- exchanger (PUCÉAT et al. 1991b) (Fig. 5). The activation of the exchanger, is associated with a band-3 like protein phosphorylation on a tyrosine site (PUCÉAT et al. 1993a, 1998). More recent work has shown that isolated cardiomyocytes express both an AE_1 and AE_3 isoform. ATP induces tyrosine phosphorylation of AE_1 while acidosis still occurs in cell in which AE_3 expression was blocked. More precisely ATP activates the tyrosine kinase Fyn and association of both Fyn and FAK with AE_1. Tyrosine kinase inhibitors and microinjection of either anti-Cst.1 antibody or recombinant CSK, both of which prevent activation of Src kinases, significantly depressed the ATP-induced activation of the anion exchanger (PUCÉAT et al. 1998b).

II. Inositol Trisphosphate, Cyclic AMP and Cyclic GMP

The effects of ATP on second messengers such as cyclic AMP (cAMP), cyclic GMP (cGMP), and IP_3 have been investigated not only in whole heart tissues

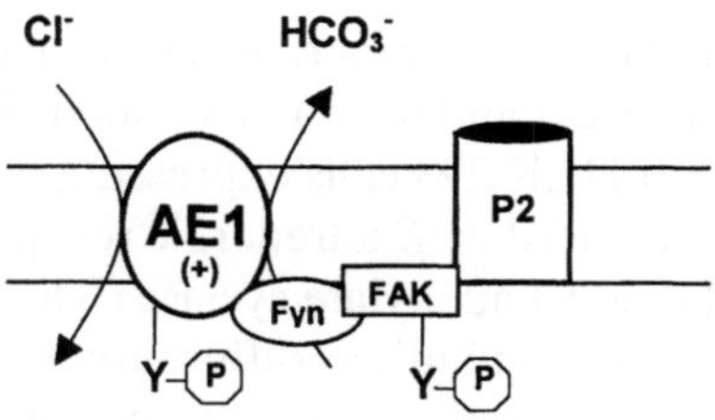

Fig. 5. Schematic description of the signal transduction pathways suggested to mediate the activation of the *AE_1* isoform of the Cl^-/HCO_3^- exchanger after stimulation of a P2-purinoceptor. The phosphorylation of tyrosine kinase *FAK* allows the docking of another tyrosine kinase of the SVC family, *Fyn*, to induce phosphorylation of the exchanger protein (PUCÉAT et al. 1998). In parallel, this signal transduction cascade activates PLCγ to produce IP3 (see Fig. 3 in PUCÉAT and VASSORT 1996)

but also, in the most recent studies, in isolated cardiomyocytes. ATP accelerates phosphatidylinositol turnover, as assessed by IP_3 formation in rat ventricles (LEGSSYER et al. 1988) and isolated fetal mouse cardiomyocytes (YAMADA et al. 1992). This pathway is not sensitive to PTX. In rat ventricular cardiomyocytes, ATP activates phospholipase Cγ_1 leading to IP_3 production by a pathway that involves a tyrosine kinase (PUCÉAT and VASSORT 1996) (Fig. 3). Simultaneously, diacylglycerol (DAG) is produced from phosphoinositide hydrolysis that should lead to activation of protein kinase C (PKC). Direct evidence in favor of an ATP-induced increase in PKC activity had been obtained. ATP triggers redistribution from cytosol to the membrane of both ε- and δ-PKC, two Ca^{2+}-insensitive PKC isoforms expressed in neonatal and adult cardiac cells. PKC also induces the phosphorylation of myristoylated alanine-rich C-kinase substrate (MARCKS) and the expression of c-*fos* in neonatal cells, two events known to be mediated by the kinase. These observations are of physiological relevance with regard to the likely specific role of PKC isoforms in cardiac function (PUCÉAT et al. 1994).

Whether or not ATP modulates intracellular cAMP has long been controversial. ZHENG et al. (1992) and SCAMPS et al. (1992) observed that ATP does not affect basal cAMP level in rat ventricular cardiomyocytes, but it facilitates the isoproterenol-induced increase in cAMP (ZHENG et al. 1992). In cardiomyocytes isolated from fetal mice, basal cAMP levels are not changed by ATP, which, however, partially antagonizes the effect of isoproterenol (YAMADA et al. 1992). While reinvestigating the effects of purinergic agonists on rat ventricular cardiomyocytes, it was found that stimulation with ATPγS in the presence of the phosphodiesterase inhibitor isobutylmethylxanthine (IBMX) increases the cAMP level twofold; at lower basal levels, a fourfold stimulation was observed (PUCÉAT et al. 1998a). The effect of ATP on cAMP production was poorly potentiated by forskolin and was additive to that of submaximal concentrations of isoproterenol. The ATP-induced activation of adenylate cyclase is mediated by a 45-kDa G_s protein, similar to that observed with isoproterenol stimulation. A paracrine effect involving PLA_2 activation and the formation of prostaglandins was excluded. Both ATP and isoproterenol increase cAMP in HEK 293 cells expressing type V adenylate cyclase while cAMP was only increased by β-adrenergic receptor stimulation of HEK expressing type IV and type VI adenylate cyclases. Thus in rat cardiomyocytes, purinergic and β-adrenergic stimulations differentially activate various cyclase isoforms; adenylate-cyclase V is the specific target of the purinergic stimulation (PUCÉAT et al. 1998b).

ATP, in the presence of IBMX, also increases basal cGMP content of isolated cardiomyocytes (SCAMPS et al. 1992). Other reports indicate that ATP activates arachidonic acid metabolism in both whole heart and isolated cardiac myocytes (DAMRON and BOND 1993). These observations could be related to the ATP-induced increase in cGMP content since arachidonic acid has been reported to activate soluble guanylate cyclase.

The increases in either cAMP or IP_3 have not yet been related to purinergic receptor subtypes in cardiac cells. A candidate would be the recently cloned $P2Y_{11}$ that is coupled to the stimulation of both phosphoinositide and adenylate cyclase pathways, although it has not been detected by Northern blot analysis in the heart (COMMUNI et al. 1997). In other cells, ATP increases cAMP preferentially through $P2Y_2$ in MDCK cells relative to $P2Y_1$ and $P2Y_{11}$ (POST et al. 1998). Most P2Y receptors found in cardiac tissues (i.e., $P2Y_1$, $P2Y_2$, $P2Y_4$, and $P2Y_6$) are assumed to be coupled to $PLC\beta$. The purinoreceptor subtype activating the tyrosine kinase cascade is unknown. However, it was previously shown (PUCÉAT et al. 1991) that the ATP-induced acidosis requires the presence of Mg^{2+} ions, a characteristic that is not generally observed with cloned purinoceptors.

H. Other General Effects of ATP in Cardiac Cells

I. Glucose Transport Inhibition

Besides increases in plasma membrane conductances for cations and anions, cytosolic Ca^{2+} and H^+ concentrations, rate of phosphoinositide hydrolysis and cyclic GMP and AMP production and force of contraction, ATP markedly inhibits glucose transport (FISCHER et al. 1998). ATP decreases the amount of glucose transporters in the plasma membrane with a concomitant increase in intracellular microsomal membranes. P2X purinergic receptors are not involved since a drastic reduction in extracellular Na^+ or Ca^{2+} ions does not alter this effect of ATP. The rank order potency of P_2 receptor agonists: ATP $\geq$ 212ATPγS $\geq$ 2-methyl-S ATP > ADP > α,β-methylene ATP does not match that of any known P2Y receptor subtype. Inhibition of transmembrane glucose transport was specific since ATP did not inhibit the rate of glycolysis or the rate of pyruvate decarboxylation. Nevertheless the physiological significance of this inhibitory effect of ATP on glucose transport is unclear.

II. Agonist-Induced Internalization of Purinergic Receptor

In the face of persistent stimulation, the response of most receptors fades away. The process of receptor desensitization is one of the many forms of the G protein coupled receptor regulation that has received much attention. Desensitization occurs generally via feedback regulation by the second messenger-stimulated kinases PKA or PKC that the Gs- and Gq-coupled receptors activate, respectively, which was first reported for the β_2-adrenergic receptor (Bouvier et al. 1988). This phosphorylation is followed by β-arrestin binding as a crucial step in internalization of most heptahelical receptors (Ferguson et al. 1996; Lefkowitz 1998).

Our knowledge of the turnover sequence activation/desensitization of the purinergic receptor is rather limited. The tagging of $P2Y_2$ receptor at its aminoterminus with a hemagglutinin epitope sequence reveals that the receptor undergoes agonist-promoted movement to an intracellular compartment. However, this internalization does not establish any functional consequence and is not required for agonist-induced desensitization (Sromek and Harden 1998).

I. Pathophysiological Effects of ATP in the Heart

I. Chronotropic and Arrhythmogenic Effects

In their seminal work, Drury and Szent-Gyorgyi (1929) reported a negative chronotropic effect of purines. Following studies suggested that ATP had dose-dependent effects; small doses producing tachycardia and relatively larger doses of ATP slowing the heart and inducing atrio-ventricular (AV) nodal conduction block (Stoner et al. 1948; Hollander and Webb 1957). However, these effects might be the result of the degradation of ATP to adenosine (Berne 1963) and the action of the latter on sino-atrial (SAN) and AV nodes (Pelleg et al. 1985). This was the case during administration of ATP into the sinus node blood supply (i.e., intracoronary) in dogs, in which model the negative chronotropic action of ATP was attenuated by theophylline, a non-selective adenosine receptor blocker (Barzu et al. 1985; Pelleg et al. 1987). On the other hand, ATP caused cardiac acceleration in 40% of the rabbit hearts using the Langendorff perfusion method (Takikawa et al. 1990). This effect of ATP was not antagonized by either theophylline or PTX pretreatment, but was almost completely blocked by apamin, neomycin and indomethacin (Takikawa et al. 1990). Based on these data it was concluded that the positive chronotropic effect of ATP was mediated by P2-purinergic receptors coupled to prostaglandins synthesis via a PTX-insensitive pathway involving the stimulation of PLC (Takikawa et al. 1990). It should be noted here that the *negative* chronotropic action of ATP is independent of prostaglandins (Pelleg et al. 1986). These observations indicate that negative chronotropy is, in part, due to P1-purinergic activation by ATP or rather by its degradation product, adenosine (Xu et al. 1993). In isolated ventricular myocytes of the guinea-pig, ATP alone did not exert any significant electrophysiological effect. However, when it was applied with compounds known to increase $[Ca^{2+}]_i$, ATP facilitated the induction of afterdepolarizations and triggered activity in ~60% of the cells (Song and Belardinelli 1994). In the presence of isoproterenol, ATP increased the amplitude of the transient inward current (I_{ti}), delayed afterdepolarizations (DADs), and I_{Ca}. In the presence of either BayK 8644 or quinidine, ATP further prolonged the action potential duration and also increased the amplitude of early afterdepolarizations (EADs) (Song and Belardinelli 1994). These findings extend earlier observations regarding the interaction between the effects of catecholamines and ATP on I_{Ca} (De Young and Scarpa 1987; Zheng et al. 1992) and support the hypothesis that the release of ATP into the extracellular space under pathophysiologic conditions could be arrhythmogenic (Scamps and Vassort 1990; Kuzmin et al. 1998).

It is difficult to anticipate the direct effect of ATP on sinus node rhythm. On the one hand, extracellular ATP can activate a time-independent, weakly inwardly rectifying current, which is non-selective for monovalent cations (Shoda et al. 1997). This current was not activated by either ADP, AMP, or adenosine, suggesting that the action of ATP is mediated by P2-purinergic receptors. On the other hand, in contrast to its effect on rat and guinea-pig ventricular myocytes, extracellular ATP inhibits, in a concentration-dependent manner, I_{Ca} in guinea-pig single sinoatrial nodal (SAN) cells (Song and Belardinelli 1994). The rank order of potency of ATP and related compounds in inhibiting I_{Ca} in SAN cells was: ATP = α,βmATP >> 2meSATP ≥ ATPγS >> UTP = AUP > AMP ≥ adenosine (Qi and Kwan 1996). This potency order has not been reported with regard to previously identified P2-purinergic receptor subtypes, suggesting mediation by a novel receptor of ATP's action. However, interpretation of a given rank order potency of ATP and its analogs should be done with great caution because of the documented high variability of ecto-nucleotidase activity among different tissues. In view of the critical role of I_{Ca} in the genesis of the action potential in SAN cells, it was proposed that extracellular ATP may play an important role in the regulation of heart rate (Qi and Kwan 1996). It is even more difficult to extrapolate data obtained in vitro (Takikawa et al. 1990; Qi and Kwan 1996) to the human heart in vivo. However, it should be noted that numerous studies in cats, dogs, and human subjects have indicated that, at least in these species, extracellular ATP exerts a negative chronotropic action on cardiac pacemakers which is mediated in part by the vagus nerve in addition to the action of adenosine (Belhassen and Pelleg 1984; Pelleg 1987; Pelleg et al. 1996). The vagal effect is due to a cardio-cardiac depressor reflex elicited by the action of ATP on vagal afferent nerve terminals in the left ventricle (Katchanov et al. 1996) similar to its action on pulmonary vagal afferent terminals (Pelleg and Hurt 1996). In both organs (i.e., heart and lungs) the triggering of the vagal reflex by ATP is mediated by P2X receptors (Pelleg and Hurt 1996; Katchanov et al. 1997).

The above-mentioned studies strongly suggest that extracellular ATP can directly affect ionic currents in sino-atrial cells. However, to what extent this action reflects a physiological role of extracellular ATP in the regulation of SAN automaticity, remains to be determined.

ATP shows positive chronotropic effects in cultured adult guinea pig myocytes that could be greatly enhanced in the presence of cardiac neurones that also possess P2-purinergic receptors (Horackova 1994). ATP increased contractile rate in intrinsic cardiac neurone-myocyte co-culture by 40% under control conditions and much more (100%) after blockade of β-adrenergic receptors by tetrodotoxin. In contrast, ATP induced much smaller effects in non-innervated myocyte cultures (26%). In isolated rat ventricular myocytes, the sudden application of ATP at micromolar concentration induces cell depolarization and triggers automaticity (Scamps and Vassort 1990). A similar ATP-triggered automaticity in rat papillary muscles requires the uncaging of ATP by UV-light flash (G. Vassort, unpublished) while the application of ATP by changing the bath solution was without significant effect.

II. Hypertrophy

Hormones and mechanical stretch can induce cardiac growth. Extracellularly applied ATP constitutes a stimulus sufficient to induce changes in the pattern of expression of immediate-early genes such as c-*fos* and *jun*-B that is mediated by a Ca^{2+}-dependent pathway in neonatal rat ventricular myocytes (Pucéat et al. 1994; Zheng et al. 1994). Similarly, extracellular ATP inhibited norepinephrine-stimulated growth of neonatal rat cardiac fibroblasts and activated *c-fos* gene expression in these cells (Zheng et al. 1996). These effects were probably mediated by P2Y receptors (Zheng et al. 1998).

However, ATP does not induce cell hypertrophy. Such an observation should be compared to the increase in expression of atrial natriuretic factor (ANF) and myosin light chain-2 (MLC_2) genes as well as cell hypertrophy by noradrenaline and angiotensin II. Similarly, it is thought that the mitogen-activated protein kinase (MAPK) plays a central role in the regulation of cell growth. However, ATP, like phenylephrine, carbachol and endothelin, activates the two p42 and p44 isoforms of MAPK; however, like carbachol, ATP neither transactivates cardiac-specific promoter/luciferase reporter genes, nor increases ANF expression (Post et al. 1996). These studies furthermore suggest that activation of *c-fos*, *jun*-B, PKC, and MAPK are not sufficient by themselves to stimulate hypertrophy, and that ATP also activates an inhibitory pathway, as suggested by the fact that it prevents phenylephrine-induced hypertrophy (Zheng et al. 1996).

The purinergic receptor subtypes that lead to MAPK activation in cardiac cells are unknown. They could be multiple and complex. Thus, in PC12 cells, MAPK activation is either by the G protein coupled $P2Y_2$ receptor (Soltoff et al. 1998) or following Ca^{2+} influx through the $P2X_2$ receptors and the involvement of the Ca^{2+}-activated tyrosine kinase, PYK_2 (SWANSON et al. 1998).

J. ATP and the Heart: Concluding Remarks

Adenosine has long been the focus of most studies concerning the effects of purines in the cardiovascular system. Adenosine is a ubiquitous biological compound released into the extracellular space by cells when oxygen supply does not meet oxygen demand or as a degradation product of ATP by ectoenzymes. At low concentrations, ATP by itself induces quite variable effects on the cardiac tissues that involve both inotropic and chronotropic effects. Positive inotropy is seen in all cardiac tissues including atria after the P1-purinergic negative inotropic effects are inhibited by P1-antagonists or PTX pretreatment (SCAMPS et al. 1990; LEGSSYER et al. 1988; MANTELLI et al. 1993). This is also true in human atria and ventricular tissues (ALVAREZ and VASSORT, unpublished). In these studies positive inotropy was mostly attributed to the increase in I_{Ca} following P2Y-purinergic receptor activation. In a more recent work positive inotropic effects of ATP and UTP in rat and guinea-pig cardiac tissues was rather attributed to P2X activation (FROLDI et al. 1997).

As discussed earlier, it is at present difficult to ascribe a given effect of ATP to a given receptor subtype and signal transduction pathway.

The $P2X_5$-receptor was cloned from a rat heart DNA library, with an abundantly expressed mRNA (GARCIA-GUZMAN et al. 1996). An $hP2X_3$ clone was also isolated from a heart cDNA library and RNA transcripts detected in human heart by RT-PCR analysis (GARCIA-GUZMAN et al. 1997). Using RT-PCR the expression of $P2Y_1$, $P2Y_2$, $P2Y_4$, and $P2Y_6$ receptor transcripts in the whole heart, neonatal cardiac fibroblasts, and myocytes and adult cardiac myocytes was reported (WEBB et al. 1996). In neonatal rat whole heart, all receptor sequences could be amplified with $P2Y_6$ being the most abundant. However, using the same procedure in adult rat myocytes $P2Y_4$ could not be detected, suggesting some changes in purinergic receptor expression during development.

In a recent study in human fetal heart, $P2X_1$, $P2X_3$, and $P2X_4$, as well as $P2Y_2$, $P2Y_4$, and $P2Y_6$ purinergic receptors have been identified using degenerated oligonucleotides (BOGDANOV et al. 1998). Indeed, very little is known about expression in cardiomyocytes per se. The need of antibodies added to the long-need of more specific and well-characterized agonists and antagonists. Moreover there might be major differences between species since, for example, the reexpression of the human and rat $P2X_4$ as well as $P2X_6$ gave different agonist/antagonist profiles (BOGDANOV et al. 1998; CHESSELL et al. 1997; WEBB et al. 1998). Nevertheless, the diversity in response to agonists and antagonists produced by the functionally expressed receptors are new findings that should help characterize the complex behavior of the cardiac tissues under purinergic stimulation, and development of reagents that might prove to have therapeutic value in various cardiac diseases.

References

Abbracchio MP, Burnstock G (1998) Purinergic signalling: Pathophysiological roles. Jpn J Pharmacol 78:113–145

Abraham EH, Okunieff P, Scala S, Vos P, Oosterveld MJ, Chen AY, Shrivastav B (1997) Cystic fibrosis transmembrane conductance regulator and adenosine triphosphate [letter]. Science 275:1324–1326

Abraham EH, Prat AG, Gerweck L, Seneveratne T, Arceci RJ, Kramer R, Guidotti G, Cantiello HF (1993) The multidrug resistance (mdr1) gene product functions as an ATP channel. Proc Natl Acad Sci USA 90:312–316

Aimond F, Rauzier J-M, Vassort G (1998) Purinergic activation of K^+ currents through activation of Gi-protein and of PLA_2 in rat ventricular myocyctes. Circulation 98:I-610

al-Awqati Q (1995) Regulation of ion channels by ABC transporters that secrete ATP [comment]. Science 269:805–806

Alvarez JL, Mongo K, Scamps F, Vassort G (1990) Effects of purinergic stimulation on the Ca current in single frog cardiac cells. Pflügers Arch 416:189–195

Alvarez JL, Vassort G (1992) Properties of the low threshold Ca current in single frog atrial cardiomyocytes. A comparison with the high threshold Ca current. J Gen Physiol 100:519–545

Babenko A, Vassort G (1997a) Enhancement of the ATP-sensitive K+ current by extracellular ATP in rat ventricular myocytes. Involvement of adenylyl cyclase-induced subsarcolemmal ATP depletion. Circ Res 80:589–600

Babenko A, Vassort G (1997b) Purinergic facilitation of ATP-sensitive potassium current in rat ventricular myocytes. Br J Pharmacol 120:631–638

Barzu T, Huerta F, Pourrias B (1985) The chronotropic effect of adenosine and ATP in dogs. The antagonism by theophylline. J Pharmacol 16:197–211

Belhassen B, Pelleg A (1984) Electrophysiologic effects of adenosine triphosphate and adenosine on the mammalian heart: clinical and experimental aspects. J Am Coll Cardiol 4:414–424

Benham CD (1989) ATP-activated channels gate calcium entry in single smooth muscle cells dissociated from rabbit ear artery. J Physiol (Lond) 419:689–701

Benham CD, Tsien RW (1987) A novel receptor-operated Ca^{2+}-permeable channel activated by ATP in smooth muscle. Nature 328:275–278

Benham CD, Bolton TB, Byrne NG, Large WA (1987) Action of externally applied adenosine triphosphate on single smooth muscle cells dispersed from rabbit ear artery. J Physiol (Lond) 387:473–488

Bergfeld GR, Forrester T (1991) Release of ATP from human erythrocytes in response to a brief period of hypoxia and hypercapnia. Cardiovasc Res 26:40–47

Berne RM (1963) Cardiac nucleotides in hypoxia: Possible role in regulation of coronary blood flow. Am J Physiol 204:317–322

Berne RM, Winn HR, Knabb RM, Ely SF, Rubio R (1983) Blood flow regulation by adenosine in heart, brain, and skeletal muscle. In: Berne RM, Rall TW, Rubio R (eds) Regulatory function of adenosine. Martins Nijihoff, Boston, pp293–313

Bjornsson OG, Monck JR, Williamson JR (1989) Identification of P2Y purinoceptors associated with voltage-activated cation channels in cardiac ventricular myocytes of the rat. Eur J Biochem 186:395–404

Boarder MR, Hourani SM (1998) The regulation of vascular function by P2 receptors: multiple sites and multiple receptors. TIPS 19:99–107

Bodin P, Burnstock G (1996) ATP-stimulated release of ATP by human endothelial cells. J Cardiovasc Pharmacol 27:872–875

Bogdanov YD, Rubino A, Burnstock G (1998) Characterisation of subtypes of the P2X and P2Y families of ATP receptors in the foetal human heart. Life Sci 62:697–703

Bogdanov YD, Wildman SS, Clements MP, King BF, Burnstock G (1998) Molecular cloning and characterization of rat P2Y4 nucleotide receptor. Br J Pharmacol 124:428–430

Bouvier M, Hausdorff WP, De Blasi A, BF OD, Kobilka BK, Caron MG, Lefkowitz RJ (1988) Removal of phosphorylation sites from the beta 2-adrenergic receptor delays onset of agonist-promoted desensitization. Nature 333:370–373

Burnstock G (1978) A basis for distinguishing two types of purinergic receptors. In: Straub RW, Bolis L (eds) Cell membrane receptors for drugs and hormones: A multidisciplinary approach. Raven Press, New York, pp 107–118

Burnstock G (1990) Noradrenaline and ATP as co-transmitters in sympathetic nerves. Neurochem Int 17:357–368

Burnstock G (1990) Dual control of local blood flow by purines. Ann NY Acad Sci 603:31–45

Cario-Toumaniantz C, Loirand G, Ladoux A, Pacaud P (1998) $P2X_7$ receptor activation-induced contraction and lysis in human saphenous vein smooth muscle. Circ Res 83:196–203

Chessell IP, Michel AD, Humphrey PP (1997) Properties of the pore-forming P2X7 purinoceptor in mouse NTW8 microglial cells. Br J Pharmacol 121:1429–1437

Christie A, Sharma VK, Sheu SS (1992) Mechanism of extracellular ATP-induced increase of cytosolic Ca2+ concentration in isolated rat ventricular myocytes. J Physiol (Lond) 445:369–388

Clemens MG, Forrester T (1981) Appearance of adenosine triphosphate in the coronary sinus effluent from isolated working rat heart in response to hypoxia. J Physiol (Lond) 312:143–158

Communi D, Govaerts C, Parmentier M, Boeynaems J-M (1997) Cloning of a human purinergic P2Y receptor coupled to phospholipase C and adenylyl cyclase. J Biol Chem 272:31969–31973

Damron DS, Bond M (1993) Modulation of Ca^{2+} cycling in cardiac myocytes by arachidonic acid. Circ Res 72:376–386

Danziger RS, Raffaeli S, Moreno-Sanchez R, Sakai M, Capogrossi MC, Spurgeon HA, Hansford RG, Lakatta EG (1988) Extracellular ATP has a potent effect to enhance cytosolic calcium and contractility in single ventricular myocytes. Cell Calcium 9:193–199

Day HJ, Holmsen H (1971) Concepts of blood platelet release reaction. Ser Haematol 4:3–27

De Mey JG, Vanhoutte PM (1981) Role of the intima in cholinergic and purinergic relaxation of isolated canine femoral arteries. J Physiol (Lond) 316:347–355

De Mey JG, Claeys M, Vanhoutte PM (1982) Endothelium-dependent inhibitory effects of acetylcholine, adenosine triphosphate, thrombin and arachidonic acid in the canine femoral artery. J Pharmacol Exp Ther 222:166–173

De Young MB, Scarpa A (1987) Extracellular ATP induces Ca^{2+} transients in cardiac myocytes which are potentiated by norepinephrine. FEBS Lett 223:53–58

De Young MB, Scarpa A (1989) ATP receptor-induced Ca2+ transients in cardiac myocytes: sources of mobilized Ca^{2+}. Am J Physiol 257:C750–C758

DiVirgilio F, Ferrari D, Chiozzi P, Falzoni S, Sanz JM, Dal Susino M, Mutini C, Hanau S, Baricordi OR (1996) Purinoceptor in the immune system. Drug Dev Res 39:319–329

Droogmans G, Callewaert G, Declerck I, Casteels R (1991) ATP-induced Ca^{2+} release and Cl^- current in cultured smooth muscle cells from pig aorta. J Physiol (Lond) 440:623–634

Drury AN, Szent-Gyorgyi A (1929) The physiological activity of adenine compounds with especial reference to their action upon mammalian heart. J Physiol (Lond) 68:213–237

Ellsworth ML, Forrester T, Ellis CG, Dietrich HH (1995) The erythrocyte as a regulator of vascular tone. Am J Physiol 269:H2155–H2161

Erlinge D, Yoo H, Edvinsson L, Reis DJ, Wahlestedt C (1993) Mitogenic effects of ATP on vascular smooth muscle cells vs. other growth factors and sympathetic cotransmitters. Am J Physiol 265:H1089–H1097

Erlinge D, You J, Wahlestedt C, Edvinsson L (1995) Characterisation of an ATP receptor mediating the mitogenesis in vascular smooth muscle cells. Eur J Pharmacol 289:135–149
Evans RJ, Kennedy C (1994) Characterization of the P_2-purinoceptors in the smooth muscle of the rat tail artery: a comparison between contractile and electrophysiological responses. Br J Pharmacol 113:853–860
Ferguson SS, Barak LS, Zhang J, Caron MG (1996) G-protein-coupled receptor regulation: role of G-protein-coupled receptor kinases and arrestins. Can J Physiol Pharmacol 74:1095–1110
Fischer TA, Singh K, DS OH, Kaye DM, Kelly RA (1998) Role of AT1 and AT2 receptors in regulation of MAPKs and MKP-1 by ANG II in adult cardiac myocytes. Am J Physiol 275:H906–H916
Forrester T (1972) An estimate of adenosine triphosphate release into the venous effluent from exercising human forearm muscle. J Physiol (Lond) 224:611–628
Forrester T, Williams CA (1977) Release of adenosine triphosphate from isolated adult heart cells in response to hypoxia. J Physiol (Lond) 268:371–390
Fredholm BB, Abbracchio MP, Burnstock G, Daly JW, Harden TK, Jacobson KA, Leff P, Williams M (1994) VI. Nomenclature and classification of purinoceptors. Pharmacol Rev 46:143–156
Friel DD, Bean BP (1988) Two ATP-activated conductances in bullfrog atrial cells. J Gen Physiol 91:1–27
Friel DD, Bean BP (1990) Dual control by ATP and acetylcholine of inwardly rectifying K^+ channels in bovine atrial cells. Pflügers Arch 415:651–657
Froldi G, Pandolfo L, Chinellato A, Ragazzi E, Caparrotta L, Fassina G (1994) Dual effect of ATP and UTP on rat atria: which types of receptors are involved?. Naunyn-Schmiedebergs Arch Pharmacol 349:381–386
Froldi G, Varani K, Chinellato A, Ragazzi E, Caparrotta L, Borea PA (1997) P2X-purinoceptors in the heart: actions of ATP and UTP. Life Sci 60:1419–1430
Fu C, Pleumsamran A, Oh U, Kim D (1995) Different properties of the atrial G protein-gated K^+ channels activated by extracellular ATP and adenosine. Am J Physiol 269:H1349–H1358
Garcia-Guzman M, Soto F, Laube B, Stuhmer W (1996) Molecular cloning and functional expression of a novel rat heart P2X purinoceptor. FEBS Lett 388:123–127
Garcia-Guzman M, Stuhmer W, Soto F (1997) Molecular characterization and pharmacological properties of the human P2X3 purinoceptor. Brain Res Mol Brain Res 47:59–66
Gerwins P, Fredholm BB (1992) ATP and its metabolite adenosine act synergistically to mobilize intracellular calcium via formation of inositol 1,4,5-trisphosphate in smooth muscle cell line. J Biol Chem 267:16081–16087
Goto M, Yatani A, Tsuda Y (1977) An analysis of the action of ATP and related compounds on membrane current and tension components in bullfrog atrial muscle. Jpn J Physiol 27:81–94
Gross GJ, Auchampach JA (1992) Blockade of ATP-sensitive potassium channels prevents myocardial preconditioning in dogs. Circ Res 70:223–233
Grover GJ, Dzwonczyk S, Parham CS, Sleph PG (1990) The protective effects of cromakalim and pinacidil on reperfusion function and infarct size in isolated perfused rat hearts and anesthetized dogs. Cardiovasc Drugs Ther 4:465–474
Grygorczyk R, Hanrahan JW (1997) Cystic fibrosis transmembrane conductance regulator and adenosine triphosphate. Science 275:1325–1326
Grygorczyk R, Tabcharani JA, Hanrahan JW (1996) CFTR channels expressed in CHO cells do not have detectable ATP conductance. J Membr Biol 151:139–148
Guibert C, Pacaud P, Loirand G, Marthan R, Savineau J-P (1996) Effect of extracellular ATP on cytosolic Ca^{2+} concentration in rat pulmonary artery myocytes. Am J Physiol 271:L450–L458

Hara Y, Nakaya H (1997) Dual effects of extracellular ATP on the muscarinic acetylcholine receptor-operated K^+ current in guinea-pig atrial cells. Eur J Pharmacol 324:295–303

Hirano Y, Abe S, Sawanobori T, Hiraoka M (1991) External ATP-induced changes in [Ca2+]i and membrane currents in mammalian atrial myocytes. Am J Physiol 260:C673–C680

Hollander PB, Webb JL (1957) Effects of adenine nucleotides on the contractility and membrane potentials of rat atrium. Circulation Research 5:349–353

Holmsen H (1985) Platelet metabolism and activation. Semin Hematol 22:219–240

Holton P (1959) The liberation of adenosine triphosphate on antidromic stimulation of sensory nerves. J Physiol (Lond) 145:494–504

Inagaki N, Gonoi T, Clement JP, Wang CZ, Aguilar-Bryan L, Bryan J, Seino S (1996) A family of sulfonylurea receptors determines the pharmacological properties of ATP-sensitive K^+ channels. Neuron 16:1011–1017

Ito H, Tung RT, Sugimoto T, Kobayashi I, Takahashi K, Katada T, Ui M, Kurachi Y (1992) On the mechanism of G protein beta gamma subunit activation of the muscarinic K^+ channel in guinea pig atrial cell membrane. Comparison with the ATP-sensitive K^+ channel. J Gen Physiol 99:961–983

Kaneda M, Fukui K, Doi K (1994) Activation of chloride current by P2-purinoceptors in rat ventricular myocytes. Br J Pharmacol 111:1355–1360

Karashima T, Takata Y (1979) The effects of ATP related compounds on the electrical activity of the rat portal vein. Gen Pharmacol 10:477–487

Katchanov G, Xu J, Hurt CM, Pelleg A (1996) Electrophysiological-anatomic correlates of adenosine 5-triphosphate-triggered vagal reflex in the dog. 3. Role of cardiac afferents. Am J Physiol 270:H1785–H1790

Katchanov G, Xu J, Clay A, Pelleg A (1997) Electrophysiological-anatomic correlates of ATP-triggered vagal reflex in the dog. 4. Stimulation of left ventricular afferent nerve terminals. Am J Physiol 272:H1899–H1903

Katsuragi T, Tokunaga T, Usune S, Sato C, Furukawa T (1991) Neurotransmitter-mediated ATP release from smooth muscles. In: Imai S, Nakazawa M (eds) Role of adenosine and adenine nucleotides in the biological system. Elsevier Science Publishers BV, pp 407–414

Katsuragi T, Tokunaga T, Ohba M, Sato C, Furukawa T (1993) Implication of ATP released from atrial, but not papillary, muscle segments of guinea pig by isoproterenol and forskolin. Life Sci 53:961–967

Kennedy C, Delbro D, Burnstock G (1985) P_2-purinoceptors mediate both vasodilation (via the endothelium) and vasoconstriction of the isolated rat femoral artery. Eur J Pharmacol 107:161–168

Kirkpatrick KA, Burnstock G (1992) Evidence that the inhibition of ATP release from sympathetic nerves by adenosine is a physiological mechanism. Gen Pharmacol 23:1045–1050

Kirsch GE, Codina J, Birnbaumer L, Brown AM (1990) Coupling of ATP-sensitive K^+ channels to A_1 receptors by G proteins in rat ventricular myocytes. Am J Physiol 259:H820–H826

Kitajima S, Ozaki H, Karaki H (1994) Role of different subtypes of P_2 purinoceptors on cytosolic Ca^{2+} levels in rat aortic smooth muscle. Eur J Pharmacol 266:263–267

Kuzmin AI, Lakomkin VL, Kapelko VI,Vassort G (1998) Interstitial ATP level and degradation in control and postmyocardial infarcted rats. Am J Physiol 275:C766–C771

Lefkowitz RJ (1998) G protein-coupled receptors. III. New roles for receptor kinases and beta-arrestins in receptor signaling and desensitization. J Biol Chem 273: 18677–18680

Legssyer A, Poggioli J, Renard D, Vassort G (1988) ATP and other adenine compounds increase mechanical activity and inositol trisphosphate production in rat heart. J Physiol (Lond) 401:185–199

Mantelli L, Amerini S, Filippi S, Ledda F (1993) Blockade of adenosine receptors unmasks a stimulatory effect of ATP on cardiac contractility. Br J Pharmacol 109:1268–1271

Matsuura H, Ehara T (1992) Activation of chloride current by purinergic stimulation in guinea pig heart cells. Circ Res 70:851–855

Matsuura H, Ehara T (1997) Selective inhibition of the slow component of delayed rectifier K^+ current in guinea-pig atrial cells by external ATP. J Physiol (Lond) 503:45–54

Matsuura H, Sakaguchi M, Tsuruhara Y, Ehara T (1996a) Activation of the muscarinic K^+ channel by P2-purinoceptors via pertussis toxin-sensitive G proteins in guinea-pig atrial cells. J Physiol (Lond) 490:659–671

Matsuura H, Tsuruhara Y, Sakaguchi M, Ehara T (1996b) Enhancement of delayed rectifier K^+ current by P2-purinoceptor stimulation in guinea-pig atrial cells. J Physiol (Lond) 490:647–658

McLaren GJ, Burke KS, Buchanan KJ, Sneddon P, Kennedy C (1998) Evidence that ATP acts at two sites to evoke contraction in the rat isolated tail artery. Br J Pharmacol 124:5–12

Mills DCB, Robb IA, Roberts GCK (1968) The release of nucleotides, 5 hydroxytryptamine and enzymes from human blood platelets during aggregation. J Physiol (Lond) 195:715–729

Noma A (1983) ATP-regulated K^+ channels in cardiac muscle. Nature 305:147–148

North RA, Barnard EA (1997) Nucleotide receptors. Current Opinion Neurobiol 7:346–357

Pacaud P, Gregoire G, Loirand G (1994) Release of Ca^{2+} from intracellular store in smooth muscle cells of rat portal vein by ATP-induced Ca^{2+} entry. Br J Pharmacol 113:457–462

Pacaud P, Loirand G (1995) Release of Ca^{2+} by noradrenaline and ATP from the same Ca^{2+} store sensitive to both $InsP_3$ and Ca^{2+} in rat portal vein myocytes. J Physiol (Lond) 484:549–555

Pacaud P, Malam-Souley R, Loirand G, Desgranges C (1995) ATP raises $[Ca^{2+}]_i$ via different P_2-receptor subtypes in freshly isolated and cultured aortic myocytes. Am J Physiol 269:H30–H36

Paddle BM, Burnstock G (1974) Release of ATP from perfused heart during coronary vasodilation. Blood Vessels 11:110–119

Parker KE, Scarpa A (1995) An ATP-activated nonselective cation channel in guinea pig ventricular myocytes. Am J Physiol 269:H789–H797

Parkinson PI (1973) The effect of graduated exercise on the concentration of adenine nucleoside in plasma. J Physiol (Lond) 234:72P–74P

Pasyk EA, Foskett JK (1997) Cystic fibrosis transmembrane conductance regulator-associated ATP and adenosine 3′-phosphate 5′-phosphosulfate channels in endoplasmic reticulum and plasma membranes. J Biol Chem 272:7746–7751

Pearson JD, Coade SB, Cusack NJ (1985) Characterization of ectonucleotidases on vascular smooth muscle cells. Biochem J 230:503–507

Pearson JD, Carleton JS, Gordon JL (1980) Metabolism of adenine nucleotides by ectoenzymes of vascular endothelial and smooth-muscle cells in culture. Biochem J 190:421–429

Pearson JD, Gordon JL (1979) Vascular endothelial and smooth muscle cells in culture selectively release adenine nucleotides. Nature 281:384–386

Pearson JD, Gordon JL (1989) P_2 purinoceptors in the blood vessel wall. Biochem Pharmacol 38:4157–4163

Pelleg A (1987) Cardiac electrophysiology and pharmacology of adenosine and ATP: modulation by the autonomic nervous system. J Clin Pharmacol 27:366–372

Pelleg A, Belardinelli L (eds) (1998) Effects of extracellular adenosine and ATP on cardiac myocytes. R.G. Landes, Austin

Pelleg A, Belhassen B, Ilia R, Laniado S (1985) Comparative electrophysiologic effects of ATP and adenosine in the canine heart: influence of atropine, propranolol, vagotomy, and aminophylline. Am J Cardiol 55:571–576
Pelleg A, Burnstock G (1990) Physiological importance of ATP released from nerve terminals and its degradation to adenosine in humans. Circulation 82:2269–2272
Pelleg A, Hurt CM (1996) Mechanism of ATP's action on canine pulmonary vagal fibre nerve terminals. J Physiol (Lond) 490:265–275
Pelleg A, Hurt CM, Wang L, Hewlett EL (1996) ATP shortens atrial action potential in the dog: Role of adenosine, the vagus nerve and G protein. Can J Physiol Pharmacol 74:15–22
Pelleg A, Mitamura H, Michelson EL, Dreifus LS (1986) Evidence against prostaglandin mediation of the electrophysiologic effects of ATP versus adenosine in the canine heart. J Cardiovasc Pharmacol 8:534–538
Pelleg A, Mitsuoka T, Michelson EL (1987) Adenosine mediates the negative chronotropic action of adenosine 5′-triphosphate in the canine sinus node. J Pharmacol Exp Ther 242:791–795
Post GR, Goldstein D, Thuerauf DJ, Glembotski CC, Brown JH (1996) Dissociation of p44 and p42 mitogen-activated protein kinase activation from receptor-induced hypertrophy in neonatal rat ventricular myocytes. J Biol Chem 271:8452–8457
Post SR, Rump LC, Zambon A, Hughes RJ, Buda MD, Jacobson JP, Kao CC, Insel PA (1998) ATP Activates cAMP production via multiple purinergic receptors in MDCK- D1 epithelial cells. Blockade of an autocrine/paracrine pathway to define receptor preference of an agonist. J Biol Chem 273:23093–23097
Pucéat M, Bony C, Jaconi M,Vassort G (1998) Specific activation of adenylyl cyclase V by a purinergic agonist. FEBS Lett 431:189–194
Pucéat M, Clement O, Scamps F, Vassort G (1991) Extracellular ATP-induced acidification leads to cytosolic calcium transient rise in single rat cardiac myocytes. Biochem J 274:55–62
Pucéat M, Clement O, Vassort G (1991) Extracellular MgATP activates the $Cl^-/HCO3^-$ exchanger in single rat cardiac cells. J Physiol (Lond) 444:241–256
Pucéat M, Clement-Chomienne O, Terzic A, Vassort G (1993) Alpha 1-adrenoceptor and purinoceptor agonists modulate Na-H antiport in single cardiac cells. Am J Physiol 264:H310–H319
Pucéat M, Clement-Chomienne O, Terzic A,Vassort G (1993) Alpha 1-adrenoceptor and purinoceptor agonists modulate Na-H antiport in single cardiac cells. Am J Physiol 264:H310–H319
Pucéat M, Hilal-Dandan R, Strulovici B, Brunton LL,Brown JH (1994) Differential regulation of protein kinase C isoforms in isolated neonatal and adult rat cardiomyocytes. J Biol Chem 269:16938–16944
Pucéat M, Roche S, Vassort G (1998) Src family tyrosine kinase regulates intracellular pH in cardiomyocytes. J Cell Biol 141:1637–1646
Pucéat M, Vassort G (1996) Purinergic stimulation of rat cardiomyocytes induces tyrosine phosphorylation and membrane association of phospholipase C gamma: a major mechanism for InsP3 generation. Biochem J 318:723–728
Qi AD, Kwan YW (1996) Modulation by extracellular ATP of L-type calcium channels in guinea- pig single sinoatrial nodal cell. Br J Pharmacol 119:1454–1462
Qu Y, Campbell DL, Strauss HC (1993) Modulation of L-type Ca2+ current by extracellular ATP in ferret isolated right ventricular myocytes. J Physiol (Lond) 471:295–317
Qu Y, Himmel HM, Campbell DL, Strauss HC (1993) Effects of extracellular ATP on ICa, [Ca2+]i, and contraction in isolated ferret ventricular myocytes. Am J Physiol 264:C702–C708
Ralevic R, Burnstock G (1991) Role of P_2-purinoceptors in the cardiovascular system. Circulation 84:1–14

Ralevic R, Milner P, Kirkpatrick KA, Burnstock G (1992) Flow-induced release of adenosine 5′-triphosphate from endothelial cells of the rat mesenteric arterial bed. Experientia 48:31–34

Rankin AC, Sitsapesan R, Kane KA (1990) Antagonism by adenosine and ATP of an isoprenaline-induced background current in guinea-pig ventricular myocytes. J Mol Cell Cardiol 22:1371–1378

Richardson PJ, Brown SJ (1987) ATP release from affinity purified rat cholinergic nerve terminals. J Neurochem 48:622–630

Scamps F, Legssyer A, Mayoux E, Vassort G (1990) The mechanism of positive inotropy induced by adenosine triphosphate in rat heart. Circ Res 67:1007–1016

Scamps F, Rybin V, Pucéat M, Tkachuk V, Vassort G (1992) A Gs protein couples P2-purinergic stimulation to cardiac Ca channels without cyclic AMP production. J Gen Physiol 100:675–701

Scamps F, Vassort G (1990) Mechanism of extracellular ATP-induced depolarization in rat isolated ventricular cardiomyocytes. Pflügers Arch 417:309–316

Scamps F, Vassort G (1994a) Effect of extracellular ATP on the Na+ current in rat ventricular myocytes. Circ Res 74:710–717

Scamps F, Vassort G (1994b) Pharmacological profile of the ATP-mediated increase in L-type calcium current amplitude and activation of a non-specific cationic current in rat ventricular cells. Br J Pharmacol 113:982–986

Schneider JA, Sperelakis N (1975) Slow Ca^{2+} and Na^{+} responses induced by isoproterenol and methylxanthines in isolated perfused guinea pig hearts exposed to elevated K^{+}. J Moll Cell Cardiol 7:249–273

Schwiebert EM, Egan ME, Hwang TH, Fulmer SB, Allen SS, Cutting GR, Guggino WB (1995) CFTR regulates outwardly rectifying chloride channels through an autocrine mechanism involving ATP. Cell 81:1063–1073

Shoda M, Hagiwara N, Kasanuki H, Hosoda S (1997) ATP-activated cationic current in rabbit sino-atrial node cells. J Moll Cell Cardiol 29:689–695

Shuba LM, Asai T, Pelzer S, McDonald TF (1996) Activation of cardiac chloride conductance by the tyrosine kinase inhibitor, genistein. Br J Pharmacol 119:335–345

Sneddon P, McLaren GJ, Kennedy C (1996) Purinergic cotransmission: sympathetic nerves. Semin Neurosci 8:201–205

Soltoff SP, Avraham H, Avraham S, Cantley LC (1998) Activation of P2Y2 receptors by UTP and ATP stimulates mitogen- activated kinase activity through a pathway that involves related adhesion focal tyrosine kinase and protein kinase C. J Biol Chem 273:2653–2660

Song Y, Belardinelli L (1994) ATP promotes development of afterdepolarizations and triggered activity in cardiac myocytes. Am J Physiol 267:H2005–H2001

Sromek SM, Harden TK (1998) Agonist-induced internalization of the P2Y2 receptor. Mol Pharmacol 54:485–494

Sprague RS, Ellsworth ML, Stephenson AH, Lonigro AJ (1996) ATP: the red blood cell link to NO and local control of the pulmonary circulation. Am J Physiol 271:H2717–H2722

Stoner HB, Green HN, Threlfall CJ (1948) Bodily reactions to trauma. A possible role of nucleotides in cardiac ischaemia. Br J Exp Pathol 29:419–446

Strøbæk D, Christophersen P, Dissing S, Olesen S-P (1996) ATP activates K and Cl channels via purinoceptor-mediated release of Ca^{2+} in human coronary artery smooth muscle. Am J Physiol 271:C1463–C1471

Su C (1985) Extracellular functions of nucleotides in heart and blood vessels. Ann Rev Physiol 47:665–676

Suzuki H (1985) Electrical responses of smooth muscle cells of the rabbit ear artery to adenosine triphosphate. J Physiol (Lond) 359:401–415

Swanson KD, Reigh C, Landreth GE (1998) ATP-stimulated activation of the mitogen-activated protein kinases through ionotrophic P2X2 purinoreceptors in PC12

Cells. Difference in purinoreceptor sensitivity in two pc12 cell lines. J Biol Chem 273:19965–19971
Takikawa R, Kurachi Y, Mashima S, Sugimoto T (1990) Adenosine-5′-triphosphate-induced sinus tachycardia mediated by prostaglandin synthesis via phospholipase C in the rabbit heart. Pflügers Arch 417:13–20
Tawada Y, Furukawa K-I, Shigekawa M (1987) ATP-induced calcium transient in cultured rat aortic smooth muscle cells. J Biochem 102:1499–1509
Terzic A, Pucéat M, Clement-Chomienne O, Vassort G (1992) Phenylephrine and ATP enhance an amiloride insensitive bicarbonate-dependent alkalinizing mechanism in rat single Naunyn-Schmiedebergs. Arch Pharmacol 346:597–600
Todorov LD, Mihaylova-Todorova S, Westfall TD, Sneddon P, Kennedy C, Bjur RA, Westfall DP (1997) Neuronal release of soluble nucleotidases and their role in neurotransmitter inactivation. Nature 387:76–79
Vials AJ, Burnstock G (1996) ATP release from the isolated perfused guinea pig heart in response to increased flow. J Vasc Res 33:1–4
Von der Weid P-Y, Serebryakov VN, Orallo F, Bergmann C, Snetkov VA, Takeda K (1993) Effects of ATP on cultured smooth muscle cells from rat aorta. Br J Pharmacol 108:638–645
Von Kugelgen I, Starke K (1991) Noradrenaline-ATP co-transmission in the sympathetic nervous system. TIPS 12:319–324
Von zur Muhlen F, Gonska BD, Kreuzer H (1997) Inhibition of the voltage-dependent calcium current by extracellular ATP in hamster ventricular cardiomyocytes. Am J Physiol 273:H250–H256
White TD, Chaudry A, Vohra MM, Webb D, Leslie R (1985) Characteristics of P_2 (nucleotide) receptors mediating contraction and relaxation of rat aortic stripes: Possible physiological relevance. Eur J Pharmacol 118:37–44
Williams CA, Forrester T (1983) Possible source of adenosine triphosphate released from rat myocytes in response to hypoxia and acidosis. Cardiovasc Res 17:301–312
Webb TE, Boluyt MO, Barnard EA (1996) Molecular biology of P2Y purinoceptors: expression in rat heart. J Auton Pharmacol 16:303–307
Webb TE, Henderson DJ, Roberts JA, Barnard EA (1998) Molecular cloning and characterization of the rat P2Y4 receptor. J Neurochem 71:1348–1357
Xiong Z, Kitamura K, Kuriyama H (1991) ATP activates cationic currents and modulates the calcium current through GTP-binding protein in rabbit portal vein. J Physiol (Lond) 440:143–165
Xu J, Wang L, Hurt CM, Pelleg A (1994) Endogenous adenosine does not activate ATP-sensitive potassium channels in the hypoxic guinea pig ventricle in vivo. Circulation 89:1209–1216
Yamada M, Hamamori Y, Akita H, Yokoyama M (1992) P2-purinoceptor activation stimulates phosphoinositide hydrolysis and inhibits accumulation of cAMP in cultured ventricular myocytes. Circ Res 70:477–485
Yang S, Cheek DJ, Westfall DP, Buxton ILO (1994) Purinergic axis in cardiac blood vessels. Agonist-mediated release of ATP from cardiac endothelial cells. Circ Res 74:401–407
Yatani A, Goto M, Tsuda Y (1978) Nature of catecholamine-like actions of ATP and other energy rich nucleotides on the bullfrog atrial muscle. Jpn J Physiol 28:47–61
Zheng JS, Boluyt MO, L ON, Crow MT, Lakatta EG (1994) Extracellular ATP induces immediate-early gene expression but not cellular hypertrophy in neonatal cardiac myocytes. Circ Res 74:1034–1034
Zheng JS, Boluyt MO, Long X, L ON, Lakatta EG, Crow MT (1996) Extracellular ATP inhibits adrenergic agonist-induced hypertrophy of neonatal cardiac myocytes. Circ Res 78:525–535
Zheng JS, Christie A, De Young MB, Levy MN, Scarpa A (1992) Synergism between cAMP and ATP in signal transduction in cardiac myocytes. Am J Physiol 262:C128–C135

Zheng JS, Christie A, Levy MN, Scarpa A (1993) Modulation by extracellular ATP of two distinct currents in rat myocytes. Am J Physiol 264:C1411–C1417

Zheng JS, O'Neill L, Long X, Webb TE, Barnard EA, Lakatta EG, Boluyt MO (1998) Stimulation of P2Y receptors activates *c-fos* gene expression and inhibits DNA synthesis in cultured cardiac fibroblasts. Cardiovasc Res 37: 718–728

Zimmermann H (1996) Extracellular purine metabolism. Drug Dev Res 39:337–352

CHAPTER 19

Roles of Purines and Pyrimidines in Endothelium

V. RALEVIC

A. Introduction

Purine nucleotides and nucleosides were shown in 1929 to have potent effects on the heart and vasculature, in the first indication that these agents can act as extracellular signalling molecules in the cardiovascular system (DRURY and SZENT-GYÖRGI 1929). The seminal discovery by FURCHGOTT and ZAWADZKI (1980) that many vasodilators require the presence of an intact endothelium to mediate vascular relaxation paved the way for the identification and characterisation of receptors for purines and pyrimidines on the endothelium. In common with a number of other endogenous vasodilators, ADP and ATP were shown to act at receptors on endothelial cells to mediate vasodilatation via the release of an endothelium-derived relaxing factor (EDRF) (DE MEY and VANHOUTTE 1981), subsequently identified as nitric oxide (NO) (FURCHGOTT et al. 1987; IGNARRO et al. 1987). Purine (ADP and ATP) and pyrimidine (UDP and UTP) nucleotides can also elicit the release of prostacyclin (GORDON 1986) and endothelium-derived hyperpolarising factor (EDHF) (MALMSJÖ et al. 1998) from endothelial cells. The vasodilator response to adenosine has also been shown to be mediated, at least in part, via receptors on the endothelium in some blood vessels.

Given the significance of these findings with respect to our understanding of circulatory physiology, it is not surprising that the literature on the actions of purines and pyrimidines in endothelium is dominated by studies elaborating their effects as mediators of endothelium-dependent vasodilatation and role in the regulation of blood flow. Partly as a consequence of the intense interest in this field, we now have a reasonably clear idea of the molecular and pharmacological profile of the different subtypes of P1 and P2 receptors that are expressed on endothelial cells, and of their signalling pathways. Vasodilatation, however, is just one of several effects that can be mediated by endothelial P1 and P2 receptors. The present chapter will consider in addition the accumulating evidence that purines and pyrimidines, acting via endothelial P1 and P2 receptors, also have other important biological roles including the regulation of angiogenesis, modulation of microvascular permeability, and modulation of inflammation.

B. Subtypes of Purine and Pyrimidine Receptors Expressed on Endothelial Cells

Receptors for purines and pyrimidines are divided into two main families known as P1 and P2 receptors. Their subclassification is reviewed comprehensively in other chapters in this book and is only briefly considered here. P1 receptors are activated by adenosine; four subclasses of the P1 receptor, all of which couple to G proteins, have been identified (A_1, A_{2A}, A_{2B} and A_3). P2 receptors are activated by ATP, ADP, UTP and UDP, and are divided into two main families of P2X receptors, which are ligand-gated cation channels, and G protein-coupled P2Y receptors (ABBRACCHIO and BURNSTOCK 1994; FREDHOLM et al. 1994). Seven mammalian P2X receptors and six P2Y receptors have been cloned and characterised to date (see RALEVIC and BURNSTOCK 1998). Receptors belonging to both the P1 and P2 receptor families are expressed on endothelial cells, although there is considerable selectivity in the subtypes that are expressed. All couple to G proteins; ionotropic P2X receptors do not appear to be expressed by endothelial cells.

Both A_{2A} and A_{2B} receptors are expressed by the endothelium. Evidence for this comes largely from studies that have defined the pharmacological profile of receptors mediating endothelium-dependent vasodilatation to adenosine in various blood vessels in vitro. A_{2A} receptors on endothelial cells in culture have also been reported (SCHIELE and SCHWABE 1994; SOBREVIA et al. 1997). Evidence implicating the A_{2A} receptor on endothelium is strong, primarily because of the availability and use of potent and highly selective agonists and antagonists. In contrast, the lack of high affinity selective agonists and antagonists for the A_{2B} receptor means that the evidence for endothelial A_{2B} receptors is largely indirect. Human aortic endothelial cells have been shown to express the mRNA of both A_{2A} and A_{2B} receptors (IWAMOTO et al. 1994), but the levels of expression of the two receptors is unclear. In most blood vessels, the indication from functional studies is that there is preferential expression of either A_{2A} or A_{2B} receptors on the endothelium, which is different in various types of blood vessels and between species (see Sect. C). The study of IWAMOTO et al. (1994), however, raises the possibility that both A_{2A} and A_{2B} receptors are expressed by the vascular endothelium, but that depending on the tissue the level of one or other of the subtypes is low. A_1 and A_3 receptors are not generally expressed by the endothelium.

The principal types of P2 receptors expressed by endothelial cells are P2Y receptors activated preferentially by ADP (ATP and UTP are weak or inactive), known as $P2Y_1$ receptors (formerly P_{2Y} receptors), and those activated equipotently by ATP and UTP, known as $P2Y_2$ receptors (formerly P_{2U} receptors). It should be noted, however, that mRNAs for $P2Y_1$, $P2Y_2$, $P2Y_4$ and $P2Y_6$ receptors have been demonstrated in endothelial cells (JIN et al. 1998). As these receptors are commonly activated by ADP, ATP and/or UTP the identity of many endothelial P2Y receptors reported to date, for which a complete agonist potency profile has not been established, is indeterminate. Diadenosine polyphosphates are endogenous signalling molecules that can act at P2Y receptors (SCHLÜTER et al. 1994; RALEVIC et al. 1995), and as they are more resistant to hydrolysis than ATP, ADP and UTP, they may have more far-reaching effects in the vasculature. $P2Y_1$ and $P2Y_2$ receptors have notably different pharmacological profiles, which has proved useful in their characterisation: 2MeSATP is a potent ligand at the $P2Y_1$, but not the $P2Y_2$ receptor; the $P2Y_1$ receptor is generally more sensitive than the $P2Y_2$ receptor to the P2 receptor antagonists suramin and pyridoxalphosphate-6-azophenyl-2′,4′-disulfonic acid (PPADS) (WILKINSON et al. 1994; BOYER et al. 1996; RALEVIC and BURNSTOCK 1996a; HANSMANN et al. 1997). A uridine nucleotide-specific receptor, activated by UTP, but not by ATP or ADP, which coexists with $P2Y_1$ and $P2Y_2$ receptors, has been identified on guinea-pig cardiac endothelial cells (YANG et al. 1996). Sensitivity to pertussis toxin and mediation of an increase in intracellular Ca^{2+} suggests that this receptor belongs to the family of G protein-coupled P2Y receptors.

Studies of single endothelial cells in culture have shown that coexpression of $P2Y_1$- and $P2Y_2$-like receptors occurs at the level of single endothelial cells (DUCHENE and TAKEDA 1997). The markedly different sensitivities of the two receptors to activation by endogenous purines and pyrimidines indicates that through their specific activation of $P2Y_1$ and $P2Y_2$ receptors, ADP, ATP and UTP may have distinct biological roles in endothelium, as well as elsewhere. It is intriguing that the $P2Y_2$ receptor is activated, equipotently, by both ATP and UTP; studies to identify the causal relationship between endogenously released ATP and UTP and activation of the $P2Y_2$ receptor are needed to resolve the physiological significance of this.

P1 and P2 receptors are also coexpressed by the endothelium. As ATP is rapidly metabolised to ADP and adenosine in biological tissues, complex response profiles to ATP at the endothelium may be generated, involving principally actions of ATP at $P2Y_2$ receptors, ADP at $P2Y_1$ receptors and adenosine at A_2 receptors. This signalling cascade is more likely to alter the temporal profile of the endothelial response than its amplitude as synergistic interactions between these endothelial receptors does not generally occur.

C. Heterogeneity of Purine and Pyrimidine Receptor Expression on Endothelium

There is considerable heterogeneity in the number and types of receptors for purines and pyrimidines expressed on endothelial cells between blood vessels and species. The functional expression of A_{2A} and A_{2B} receptors by endothelium, for instance, differs widely; A_{2A} receptors account for adenosine-mediated endothelium-dependent vasodilatation in rat aorta (CONTI et al. 1993; LEWIS et al. 1994; MONOPOLI et al. 1994; PRENTICE and HOURANI 1996) and porcine coronary artery (ABEBE et al. 1994), whereas A_{2B} receptors appear to be more significant as mediators of endothelium-dependent vasodilatation in rabbit corpus cavernosum (CHIANG et al. 1994) and rat renal artery (MARTIN and POTTS 1994). Adenosine releases NO from endothelial cells cultured from porcine carotid artery, but not from human saphenous vein (LI et al. 1995), although whether this is because of a difference in the expression of adenosine receptors by the arterial and venous endothelial cells, or due to differences in their ability to generate NO, is unknown. The differential expression of A_{2A} and A_{2B} receptors by the endothelium between blood vessels has implications for their sensitivity to adenosine, as A_{2A} receptors have a high affinity for adenosine, binding in the nanomolar level, and A_{2B} receptors a low affinity, requiring micromolar concentrations of adenosine for activation.

Heterogeneity of receptor expression has also been described for endothelial P2 receptors. Radioligand binding studies have shown that rat brain capillary cell B10 cells express a single type of P2 receptor, namely the $P2Y_1$, and not the $P2Y_2$ receptor (WEBB et al. 1996). Only 17% of cells from an immortalised rat brain microvascular endothelial cell line (RBE4) were shown to express $P2Y_1$ receptors, i.e. were activated by 2MeSATP to cause elevation of intracellular $[Ca^{2+}]$, whereas all of the cells express $P2Y_2$ receptors, as evidenced by their responsiveness to ATP and UTP (NOBLES et al. 1995). Purinergic responses of the RBE4 cells were shown to be quantitatively and qualitatively similar to those of primary cultured rat brain endothelial cells (NOBLES et al. 1995), which is important as there may be considerable plasticity of purine receptor expression by endothelial cells in culture (MATEO et al. 1996; ALBERT et al. 1997). For example, $P2Y_1$-like receptors, but not $P2Y_2$-like receptors, in bovine adrenomedullary endothelial cells are lost in culture with the first passage of the cells (MATEO et al. 1996).

Pharmacological studies have shown that, in the golden hamster mesenteric arterial bed, whereas ATP and UTP mediate potent vasodilatation via endothelial $P2Y_2$ receptors, ADP and 2MeSATP are weak or inactive, indicating a lack of endothelial $P2Y_1$ receptors (RALEVIC and BURNSTOCK 1996b) (Fig. 1). A lack of responsiveness to ADP of the mouse mesenteric arterial bed similarly indi-

Fig. 1a,b. Species difference in vasodilator responsiveness of hamster and rat mesenteric arterial beds to 2-methylthio ATP (2MeSATP) at endothelial $P2Y_1$ receptors. **a** Hamster mesenteric arterial bed. *Left*: Representative trace showing dose-dependent vasodilator responses to ATP (0.005–500 nmol) and UTP (0.005–500 nmol) at $P2Y_2$ receptors in a preparation with tone raised with methoxamine (15 μmol/l). 2MeSATP (0.005–50 nmol) had no vasodilator effect in this preparation, and at the highest dose elicited vasoconstriction. In other preparations, 2MeSATP gave weak vasodilator responses. *Right*: Vasodilator responses in preparations with tone raised with methoxamine (10–80 μmol/l) to: ATP (0.005–500 nmol, $n = 18$); UTP (0.005–500 nmol, $n = 16$); 2MeSATP (0.005–50 nmol, $n = 16$) and ADP (0.005–500 nmol, $n = 5$). At the highest dose of 2MeSATP vasoconstriction and no vasodilatation was typically observed. **b** Rat mesenteric arterial bed. *Left*: Representative trace of preparation with tone raised with methoxamine (20 μmol/l), where 2MeSATP (0.0005–5 nmol) is more potent as a vasodilator at $P2Y_1$ receptors than ATP (0.005–5 nmol) and UTP (0.5–500 nmol) at $P2Y_2$ receptors. *Right*: Vasodilator dose-response curves to 2MeSATP ($n = 8$), ADP ($n = 8$), ATP ($n = 7$) and UTP ($n = 8$). (Modified from RALEVIC and BURNSTOCK 1996a,b)

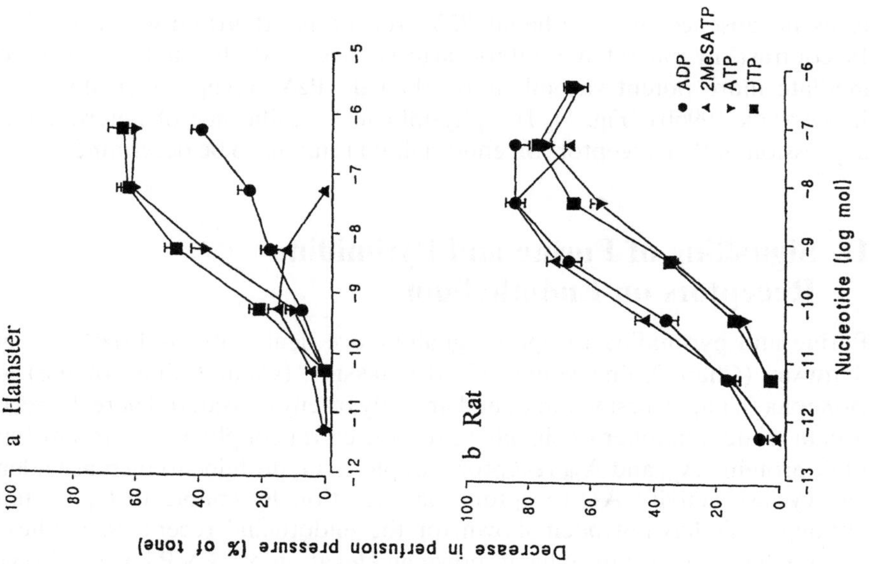
a Hamster
b Rat
Decrease in perfusion pressure (% of tone)
Nucleotide (log mol)
ADP
2MeSATP
ATP
UTP

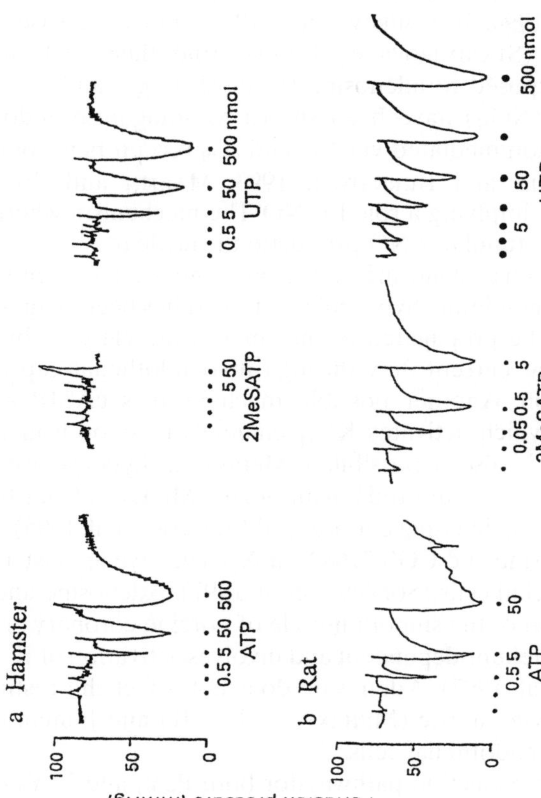
a Hamster
b Rat
Perfusion pressure (mmHg)
0.5 5 50 500
ATP
5 50
2MeSATP
0.5 5 50 500 nmol
UTP
0.5 5 50
ATP
0.05 0.5 5
2MeSATP
5 50 500 nmol
UTP

cates an absence of endothelial $P2Y_1$ receptors (BERTHIAUME et al. 1997). In contrast, in the rat mesenteric arterial bed, endothelial $P2Y_1$ receptors mediate more potent vasodilatation than do $P2Y_2$ receptors (RALEVIC and BURNSTOCK 1996b) (Fig. 1). The physiological significance of heterogeneous expression of P2Y receptors on endothelium remains to be determined.

D. Signalling of Purine and Pyrimidine Receptors on Endothelium

Purine and pyrimidine receptor signalling are dealt with by LORENZEN and SCHWABE (Chap. 2, first volume) and CHESSELL (Chap. 3, first volume) and BOARDER (Chap. 4, first volume) and are only briefly considered here. P1 receptors mediate a number of signalling responses via coupling to different types of G proteins. A_{2A} and A_{2B} receptors couple to G_s proteins to increase adenylate cyclase activity. A_{2B} receptors may additionally couple to G_q proteins, although this has not been shown for the endothelial receptors. Studies of endothelial cells in culture have shown an elevation of cAMP levels following activation of A_2 receptors (DES ROSIERS and NEES 1987; WATANABE et al. 1992; SCHIELE and SCHWABE 1994; STANIMIROVIC et al. 1994), and a decrease in cAMP levels via A_1 receptors (STANIMIROVIC et al. 1994). Endothelial cells in culture generate NO when exposed to adenosine (LI et al. 1995), and inhibitors of nitric oxide synthase (NOS) have been shown to antagonise endothelium-dependent vasodilatation mediated via A_{2A} and A_{2B} receptors in some vascular preparations (VIALS and BURNSTOCK 1993; MARTIN and POTTS 1994; STEINHORN et al. 1994), implying a role for NO. The mechanism whereby activation of A_2 receptors stimulates NO production is unclear.

In some blood vessels, adenosine and its analogues have been shown to cause endothelial cell membrane hyperpolarisation. It has been suggested that, in the heart, this may be propagated to the smooth muscle cells by electrotonic coupling and that current flow through myoendothelial gap junctions mediates vasodilatation. Another possible mechanism is cAMP activation of protein kinase A, which activates K^+_{ATP} channels in smooth muscle. The generation of EDHF is also a possibility. Membrane hyperpolarisation by adenosine has been shown in cultured bovine aortic (MEHRKE et al. 1991; SEISS-GEUDER et al. 1992) and guinea-pig endocardial (MANABE et al. 1995) endothelial cells, and by adenosine and CGS 21680, an A_{2A} selective agonist, in human umbilical vein endothelial cells (SOBREVIA et al. 1997). Adenosine and NECA cause hyperpolarisation of the smooth muscle of porcine coronary artery that is at least partly endothelium-dependent and involves activation of K^+_{ATP} channels (OLANREWAJU et al. 1997). Adenosine does not affect the resting membrane potential of bovine aortic (MEHRKE et al. 1991) and human coronary (ZUNKLER et al. 1995) endothelial cells.

The main signal transduction pathway for both $P2Y_1$ and $P2Y_2$ receptors is activation of PLC via coupling with G_q ($P2Y_1$ and $P2Y_2$) and G_{io} ($P2Y_2$) proteins. Subsequent IP_3 and diacylglycerol (DAG) formation and Ca^{2+} mobilisation can lead to activation of a variety of signalling pathways. Elevation of intracellular Ca^{2+} stimulates NOS and leads to NO formation. This occurs in response to activation of $P2Y_1$ and $P2Y_2$ receptors on the endothelium of most, if not all, blood vessels. The main physiological target of DAG is activation of protein kinase C (PKC), which may subsequently cause tyrosine phosphorylation of mitogen-activated protein kinases (MAPK). Both $P2Y_1$ and $P2Y_2$ receptors are linked, albeit differentially, to tyrosine phosphorylation of MAPK, and activation of this pathway is a requirement for stimulation of endothelial prostacyclin (PGI_2) production (BOWDEN et al. 1995; PATEL et al. 1996; ALBERT et al. 1997). $P2Y_1$ and $P2Y_2$ receptors show a differential sensitivity to pharmacological manipulations of PKC; the $P2Y_2$ receptor is generally less sensitive, indicating that PKC provides greater inhibitory feedback control of $P2Y_1$ receptor signalling (PURKISS et al. 1994; COMMUNI et al. 1995; CHEN et al. 1996). Albeit subtle, these differences in intracellular signalling lend support to the concept of distinct biological roles of $P2Y_1$ and $P2Y_2$ receptors. Endothelial cell hyperpolarisation and potassium channel activation by P2Y receptors has been described (IKEUCHI and NISHIZAKI 1995; MANABE et al. 1995). EDHF generation in response to activation of endothelial $P2Y_1$ and $P2Y_2$ receptors has been shown in rat carotid and mesenteric arteries (MALMSJÖ et al. 1998).

$P2Y_1$ receptor-mediated inhibition of adenylate cyclase has been described in a clonal population of rat brain capillary endothelial cells (B10 cells) and may be due to different G protein coupling of the $P2Y_1$ receptor (WEBB et al. 1996).

E. Endogenous Sources of Purines and Pyrimidines

A consideration of the endogenous sources and stimuli for the release of purines and pyrimidines is essential for an understanding of their specific roles at receptors on the vascular endothelium. Purines and pyrimidines are metabolised rapidly by ectonucleotidases (ZIMMERMANN, Chap. 8, first volume). Thus their accumulation to biologically active concentrations at the endothelial cell surface depends on local release, which may be from endothelium, erythrocytes, platelets and adrenal chromaffin cells.

Many biological tissues, including endothelial cells (MEININGER et al. 1988), release adenosine during hypoxic conditions. An increase in pCO_2 or the addition of thrombin to the culture medium evoke the release of adenosine from guinea-pig coronary endothelial cells (KRATZER et al. 1996). The endothelium is also a source of ATP and UTP, which can be released by hypoxia, shear stress and vasoactive agents (MILNER et al. 1990; BODIN et al. 1992; RALEVIC et al. 1992; YANG et al. 1994; SAIAG et al. 1995; SHINOZUKA et al. 1997), or by injury in pathophysiological conditions. ATP and UTP may act on P2 receptors on endothelial cells adjacent to, or downstream from, the site of nucleotide release to elicit local vasodilatation and an increase in blood flow. In addition, they may release further ATP, and presumably UTP, in an autocrine feedback loop via actions on endothelial P2Y receptors (YANG et al. 1994). ATP and UTP are rapidly broken down to ADP and UDP respectively, and ADP is further degraded to adenosine; thus, multiple endothelial P1 and P2 receptors may be activated in a complex signalling cascade.

Erythrocytes are sources of ATP, and of ADP and adenosine following metabolic degradation of ATP (FORRESTER 1990; ELLSWORTH et al. 1995). It has been shown that ATP is released from erythrocytes when these cells are exposed to a low pO_2 or low-pH environment (ELLSWORTH et al. 1995). Thus, erythrocytes may act as sensors of oxygen demand and regulators of blood supply via the release of ATP and its interaction with specific endothelial purine receptors, leading to vasodilatation and an increase in blood flow.

Endogenous vasoactive agents including NA and ATP can induce the release of ATP, and presumably other purines and pyrimidines, from smooth muscles, but apparently not significantly in blood vessels (HASHIMOTO et al. 1997). ATP, ADP, UTP and the diadenosine polyphosphates are also released from activated platelets during aggregation, but this process occurs when there is damage to the function or integrity of the endothelium and may be more relevant to activation of P2 receptors on the underlying smooth muscle cells, leading to vasoconstriction or vasospasm. As purines and pyrimidines are present in the cytoplasm of all types of cells, injury can potentially lead to their accumulation extracellularly in concentrations sufficient to activate endothelial P2 receptors.

F. Roles of Purines and Pyrimidines on Endothelium

I. Vasodilatation

ATP, ADP and UTP act at endothelial P2Y receptors to mediate potent vasodilatation in most blood vessels, which generally occurs, at least in part, via generation of NO (RALEVIC and BURNSTOCK 1988, 1991; RALEVIC et al. 1997; YOU et al. 1997; MALMSJÖ et al. 1998; RUMP et al. 1998) (Fig. 2). In rat carotid and mesenteric arteries, EDHF, in addition to NO, is generated as a vasodilator mediator in response to activation of $P2Y_1$ and $P2Y_2$ receptors on the endothelium (MALMSJÖ et al. 1998). As NO does not account entirely for endothelium-dependent vasodilatation in many blood vessels, especially in resistance vessels, it is likely that a more widespread involvement of EDHF in endothelium-dependent relaxation by purines and pyrimidines will be revealed when this is specifically investigated. In general, PGI_2 does not contribute significantly to P2 receptor-mediated endothelium-dependent vasodilatation; notable exceptions are its contribution to ADPβS-mediated vasodilatation via $P2Y_1$-like receptors in rat thoracic aorta and pancreatic vascular bed (SAIAG et al. 1996) and endothelium-dependent vasodilatation

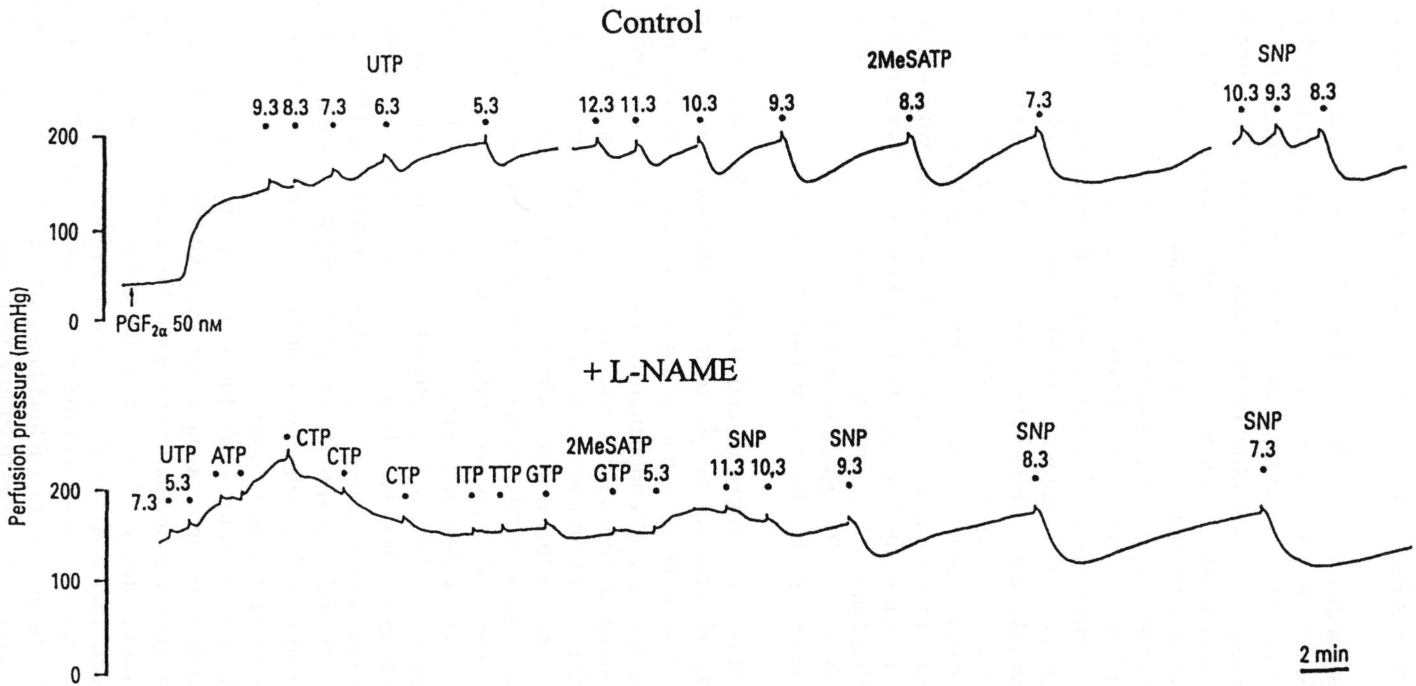

Fig. 2. Nitric oxide (NO) mediates vasodilatation of human placental cotyledons by purine and pyrimidine nucleotides at endothelial P2Y receptors. Representative trace showing vasodilator responses (decrease in perfusion pressure, mmHg) of a single human placental cotyledon to purine and pyrimidine nucleotides and to the endothelium-independent vasodilator sodium nitroprusside (SNP) in the absence (*top*) and presence (*bottom*) of a NO synthase inhibitor, N^G-nitro-L-arginine methyl ester (L-NAME; 100 μmol/l). Doses are given as –log mol. The *upper trace* shows responses to doses of UTP, 2-methylthio ATP (2MeSATP) and SNP. In the *lower trace*, UTP, ITP and TTP were all applied as doses of 5.3 (–log mol; equivalent to 5 μmol). ATP was applied at doses of 6.3 and 5.3; CTP at 7.3, 6.3 and 5.3; GTP at 6.3 and 5.3. Tone of the preparations was raised with 50 nmol/l $PGF_{2\alpha}$ in the absence of L-NAME (basal perfusion pressure 42 mmHg) and 30 nmol/l $PGF_{2\alpha}$ in the presence of L-NAME. (From Ralevic et al. 1997)

mediated by $P2Y_1$, but not $P2Y_2$, receptors in intercostal artery rings (WILKINSON et al. 1994).

Adenosine does not mediate endothelium-dependent vasodilatation in all blood vessels, and where this does occur it is generally less potent than that of the purine and pyrimidine nucleotides at endothelial P2Y receptors. In contrast, vasodilatation via P1 receptors on the underlying vascular smooth muscle contributes significantly to adenosine's relaxation response. In some vessels NO mediates endothelium-dependent vasodilatation to adenosine, as reported for A_{2A} receptors in porcine coronary artery (ABEBE et al. 1995), A_{2B} receptors in rat renal artery (MARTIN and POTTS 1994), and A_2-like receptors in rabbit pulmonary arteries and veins (STEINHORN et al. 1994). Inhibitors of NOS have also been shown to antagonise vasodilatation to adenosine in dog heart (ZANZINGER and BASSENGE 1993) and human forearm (SMITS et al. 1995), and vasodilatation via A_{2A} receptors in the guinea-pig isolated heart (VIALS and BURNSTOCK 1993), but not that mediated by A_{2A} and A_{2B} receptors in the rat isolated heart (LEWIS and HOURANI 1997). It is unclear whether the rat heart adenosine receptors mediate endothelium-dependent vasodilatation by a mechanism other than NO generation, or whether they are located on the vascular smooth muscle. P1 receptors in rabbit renal arteries mediate endothelium-dependent vasodilatation independently of NO (RUMP et al. 1998).

Pharmacological demonstrations of endothelium-dependent vasodilatation by purines and pyrimidines do not *per se* define a role for these agents in the control of blood flow. Direct evidence for this is hard to obtain as it requires the demonstration that endogenously released purines and pyrimidines mediate endothelium-dependent vasodilatation. One approach is to use selective P1 and P2 receptor antagonists to modulate the purinergic contribution to vascular tone. In this respect, whilst inhibitors of NO formation clearly cause a resetting of basal vascular responsiveness in vitro (i.e. an increase in vascular tone and/or an augmentation of responsiveness to vasoconstrictors), this generally does not occur using antagonists of P1 and P2 receptors. This suggests that there is no significant tonic release of endogenous purines and pyrimidines in blood vessels under normal conditions. However, an increase in perfusate flow rate evokes the release of ATP and UTP from endothelial cells in situ and in culture (BODIN et al. 1992; RALEVIC et al. 1992; SAIAG et al. 1995), and thus their role may be in the regulation of blood flow under conditions of increased shear stress.

Substantial evidence exists for a role for purines and pyrimidines in the regulation of blood flow during hypoxia. It has been known for many years that adenosine is an important mediator of vasodilatation during hypoxia following its release from a variety of hypoxic tissues, and part of this effect may be via actions on endothelial A_2 receptors. In their position at the interface between the rest of the vessel wall and the blood, endothelial cells are in a prime location to act as sensors of local fluctuations in oxygenation and blood flow; stimulation of endothelial cells in culture with hypoxia causes an elevation in the levels of extracellular ATP (BODIN et al. 1992), pointing to a possible role of ATP in regulation of blood flow under such conditions. In the rat isolated heart, the P2Y receptor antagonist, reactive blue 2 reduced by about 40% the vasodilator response to hypoxic perfusion. Furthermore, hydroquinone treatment reduced the hypoxic vasodilatation of the rat coronary vascular bed by over 60%, indicating the contribution of endothelial NO (HOPWOOD et al. 1989).

Evidence of a specific role for A_{2A} receptors in functional hyperaemia is provided with the demonstration that ZM 241385, a potent and selective A_{2A} receptor antagonist, blocks by up to 30% the functional hyperaemia response in the hindlimb of anaesthetised cats (POUCHER 1996), although it was not determined whether the receptors are located on the endothelium or smooth muscle. Release of purine nucleotides has been demonstrated under similar conditions.

Erythrocytes, platelets and adrenal chromaffin cells are significant sources of releasable purines, pyrimidines and the diadenosine polyphosphates, but, as with the endothelium as a source of these agents, a causal link between their release from these endogenous sources and activation of endothelial P1 and P2 receptors has yet to be demonstrated. This may help to determine why there are so many different types of purine and pyrimidine receptors on the endothelium that commonly mediate vasodilatation.

II. Angiogenesis

Angiogenesis, the formation of new microvessels, is stimulated by hypoxia as part of a physiological defence mechanism to match tissue vascularity with oxygen demands. Adenosine is a recognised extracellular signal of tissue hypoxia and plays several roles in initiation and/or modulation of angiogenesis. Evidence for the link between tissue hypoxia, the release of adenosine and mitogenesis has been provided by studies of endothelial cells in culture and adenosine-mediated neovascularisation in vivo (DUSSEAU et al. 1986; MEININGER et al. 1988; HUDLICKA et al. 1992; ETHIER et al. 1993). Adenosine has been shown to stimulate the proliferation of human umbilical vein endothelial cells, and adenosine deaminase inhibited proliferation in control cultures, indicating a mitogenic role for endogenous adenosine (ETHIER et al. 1993). Adenosine mimics the effects of hypoxia as a stimulant of the proliferation and migration of bovine aortic or coronary venular endothelial cells in culture, and the effects of both stimuli are blocked by 8-phenyltheophylline, suggesting that adenosine, released as a result of tissue hypoxia, is an angiogenic stimulus for the growth of new vessels via actions at extracellular P1 receptors (MEININGER et al. 1988). Adenosine also stimulates endothelial cell chemotaxis (MEININGER et al. 1988).

In general, the mitogenic and angiogenic effects of adenosine appear to be mediated by A_2 receptors, although an additional, but smaller, involvement of A_1 receptors has been reported in microvascular retinal endothelial cell migration and tube formation (LUTTY et al. 1998). Adenosine-induced proliferation of bovine coronary venular endothelial cells, via A_{2A}-like receptors, requires stimulation of G_s and activation of adenylate cyclase (MEININGER and GRANGER 1990). In contrast, A_{2A} receptor stimulation of human umbilical vein endothelial cell DNA synthesis and proliferation occurs via a mechanism that is independent of G_s and G_i and PKC, and appears to operate via *p21(ras)* and mitogen-activated protein kinase kinase 1 (MEK1) (SEXL et al. 1997). However, receptor-independent stimulation by adenosine of DNA synthesis in human umbilical vein endothelial cells, which was blocked by inhibitors of the Na^+/H^+ exchanger or phospholipase A_2, has also been reported (ETHIER and DOBSON 1997).

Some of the mitogenic effects of adenosine may be mediated indirectly via modulation of vascular endothelial growth factor (VEGF) signalling. Augmentation by adenosine of the expression of VEGF has been described in cerebral (FISCHER et al. 1995) and retinal (TAKAGI et al. 1996b) microvascular endothelial cells. Hypoxic accumulation of adenosine stimulates VEGF expression in retinal microvascular endothelial cells and appears to involve A_{2A} receptors and activation of the cAMP-dependent protein kinase A pathway (TAKAGI et al. 1996b). An initial decline in mRNA for KDR receptors for VEGF and of VEGF binding sites during hypoxia was also shown to be antagonised by A_2 receptor blockade (TAKAGI et al. 1996a). A combination of VEGF and adenosine elicited a greater stimulation of retinal microvascular endothelial cell migration than either agent alone (LUTTY et al. 1998). It is possible that this occurs through synergistic activation of NOS, as many of the actions of VEGF on endothelium are NO-dependent, although it remains to be determined whether the angiogenic effects of adenosine involve generation of endothelial NO.

The evidence for a role of extracellular purine and pyrimidine nucleotides in angiogenesis is less extensive than that for adenosine, and their effects may be indirect, via actions on growth factors. Proliferation of human brain capillary endothelial cells is stimulated by the P2 receptor agonists α,β-meATP, 2MeSATP and ATP (as well as by adenosine acting via A_2 receptors) (RATHBONE et al. 1992). In bovine aortic endothelial cells ATP, ADP, ATPγS and ADPβS, but not α,β-meATP, increased proliferation, implying involvement of a P2Y receptor (VAN DAELE et al. 1992). These compounds also inhibited the growth-promoting effect of basic fibroblast growth factor, suggesting that their effects may be dependent on the presence of other locally released compounds (VAN DAELE et al. 1992). Interestingly, adenosine also induced a mitogenic response in bovine aortic endothelial cells, which was of similar magnitude to that of ATP, but the receptor subtype is unclear as this effect was not mimicked by A_1 and A_2 selective agonists, nor was it blocked by 8-phenyltheophylline (VAN DAELE et al. 1992).

III. Microvascular Permeability

Adenosine has been shown to modulate endothelial permeability and the general consensus is that this occurs via activation of A_2-like receptors (WATANABE et al. 1992; HASELTON et al. 1993), although a role for adenosine A_1 receptors has also been reported (PATY et al. 1992). Both an increase (WATANABE et al. 1992; HEMPEL et al. 1996) and a decrease (HASELTON et al. 1993; LENNON et al. 1998) in endothelial permeability mediated by A_2 receptors has been described. In coronary endothelial monolayers the increase in permeability mediated by the A_2-receptor selective agonist NECA was accompanied by an increase in cAMP and could be antagonised by agents that reduced cAMP (HEMPEL et al. 1996). In the eye, a breakdown of the blood-retinal barrier by adenosine and NECA is due, at least in part, to the opening of tight junctions between retinal vascular endothelial cells (VINORES et al. 1992). In contrast, a decrease in endothelial cell permeability by adenosine was suggested to occur via A_{2B} receptors (LENNON et al. 1998). Activation of A_1 receptors by adenosine released from platelets has been shown to mediate a reduction in permeability of bovine pulmonary artery endothelial monolayers (PATY et al. 1992). All purines and pyrimidines may contribute to modulation of microvascular permeability indirectly via effects on blood flow as increases in blood flow promote capillary filtration and oedema formation.

IV. Inflammation

The vascular endothelium is an important target for the anti-inflammatory actions of adenosine. Adenosine inhibits cytokine (IL-6 and IL-8) production and the expression of adhesion molecules by human endothelial cells stimulated with IL-1, tumour necrosis factor-α, or endotoxin (BOUMA et al. 1996). A_{2A} receptor agonists decrease oxygen radical generation and adhesion of activated neutrophils to endothelial cells, whilst A_1 receptor agonists enhance neutrophil adhesion to the endothelium (CRONSTEIN 1994; ZAHLER et al. 1994; FELSCH et al. 1995; MONTESINO and CRONSTEIN, Chap. 24, this volume). However, as neutrophils express both A_1 and A_{2A} adenosine receptors the contribution of adenosine modulation of the adhesive capacity of vascular endothelium to these anti- and pro-inflammatory effects is unclear.

V. Regulation of Thrombus Formation

Von Willebrand factor is a blood glycoprotein that is crucially involved in platelet adhesion and aggregation and thus in the maintenance of normal haemostasis. It mediates the adhesion of platelets to sites of vascular damage by binding to specific platelet membrane glycoproteins and to constituents of exposed connective tissue. Von Willebrand factor is stored in and released from endothelial cells and its release can be induced by a variety of mediators of thrombosis or inflammation. ATP and ADP, which can be released from platelets, have been shown to induce the release of von Willebrand factor from endothelial cells by an action that appears to require dual activation of P2Y and A_2 receptors (VISCHER and WOLLHEIM 1998). This indicates a possible role for endothelial P2Y and A_2 receptors in regulation of thrombus formation (see also HOURANI, Chap. 20, this volume).

VI. Preconditioning

Pretreatment of coronary endothelial cells with the A_{2A}-selective agonist CGS 21680 has been shown to mediate protection against anoxia and reoxygenation injury similar to that seen with anoxic preconditioning (ZHOU et al. 1996; BROAD and LINDEN, Chap. 16, this volume). This effect appears to involve protein kinase C, which is not consistent with current indications that A_{2A} receptors couple exclusively to adenylate cyclase. A role of adenosine in preconditioning is, however, consistent with its release from ischaemic tissues.

VII. Stimulation of L-Arginine Transport

Activation of human umbilical vein endothelial cell A_{2A}-like receptors, but not A_1 receptors, stimulates L-arginine transport via the y^+ basic amino acid transporter, and NO biosynthesis (SOBREVIA et al. 1997). The transporter is known to be activated by hyperpolarisation, which occurs with activation of A_{2A} receptors on the umbilical vein endothelial cells. Thus, endothelial cell hyperpolarisation by adenosine and other endogenous agents may be important to sustain or replenish levels of L-arginine.

VIII. Apoptosis

Extracellular adenosine and ATP (following metabolism to adenosine) have been shown to cause apoptosis of pulmonary artery endothelial cells and this action is mediated intracellularly rather than by activation of cell surface purine receptors (DAWICKI et al. 1997; ROUNDS et al. 1998). Stimulation of endothelial cells with extracellular ATP and ADP has also been shown to induce activation of transcription factor NFκB, expression of E-selectin protein on the cell membranes and apoptosis (VON ALBERTINI et al. 1998). This may have a possible pathophysiological significance in inflammation or in acute vascular injury associated with the release of purines locally within the vasculature.

G. Conclusions

Extracellular purines and pyrimidines mediate potent and diverse biological effects via activation of specific P1 and P2 receptors on the vascular endothelium; the principal subtypes are A_{2A} and A_{2B} receptors for adenosine, $P2Y_1$

receptors for ADP, and P2Y_2 receptors for ATP and UTP. Considerable attention has been directed towards the roles of purines and pyrimidines as mediators of endothelium-dependent vasodilatation and in the regulation of blood flow. Adenosine is a relatively weak endothelium-dependent vasodilator compared to ADP, ATP and/or UTP, which indicates selectivity for nucleotides and P2Y receptors in this effect. The integrity of the endothelium may have especial significance for the actions of endogenous ATP, as impairment of endothelial vasodilator function may lead to vasoconstriction by ATP at P2X receptors on the underlying vascular smooth muscle (BURNSTOCK 1987). Other effects of extracellular purines and pyrimidines on the endothelium include modulation of angiogenesis, microvascular permeability and inflammation; of the different types of P1 and P2 receptors expressed on endothelial cells, available evidence links these effects most closely with activation of adenosine A_2 receptors.

The above-described diverse roles of purines and pyrimidines in endothelium have been proposed after careful consideration of the correlations between their pharmacological/ biochemical actions, endogenous sources and the stimuli that evoke their release. More studies using selective antagonists are needed to establish a causal relationship between endogenous release of purines and pyrimidines and responses evoked at endothelial P1 and P2 receptors. This would help to clarify why multiple different P1 and P2 receptors are coexpressed by endothelial cells.

Acknowledgements. V.R. is supported by a Royal Society University Research Fellowship.

References

Abbracchio MP, Burnstock G (1994) Purinoceptors: are there families of P2X and P2Y purinoceptors. Pharmacol Ther 64:445–475

Abebe W, Hussain T, Olanrewaju H, Mustafa SJ (1995) Role of nitric oxide in adenosine receptor-mediated relaxation of porcine coronary artery. Am J Physiol 269:H1672–H1678

Abebe W, Makujina SR, Mustafa SJ (1994) Adenosine receptor-mediated relaxation of porcine coronary artery in presence and absence of endothelium. Am J Physiol 266:H2018–H2025

Albert JL, Boyle JP, Roberts JA, Challis RAJ, Gubby SE, Boarder MR (1997) Regulation of brain capillary endothelial cells by P2Y receptors coupled to Ca^{2+}, phospholipase C and mitogen-activated protein kinase. Br J Pharmacol 122:935–941

Berthiaume N, Hess F, Chen A, Regoli D, D'Orleans-Juste P (1997) Pharmacology of kinins in the arterial and venous mesenteric bed of normal and B2 knockout transgenic mice. Eur J Pharmacol 333:55–61

Bodin P, Milner P, Winter R, Burnstock G (1992) Chronic hypoxia changes the ratio of endothelin to ATP release from rat aortic endothelial cells exposed to high flow. Proc R Soc Lond B Biol Sci 247:131–135

Bouma MG, Van den Wildenberg FAJM, Buurman WA (1996) Adenosine inhibits cytokine release and expression of adhesion molecules by activated human endothelial cells. Am J Physiol 270:C522–C529

Bowden A, Patel V, Brown C, Boarder MR (1995) Evidence for requirement of tyrosine phosphorylation in endothelial P_{2Y}- and P_{2U}-purinoceptor stimulation of prostacyclin release. Br J Pharmacol 116:2563–2568
Boyer JL, Romero-Avila T, Schachter JB, Harden TK (1996) Identification of competitive antagonists of the $P2Y_1$ receptor. Mol Pharmacol 50:1323–1329
Burnstock G (1987) Local control of blood pressure by purines. Blood Vessels 24:156–160
Chen BC, Lee C-M, Lee YT, Lin W-W (1996) Characterization of signaling pathways of P_{2Y} and P_{2U} purinoceptors in bovine pulmonary artery endothelial cells. J Cardiovasc Pharmacol 28:192–199
Chiang PH, Wu SN, Tsai EM, Wu CC, Shen MR, Huang CH, Chiang CP (1994) Adenosine modulation of neurotransmission in penile erection. Br J Clin Pharmacol 38:357–362
Communi D, Raspe E, Pirroton S, Boeynaems JM (1995) Coexpression of P_{2Y} and P_{2U} receptors on aortic endothelial cells: comparison of cell localization and signaling pathways. Circ Res 76:191–198
Conti A, Monopoli A, Gamba M, Borea PA, Ongini E (1993) Effects of selective A_1 and A_2 adenosine receptor agonists on cardiovascular tissues. Naunyn Schmiedebergs Arch Pharmacol 348:108–122
Cronstein BN (1994) Adenosine, an endogenous anti-inflammatory agent. J Appl Physiol 76:5–13
Dawicki DD, Chatterjee D, Wyche J, Rounds S (1997) Extracellular ATP and adenosine cause apoptosis of pulmonary artery endothelial cells. Am J Physiol 273:L485–L494
De Mey JG, Vanhoutte PM (1981) Role of the intima in cholinergic and purinergic relaxation of isolated canine femoral arteries. J Physiol (Lond) 316:347–355
Des Rosiers S, Nees S (1987) Functional evidence for the presence of adenosine A_2 receptors in cultured coronary endothelial cells. Naunyn Schmiedebergs Arch Pharmacol 336:94–98
Drury AN, Szent-Györgyi A (1929) The physiological activity of adenine compounds with especial reference to their action upon the mammalian heart. J Physiol (Lond) 68:213–237
Duchene AD, Takeda K (1997) P_{2Y}- and P_{2U}-mediated increases in internal calcium in single bovine aortic endothelial cells in primary culture. Endothelium 5:277–286
Dusseau JW, Hutchins PM, Malbasa DS (1986) Stimulation of angiogenesis by adenosine on the chick chorioallantoic membrane. Circ Res 59:163–170
Ellsworth ML, Forrester T, Ellis CG, Dietrich HH (1995) The erythrocyte as a regulator of vascular tone. Am J Physiol 269:H2155–H2161
Ethier MF, Chander V, Dobson JG Jr (1993) Adenosine stimulates proliferation of human endothelial cells in culture. Am J Physiol 265:H131–H138
Ethier MF, Dobson JG (1997) Adenosine stimulation of DNA synthesis in human endothelial cells. Am J Physiol 41:H1470–H1479
Felsch A, Stocker K, Borchard U (1995) Phorbol ester-stimulated adherence of neutrophils to endothelial cells is reduced by adenosine A_2 receptor agonists. J Immunol 155:333–338
Fischer S, Sharma HS, Karliczek GF, Schaper W (1995) Expression of vascular permeability factor/vascular endothelial growth factor in pig cerebral microvascular endothelial cells and its upregulation by adenosine. Mol Brain Res 28:141–148
Forrester T (1990) Release of ATP from heart: presentation of a release model using human erythrocyte. Ann NY Acad Sci 603:335–352
Fredholm BB, Abbracchio MP, Burnstock G, Daly JW, Harden KT, Jacobson KA, Leff P, Williams M (1994) Nomenclature and classification of purinoceptors. Pharmacol Rev 46:143–156
Furchgott RF, Zawadzki JV (1980) The obligatory role of endothelial cells in the relaxation of arterial smooth muscle by acetylcholine. Nature 288:373–376

Furchgott RF, Khan MT, Jothianandan D (1987) Comparison of endothelium dependent relaxation and nitric oxide induced relaxation in rabbit aorta. Fed Proc (Abstr) 46:385

Gordon JL (1986) Extracellular ATP: effects, sources and fate. Biochem J 233:309–319

Hansmann G, Bultmann R, Tuluc F, Starke K (1997) Characterization by antagonists of P2 receptors mediating endothelium-dependent relaxation in the rat aorta. Naunyn Schmiedebergs Arch Pharmacol 356:641–652

Haselton FR, Alexander JS, Mueller SN (1993) Adenosine decreases permeability of in vitro endothelial monolayers. J Appl Physiol 74:1581–1590

Hashimoto M, Shinozuka K, Tanaka Y, Honda M, Ishida Y, Bjur RA, Westfall DP, Hattori K, Masumura S (1997) Source of ATP, ADP, AMP and adenosine released from isolated rat caudal artery exposed to noradrenaline. J Smooth Muscle Res 33:127–133

Hempel A, Noll T, Muhs A, Piper HM (1996) Functional antagonism between cAMP and cGMP on permeability of coronary endothelial monolayers. Am J Physiol 270:H1264–H1271

Hopwood AM, Lincoln J, Kirkpatrick KA, Burnstock G (1989) Adenosine 5′-triphosphate, adenosine and endothelium-derived relaxing factor in hypoxic vasodilatation of the heart. Eur J Pharmacol 165:323–326

Hudlicka O, Brown M, Egginton S (1992) Angiogenesis in skeletal and cardiac muscle. Physiol Rev 72:369–417

Ignarro LJ, Byrns RE, Buga GM, Wood KS (1987) Endothelium-derived relaxing factor from pulmonary artery and vein possess pharmacologic and chemical properties identical to those of nitric oxide radical. Circ Res 61:866–879

Ikeuchi Y, Nishizaki T (1995) ATP activates the potassium channel and enhances cytosolic Ca^{2+} release via a P_{2y} purinoceptor linked to pertussis toxin-insensitive G protein in brain artery endothelial cells. Bioch Biophys Res Commun 215:1022–1028

Iwamoto T, Umemura S, Toya Y, Uchibori T, Kogi K, Takagi N, Ishii M (1994) Identification of adenosine A_2 receptor-cAMP system in human aortic endothelial cells. Biochem Biophys Res Commun 199:905–910

Jin J, Dasari VR, Sistare FD, Kunapuli SP (1998) Distribution of P2Y receptors on haematopoietic cells. Br J Pharmacol 123:789–794

Kratzer MAA, Nees S, Schrader J (1996) Coronary endothelial cells do not release ATP upon stimulation with thrombin or CO_2. Pharmaceut Pharmacol Lett 6:56–59

Lennon PF, Taylor CT, Stahl GL, Colgan SP (1998) Neutrophil-derived 5′-adenosine monophosphate promotes endothelial barrier function via CD73-mediated conversion to adenosine and endothelial A_{2B} receptor activation. J Exp Med 188:1433–1443

Lewis CD, Hourani SMO (1997) Involvement of functional antagonism in the effects of adenosine antagonists and L-NAME in the rat isolated heart. Gen Pharmacol 29:421–427

Lewis CD, Hourani SM, Long CJ, Collis MG (1994) Characterization of adenosine receptors in the rat isolated aorta. Gen Pharmacol 25:1381–1387

Li J-M, Fenton RA, Cutler BS, Dobson JG Jr (1995) Adenosine enhances nitric oxide production by vascular endothelial cells. Am J Physiol 269:C519–C523

Lutty GA, Mathews, MK, Merges C, McLeod DS (1998) Adenosine stimulates canine retinal microvascular endothelial cell migration and tube formation. Curr Eye Res 17:594–607

Malmsjö M, Edvinsson L, Erlinge D (1998) P2U-receptor mediated endothelium-dependent but nitric oxide-independent vascular relaxation. Br J Pharmacol 123:719–729

Manabe K, Ito H,Matsuda H, Noma A (1995) Hyperpolarization induced by vasoative substances in intact guinea-pig endocardial endothelial cells. J Physiol 484:25–40

Martin PL, Potts AA (1994) The endothelium of the rat renal artery plays an obligatory role in A_2 adenosine receptor-mediated relaxation induced by 5′-*N*-

ethylcarboxamidoadenosine and N^6-cyclopentyladenosine. J Pharmacol Exp Ther 270:893–899

Mateo J, Miras-Portugal MT, Castro E (1996) Co-existence of P_{2Y}- and PPADS-insensitive P_{2U}-purinoceptors in endothelial cells from adrenal medulla. Br J Pharmacol 119:1223–1232

Mehrke G, Pohl U, Daut J (1991) Effects of vasoactive agonists on the membrane potential of cultured bovine aortic and guinea-pig coronary endothelium. J Physiol 439:277–299

Meininger CJ, Granger HJ (1990) Mechanisms leading to adenosine-stimulated proliferation of microvascular endothelial cells. Am J Physiol 258:H198–H206

Meininger CJ, Schelling ME, Granger HJ (1988) Adenosine and hypoxia stimulate proliferation and migration of endothelial cells. Am J Physiol 255:H554–H562

Milner P, Bodin P, Loesch A, Burnstock G (1990) Rapid release of endothelin and ATP from isolated aortic endothelial cells exposed to increased flow. Biochem Biophys Res Commun 170:649–656

Monopoli A, Conti A, Zocchi C, Casati C, Volpini R, Cristalli G, Ongini E (1994) Pharmacology of the new selective A_{2A} adenosine receptor agonist 2-hexynyl-5′-*N*-ethylcarboxamidoadenosine. Arzneimittelforschung 44:1296–1304

Nobles M, Revest PA, Couraud P-O, Abbott NJ (1995) Characteristics of nucleotide receptors that cause elevation of cytoplasmic calcium in immortalized rat brain endothelial cells (RBE4) and in primary cultures. Br J Pharmacol 115:1245–1252

Olanrewaju HA, Hargittai PT, Lieberman EA, Mustafa SJ (1997) Effect of ouabain on adenosine receptor-mediated hyperpolarization in porcine coronary artery smooth muscle. J Cardiovasc Pharmacol 25:234–239

Patel V, Brown C, Goodwin A, Wilkie N, Boarder MR (1996) Phosphorylation and activation of p42 and p44 mitogen-activated protein kinase are required for the P2 purinoceptor stimulation of endothelial prostacyclin production. Biochem J 320:221–226

Paty PSK, Sherman PF, Shepard JM, Malik AB, Kaplan JE (1992) Role of adenosine in platelet-mediated reduction in pulmonary vascular permeability. Am J Physiol 262:H771–H777

Poucher SM (1996) The role of the A_{2A} adenosine receptor in functional hyperemia in the hindlimb of anaesthetized cats. J Physiol 492:495–503

Prentice DJ, Hourani SMO (1996) Activation of multiple sites by adenosine analogs in the rat isolated aorta. Br J Pharmacol 118:1509–1517

Purkiss JR, Wilkinson GF, Boarder MR (1994) Differential regulation of inositol 1,4,5-trisphosphate by co-existing P_{2Y}-purinoceptors and nucleotide receptors on bovine aortic endothelial cells. Br J Pharmacol 111:723–728

Ralevic V, Burnstock G (1988) Actions mediated by P_2-purinoceptor subtypes in the isolated perfused mesenteric bed of the rat. Br J Pharmacol 95:637–645

Ralevic V, Burnstock G (1991) Roles of P_2-purinoceptors in the cardiovascular system. Circulation 84:1–14

Ralevic V, Hoyle CHV, Burnstock G (1995) Pivotal role of phosphate chain length in vasoconstrictor versus vasodilator actions of adenine dinucleotides in rat mesenteric arteries. J Physiol (Lond) 483:703–713

Ralevic V, Milner P, Kirkpatrick KA, Burnstock G (1992) Flow-induced release of adenosine 5′-triphosphate from endothelial cells of the rat mesenteric arterial bed. Experientia 48:31–34

Ralevic V, Burnstock G (1996a) Discrimination by PPADS between endothelial P_{2Y}- and P_{2U}-purinoceptors in the rat isolated mesenteric arterial bed. Br J Pharmacol 118:428–434

Ralevic V, Burnstock G (1996b) Relative contribution of P_{2U}- and P_{2Y}-purinoceptors to endothelium-dependent vasodilatation in the golden hamster isolated mesenteric arterial bed. Br J Pharmacol 117:1797–1802

Ralevic V, Burnstock G (1998) Receptors for purines and pyrimidines. Pharmacol Rev 50; 413–492

Ralevic V, Burrell S, Kingdom J, Burnstock G (1997) Characterization of P2 receptors for purine and pyrimidine nucleotides in human placental cotyledons. Br J Pharmacol 121:1121–1126

Rathbone MP, Middlemiss PJ, Gysbers JW, DeForge S, Costello P, Del Maestro RF (1992) Purine nucleosides and nucleotides stimulate proliferation of a wide range of cell types. In vitro cell Dev Biol 28:529–536

Rounds S, Yee WL, Dawicki DD, Harrington E, Parks N, Cutaia MV (1998) Mechanism of extracellular ATP- and adenosine-induced apoptosis of cultured pulmonary artery endothelial cells. Am J Physiol 275:L379–L388

Rump LC, Oberhauser V, Von Kügelgen I (1998) Purinoceptors mediate renal vasodilatation by nitric oxide dependent and independent mechanisms. Kidney Int 54:473–481

Saiag B, Bodin P, Shacoori V, Catheline M, Rault B, Burnstock G (1995) Uptake and flow-induced release of uridine nucleotides from isolated vascular endothelial cells. Endothelium 2:279–285

Saiag B, Hillaire-Buys D, Chapal J, Petit P, Pape D, Rault B, Allain H, Loubatieres-Mariani MM (1996) Study of the mechanisms involved in adenosine-5′-*O*-(2-thiodiphosphate) induced relaxation of rat thoracic aorta and pancreatic vascular bed. Br J Pharmacol 118:804–810

Schiele JO, Schwabe U (1994) Characterization of the adenosine receptor in microvascular coronary endothelial cells. Eur J Pharmacol Mol Pharmacol 269:51–58

Schlüter H, Offers E, Brüggemann G, van der Giet M, Tepel M, Nordhoff E, Karas M, Spieker C, Witzel H, Zidek W (1994) Diadenosine polyphosphates and the physiological control of blood pressure. Nature 367:186–188.

Seiss-Geuder M, Mehrke G, Daut J (1992) Sustained hyperpolarization of cultured guinea-pig coronary endothelial cells induced by adenosine. J Cardiovasc Pharmacol 20:S97–S100

Sexl V, Mancusi G, Holler C, Gloria-Maercker E, Schutz W, Freissmuth M (1997) Stimulation of the mitogen-activated protein kinase via the A_{2A} adenosine receptor in primary human endothelial cells. J Biol Chem 272:5792–5799

Shinozuka K, Hashimoto M, Kwon YM, Masumara S, Bjur RA, Westfall DP, Kunitomo M (1997) Regional differences in noradrenaline-induced release of adenosine triphosphate from rat vascular endothelium. J Smooth Muscle Res 33:135–143

Smits P, Williams SB, Lipson DE, Banitt P, Rongen GA, Creager MA (1995) Endothelial release of nitric oxide contributes to the vasodilator effect of adenosine in humans. Circulation 92:2135–2141

Sobrevia L, Yudilevich DL, Mann GE (1997) Activation of A_2 purinoceptors by adenosine stimulates L-arginine transport (system y^+) and nitric oxide synthesis in human fetal endothelial cells. J Physiol 499:135–140

Stanimirovic DB, Bertrand N, Merkel N, Bembry J, Spatz M (1994) Interaction between histamine and adenosine in human cerebromicrovascular endothelial cells: modulation of second messengers. Metab Brain Disease 9:275–289

Steinhorn RH, Morin FC III, Van Wylen DGL, Gugino SF, Giese EC, Russell JA (1994) Endothelium-dependent relaxations to adenosine in juvenile rabbit pulmonary arteries and veins. Am J Physiol 266:H2001–H2006

Takagi H, King GL, Ferrara N, Aiello LP (1996a) Hypoxia regulates vascular endothelial growth factor receptor KDR/Flk gene expression through adenosine A2 receptors in retinal capillary endothelial cells. Invest Ophthalmol Vis Sci 37:1311–1321

Takagi H, King GL, Robinson GS, Ferrara N, Aiello LP (1996b) Adenosine mediates hypoxic induction of vascular endothelial growth factor in retinal pericytes and endothelial cells. Invest Ophthalmol Vis Sci 37:2165–2176

Van Daele P, Van Coevorden A, Roger PP, Boeynaems J-M (1992) Effects of adenine nucleotides on the proliferation of aortic endothelial cells. Circ Res 70:82–90

Vials A, Burnstock G (1993) A_2-purinoceptor-mediated relaxation in the guinea-pig coronary vasculature: a role for nitric oxide. Br J Pharmacol 109:424–429
Vinores SA, Sen H, Campochiaro PA (1992) An adenosine agonist and prostaglandin E1 cause breakdown of the blood-retinal barrier by opening tight junctions between vascular endothelial cells. Invest Ophthalmol Vis Sci 33:1870–1878
Vischer UM, Wollheim CB (1998) Purine nucleotides induce regulated secretion of von Willebrand factor: involvement of cytosolic Ca^{2+} and cyclic adenosine monophosphate-dependent signaling in endothelial exocytosis. Blood 91:18–127
Von Albertini M, Palmetshofer A, Kaczmarek E, Koziak K, Stroka D, Grey ST, Stuhlmeier KM, Robson SC (1998) Extracellular ATP and ADP activate transcription factor NF-kappaB and induce endothelial cell apoptosis. Biochem Biophys Res Commun 248:822–829
Watanabe H, Kuhne W, Schwartz P, Piper HM (1992) A_2 adenosine receptor stimulation increases macromolecular permeability of coronary endothelial cells. Am J Physiol 262:H1174–H1181
Webb TE, Foeolde E, Vigne P, Neary JT, Runberg A, Frelin C, Barnard EA (1996) The P2Y purinoceptor in rat brain microvascular endothelial cells couple to inhibition of adenylate cyclase. Br J Pharmacol 119:1385–1392
Wilkinson GF, McKechnie K, Dainty IA, Boarder MR (1994) P_{2Y} purinoceptor and nucleotide receptor-induced relaxation of precontracted bovine aortic collateral artery rings: differential sensitivity to suramin and indomethacin. J Pharmacol Exp Ther 268:881–887
Yang S, Burton ILO, Probert CB, Talbot JN, Bradley ME (1996) Evidence for a discrete UTP receptor in cardiac endothelial cells. Br J Pharmacol 117:1572–1578
Yang S, Cheek DJ, Westfall DP, Buxton ILO (1994) Purinergic axis in cardiac blood vessels: agonist-mediated release of ATP from cardiac endothelial cells. Circ Res 74:401–407
You J, Johnson TD, Childres WF, Bryan RM Jr (1997) Endothelial-mediated dilations of rat middle cerebral arteries by ATP and ADP. Am J Physiol 273:H1472–H1477
Zahler S, Becker BF, Raschke P, Gerlach E (1994) Stimulation of endothelial adenosine A_1 receptors enhances adhesion of neutrophils in the intact guinea pig coronary system. Cardiovasc Res 28:1366–1372
Zanzinger J, Bassenge E (1993) Coronary vasodilatation to acetylcholine, adenosine and bradykinin in dogs: Effects of inhibition of NO synthesis and captopril. Eur Heart J 14 [Suppl 1]:164–168
Zhou X, Xhai X, Ashraf M (1996) Preconditioning of bovine endothelial cells: the protective effect is mediated by an adenosine A_2 receptor through a protein kinase C signaling pathway. Circ Res 78:73–81
Zunkler BJ, Henning B, Grafe M, Bass R, Hildebrandt AG, Fleck E (1995) Electrophysiological properties of human coronary endothelial cells. Basic Res Cardiol 90:435–442

CHAPTER 20

P1 and P2 Receptors in Platelets

S.M.O. HOURANI

A. Introduction

Because of the importance of platelets in haemostasis and thrombosis, these cell fragments have been extensively studied for many years and the existence of receptors for adenosine and adenine nucleotides on platelets has been recognised since the early 1960s. In 1961 adenosine 5′-diphosphate (ADP) was identified as the factor in red blood cells which caused the development of adhesiveness in platelets, and this effect was shown to be highly specific for ADP (GAARDER et al. 1961). The study of platelets was greatly facilitated by the invention of the aggregometer, which allowed rapid and highly reproducible measurement of this ability of platelets to adhere to each other ("aggregate") (BORN 1962). This seminal paper also showed that ADP could aggregate platelets and that this aggregation was inhibited by ATP, and suggested that these substances were competing for specific sites, a suggestion which was later confirmed (MACFARLANE and MILLS 1975). Adenosine was also shown to inhibit aggregation (BORN and CROSS 1963), and it was later demonstrated that it acted via stimulation of adenylate cyclase and that caffeine and theophylline could antagonise this effect (HASLAM and ROSSON 1975). HASLAM and ROSSON (1975) also showed that ADP could inhibit adenylate cyclase. By the early 1980s a very large number of analogues of adenosine and adenine nucleotides had been tested for their effects on platelets, and the possible importance of extracellular enzymic degradation and adenosine uptake had been recognised (see HASLAM and CUSACK 1981). At that time it was clear that the platelet receptor was of the A_2 subtype, and GORDON (1986) gave the name of P_{2T} to the ADP receptor on platelets. Although this was useful in so far as it brought platelets to the attention of those working in the purinergic field, it was perhaps unfortunate in so far as it is clear now that there is not one single receptor responsible for all the effects of ADP on platelets (see below).

B. P1 Receptors in Platelets

Early studies showed that N^6-substituted analogues of adenosine, like N^6-cyclopentyladenosine (CPA), were less active than adenosine in inhibiting platelet aggregation (Kikugawa et al. 1973), whereas 2-chloroadenosine (2-CADO) was more active (Born 1963). In general the potency of adenosine analogues in inhibiting aggregation was shown to parallel their potency in causing vasodilatation, a finding noted to be likely to limit the potential therapeutic value of such compounds as antithrombotics (Born et al. 1965). 5′-*N*-Ethylcarboxamidoadenosine (NECA), known to be a very powerful vasodilator (Raberger et al. 1977), was shown also to be more potent than adenosine as an inhibitor of aggregation and as a simulator of adenylate cyclase (Cusack and Hourani 1981a), and this potency order of NECA > 2-CADO > CPA is characteristic of an A_2 receptor (see, e.g. Collis and Hourani 1993; Fredholm et al. 1994).

More recently, use of newer analogues such as the A_{2A}-selective agonist CGS 21680 has shown that the adenosine receptor on platelets is of the A_{2A} subtype, and indeed a very good correlation was found between the potencies of analogues to inhibit human and rabbit platelet aggregation and to bind A_{2A} receptors in the rat striatum (Dionisotti et al. 1992). A comparison of human platelets and human erythroleukemia (HEL) cells, a cell line thought to be similar to megakaryocytes, the precursors of platelets, found a different order of potency in these two cell types, with that in platelets [CGS 21680 ≥ NECA > N^6-*R*-phenylisopropyladenosine (R-PIA)] being characteristic of A_{2A} and that in HEL cells (NECA > R-PIA > CGS 21680) characteristic of A_{2B} receptors (Feoktistov and Biaggioni 1993). In a comparative functional study of a range of tissues the rank order of potencies for adenosine analogues in platelets was also found to be similar to those found in tissues (such as the dog coronary artery) thought to contain A_{2A} receptors (Gurden et al. 1996). However, in another study on human platelets, CGS 21680 achieved a lower maximal response than NECA both for accumulation of adenosine 3′,5′-cyclic monophosphate (cyclic AMP) and for inhibition of collagen-induced 5-hydroxytryptamine (5-HT) secretion from platelets, without appearing to be a partial agonist as it did not itself antagonise responses to NECA (Cooper et al. 1995). The responses to CGS 21680 were enhanced to the level of those of NECA by the inclusion of a phosphodiesterase inhibitor in either assay, and these authors suggested that as well as acting via an A_{2A} receptor, NECA might act by an additional mechanism. The study by Feoktistov and Biaggioni (1993) had also shown a greater maximal response for NECA than for CGS 21680 or R-PIA. A group of 2-substituted adenosine analogues were also reported to be partial agonists compared to NECA for stimulation of adenylate cyclase, in that they achieved lower maximal responses, although their ability to act as antagonists was not investigated (Cristalli et al. 1994). These analogues all behaved as full agonists for inhibition of aggregation, and on the basis of this difference and because the correlation between antiaggregatory activity and inhibition of [^{3}H] NECA binding to the rat striatum was low it was again suggested that the responses of platelets to adenosine analogues might not be simply explained in terms of an A_{2A} receptor.

Interpretation of functional studies with adenosine and its analogues in platelets is complicated by their potential to inhibit adenylate cyclase directly by interacting with the intracellular "P" site on the enzyme (HASLAM and LYNHAM 1972; HASLAM and ROSSON 1975; JAKOBS et al. 1979). The extent to which this occurs clearly depends on their affinity for the P site and their ability to enter the platelets, which is time-dependent and can occur both by the active uptake system and by passive diffusion. The concentration used also plays a part, as the P site requires high concentrations of agonists, so concentration-response curves for adenosine may appear bell-shaped (see, e.g. HASLAM and ROSSON 1975; JAKOBS et al. 1979). Ribose-modified adenosine analogues, including 2′-5′-dideoxyadenosine and a series of 9-substituted adenine derivatives, are selective for this P site and can be used as adenylate cyclase inhibitors (see HASLAM and CUSACK 1981; HARRIS et al. 1979).

Use of adenosine receptor antagonists has also confirmed the presence of A_2 rather than A_1 receptors in platelets, with A_1 selective antagonists being less potent as inhibitors of the response to NECA than non-selective antagonists (SCHWABE et al. 1985; MARTINSON et al. 1987). That the A_2 receptors are of the A_{2A} subtype is again confirmed by the use of antagonists, with the A_1/A_{2A} selective antagonist PD 115199 antagonising the effects of NECA with a K_B in the nanomolar range (COOPER et al. 1995). The highly selective A_{2A} antagonist SCH 58261 also inhibits the effects of CGS 21680 in rabbit platelets (ZOCCHI et al. 1996). That an A_{2B} receptor is unlikely to be involved in the effects of NECA is shown by the complete inhibition of the responses to NECA by PD 115199 (COOPER et al. 1995), and the fact that enprofylline, now demonstrated to act as an A_{2B} antagonist (FEOKTISTOV and BIAGGIONI 1995), enhanced (due to phosphodiesterase inhibition) rather than inhibited the responses to NECA (UKENA et al. 1985).

[^{3}H] NECA and [^{3}H] CGS 21680 have been used as a radioligands for the A_{2A} receptors in platelets, but their usefulness is limited by the presence of a non-receptor NECA-binding site ("adenotin") which can however be separated from the receptor itself (HUTTEMANN et al. 1984; LOHSE et al. 1988; VARANI et al. 1994; FEIN et al. 1994; VARANI et al. 1996; LORENZEN and SCHWABE, Chap. 2, first volume). More recently the A_{2A}-selective [^{3}H] SCH 58261 has been shown to bind and to be displaced by adenosine receptor agonists and antagonists with affinities similar to those observed in the rat striatum (DIONISOTTI et al. 1996; for review see GESSI et al. 2000). The presence of A_{2A} receptors in human platelets has also been demonstrated using an antibody raised against the A_{2A} receptor (MARALA and MUSTAFA 1998).

That the A_{2A} receptor indeed mediates inhibition of platelet aggregation was shown by the development of genetically-modified mice lacking the A_{2A} receptor (LEDENT et al. 1997). In these mice inhibitory responses to both NECA and CGS 21680 were abolished, showing that in mouse platelets at least they act entirely through the A_{2A} receptor. Interestingly, aggregation to ADP in the absence of adenosine receptor agonists was enhanced, suggesting that in normal mice ADP-induced aggregation is tonically inhibited by endogenous adenosine (LEDENT et al. 1997). A similar enhancement of ADP-induced platelet aggregation is observed in the presence of adenosine deaminase, which degrades any adenosine produced, and the inhibition of platelet aggregation by the drug dipyridamole is thought to be due to inhibition of adenosine uptake enhancing the inhibitory effects of endogenous adenosine (AGARWAL 1991). A recent study confirmed that breakdown of ADP to adenosine in plasma can result in activation of platelet adenylate cyclase and thus inhibition of aggregation (GLENN and HEPTINSTALL 1998). However, the A_{2A} antagonist SCH 58261 was reported not to affect ADP-induced aggregation of rabbit platelets (ZOCCHI et al. 1996), although the limited solubility of this compound may not have allowed a high enough concentration to be used. In vivo adenosine is likely to be an important modulator of platelet function, being produced by breakdown of adenine nucleotides released both from platelets themselves and from other cells in the cardiovascular system (see, e.g. BOARDER and HOURANI 1998).

C. P2 Receptors in Platelets

ADP is just one of the many aggregating agents which can stimulate platelets by acting at specific receptors (see, e.g. HOURANI and CUSACK 1991), but it plays a very important physiological and pathological role because it is one of the positive feedback mechanisms which acts to spread and enhance platelet aggregation. When platelets are activated by any strong aggregating agent the functional response observed consists of a rapid change in shape (from a disc to a spiny sphere), exposure of receptors for fibrinogen and "primary" aggregation due to a calcium-dependent binding of fibrinogen to these receptors. This is accompanied by the release from platelets of the aggregating agent thromboxane (TXA_2) and the contents of dense granules (ATP, ADP and 5-HT, a weak aggregating agent) and α granules (growth factors and prothrombotic proteins including fibrinogen). The release of TXA_2, ADP and 5-HT results in a "secondary" wave of aggregation which is irreversible due to cross-linking of platelets by proteins released from the α granules. Lysosomal enzymes can also be released, but their functional significance during aggregation is unknown. ADP can induce shape change and primary aggregation directly, but release of granule contents is dependent on the synthesis of TXA_2 which is stimulated by the platelet-platelet contact during aggregation and only occurs when extracellular calcium levels have been reduced from the physiological 1.5 mmol/l to around 100 μmol/l, as found in plasma anticoagulated by a chelating agent such as the widely-used citrate (MUSTARD et al. 1975; HEPTINSTALL et al. 1979; PACKHAM et al. 1989). Shape change but not aggregation can occur in the absence of extracellular calcium, and indeed a calcium-free medium is commonly used for such experiments. Washing platelets free of plasma allows the calcium concentration to be returned to mmol/l levels, but runs the risk of reducing platelet sensitivity, particularly to ADP, which may be released by platelets during the washing procedure. Thus although the functional responses of platelets of shape change, aggregation and release are technically easy to investigate, results have to be interpreted with caution because the experimental conditions used can greatly affect the results obtained. There is a very large body of literature concerned with platelet responses and the general mechanisms of platelet activation (see, e.g. chapters in VON BRUCHHAUSEN and WALTER 1997; for a brief review see HOLMSEN 1996), but many uncertainties still remain.

Increases in intracellular Ca^{2+} clearly play some role in platelet responses, as the ionophore A23187 can induce shape change, aggregation and release (WHITE et al. 1974; MASSINI and LUSCHER 1976). In the case of shape change, if this is triggered by increases in Ca^{2+} then sufficient must come from internal stores since extracellular Ca^{2+} is not required for this response. ADP, like the other aggregating agents, clearly causes an increase in Ca^{2+}, as a result of both release from internal stores and influx (HALLAM and RINK 1985). A comparison of the ability of aggregating agents to increase Ca^{2+} showed that it was less potent in this regard than thrombin, platelet-activating factor, vasopressin or a TXA_2 analogue but considerably more potent than 5-HT, while adrenaline caused no detectable increase. However, when the ability of these agonists to induce phosphoinositide metabolism was similarly compared, while for the other agonists their potency was similar to that for increasing Ca^{2+}, no effect of ADP (or adrenaline) could be detected (MACINTYRE et al. 1985). Over many years indeed while some authors have reported stimulation of phosphoinositide metabolism by ADP (LLOYD et al. 1972, 1973; DANIEL et al. 1986; OLBRICH et al. 1989; HEEMSKERK et al. 1993; RAHA et al. 1993; VANAGS et al. 1998), others have failed to detect it (FISHER et al. 1985; VICKERS et al. 1990; PACKHAM et al. 1993; PULCINELLI et al. 1995). In the studies by RAHA et al. (1993) and VANAGS et al. (1998) the responses induced by ADP were rather small and very transient: also, in the former study they were not different to responses induced by shear stress (which did not result in an increase in Ca^{2+} levels), and in the latter study no clear concentration-dependence was observed. The conflicting results obtained in all these studies probably reflects the difficulty of ensuring that it is only the primary response to ADP itself that is measured, not secondary events triggered as a result of the complex amplification systems of platelets. Overall, however, it is clear that while ADP may induce small, transient and rather variable phospholipase C (PLC) activation resulting in phosphoinositide breakdown, inositol 1,4,5-trisphosphate (IP_3) formation and Ca^{2+} mobilisation by this method, it is very weak agonist in this respect and other mechanisms must exist to account for its robust ability to increase Ca^{2+} and to cause aggregation.

As well as releasing Ca^{2+} from internal stores (presumably as a result of small amounts of IP_3 formed) ADP causes an influx of Ca^{2+}. While the majority of this influx, as with other aggregating agents, is due to store depletion, ADP has the unique ability to cause directly Ca^{2+} influx via receptor-operated channels (SAGE and RINK 1987; SAGE et al. 1989, 1990; for reviews see SAGE et al. 1992, 1997). Influx through the receptor-operated channel, unlike calcium mobilisation and store-mediated influx, is resistant to inhibition by agents which cause an increase in cyclic nucleotides in platelets (SAGE and RINK 1987). Patch-clamp studies in human platelets showed the ability of ADP to evoke a rapid influx of monovalent or divalent cations, with estimates of the number of channels per platelets ranging between 13 and 130 (MAHAUT-SMITH et al. 1990, 1992; MACKENZIE et al. 1996). The characteristics of these channels were noted by these authors to be similar to those evoked by ATP in vascular smooth muscle, and indeed it is now clear that they are also due to activation of $P2X_1$ receptors (see SAGE et al. 1997). ATP can also induce rapid Ca^{2+} entry into platelets (MACKENZIE et al. 1996), and the presence of $P2X_1$ receptors in platelets has been directly demonstrated by molecular techniques (VIAL et al. 1997; CLIFFORD et al. 1998; SCASE et al. 1998; SUN et al. 1998). The functional importance of this is however unclear, as ATP has never been reported directly to cause aggregation, shape change or release, and indeed competitively inhibits all the effects of ADP (MACFARLANE and MILLS 1985). It should be noted here that platelets, unlike smooth muscle, do not possess voltage-operated Ca^{2+} channels (see SAGE et al. 1992, 1997), so influx of cations through the $P2X_1$ receptor will not result in a larger influx of Ca^{2+} through such channels. Also, although ATP has a higher potency at the cloned $P2X_1$ receptor than ADP (VALERA et al. 1994; SUN et al. 1998), responses to ATP in platelets can only be detected if the platelets are pre-treated with the adenine nucleotide-degrading enzyme apyrase first (MACKENZIE et al. 1996), suggesting that in vitro these receptors may normally be desensitised. In vivo they may be more important, although no bleeding abnormalities were reported in genetically-modified mice lacking the $P2X_1$ receptor (MULRYAN et al. 2000).

As well as increasing Ca^{2+} levels, ADP has been known for a long time to inhibit adenylate cyclase (MILLS and SMITH 1972; COOPER and RODBELL 1979), and as stimulation of adenylate cyclase inhibits platelet stimulation it seems likely that this inhibition of adenylate cyclase by ADP has some functional significance, but exactly what is still the subject of discussion. Inhibition of adenylate cyclase by adenine derivatives does not induce aggregation, and their reported lack of effect on ADP-induced aggregation suggested that such inhibition of adenylate cyclase does not normally enhance aggregation (HARRIS et al. 1979). However, inhibition of adenylate cyclase may offset the inhibitory effects of stimulators of adenylate cyclase such as adenosine to which platelets may be exposed in vivo (CUSACK and HOURANI 1982a), and, as discussed below, there is increasing evidence that some process linked to inhibition of adenylate cyclase is involved in aggregation. The ability of ADP to inhibit adenylate cyclase suggests that it acts via a G_i coupled receptor, and indeed ADP has been shown to increase binding of the non-hydrolysable guanosine 5′-triphosphate (GTP) analogue [^{35}S]GTPγS to platelet membranes, as expected for a G protein-coupled response (GACHET et al. 1992). Immunoprecipitation with subtype-specific antibodies showed in humans that ADP increased the binding of a radiolabelled GTP photoaffinity ligand to $G\alpha_{i2}$, and that this effect was inhibited by ATP (Ohlmann et al. 1995). These authors also reported that ADP did not stimulate labelling of $G\alpha_q$ whereas a TXA_2 analogue did, and interpreted this as evidence against PLC stimulation by ADP. However in genetically-modified mice lacking $G\alpha_q$ both shape change and aggregation in response to ADP were abolished, suggesting that coupling via this G protein, presumably to PLC, does occur and is essential for both responses (Offermanns et al. 1997).

It is clear from the above that, setting aside for the moment the $P2X_1$ receptor, ADP must act on platelets via a G protein coupled receptor, and that there is evidence for coupling both via G_{i2} to adenylate cyclase and via G_q probably to PLC. By definition the P2 receptor by which it acts must be of the P2Y type, but the question remains as to which subtype it belongs. Early studies on the effects of ADP analogues on platelets showed that substitutions on the 2 position enhanced potency, while modification of the phosphate chain reduced it (for reviews of structure-activity relationships see Haslam and Cusack 1981; Hourani and Cusack 1991; Mills 1996). Thus 2-chloroadenosine 5′-diphosphate (2-ClADP) and 2-methylthioadenosine 5′-diphosphate (2-MeSADP) are more potent than ADP as aggregating agents, while adenosine 5′-(α,β-methylene)diphosphonate (α,β-me-ADP) is less potent (Maguire and Michal 1968; Gough et al. 1969, 1972). Similarly, 2-chloroadenosine 5′-triphosphate (2-ClATP) is a more powerful antagonist of ADP-induced aggregation than is ATP, whereas adenosine 5′-(β,γ-methylene)triphosphonate (β,γ-me-ATP) is less potent (Born and Foulks 1979; Cusack and Hourani 1982c). In the original classification of Burnstock and Kennedy (1985) of P2 into P_{2X} and P_{2Y} overall these structure-activity relationships are clearly most similar to those of the P_{2Y} receptor, although as discussed above the platelet ADP receptor was given the name of P_{2T} (Gordon 1986) and thought to be unique in that ATP is an antagonist whereas in functional studies on, for example, smooth muscle preparations, ADP and ATP are equipotent agonists. Of the cloned P2Y receptors it is most similar therefore to the $P2Y_1$, apart from the fact that in most studies on the cloned $P2Y_1$, as with the functionally-defined P_{2Y}, both ADP and ATP appeared to be agonists (Webb et al. 1993; Schachter et al. 1996). However, in studies in which great care was taken to use only pure compounds and to avoid degradation of triphosphate derivatives to their corresponding diphosphates, ATP and its analogues adenosine 5′-(*O*-1-thiotriphosphate) (ATPαS), adenosine 5′-(α,β-methylene)triphosphonate (α,β-meATP), 2-ClATP and 2-methylthioadenosine 5′-triphosphate (2-MeSATP) were clearly shown to act as antagonists rather than agonists at the cloned human $P2Y_1$ receptor (Leon et al. 1997; Hechler et al. 1998b). However, while the $P2Y_1$ receptor is known to couple to PLC, its ability to couple via G_i to inhibition of adenylate cyclase is not so widely accepted (Schachter et al. 1996; Webb et al. 1996). If it is indeed unable to couple to adenylate cyclase then there must be another ADP receptor responsible for this effect.

The first suggestion that there might be two receptors on platelets, one responsible for shape change and aggregation and one for inhibition of adenylate cyclase, was based on the observation that 2-MeSADP was more potent as an inhibitor of adenylate cyclase than as an aggregating agent, although ADP itself and 2-ClADP are roughly equipotent for both effects (Macfarlane et al. 1983; Cusack and Hourani 1982d). Further evidence in support of this hypothesis came from a series of studies using the affinity reagent 5′-fluorosulphonylbenzoyladenosine (FSBA), which inhibits aggregation but not the ability of ADP to inhibit adenylate cyclase (see Colman 1990, 1992 for reviews). This reagent and related ones later developed by this group labelled a 100-kDa protein which was named "aggregin" and suggested to be the ADP receptor responsible for aggregation, although this was not universally accepted (see Colman 1990, 1992; and Hall 1994; Mills 1996 for discussion). Arguments against this hypothesis came from a comparison of the ability of a series of adenine nucleotide analogues (including 2-ClATP, ATPαS and α,β-meATP) to inhibit competitively aggregation and the effects of ADP on adenylate cyclase, in which a good correlation was found between their apparent affinity for both putative receptors (Cusack and Hourani 1982c). The non-selective P2 antagonist suramin also inhibits both these effects of ADP and also the increases in Ca^{2+} induced by ADP, again suggesting that they are all mediated by the same receptor (Hourani et al. 1992; Hall and Hourani 1994). The structure-activity relationships for a number of adenine nucleotide analogues in causing or inhibiting ADP-induced increases in Ca^{2+} were also in close agreement with their effects on both aggregation and adenylate cyclase (Hall and Hourani 1993; Hourani and Hall 1996) whereas FSBA was reported not to inhibit ADP-induced increases in Ca^{2+} (Rao and Kowalska 1987). However, differences were noted in the responses to some adenine nucleotide analogues. In particular, although adenosine 5′-(*O*-2-thiodiphosphate) (ADPβS) and both isomers of adenosine 5′-(*O*-1-thiodiphosphate (ADPαS) appeared to be partial agonists relative to ADP for aggregation and for increases in Ca^{2+}, when inhibition of adenylate cyclase was measured only ADPβS retained its partial agonist activity, while ADPαS was an antagonist (Cusack and Hourani 1981b,c; Hall and Hourani 1993). Both compounds are full agonists for shape change, being nearly as potent as ADP for this response (Park and Hourani 1999). For ADPαS the potency as agonists of both isomers of ADPαS for aggregation was similar to their apparent affinity as antagonists for inhibition of adenylate cyclase, which was taken to imply that although they had a different efficacy for each response, they were mediated by the same receptor. Another explanation of course is that inhibition of adenylate cyclase is mediated through a receptor different to the $P2Y_1$ receptor and at which ADPαS is an antagonist rather than an agonist.

An important compound in the classification of P2 receptors is 2-MeSATP, which characteristically is shown to have a higher potency than ATP at $P2Y_1$ receptors but not at P2X or the other P2Y subtypes (see, e.g. Harden et al. 1998). This, like ATP, was shown to be a competitive antagonist at the cloned human $P2Y_1$ receptor and had an apparent K_B value of 38 μmol/l, compared to an apparent K_B value for ATP of 25 μmol/l (Hechler et al. 1998b). This suggests that in fact 2-MeSATP has a rather lower affinity for the $P2Y_1$ receptor than does ATP, and its apparently higher potency in most studies is due to the higher potency of its breakdown product, 2-MeSADP, compared to ADP, the breakdown product of ATP. MeSATP and related 2-alkylthio analogues of AMP and ATP have however rather intriguing effects on platelets. They are highly specific inhibitors of ADP-induced aggregation, but do not display competitive kinetics, being unable to inhibit completely this response to ADP in platelets in plasma. Even at 100 μmol/l MeSATP and its stable analogue 2-methylthioadenosine 5′-(β,γ-methylene)triphosphonate (2-MeS-β,γ-meATP) can only achieve around 50% inhibition of any concentration of ADP, although their IC_{50} values for this inhibition are rather lower than one might expect from their affinity at $P2Y_1$ receptors, around 3 μmol/l and 0.3 μmol/l respectively (CUSACK and HOURANI 1982b; HOURANI et al. 1996). They do, however, powerfully and competitively inhibit the effect of ADP on adenylate cyclase, with apparent K_B values of 63 nmol/l and 50 nmol/l respectively (HOURANI et al. 1996; PARK and HOURANI 1999). These observations led to the suggestion that ADP may induce aggregation by interacting with two receptors or two forms of one receptor, only one of which is coupled to inhibition of adenylate cyclase and at which 2-alkylthio analogues of ATP and AMP are specific and competitive antagonists (HOURANI et al. 1996). Closely related compounds including ARL 66096 also strongly inhibit ADP-induced platelet aggregation and the effect of ADP on adenylate cyclase but are very weak antagonists at the cloned $P2Y_1$ receptor and have little effect on the ADP-induced increases in Ca^{2+} or PLC activation in platelets (HUMPHRIES et al. 1994; FAGURA et al. 1998; DANIEL et al. 1998). Interestingly, MeSATP is a rather weak inhibitor of ADP-induced shape change, and its K_B value for this effect (6.3 μmol/l) is similar to its affinity for the $P2Y_1$ receptor (PARK et al. 1999), and ARL 66096 has been reported not to inhibit shape change (SANDERSON et al. 1996). The mechanism of action of these compounds is of more than academic interest as stable 2-alkylthio analogues of ATP are being developed as antithrombotic drugs (HUMPHRIES et al. 1995; INGALL et al. 1999).

Rather similar results have been obtained over many years with the clinically-used antithrombotic drugs ticlopidine and clopidogrel, which are only active in vivo and are therefore harder to use as rigorous pharmacological tools. These compounds also specifically inhibit ADP-induced aggregation and inhibition of adenylate cyclase, while having little effect on shape change or on Ca^{2+} increases (see MILLS 1996; GACHET et al. 1997; SAVI et al. 1998 for reviews). 2-MeSADP radiolabelled in various ways has been successfully used in binding studies with platelets (MACFARLANE et al. 1983; CRISTALLI and MILLS 1993), and ticlopidine and clopidogrel apparently remove selectively and irreversibly approximately 70% of the binding sites for this ligand, which have been suggested to correspond to an ADP receptor coupled to adenylate cyclase (MILLS et al. 1992; SAVI et al. 1994; GACHET et al. 1995; see MILLS 1996; GACHET et al. 1997; SAVI et al. 1998 for reviews). In addition, some patients have been described with a bleeding disorder and impaired platelet responses to ADP, similar to those seen after treatment with these drugs. This abnormality has been attributed to a hereditary lack of the ADP receptor coupled to adenylate cyclase (see CATTANEO and GACHET 1999 for review).

The differential effects of the 2-alkylthio analogues on the $P2Y_1$ receptor and the effect of ADP on adenylate cyclase suggests that there is indeed a separate G_i-coupled P2Y receptor on platelets. Further evidence comes from the use of a series of adenosine 3′, 5′-bisphosphate derivatives such as adenosine 3′-phosphate, 5′-phosphosulfate (A3P5PS) which have been shown to act as selective $P2Y_1$ antagonists (BOYER et al. 1996). The significance of these antagonists for the study of platelets was clear as soon as the $P2Y_1$ receptor was shown to exist on platelets (LEON et al. 1997), and they were immediately tested by a number of independent groups, who all found that they inhibited ADP-induced shape change, aggregation and increases in Ca^{2+}, but not the effect of ADP on adenylate cyclase (HECHLER et al. 1998a; JIN and KUNAPULI 1998; JIN et al. 1998; SAVI et al. 1998; PARK et al. 1998; JANTZEN et al. 1999). This led to the development of a model for ADP-induced platelet activation in which increases in Ca^{2+} and shape change are mediated via the $P2Y_1$ receptor coupled via G_q to PLC, while G_{i2}-mediated inhibition of adenylate cyclase is via another P2Y receptor (variously called P_{2T}, $P2Y_{ADP}$ or $P2Y_{AC}$) with activation of both receptors being required for full aggregation (HECHLER et al. 1998a; KUNAPULI 1998; SAVI et al. 1998). Overall, the $P2Y_1$ receptor is the target for A3P5PS and related compounds, and ADPαS is an agonist here, while the adenylate cyclase-coupled P2Y receptor is the target for both the 2-alkylthio analogues of ATP and AMP and for ticlopidine and clopidogrel, with ADPαS being an antagonist here (Fig. 1). MeSADP has a higher potency at the adenylate cyclase coupled receptor than at the $P2Y_1$, which might suggest a higher affinity, but binding studies combined with the use of ticlopidine and clopidogrel to reduce the number of binding sites have in some (SAVI et al. 1994) but not all (MILLS et al. 1992; GACHET et al. 1995) studies reported that the drug-resistant sites (presumably the $P2Y_1$) have a higher affinity than the ones affected (presumably the cyclase-coupled receptors). A comparison of the divalent cation-dependence of the effects of ADP and ATP on shape change and adenylate cyclase suggests that both these receptors recognise the uncomplexed forms of the nucleotides (HALL et al. 1994).

The importance of the $P2Y_1$ receptor in ADP-induced aggregation has recently been unequivocally demonstrated by a study using genetically-modified mice lacking this receptor (FABRE et al. 1999). Platelets from these mice showed greatly reduced shape change, calcium increases and aggregation in response to ADP, and had an increased bleeding time. This study also demonstrated clearly the existence of another ADP receptor on platelets coupled to inhibition of adenylate cyclase, as this response was unchanged. A rather similar phenomenon was observed in mice lacking CD39, the main vascular ectoATPdiphosphohydrolase, in which the platelet $P2Y_1$ receptor appeared to be desensitised in vivo (ENJYOJI et al. 1999). The adenylate cyclase

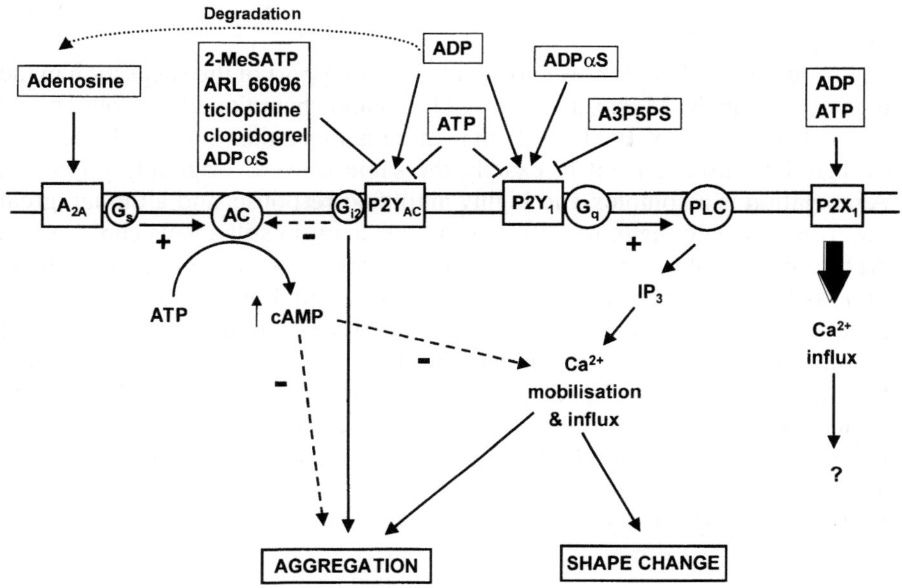

Fig. 1. P1 and P2 receptors on platelets and the possible pathways by which they act

coupled P2Y receptor has recently been cloned and designated as the $P2Y_{12}$ receptor (HOLLOPETER et al. 2000).

Whether or not activation of the $P2Y_1$ receptor alone is able to induce aggregation at all may depend on the experimental conditions, as 2-MeSATP and ARL 66096 are apparently competitive and can abolish ADP-induced aggregation in platelets washed free of plasma (HUMPHRIES et al. 1994; PARK and HOURANI 1999), while as discussed above the 2-alkylthio analogues can only partially inhibit ADP-induced aggregation in platelets in plasma (CUSACK and HOURANI 1982b; HOURANI et al. 1996). In addition, ADPαS, which is a full agonist at the $P2Y_1$ receptor but an antagonist at the adenylate cyclase-linked receptor, is incapable of inducing aggregation at all in washed platelets, instead acting as an antagonist (PARK and HOURANI 1999). This suggests that in plasma the $P2Y_1$ receptor can induce aggregation to a limited extent, but that this ability is abolished in washed platelets, possibly due to desensitisation of the $P2Y_1$ receptor. The role of the adenylate cyclase-coupled P2Y receptor may therefore normally be to enhance aggregation, and a study of the kinetics of aggregation and its inhibition concluded that the $P2Y_1$ receptor causes the immediate but transient effect of ADP while the adenylate cyclase-coupled receptor determines the final extent of aggregation (JARVIS et al. 1998). The way in which the $P2Y_{12}$ adenylate cyclase-linked receptor enhances effects mediated through the $P2Y_1$ receptor is still to be investigated, but as discussed above the mechanism is unlikely to be simply inhibition of adenylate cyclase although this may play some role, but probably involves some other G_i-mediated effect.

In summary, there is now good evidence in support of the suggestion made many years ago by Macfarlane et al. (1983) and championed by Colman and co-workers (Colman 1990, 1992) that there is more than one ADP receptor on platelets, although not in exactly the same form as originally proposed. Aggregation is a complex and highly amplified response, and although great progress has been made recently the full mechanism still awaits clarification. ADP is clearly of physiological and pathological importance in platelet function, as shown by the studies in patients (Cattaneo and Gachet 1999) and genetically-modified mice (Fabre et al. 1999; Enjyoji et al. 1999), and by the antithrombotic effects in animal models of ADP-removing enzymes (Zawilska et al. 1982) and ADP antagonists (Mcclure et al. 1988; Humphries et al. 1995; Ingall et al. 1999) as well as the established clinical benefits of ticlopidine and clopidogrel. The identification of the $P2Y_{12}$ receptor and the availability of potent and selective $P2Y_{12}$ receptor antagonists are likely to soon lead to exciting and significant advancements in the field.

List of Abbreviations

ADP	adenosine 5′-diphosphate
A3P5PS	adenosine 3′-phosphate,5′-phosphosulfate
ARL 66096	2-propylthioadenosine 5′-(β,γ-difluoromethylene) triphosphonate
ATP	adenosine 5′-triphosphate
ATPαS	adenosine 5′-(*O*-1-thiotriphosphate)
CGS 21680	2-[*p*-(2-carboxyethyl)phenethylamino]-5′-*N*-ethylcarboxamidoadenosine
2-ClADP	2-chloroadenosine 5′-diphosphate
2-ClATP	2-chloroadenosine 5′-triphosphate
CPA	N^6-cyclopentyladenosine
cyclic AMP	adenosine 3′,5′-cyclic monophosphate
GTP	guanosine 5′-triphosphate
HEL	human erythroleukemia
5-HT	5-hydroxytryptamine
IP_3	inositol 1,4,5-trisphosphate
2-MeSADP	2-methylthioadenosine 5′-diphosphate
2-MeSATP	2-methylthioadenosine 5′-triphosphate
2-MeS-β,γ-meATP	2-methylthioadenosine 5′-(β,γ-methylene) triphosphonate
α,β-me-ADP	adenosine 5′-(α,β-methylene)diphosphonate
α,β-meATP	adenosine 5′-(α,β-methylene)triphosphonate
β,γ-me-ATP	adenosine 5′-(β,γ-methylene)triphosphonate
NECA	5′-*N*-Ethylcarboxamidoadenosine

PD 115199	*N*-[2-(dimethylamino)-*N*-methyl-4-(2,3,6,7-tetrahydro-2,6-dioxo-1,3-dipropyl-1H-purin-8-yl)benzene sulphonamide
PLC	phospholipase C
R-PIA	N^6-R-phenylisopropyladenosine
SCH 58261	5-amino-7-(2-phenylethyl)-2-(2-furyl)-pyrazolo [4,3-*e*]-1,2,4-triazolo [1,5-*c*] pyrimidine
TXA_2	thromboxane A_2

References

Agarwal KC (1991) Modulation of platelet function by plasma adenosine levels. In: Imai S, Nakazawa M (eds) Role of adenosine and adenine nucleotides in the biological system. Elsevier, Amsterdam, pp 457–468

Boarder MR, Hourani SMO (1998) The regulation of vascular function by P2 receptors: multiple sites and multiple receptors. Trends in Pharmacol Sci 19:99–107

Boeynaems J-P, Communi D, Savi P, Herbert J-M (2000) P2Y receptors: in the middle of the road. Trends Pharmacol Sci 21:1–3

Born GVR (1962) Aggregation of blood platelets by adenosine diphosphate and its reversal. Nature 194:927–929

Born GVR (1963) Strong inhibition by 2-chloroadenosine of the aggregation of blood platelets by adenosine diphosphate. Nature 202:95–96

Born GVR, Foulks, JG (1977) Inhibition by a stable analogue of adenosine triphosphate of platelet aggregation by adenosine diphosphate. Br J Pharmacol 61:87–89

Born GVR, Haslam RJ, Goldman M, Lowe RD (1965) Comparative effectiveness of adenosine analogues as inhibitors of blood platelet aggregation and as vasodilators in man. Nature 205:678–680

Born GVR, Cross MJ (1963) The aggregation of blood platelets. J Physiol 168:178–195

Boyer JL, Romero-Avila T, Schachter JB, Harden TK (1996) Identification of competitive antagonists of the $P2Y_1$ receptor. Mol Pharmacol 50:1323–1329

Von Bruchhausen F, Walter U (1997) Platelets and their factors (Handbook of Experimental Pharmacology, vol 126). Springer, Berlin Heidelberg New York

Burnstock G, Kennedy C (1985) Is there a basis of distinguishing two types of P2-purinoceptor? Gen Pharmacol 16:433–440

Cattaneo M, Gachet C (1999) ADP receptors and clinical bleeding disorders. Arterioscler Thromb Vasc Biol 19:2281–2285

Clifford EE, Parker K, Humphreys BD, Kertesy SB, Dubyak GR (1998) The $P2X_1$ receptor, an adenosine-triphosphate-gated cation channel, is expressed in human platelets but not in human blood leukocytes. Blood 91:3172–3181

Collis MG, Hourani SMO (1993) Adenosine receptor subtypes. Trends in Pharmacol Sci 14:360–366

Colman RW (1990) Aggregin: a platelet ADP receptor that mediates aggregation. FASEB J 3:1425–1435

Colman RJ (1992) Platelet ADP receptors stimulating shape change and inhibiting adenylate cyclase. News Physiol Sci 7:274–278

Cooper DMF, Rodbell M (1979) ADP is a potent inhibitor of human platelet plasma membrane adenylate cyclase. Nature 282:517–518

Cooper JA, Hill SJ, Alexander SPH, Rubin PC, Horn EH (1995) Adenosine receptor-induced cyclic AMP generation and inhibition of 5-hydroxytryptamine release in human platelets. Br J Clin Pharmacol 40:43–50

Cristalli G, Mills DCB (1993) Identification of a receptor for ADP on blood platelets by photoaffinity labelling. Biochem J 291:875–881

Cristalli G, Vittori S, Thompson RD, Padgett WL, Shi D, Daly JW, Olsson RA (1994) Inhibition of platelet aggregation by adenosine receptor agonists. Naunyn Schmiedebergs Arch Pharmacol 349:644–650

Cusack NJ, Hourani SMO (1981a) 5-*N*-Ethylcarboxamidoadenosine: a potent inhibitor of human platelet aggregation. Br J Pharmacol 72:443–447

Cusack NJ, Hourani SMO (1981b) Partial agonist behaviour of adenosine 5′-*O*-(2-thiodiphosphate) on human platelets. Br J Pharmacol 73:405–408

Cusack NJ, Hourani SMO (1981c) Effects of Rp and Sp diasteroisomers of adenosine 5′-*O*-(1-thiodiphosphate) on human platelets. Br J Pharmacol 73:409–412

Cusack NJ, Hourani SMO (1982a) Differential inhibition by adenosine or by prostaglandin E_1 of human platelet aggregation induced by adenosine 5′-*O*-(1-thiodiphosphate) and adenosine 5′-*O*-(2-thiodiphosphate). Br J Pharmacol 75:257–259

Cusack NJ, Hourani SMO (1982b) Specific but noncompetitive inhibition by 2-alkylthio analogues of adenosine 5′-monophosphate and adenosine 5′-triphosphate of human platelet aggregation induced by adenosine 5′-diphosphate. Br J Pharmacol 75:397–400

Cusack NJ, Hourani SMO (1982c) Adenosine 5′-diphosphate antagonists and human platelets: no evidence that aggregation and inhibition of stimulated adenylate cyclase are mediated by different receptors. Br J Pharmacol 76:221–227

Cusack NJ, Hourani SMO (1982d) Inhibition by adenosine 5′-triphosphate of the actions on human platelets of 2-chloroadenosine 5′-diphosphate, 2-azidoadenosine 5′-diphosphate and 2-methylthioadenosine 5′-diphosphate. Br J Pharmacol 77:329–333

Daniel JL, Dangelmaier CA, Jin J, Ashby B, Smith JB, Kunapuli SP (1998) Molecular basis for ADP-induced platelet activation. J Biol Chem 273:2024–2029

Daniel JL, Dangelmaier CA, Smith JB (1986) ADP stimulates IP_3 formation in human platelets. FEBS Letters 206:299–303

Dionisotti S, Ferrara S, Molta C, Zocchi C, Ongini E (1996) Labeling of A_{2A} adenosine receptors in human platelets by use of the new nonxanthine antagonist radioligand [^{3}H] SCH 58261. J Pharmacol Exp Ther 278:1209–1214

Dionisotti S, Zocchi C, Varani K, Borea PA, Ongini E (1992) Effects of adenosine derivatives on human and rabbit platelet aggregation. Naunyn Schmiedebergs Arch Pharmacol 346:673–676

Enjyoji K, Sévigny J, Lin Y, Frenette PS, Christie PD, Schulte am Esch J, Imai M, Edelberg JM, Rayburn H, Lech M, Beeler DL, Csizmadia E, Wagner DD, Robson SC, Rosenberg RD (1999) Targeted disruption of cd39/ATPdiphosphohydrolase results in disordered hemostasis and thromboregulation. Nature Med 5:1010–1017

Fabre J-E, Nguyen MT, Latour A, Keifer JA, Audoly LP, Coffman TM, Koller BH (1999) Decreased platelet aggregation, increased bleeding time and resistance to thromboembolism in $P2Y_1$-deficient mice. Nature Med 5:1199–1202

Fagura MS, Dainty IA, McKay GD, Kirk IP, Robertson MJ, Dougall IG, Leff P (1998) $P2Y_1$-receptors in human platelets which are pharmacologically distinct from $P2Y_{ADP}$-receptors. Br J Pharmacol 124:157–164

Fein T, Schulze E, Bär J, Schwabe U (1994) Purification and characterization of an adenotin-like adenosine binding protein from human platelets. Naunyn Schmiedebergs Arch Pharmacol 349:374–380

Feoktistov I, Biaggioni I (1993) Characterization of adenosine receptors in human erythroleukemia cells and platelets: further evidence for heterogeneity of adenosine A_2 receptor subtypes. Mol Pharmacol 43:909–914

Feoktistov I, Biaggioni I (1995) Adenosine A_{2b} receptors evoke interleukin-8 secretion in human mast cells. An enprofylline-sensitive mechanism with implications for asthma. J Clin Invest 96:1979–1986

Fisher GJ, Bakshian S, Baldassare JJ (1985) Activation of human platelets by ADP causes a rapid rise in cytosolic free calcium without hydrolysis of

phosphatidylinositol-4,5-bisphosphate. Biochem Biophys Res Commun 129: 958–964
Fredholm BB, Abbracchio MP, Burnstock G, Daly JW, Harden T., Jacobson KA, Leff P, Williams M (1994). Nomenclature and classification of purinoceptors. Pharmacol Rev 46:143–156
Gaarder A, Jonsen L, Laland S, Hellem A, Owren PA (1961) Adenosine diphosphate in red cells as a factor in the adhesiveness of human blood platelets. Nature 192:531–532
Gachet C, Cattaneo M, Ohlmann P, Lecchi A, Chevalier J, Cassel D, Mannucci P M, Cazenave J-P (1995) Purinoceptors on blood platelets: further pharmacological and clinical evidence to suggest the presence of two ADP receptors. Br J Haematol 91:434–444
Gachet C, Cazenave J-P, Ohlmann P, Hilf G, Wieland T, Jakobs KH (1992) ADP receptor-induced activation of guanine-nucleotide-binding proteins in human platelet membranes. Eur J Biochem 207:259–263
Gachet C, Hechler B, Léon C, Vial C, Leray C, Ohlmann P, Cazenave J-P (1997) Activation of ADP receptors and platelet function. Thrombos Haemostas 78:271–275
Gessi S, Varani K, Merighi S, Ongini E, Borea PA (2000) A_{2A} adenosine receptors in human peripheral blood cells. Br J Pharmacol 129:2–11
Glenn JR, Heptinstall S (1998) Evidence that adenosine diphosphate can activate adenylate cyclase via conversion to adenosine in platelet-rich plasma containing magnesium. Thrombos Haemostas 80:321–325
Gordon JL (1986) Extracellular ATP: effects, sources and fate. Biochem J 233:309–319
Gurden MF, Coates J, Ellis F, Evands F, Evans B, Foster M, Hornby E, Kennedy I, Martin DP, Strong P, Vardey CJ, Wheeldon A (1993) Functional characterization of 3 adenosine receptor types. Br J Pharmacol 109:693–698
Hall DA, Hourani SMO (1993) Effects of analogues of adenine nucleotides on increases in intracellular calcium mediated by P_{2T}-purinoceptors on human blood platelets. Br J Pharmacol 108:728–733
Hall DA, Frost V, Hourani SMO (1994) Effects of extracellular divalent cations on responses of human blood platelets to adenosine 5′-diphosphate. Biochem Pharmacol 48:1319–1326
Hall DA, Hourani SMO (1994) Effects of suramin on increases in cytosolic calcium and on inhibition of adenylate cyclase induced by adenosine 5′-diphosphate in human platelets. Biochem Pharmacol 47:1013–1018
Hallam TJ, Rink TJ (1985) Responses to adenosine diphosphate in human platelets loaded with the fluorescent calcium indicator quin2. J Physiol 368:131–146
Harden TK, Nicholas RA, Schachter JB, Lazarowski ER, Boyer JL (1998) Pharmacological selectivities of molecularly defined subtypes of P2Y receptors. In: Turner JT, Weisman GA, Fedan JS (eds) The P2 nucleotide receptors. Humana Press, Totowa NJ, pp 109–134
Harris DN, Asaad MM, Phillips MB, Goldenberg HJ, Antonaccio MJ (1979) Inhibition of adenylate cyclase in human blood platelets by 9-substituted adenine derivatives. J Cyclic Nuc Res 5:125–134
Haslam RJ, Cusack NJ (1981) Blood platelet receptors for ADP and for adenosine. In: Burnstock G (ed) Purinergic receptors: receptors and recognition, Series B, vol 12. Chapman and Hall, London, pp 222–285
Haslam RJ, Lynham JA (1972) Activation and inhibition of blood platelet adenylate cyclase by adenosine or by 2-chloroadenosine. Life Sciences 11:1143–1154
Haslam RJ, Rosson GM (1975) Effects of adenosine on levels of adenosine cyclic 3′,5′-monophosphate in human blood platelets in relation to adenosine incorporation and platelet aggregation. Mol Pharmacol 11:528–544
Hechler B, Léon C, Vial C, Vigne P, Frelin C, Cazenave J-P, Gachet C, (1998a) The $P2Y_1$ receptor is necessary for adenosine 5′-diphosphate-induced platelet aggregation. Blood 92:152–159

Hechler B, Vigne P, Léon C, Breittmayer J-P, Gachet C, Frelin C (1998b) ATP derivatives are antagonists of the $P2Y_1$ receptor: similarities to the platelet ADP receptor. Mol Pharmacol 53:727–733
Heemskerk JWM, Vis P, Feijge MAH, Hoyland J, Mason WT, Sage SO (1993) Roles of phospholipase C and Ca^{2+}-ATPase in calcium responses of single, fibrinogen-bound platelets. J Biol Chem 268:356–363
Heptinstall S, Taylor, PM (1979) The effects of citrate and extracellular calcium ions on the platelet release reaction induced by adenosine diphosphate and collagen. Thrombos Haemostas 42:778–793
Hollopeter G, Jantzen H-M, Vincent D, Li G, England L, Ramakrishnan V, Yang R-B, Nurden P, Nurden A, Julius D, Conley PB (2000) Identification of the Platelet ADP Receptor Targeted by Antithrombotic Drugs. Nature, in press
Holmsen H (1996) Activation of human platelets in vitro. Platelets 7:336–341
Hourani SMO, Cusack NJ (1991) Pharmacological receptors on blood platelets. Pharmacol Rev 43:243–298
Hourani SMO, Hall DA (1994) Platelets for ADP on human blood platelets. Trends in Pharmacol Sci 15:103–108
Hourani SMO, Hall DA (1996) P_{2T} purinoceptors: ADP receptors on platelets. Ciba Foundation Symposium 198:52–69
Hourani SMO, Hall DA, Nieman CJ (1992) Effects of the P2-purinoceptor antagonist, suramin, on human platelet aggregation induced by adenosine 5′-diphosphate. Br J Pharmacol 105:453–457
Hourani, SMO, Welford LA, Cusack NJ (1996) Effects of 2-methylthioadenosine 5′-β,γ-methylenetriphosphonate and 2-ethylthioadenosine 5′-monophosphate on human platelet aggregation induced by adenosine 5′-diphosphate. Drug Dev Res 38:12–28
Humphries RG, Tomlinson W, Ingall AH, Cage PA, Leff P (1994) FPL 66096: a novel, highly potent and selective antagonist at human platelet P_{2T}-purinoceptors. Br J Pharmacol 113:1057–1063
Humphries RG, Robertson MJ, Leff P (1995) A novel series of P_{2T} purinoceptor antagonists: definition of the role of ADP in arterial thrombosis. Trends in Pharmacol Sci 16, 179–181
Hütteman E, Ukena D, Lenschow V, Schwabe U (1984) R_a Adenosine receptors in human platelets. Characterization by 5′-*N*-ethylcarboxamido [^{3}H] adenosine binding in relation to adenylate cyclase activity. Naunyn Schmiedebergs Arch Pharmacol 325:226–233
Ingall AH, Dixon J, Bailey A, Coombs ME, Cox D, McInally JI, Hunt SF, Kindon ND, Teobald BJ, Willis PA, Humphries RG, Leff P, Clegg JA, Smith JA, Tomlinson W (1999) Antagonists of the platelet P2T receptor: a novel approach to antithrombotic therapy. J Med Chem 42:213–220
Jakobs KH, Saur W, Johnson RA (1979) Regulation of platelet adenylate cyclase by adenosine. Biochim Biophys Acta 583:409–421
Jantzen H-M, Goussent L, Bhaskar V, Vincent D, Tai A, Reynolds EE, Conley PB (1999) Evidence for two distinct G-protein-coupled ADP receptors mediating platelet activation. Thromb Haemost 81:111–117
Jarvis GE, Humphries RG, Robertson MJ, Leff P (1998) ADP can induce aggregation of human platelets via both $P2Y_1$ and P_{2T} receptors. Br J Pharmacol 124;32P
Jin J, Daniel JL, Kunapuli SP (1998) Molecular basis for ADP-induced platelet activation. II. The $P2Y_1$ receptor mediates ADP-induced intracellular calcium mobilization and shape change in platelets. J Biol Chem 273:2030–2034
Jin J, Kunapuli SP (1998) Coactivation of two different G protein-coupled receptors in essential for ADP-induced platelet aggregation. Proc Natl Acad Sci USA 95:8070–8074
Kikugawa K, Iizuka K, Ichino M (1973) Platelet aggregation inhibitors. 4. N^6-substituted adenosine. J Med Chem 16:358–364
Kunapuli SP (1998) Multiple P2 receptor subtypes on platelets; a new interpretation of their function. Trends in Pharmacol Sci 19:391–394

Ledent C, Vaugeois J-M, Schiffmann SN, Pedrazzini T, El Yacoubi M, Vanderhaegen J-J, Costentin J, Heath JK, Vassart G, Parmentier M (1997) Aggressiveness, hypoalgesia and high blood pressure in mice lacking the adenosine A_{2a} receptor. Nature 388:674–678

Léon C, Hechler B, Vial C, Leray C, Cazenave J-P, Gachet C (1997) The $P2Y_1$ receptor is an ADP receptor antagonized by ATP and expressed in platelets and megakaroblastic cells. FEBS Letters 403:26–30

Lloyd JV, Nishizawa EE, Haldar J, Mustard JF (1972) Changes in ^{32}P-labelling of platelet phospholipids in response to ADP. Br J Haematol 23:571–585

Lloyd JV, Nishizawa EE, Mustard JF (1973) Effect of ADP-induced shape change on incorporation of ^{32}P into platelet phosphatidic acid and mono-, di- and triphosphatidyl inositol. Br J Haematol 25:77–99

Lohse MJ, Elger B, Lindenborn-Fotinos J, Klotz K-N, Schwabe U (1988) Separation of solubilized A_2 adenosine receptors of human platelets from non-receptor [^{3}H] NECA binding sites by gel filtration. Naunyn Schmiedebergs Arch Pharmacol 337:64–68

Macfarlane DE, Mills DCB (1975) The effects of ATP on platelets: evidence against the central role of released ADP in primary aggregation. Blood 46:309–320

Macfarlane DE, Srivastava PC, Mills DCB (1983) 2-Methylthioadenosine [β-^{32}P]diphosphate. An agonist and radioligand for the receptor that inhibits the accumulation of cyclic AMP in intact blood platelets. J Clin Invest 71:420–425

MacIntyre DE, Pollock WK, Shaw AM, Bushfield M, MacMillan LJ, McNicol A (1985) Agonist-induced inositol phospholipid metabolism and Ca^{2+} flux in human platelet activation. Adv Exp Biol Med 192:127–144

MacKenzie AB, Mahaut-Smith MP, Sage SO (1996) Activation of receptor-operated cation channels via P_{2X1} not P_{2T} purinoceptors in human platelets. J Biol Chem 271:2879–2881

Mahaut-Smith MP, Sage SO, Rink TJ (1990) Receptor-activated single channels in intact human platelets. J Biol Chem 265:10479–10483

Mahaut-Smith MP, Sage SO, Rink TJ (1992) Rapid ADP-evoked currents in human platelets recorded with the nystatin permeabilized patch technique. J Biol Chem 267:3060–3065

Marala RB, Mustafa SJ (1998) Immunological characterization of adenosine A_{2A} receptors in human and porcine cardiovascular tissues. J Pharmacol Exp Ther 286:1051–1057

Martinson EA, Johnson RA, Wells JN (1987) Potent adenosine receptor antagonists that are selective for the A_1 receptor subtype. Mol Pharmacol 31:247–252

Massini P, Lüscher EF (1976) On the significance of the influx of calcium ions into stimulated human blood platelets. Biochim Biophys Acta 436:652–663

McClure MO, Kakkar A, Cusack NJ, Born GCR (1988) Evidence for the dependence of arterial haemostasis on ADP. Proc R Soc Lond B 234:255–262

Mills DCB (1996) ADP receptors on platelets. Blood 76:835–856

Mills DCB, Puri R, Hu C-J, Minniti C, Grana G, Freedman MD, Colman RF, Colman RW (1992) Clopidogrel inhibits the binding of ADP analogues to the receptor mediating inhibition of platelet adenylate cyclase. Arteriosclerosis Thrombosis 12:430–436

Mills DCB, Smith JB (1972) The control of platelet responsiveness by agents that influence cyclic AMP metabolism. Ann N Y Acad Sci 201:391–399

Mulryan K, Gitterman DP, Lewis CJ, Vial C, Leckie BJ, Cobb AL, Brown JE, Conley EC, Buell G, Pritchard CA, Evans RJ (2000) Reduced vas deferens contraction and male infertility in mice lacking $P2X_1$ receptors. Nature 403:86–89

Mustard JF, Perry DW, Kinlough-Rathbone RL, Packham MA (1975) Factors responsible for ADP-induced release reaction of human platelets. Am J Physiol 228:1757–1765

Offermanns S, Toombs CF, Hu Y-H, Simon MI (1997) Defective platelet activation in $G\alpha_q$-deficient mice. Nature 389:183–186

Ohlmann P, Laugwitz K-L, Nürnberg B, Spicher K, Schultz G, Cazenave J-P, Gachet C (1995) The human platelet ADP receptor activates G_{i2} proteins. Biochem J 312:775–779

Olbrich C, Aepfelbacher M, Siess W (1989) Epinephrine potentiated calcium mobilization and activation of protein kinases in platelets stimulated by ADP through a mechanism unrelated to phospholipase C. Cell Signal 1:483–492

Packham MA, Bryant NL, Guccione MA, Kinlough-Rathbone RL, Mustard JF (1989) Effect of the concentration of Ca^{2+} in the suspending medium on the responses of human and rabbit platelets to aggregating agents. Thrombos Haemostas 62:978–976

Packham MA, Livne A-A, Ruben DH, Rand ML (1993) Activation of phospholipase C and protein kinase C has little involvement in ADP-induced primary aggregation of human platelets: effects of diacylglycerols, the diacylglycerol kinase inhibitor R59022, staurosporine and okadaic acid. Biochem J 290:849–856

Park H-S, Hourani SMO (1999) Different effects of adenine nucleotide analogues on shape change and aggregation induced by adenosine 5′-diphosphate (ADP) in human platelets. Br J Pharmacol 127:1359–1366

Park H-S, Tennant JP, Waktolla GF, Sarkardei S, Kass GEN, Hourani SMO (1998) Effects of adenosine 3′-phosphate 5′-phosphosulfate on P2 receptors in platelets and smooth muscle preparations. Drug Dev Res 45:67–73

Pulchinelli FM, Ashby B, Gazzaniga PP, Daniel JL (1995) Protein kinase C activation is not a key step in ADP-mediated exposure of fibrinogen receptors on human platelets. FEBS Letters 364:87–90

Raberger G, Schütz W, Kraupp O (1977) Coronary dilator action of adenosine analogues: a comparative study. Arch Int Pharmacodyn 230:140–149

Raha S, Jones GD, Gear ARL (1993) Sub-second oscillations of inositol 1,4,5-trisphosphate and inositol 1,3,4,5-tetrakisphosphate during platelet activation by ADP and thrombin: lack of correlation with calcium kinetics. Biochem J 292:643–646

Rao AK, Kowalsa MA (1987) ADP-induced platelet shape change and mobilization of cytoplasmic ionised calcium are mediated by distinct binding sites on platelets: 5′-*p*-fluorosulfonylbenzoyladenosine is a weak platelet agonist. Blood 70:751–756

Sage SO, MacKenzie AB, Jenner S, Mahaut-Smith MP (1997) Purinoceptor-evoked calcium signalling in human platelets. Prostaglandins Leukotrienes Essential Fatty Acids 57:435–438

Sage SO, Mahaut-Smith MP, Rink TJ (1992) Calcium entry in nonexcitable cell: lessons from human platelets. News in Physiol Sci 7:108–113

Sage SO, Merritt JE, Hallam TJ, Rink TJ (1989) Receptor-mediated calcium entry in fura-2-loaded human platelets stimulated with ADP and thrombin. Biochem J 258:923–926

Sage SO, Reast R, Rink TJ (1990) ADP evokes biphasic Ca^{2+} influx in fura-2-loaded human platelets. Evidence for Ca^{2+} entry regulated by the intracellular Ca^{2+} store. Biochem J 265:675–680

Sage SO, Rink TJ (1987) The kinetics of changes in intracellular calcium concentration in fura-2-loaded human platelets. J Biol Chem 262:16364–16369

Sanderson HM, Heptinstall S, Vickers J, Lösche W (1996) Studies on the effects of agonists and antagonists on platelet shape change and platelet aggregation in whole blood. Blood Coag Fibrinol 7:245–248

Savi P, Beauverger P, Labouret C, Delfaud M, Salel V, Kaghad M, Herbert JM (1998) Role of P2Y1 in ADP-induced platelet activation. FEBS Letters 422:291–295

Savi P, Nurden P, Nurden AT, Levy-Toledano S, Herbert J-M (1998) Clopidogrel: a review of its mechanism of action. Platelets 9:251–255

Scase TJ, Heath MF, Allen JM, Sage SO, Evans RJ (1998) Identification of a P_{2X1} purinoceptor expressed on human platelets. Biochem Biophys Res Commun 242:525–528

Schachter JB, Li Q, Boyer JL, Nicholas RA, Harden TK (1996) Second messenger cascade specificity and pharmacological selectivity of the human P_{2Y1}-purinoceptor. Br J Pharmacol 118:167–173
Schwabe U, Ukena D, Lohse MJ (1985) Xanthine derivatives as antagonists at A_1 and A_2 adenosine receptors. Naunyn Schmiedebergs Arch Pharmacol 330:212–221
Sun B, Okahara K, Kambayashi J (1998) $P2X_1$ purinoceptor in human platelets. J Biol Chem 273:11544–11547
Ukena D, Schirren CG, Schwabe U (1985) Effects of enprofylline on A_1 and A_2 adenosine receptors. Eur J Pharmacol 117:25–33
Valera S, Hussy N, Evans RJ, Adami N, North RA, Surprenant A, Buell G (1994) A new class of ligand-gated ion channel defined by P_{2X} receptor for extracellular ATP. Nature 371:516–519
Vanags DM, Lloyd JV, Rodgers SE, Bochner F (1998) ADP, adrenaline and serotonin stimulate 1,4,5-trisphosphate production in human platelets. Eur J Pharmacol 358:93–100
Varani K, Borea PA, Guerra L, Dionisotti S, Zocchi C, Ongini E (1994) Binding characteristics of the adenosine A_2 receptor ligand [^{3}H] CGS 21680 to human platelet membranes. Biochem Pharmacol 48:1658–1661
Varani K, Gessi S, Dalpiaz A, Borea PA (1996) Pharmacological and biochemical characterization of purified A_{2A} adenosine receptors in human platelet membranes by [^{3}H]- CGS 21680 binding. Br J Pharmacol 117:1693–1701
Vial C, Hechler B, Leon C, Cazenave J-P, Gachet C (1997) Presence of $P2X_1$ purinoceptors in human platelets and megakaryoblastic cell lines. Thrombos Haemostas 78:1500–1504
Vickers JD, Kinlough-Rathbone RL, Packham MSA, Mustard JF (1996) Inositol phospholipid metabolism in human platelets stimulated by ADP. Eur J Biochem 193:521–528
Webb TE, Simon J, Krishek BJ, Bateson AN, Smart TG, King BF, Burnstock G, Barnard EA (1993) Cloning and functional expression of a brain G-protein coupled ATP receptor. FEBS Letters 324:219–225
Webb TE, Feolde E, Vigne P, Neary JT, Runberg A, Frelin C, Barnard EA (1996) The P2Y purinoceptor in rat brain microvascular endothelial cells couples to inhibition of adenylate cyclase. Br J Pharmacol 119:1385–1392
White JG, Rao GHR, Gerrard JM (1974) Effects of the ionophore A23187 on blood platelets. I. Influence on aggregation and secretion. Am J Pathol 77:135–149
Zawilska KM, Born GVR, Begent NA (1982) Effect of ADP-utilizing enzymes on the arterial bleeding time in rats and rabbits. Br J Haematol 50:317–325
Zocchi C, Ongini E, Conti A, Monopoli A, Negretti A, Baraldi PG, Dionisotti S (1996) The non-xanthine heterocyclic compound SCH 58261 is a new potent and selective A_{2A} adenosine receptor antagonist. J Pharm Exp Ther 276:398–404

CHAPTER 21

Purinergic Signalling in Gut

G. BURNSTOCK

A. Introduction

Purines can influence motility secretion and absorption in a variety of direct and indirect ways. They can be released from intrinsic enteric nerves, sympathetic nerves, or sensory-motor nerves during axon reflexes, to act directly on smooth muscle purinoceptors mediating relaxation or contraction or on epithelial cell receptors. They can act on prejunctional nerve terminals to modify transmitter release from motor and inhibitory neural control pathways. They can participate in synaptic transmission in myenteric and submucosal ganglia that are involved in control of gastrointestinal motility, mucosal secretion and absorption. They can act on blood vessels or interstitial cells of Cajal thereby indirectly modulating motility patterns. They can act on sensory nerve endings in the gut wall to initiate local and/or central reflex activity that alters gastrointestinal motility and secretory patterns.

A signalling role for ATP was first proposed for transmission from non-adrenergic, non-cholinergic (NANC) nerves and the smooth muscle of the intestine in 1970 (BURNSTOCK et al. 1970, 1972; BURNSTOCK 1972) (Fig. 1). This hypothesis was resisted by many for over two decades, first vasoactive intestinal polypeptide (VIP) (see GRIDER and MAKHLOUF 1986; FURNESS and COSTA 1987; HOYLE and BURNSTOCK 1989) and later nitric oxide (NO) (see RAND 1992; LINCOLN et al. 1997) being more popular as putative NANC transmitters. However, the evidence that ATP acts as a neuromuscular transmitter in the gut is now convincing (see reviews HOYLE and BURNSTOCK 1989; HOYLE 1992, 1996; BAUER 1993; ITOH et al. 1995; BURNSTOCK 1996). The current consensus is that ATP, NO and VIP or its related peptide, pituitary adenylyl cyclase activating peptide (PACAP) (MCCONALOGUE et al. 1995a), are cotransmitters in NANC inhibitory nerves, although their proportions vary considerably between different regions of the gastrointestinal tract and different species (COSTA et al. 1986a; BURNSTOCK 1990; SHUTTLEWORTH et al. 1991; CRIST et al. 1992; SAFFREY et al. 1992; SANDERS and WARD 1992; BELAI and BURNSTOCK 1994; GERSHON and WADE 1994; FURNESS et al. 1995; KNUDSEN et al. 1995; MAGGI and GIULIANI 1996; RAE and MUIR 1996; SMITS and LEFEBVRE 1996; BOECKXSTAENS and PELCKMANS 1997; SANDERS 1998; BÖRJESSON et al. 1999) and

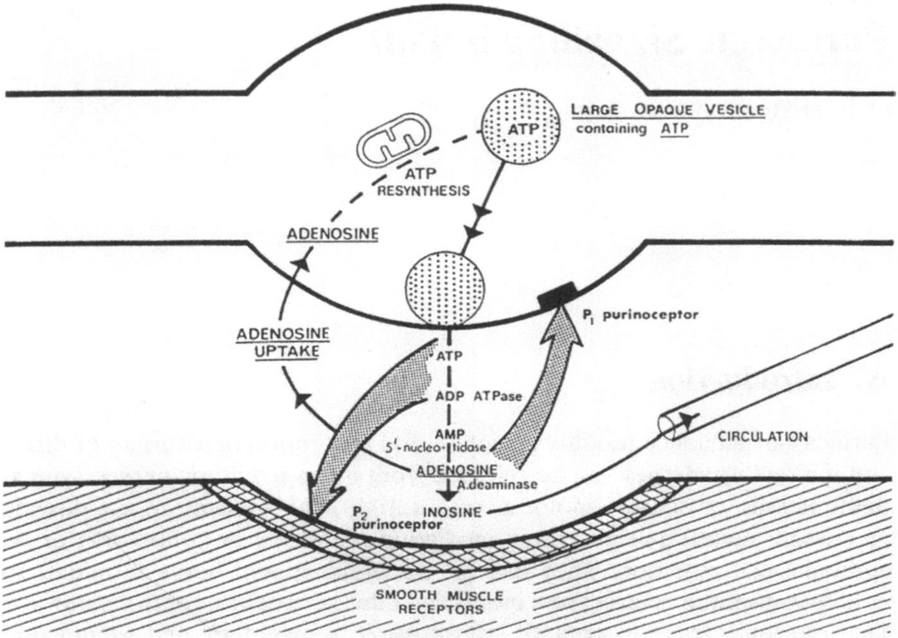

Fig. 1. Schematic representation of purinergic neuromuscular junctions depicting the synthesis, storage, release and inactivation of ATP as an autonomic neurotransmitter (modified from BURNSTOCK 1972)

the evidence is presented in detail in Sect. B. Signalling roles for nucleosides and nucleotides in the gut is a primitive mechanism and has been documented from early evolutionary times in most invertebrates as well as in lower vertebrate animals (see BURNSTOCK 1996).

Other signalling roles for ATP have emerged through the years, including synaptic transmission between neurones in myenteric and submucosal plexuses (Sect. C.), control of epithelial cell secretion and absorption (Sect. D.), and as a sympathetic nerve cotransmitter in controlling intestinal vascular tone (Sect. E.).

The possible roles of purinergic signalling in various disease conditions are discussed in Sect. F. Speculations are made about the role of purines in gut pain and inflammation and about the long-term (trophic) actions of purines in the final section (Sect. G.).

B. Purinergic Neuromuscular Transmission

NANC *inhibitory* nerves are dominant in many regions of the gut (Fig. 2) (see HOYLE and BURNSTOCK 1989; BURNSTOCK 1992), but NANC *excitatory* nerves have also been described, notably in the guinea-pig ileum, and in the gastrointestinal tract of lower vertebrates (Fig. 3) (Burnstock 1969, 1996; Burnstock et al. 1972; Sneddon et al. 1973), but are also found in neonatal development (see Burnstock 1996; and Chap. 5, first volume).

While ATP, NO and VIP appear to be cotransmitters in many of these NANC nerves (Fig. 4), there is much variability in their proportions in different regions of the gut and between species. In general, it seems that in most species, NO is the dominant cotransmitter in anterior regions of the gut, while ATP is more prominent in the posterior regions.

I. Oesophagus

The main body of experimental evidence supports the view that, at least in healthy conditions, NO is the dominant NANC transmitter in the smooth muscle of the oesophagus and lower oesophageal sphincter (LES). Both pharmacological (Murray et al. 1991; Tøttrup et al. 1991a,b, 1993) and histochemical (McKirdy et al. 1992; Ny et al. 1995; Rodrigo et al. 1998) evidence has been presented to support this view. Nitrergic co-innervation of striatal muscle fibres in the oesophagus of the rat has also been claimed (Wörl et al. 1994).

VIP has also been reported to participate in control of oesophageal smooth muscle in some species (Goyal et al. 1980; Biancani et al. 1984; Behar et al. 1989; Szewczak et al. 1990), although evidence against a role for VIP in the opossum oesophagus has also been presented (Daniel et al. 1983, 1989; Akbarali et al. 1986).

The possibility that ATP is also involved as a neurotransmitter in the oesophagus has also been raised (Castell 1975; de Carle and Christensen 1976) but no hard evidence has emerged to support this possibility (e.g. Rattan and Goyal 1980; Daniel et al. 1984). ATP and ADP produce about 40% maximal contractions of rabbit oesophageal muscularis mucosae (Percy et al. 1997).

Purinergic receptors have been identified, however, in the smooth muscle of the rat oesophageal muscularis mucosae (Will et al. 1990). The results indicated that adenosine analogues can modulate oesophageal smooth muscle tone both *indirectly* via the activation of an A_3-like receptor leading to the release of 5-hydroxytryptamine (5-HT) and histamine from tissue-bound mast cells and *directly* by activation of excitatory, suramin sensitive, P2 receptors associated with the smooth muscle cells. The possibility that ATP (or its breakdown product adenosine) released as a cotransmitter from sympathetic or parasympathetic nerves may act prejunctionally to modulate the release of the NANC transmitter has also been raised (Kamikawa and Shimo 1982; Lundgren 1983; Mishra and Raviprakesh 1987). ATP and adenosine are reported to augment the contractions of the guinea-pig oesophagus both to cholinergic nerve stimulation and applied acetylcholine (ACh) (Kamikawa and Shimo 1982).

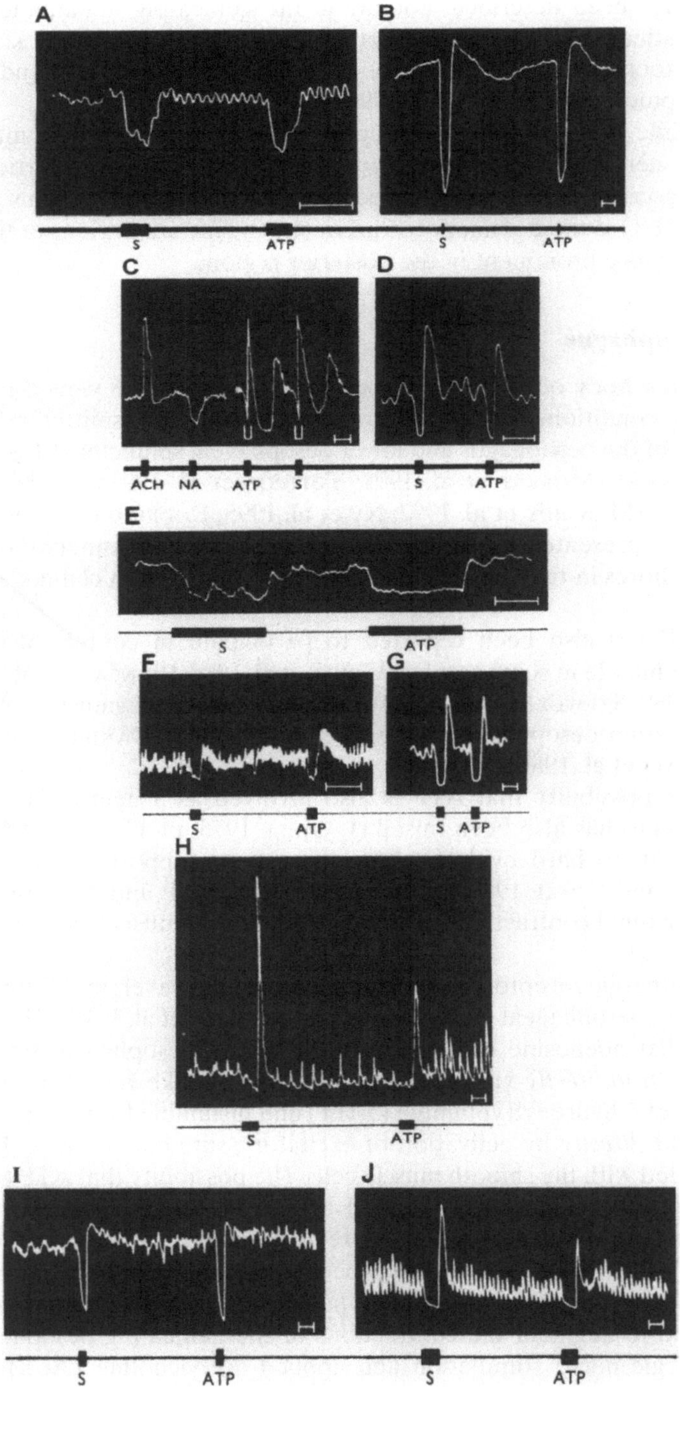
A
B
S
ATP
S
ATP
C
D
ACH
NA
ATP
S
S
ATP
E
S
ATP
F
G
S
ATP
S
ATP
H
S
ATP
I
J
S
ATP
S
ATP

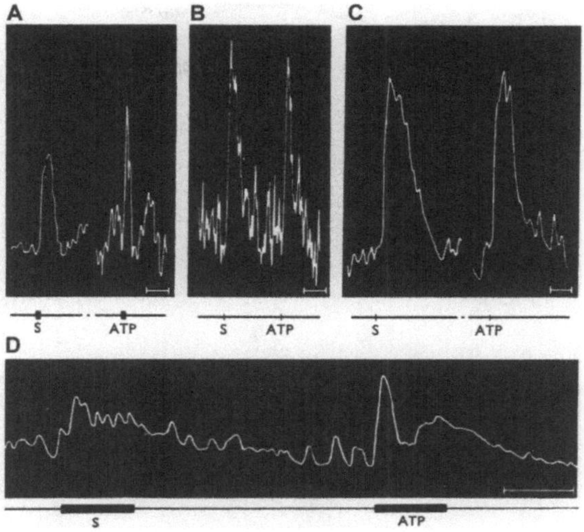

Fig. 3A–D. Mimicry *excitatory* responses of gut segments from lower vertebrates to transmural stimulation and ATP. Hyoscine (1.3 μmol/l) and guanethidine (3.5 μmol/l) were present throughout. **A** Lizard ileum; transmural stimulation (S; 10 Hz for 1 min), ATP (10 μmol/l for 1 min). **B** Toad duodenum, transmural stimulation (5 Hz for 15 s), ATP (10 μmol/l for 15 s). **C** Toad ileum, transmural stimulation (5 Hz for 15 s), ATP (25 μmol/l for 15 s). **D** Goldfish large intestine, transmural stimulation (10 Hz for 1 min), ATP (12 μmol/l for 1 min). Time markers: **A**,**B**,**C**, 5 min; **D**, 1 min (from BURNSTOCK et al. 1972)

◄

Fig. 2A–J. Mimicry *inhibitory* responses of various gastrointestinal smooth muscle preparations to transmural stimulation and ATP, often followed by rebound contractions. Hyoscine (1.3 μmol/l) and guanethidine (3.5 μmol/l) were present except where stated. **A** Guinea-pig stomach preparations consisting of strips (4 × 40 mm) cut as a spiral around the mid portion of the stomach; transmural stimulation (S, 5 Hz for 30 s), ATP (5 μmol/l for 30 s). **B** Guinea-pig taenia coli, transmural stimulation (S, 5 Hz for 15 s), ATP (1 μmol/l for 15 s). **C** Guinea-pig ileum, acetylcholine (ACh, 0.006 μmol/l for 30 s, hyoscine omitted), noradrenaline (NA, 0.17 μmol/l for 30 s, hyoscine omitted), ATP (5 μmol/l for 30 s), transmural stimulation (S, 5 Hz for 30 s); **D** Guinea-pig colon, transmural stimulation (S, 5 Hz for 15 s), ATP (5 μmol/l for 15 s). **E** Biopsy specimen of human colon cut as 10 × 5 × 4 mm strips; transmural stimulation (S, 5 Hz for 2 min), ATP (400 μmol/l for 2 min). **F** Rat duodenum, transmural stimulation (S, 5 Hz for 20 s), ATP (10 μmol/l for 20 s). **G** Rat ileum, transmural stimulation (S, 5 Hz for 30 s), ATP (50 μmol/l for 30 s). **H** Rat rectum, transmural stimulation (S, 3 Hz for 1 min), ATP (200 μmol/l for 1 min). **I** Mouse colon, transmural stimulation (S, 5 Hz for 30 s), ATP (40 μmol/l for 30 s). **J** Mouse rectum, transmural stimulation (S, 5 Hz for 1 min), ATP (40 μmol/l for 1 min). Time markers, 1 min (from BURNSTOCK et al. 1972)

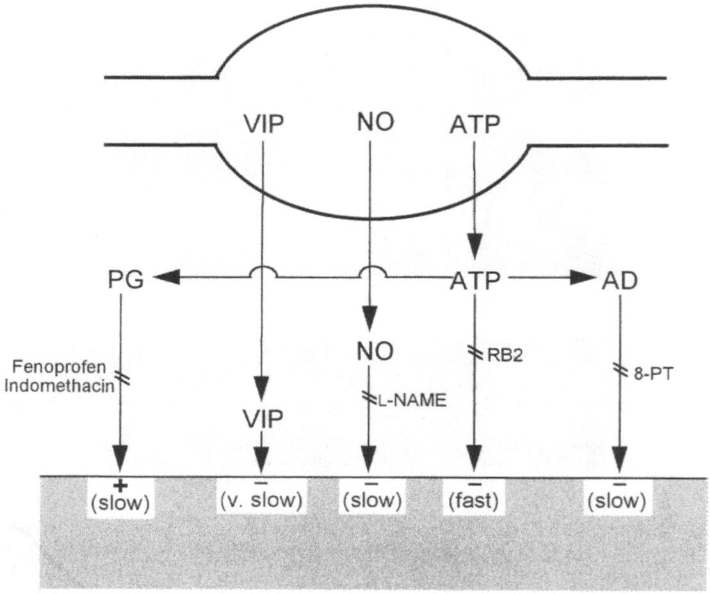

Fig. 4. Schematic representation of nonadrenergic noncholinergic (NANC) inhibitory nerves in the gut. Neurotransmitters and/or agonists: VIP, vasoactive intestinal polypeptide; NO, nitric oxide; ATP, adenosine 5′-triphosphate; AD, adenosine; PG, prostaglandins. Antagonists or inhibitors: L-NAME, N^G-nitro-L-aginine methyl ester; RB2, Reactive blue 2; 8-PT, 8-phenyltheophylline. Responses: +, excitatory; –, inhibitory

There are some indications that in pathological conditions, such as achalasia or symptomatic diffuse oesophageal spasm, that purinergic signalling might be involved (DiMarino 1974; Fisher and Cohen 1975).

II. Stomach

There is growing evidence that ATP is involved in NANC inhibitory transmission in the stomach of most species in concert with NO and to a lesser extent, VIP (Okwuasaba et al. 1977; Kasakov and Milenov 1979; Minker and Matejka 1981; Delbro and Fändriks 1982; Grider et al. 1982; Brizzi et al. 1985; Baccari et al. 1991; Ohno et al. 1993; Mashimo et al. 1996; Otsuguro et al. 1996, 1998; Zagorodnyuk et al. 1996; Beck et al. 1998; Currò and Preziosi 1998; Glasgow et al. 1998; Xue et al. 1998), although many early papers did not favour purinergic involvement (e.g. Heazell 1975; Ohga and Taneiki 1977; Hunt et al. 1978; Baer and Frew 1979; Lefebvre and Willems 1979; Frew and Lundy 1982; Andrews and Lawes 1985; Lefebvre 1986).

As early as 1970, the involvement of ATP in NANC neurotransmission was supported by evidence of its release from bullfrog stomach upon electrical field stimulation of the vagal nerves evoking a NANC relaxation of the smooth muscle of the stomach (BURNSTOCK et al. 1970; SATCHELL and BURNSTOCK 1971). In the presence of α- and β-adrenergic blockers, ATP and related substances were shown to produce inhibitory effects on the spike activity and generation of slow waves in isolated muscle preparation of the guinea-pig antrum, suggesting direct action of ATP and related substances on the antrum muscle (OHKAWA and WATANABE 1976). Subsequent studies give further supporting evidence for purinergic NANC inhibitory transmission in the stomach, including: the fast component of the inhibitory junction potentials (IJPs) is blocked by apamin and suramin (OHNO et al. 1996; XUE et al. 1996); ATP and analogues mimic the relaxation and hyperpolarisation produced by NANC nerve stimulation. An early paper examined the effect of theophylline on IJPs and the responses to ATP and adenosine (HUIZINGA and DEN HERTOG 1980). Perhaps not surprisingly, in retrospect, theophylline (an adenosine antagonist) did not inhibit either the effects of NANC inhibitory stimulation or the relaxation to ATP.

Vagally induced NANC gastric relaxation of cat stomach is inhibited by P2 receptor desensitisation with α,β-methylene ATP ($\alpha\beta$-meATP), but it is likely that this is due to interference with ganglionic transmission in the vagal pathway, rather than neuromuscular blockade (DELBRO and FÄNDRIKS 1982, 1984).

Initially, BURNSTOCK et al. (1970) described the effect of ATP on rat stomach as a biphasic response involving a rapid and brief relaxation followed by a sustained contraction which was later confirmed by other investigators (HEAZELL 1975; LEFEBVRE 1986; LEFEBVRE and BURNSTOCK 1990; LEFEBVRE et al. 1991; OTSUGURO et al. 1996). Apamin, a polypeptide derived from bee venom, which blocks small conductance Ca^{2+}-activated K^+ channels, was reported to inhibit the relaxant effect of ATP in various gastrointestinal preparations (BANKS et al. 1979; SHUBA and VLADIMIROVA 1980; COSTA et al. 1986b; LEFEBVRE et al. 1991). Responses to stimulation of enteric inhibitory neurones were reported to be substantially reduced by apamin in the circular muscle coat of the antrum, but not fundus (COSTA et al. 1986b).

The excitatory effects of ATP and related compounds on the stomach may indicate the presence of some NANC excitatory transmission, mediated by P2 receptors (RHEE et al. 1996). Another explanation is that the excitatory effects of ATP are indirect, due to stimulation of prostaglandin biosynthesis (Fig. 5) (BURNSTOCK et al. 1975; HUIZINGA et al. 1981; SAKAMOTO et al. 1987; LEFEBVRE and BURNSTOCK 1990; BACCARI et al. 1990b, 1991, 1992, 1996).

Some of the actions of ATP, usually mediated by its breakdown product adenosine acting on prejunctional $P1(A_1)$ receptors, involve modulation of release of enteric transmitters, including somatostatin (SMS) (KWOK et al. 1990) and VIP (SJÖQVIST et al. 1985) and there is also a report that ATP depresses the contractile activity probably mediated by ACh elicited by vagal stimulation in vivo (BACCARI et al. 1990a, 1994).

An area of ongoing debate concerns the identity of the P2 receptor subtypes involved in gastric motility (MATHARU and HOLLINGSWORTH 1992;

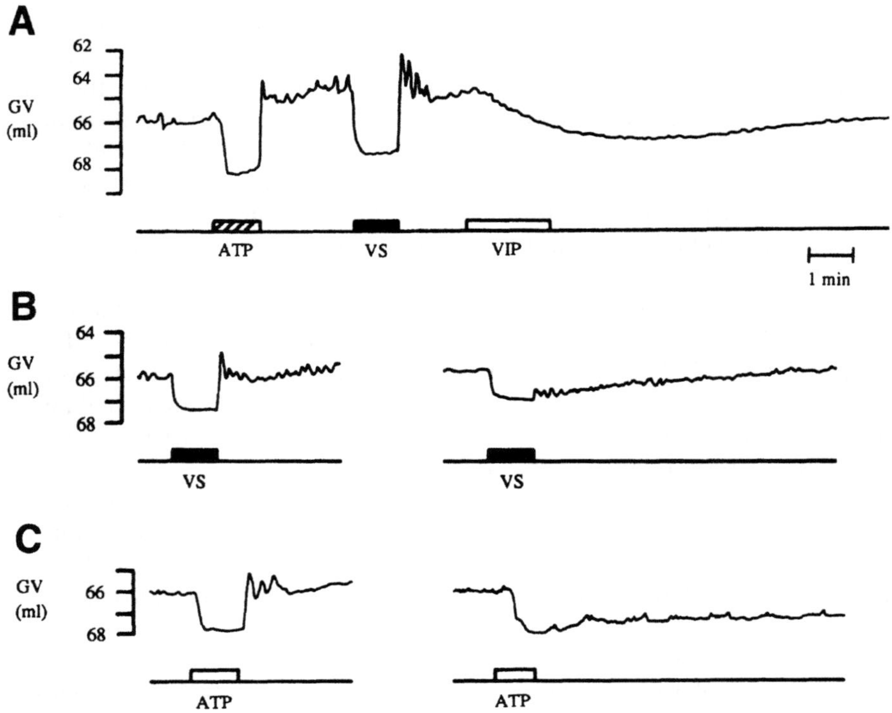

Fig. 5A–C. Recordings of gastric volume (GV) from anaesthetised rabbits. **A** Gastric inhibitory responses elicited by ATP (15 μmol/kg/min for 1 min), vagal stimulation (VS, 15 V, 20 Hz, 3 ms for 1 min) and vasoactive intestinal polypeptide (VIP, 3 μg/kg/min for 2 min). **B**,**C** Effects of fenoprofen (20 mg/kg) on the rebound contractions caused by VS and ATP. After 35 min (**B**) and 55 min (**C**) from the administration the drug, the rebound contractions were completely abolished. All agents were administered by close arterial injection. Recordings are from three different experiments (from Baccari et al. 1990b)

Otsuguro et al. 1996, 1998). For relaxation, most reports suggest a P2Y receptor is involved, although there are indications that it may be of a yet undiscovered subtype. For example, one component of relaxation of rat gastric circular muscle is sensitive to ATP and adenosine 5′-*O*-2-thiodiphosphate (ADPβS) and resistant to block by pyridoxalphosphate-6-azophenyl-2′,4′-disulphonic acid (PPADS) and 4,4′-diisothiocyanatostilbene-2,2′-disulphonate (DIDS), while αβ-meATP causes relaxation by distinct receptors sensitive to PPADS and DIDS (Otsuguro et al. 1998). For contraction, it seems more likely that a P2X receptor is involved (Murthy and Makhlouf 1998); immunostaining for P2X receptors is positive in the muscularis externa and muscularis mucosa in some regions of the gastrointestinal tract in some species (Burnstock, unpublished data). A recent study of the muscularis mucosae of the rabbit stomach showed that it contracted in response to ATP and ADP (Fig. 6) (Percy et al. 1999).

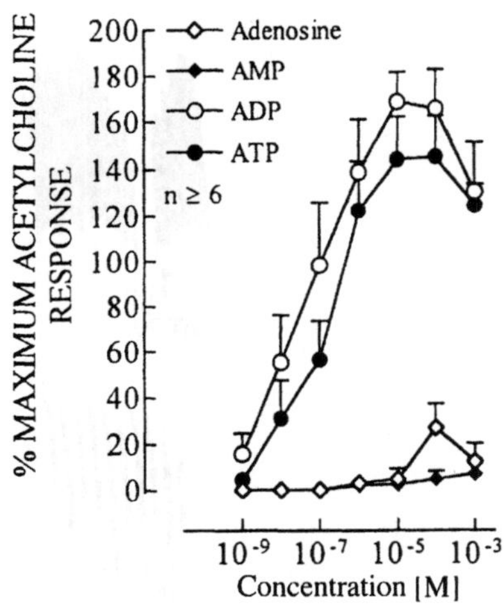

Fig. 6. Contractile effects of adenosine, AMP, ADP and ATP on rabbit isolated gastric muscularis mucosae strips from the fundic region of gastric corpus. Both ADP and ATP elicited large contractions, whereas AMP and adenosine were essentially without effect. Data are expressed as percentage of maximum contraction to ACh of each tissue and are means ± SEM of number of observations (n) indicated (from PERCY et al. 1999)

III. Small Intestine

There is persuasive evidence for purinergic *inhibitory* neuromuscular transmission in the duodenum and jejunum (OHKAWA 1974; BARTLETT et al. 1979; YAGASAKI et al. 1983; MAGGI et al. 1984, 1986; MANZINI et al. 1985, 1986; IRIEET et al. 1994; WINDSCHEIF et al. 1995; XUE et al. 1999) (Fig. 7). A recent study of the human jejunal circular smooth muscle showed that the fast IJP was reduced by suramin and apamin, while desensitisation with the P2Y receptor agonist ADPβS completely abolished the IJP; ATP-evoked hyperpolarisations modified the IJP and were reduced or blocked by the agents effective against the IJP; the authors therefore concluded that the IJPs in human jejunum were mediated in part through an ADPβS-sensitive P2 receptor (XUE et al. 1999). However, another study of human jejunal circular muscle did not find evidence to support ATP as an inhibitory transmitter in this preparation (MURR et al. 1999). There is also recent support for purinergic NANC transmission in the ileum of pig (FERNÁNDEZ et al. 1998) (Fig. 8) and humans (ZAGORODNYUK and SHUBA 1986) where ATP as well as $\alpha\beta$-meATP cause hyperpolarisations mimicking responses to NANC nerve stimulation; suramin and apamin, as well as desensitisation of P2 receptors with $\alpha\beta$-meATP, inhibit IJPs and the relaxations produced by ATP. ATP has also been claimed to be an inhibitory transmitter together with VIP in the circular muscle of the guinea-pig ileum (CRIST et al. 1992). SMITS and LEFEBVRE (1996) presented evidence for ATP and NO as NANC cotransmitters in rat ileum.

Earlier papers, using other species, were mostly negative about purinergic signalling in the small intestine (e.g. SMALL and WESTON 1979; BAUER and KURIYAMA 1982; YAGASAKI et al. 1983; OHKAWA 1984; FOX et al. 1986; MATUSÁK and BAUER 1986; GUIMARAES et al. 1988). Two papers claim that, while P2 receptors are present in muscle of the rat duodenum mediating relaxation followed by rebound contraction, comparable to the responses produced

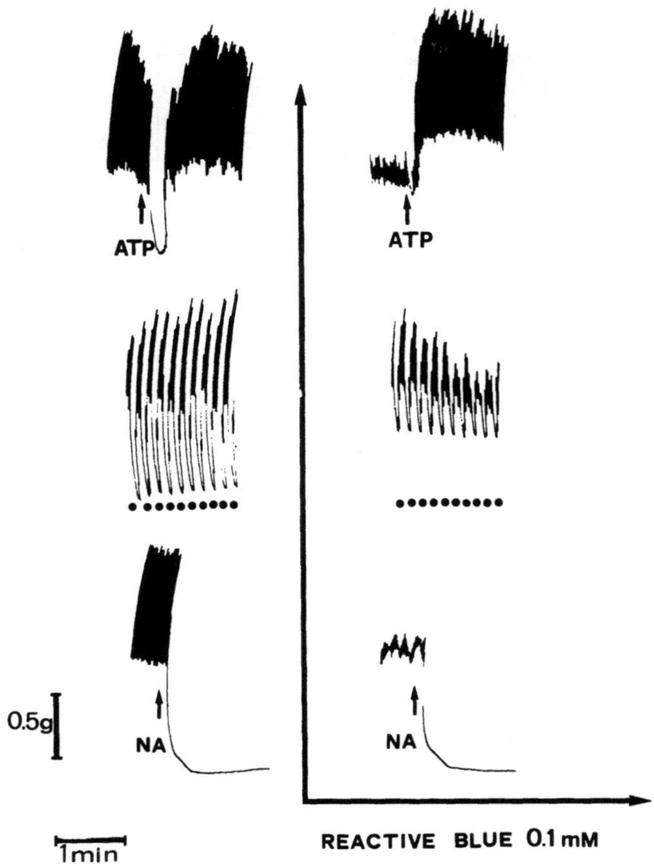

Fig. 7. Typical tracings showing effect of Reactive blue 2 (0.1 mmol/l) on relaxation of rat duodenum induced by ATP (1 mmol/l), field stimulation (0.1 Hz), and noradrenaline (1 μmol/l; NA). *Black dots*, application of single electrical pulse (60 V, 2 ms), which stimulates nonadrenergic, noncholinergic inhibitory nerves (from MANZINI et al. 1985)

by NANC nerve stimulation, ATP was unlikely to be involved in the nerve-mediated responses (SERIO et al. 1990; POSTERINO et al. 1990); however, their evidence was not strong, being based on the use of non-selective antagonists.

Several groups have reported that ATP or $\alpha\beta$-meATP *contract* guinea-pig ileum (KAMIKAWA et al. 1977; KAZIC and MILOSAVLJEVIC 1977; MOODY and BURNSTOCK 1982; WIKLUND and GUSTAFSSON 1988a,b; SATO et al. 1999). KAMIKAWA et al. (1977), following the earlier proposal of BURNSTOCK et al. (1975) in taenia coli, suggested that this was due to prostaglandins produced by ATP. ATP and $\alpha\beta$-meATP produced contractions that were antagonised by atropine, suggesting that P2 receptors mediated release of ACh from cholin-

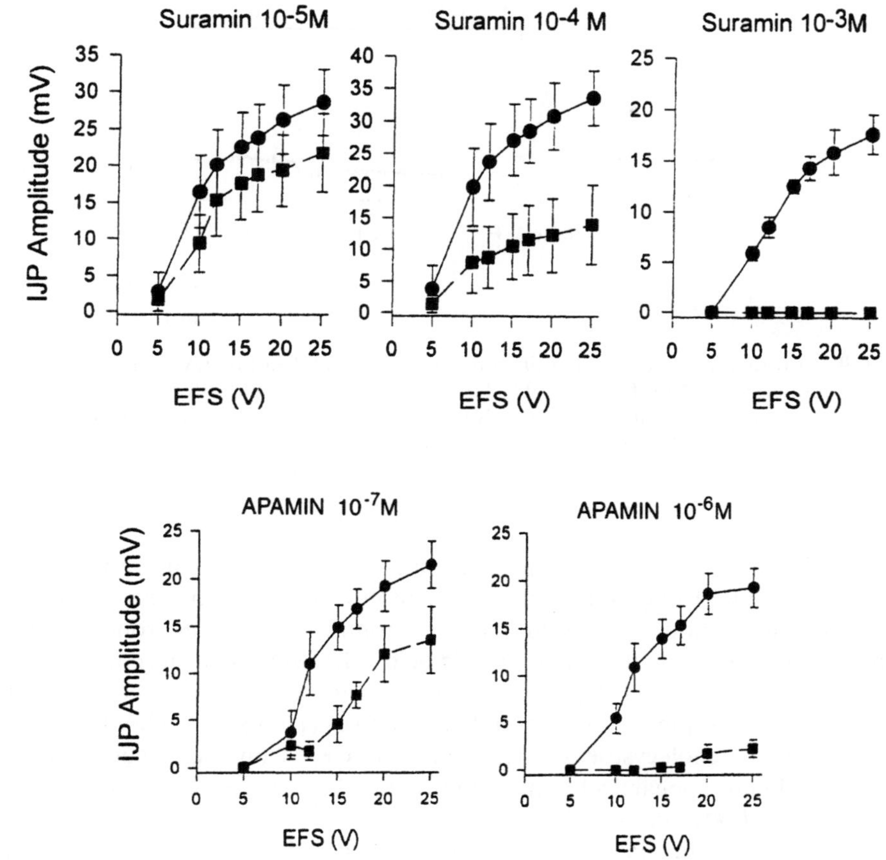

Fig. 8. Recordings showing the IJPs elicited in porcine ileum by EFS (5, 10, 12, 17, 20 and 25 V) under control conditions (●) and in the presence of increasing concentrations of suramin (10^{-5} to 10^{-3} mol/l, top panel, ■) or apamin (10^{-7} to 10^{-6} mol/l, lower panel, ■) (from FERNÁNDEZ et al. 1998)

ergic enteric nerves (NORTHWAY and BURKS 1980; MOODY and BURNSTOCK 1982) (Fig. 9). This was confirmed in later studies (SPERLÁGH and VIZI 1990, 1991; KENNEDY and HUMPHREY 1994). It has also been suggested that ATP causes a fast contraction of rat ileum by a stimulation of neuronal elements in the myenteric plexus involving cholinergic interneurones (SAKAI et al. 1979). In a more recent study (KENNEDY and HUMPHREY 1994), evidence for two types of P2 receptor in guinea-pig ileum was presented, one where $\alpha\beta$-meATP and 2-methylthioATP (2-meSATP) were equipotent in producing direct contraction of smooth muscle, while UTP was inactive (perhaps the receptor mediating NANC inhibitory transmission), the other where $\alpha\beta$-meATP, but not 2-meSATP, produced contractions by activating cholinergic nerves as suggested by MOODY and BURNSTOCK (1982).

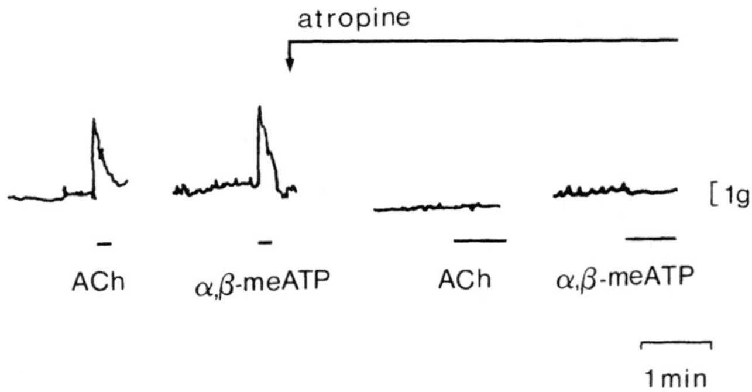

Fig. 9. Contractile responses of guinea-pig ileum to acetylcholine (ACh, 2×10^{-9} mol/l) and α,β-methyleneATP (α,β-meATP, 4×10^{-6} mol/l) before and after equilibration of the tissues with 3×10^{-7} mol/l atropine for 15 min. Atropine abolished the responses to both compounds (from MOODY and BURNSTOCK 1982)

Evidence has been presented to suggest that calcitonin gene-related peptide (CGRP) released from sensory-motor nerves in the rat duodenum releases endogenous ATP by stimulating intramural NANC neurones (MAGGI et al. 1986). Actions of γ-aminobutyric acid (GABA) on cat terminal ileum, mediated through $GABA_A$ receptors, were suggested to involve concomitant activation of postjunctional P1 and P2Y receptors (PENCHEVA 1997). The relaxing effects of cholecystokinin-octapeptide (CCK-8) on the circular muscle of avian ileum appear to be mediated by via ATP released from purinergic neurones (MARTIN et al. 1998).

Analysis of the P2 receptor subtypes involved in motility in the small intestine is still under debated (MAHMOD and HUDDART 1993; KENNEDY and HUMPHREY 1994; BLOTTIÉRE et al. 1996; JOHNSON et al. 1996; PACAUDET al. 1996; NICHOLLS and HOURANI 1997; MATSUO et al. 1997). The general consensus is that:

1. The P2 receptors in intestinal smooth muscle mediating NANC inhibitory transmission are probably P2Y receptors, perhaps a mixture of $P2Y_1$ and another P2Y receptor of as yet unknown subtype, where $\alpha\beta$-meATP has a potent relaxant action in some preparations (JOHNSON and HOURANI 1994; JOHNSON et al. 1996; PACAUD et al. 1996). PPADS is not a competitive antagonist of $P2Y_1$ receptors for the responses of rat ileal myocytes to ADP (VIGNE et al. 1998). Human small intestinal muscle contains appreciable levels of mRNA for the $P2Y_6$ receptor (COMMUNI et al. 1996).
2. The P2 receptors mediating smooth muscle contractions, at least in guinea-pig and in the small intestine of most lower vertebrates (BURNSTOCK 1969; SNEDDON et al. 1973) (perhaps in response to ATP released by NANC excitatory nerves) are probably also P2Y receptors, perhaps $P2Y_2$ or $P2Y_4$, since they are activated by UTP as well as by ATP (KENNEDY and HUMPHREY

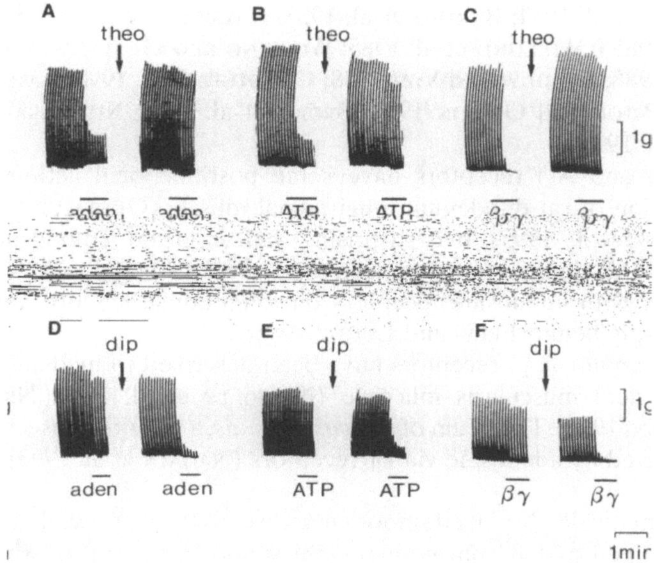

Fig. 10A–F. Cholinergic twitch responses of guinea-pig ileum to transmural electrical stimulation (0.5 Hz, 25–30 V, 0.5 ms). **A–C** Inhibitory responses to 2×10^{-6} mol/l adenosine (aden), 2×10^{-6} mol/l ATP and 2×10^{-6} mol/l β,γ-methylene ATP (β,γ) before and after equilibration of the tissues with 5×10^{-5} mol/l theophylline (theo) for 30 min. Theophylline reduced the responses to each inhibitory agonist. **D–F** Inhibitory responses to 4×10^{-6} mol/l adenosine, 4×10^{-6} mol/l ATP and 4×10^{-7} mol/l β,γ-methylene ATP before and after equilibration of the tissues with 5×10^{-8} mol/l dipyridamole (dip) for 30 min. Dipyridamole potentiated the responses to each inhibitory agonist (from Moody and Burnstock 1982)

1994; Johnson and Hourani 1994; Johnson et al. 1996; Windscheif et al. 1995).

3. Contraction of rat duodenal muscularis mucosae is mediated by P2X receptors (Johnson et al. 1996), which have been identified immunohistochemically (G. Burnstock, unpublished data).
4. The P2 receptors responsive to $\alpha\beta$-meATP on cholinergic nerve terminals, mediating release of ACh are probably P2X receptors (Kennedy and Humphrey 1994; Barthó et al. 1997; Ziganshin et al. 1995).

For receptors to adenosine the consensus is:

1. P1 (A_1) purinergic receptors on cholinergic and other enteric nerve types, including NANC inhibitory nerves, mediate prejunctional inhibition of release of ACh and other transmitters including tachykinins (Fig. 10) (Sawynok and Jhamandas 1976; Vizi and Knoll 1976; Gustafsson et al. 1978; Hayashi et al. 1978, 1985; Moritoki et al. 1978; Okwuasaba et al. 1978; Dowdle and Maske 1980; Gustafsson et al. 1981; Vizi et al. 1981; Hayashi et al. 1982; Moody and Burnstock 1982; Paul et al. 1982; Gustafsson 1984;

MOODY et al. 1984; BARTHÓ et al. 1985; KATSURAGI et al. 1985; SHINOZUKA et al. 1985a; WIKLUND et al. 1985; WIKLUND and GUSTAFSSON 1987; GAION et al. 1988; SOMOGYI and VIZI 1988; CHRISTOFI et al. 1990; KATSURAGI et al. 1990; PATON and OLSSON 1991; BROAD et al. 1992; NITAHARA et al. 1995; COUPAR 1999).
2. P1 (A_1 and A_2) receptors have some postjunctional actions mediating relaxation of rat duodenum longitudinal muscle (GUSTAFSSON et al. 1985; MURTHY et al. 1995). It is surprising that arylazido-aminopropionyl ATP ($ANAPP_3$), claimed to be a selective P2 receptor antagonist, was shown to interact with adenosine receptors on longitudinal smooth muscle of the guinea-pig ileum (FREW and LUNDY 1986).
3. P1 (A_1 and/or A_{2B}) receptors have been described to mediate contraction of rat ileal muscularis mucosae (NICHOLLS et al. 1996; NICHOLLS and HOURANI 1997). The ileum of *Suncus murinus*, a primitive insectivore, is also contracted by adenosine via P1 receptors (NAGATA et al. 1993).

Enzymatically dispersed smooth muscle cells from mouse ileum have been patch-clamped and P2 purinergic agonists shown to activate small conductance Ca^{2+}-dependent K^+-channels, responses decreased by apamin (VOGALIS and GOYAL 1997). Isolated myocytes from the longitudinal muscle of jejunum and ileum showed a slow transient increase in intracellular Ca^{2+} concentration ($[Ca^{2+}]_i$) in response to ATP and 2-meSATP, taken to indicate a P2Y receptor (BLOTTIÈRE et al. 1996).

Incubation of the guinea-pig ileum with adenosine or 2-CADO has been claimed to induce a novel form of drug dependence, made manifest by withdrawal of the inducing drug, but not by antagonists of opiates or clonidine (GINTZLER and MUSACCHIO 1975; COLLIER and TUCKER 1983).

Responses of the isolated rat duodenum to adenine nucleotides have been examined in neonatal rats and it was concluded that postganglionic neurones produce contractile responses upon ganglionic stimulation with nicotine during the first few weeks after birth and that thereafter NANC purinergic inhibitory neurones gradually develop and become predominant (IRIE et al. 1994).

A more recent study of the ontogeny of P2 receptors in the rat duodenum (BROWNHILL et al. 1997) has shown that in the neonatal longitudinal muscle P2Y receptor mediate relaxation and that the receptor population is fully developed by day 25, in the neonatal muscularis mucosae P2Y receptors mediate contractions before day 20, while after day 20 P2X receptors mediate this effect.

IV. Caecum and Taenia Coli

The taenia coli is a misnomer for taenia caeci, since it consists of three strips of longitudinal muscle overlying the caecum, but for historical reasons is still retained by most workers. The guinea-pig taenia coli was the original gut preparation where NANC inhibitory neural control of smooth muscle was demonstrated (BURNSTOCK et al. 1963, 1964, 1966; BENNETT et al. 1966), but it was not until later that evidence began to accumulate that ATP was a NANC inhibitory transmitter in the taenia coli (BURNSTOCK et al. 1970; SU et al. 1971; SATCHELL et al. 1972, 1973; TOMITA and WATANABE 1973; JAGER and SCHEVERS 1980), although some early papers concluded that this was unlikely (RIKIMARU et al. 1971; KUCHII et al. 1973; WESTON 1973; SPEDDING et al. 1975). BURNSTOCK et al. (1970) showed a block of responses of the guinea-pig taenia coli and ileum to exogenously applied ATP and NANC inhibitory nerve stimulation by high concentrations of quinidine. A high concentration of phentolamine was also effective in blocking IJPs and ATP actions (TOMITA and WATANABE 1973). 2,2′-Pyridylisatogen tosylate (PIT) was claimed as an antagonist of the inhibitory affect of ATP on smooth muscle of the taenia coli (HOOPER et al. 1974). Later publications from this group reported that PIT did not antagonise the relaxations produced by adenosine (SPEDDING and WEETMAN 1976; FOSTER et al. 1978) and in a study of a series of 6-substituted 2-phenylisatogens they claimed that 6-methoxy-2-phenylisatogen was the most effective ATP receptor antagonist in the guinea-pig taenia coli (FOSTER et al. 1983). The enzyme nucleotide pyrophosphatase that converts ATP to AMP reduced both the responses of taenia coli to NANC inhibitory stimulation and ATP, while responses to perivascular sympathetic nerve stimulation and noradrenaline (NA) were unaffected; this was taken as support for the purinergic neurotransmission hypothesis (SATCHELL 1981).

Responses of the guinea-pig taenia coli to ultraviolet light (340–380 nm), in the presence of sodium nitrite, consisted of fast relaxations that closely resembled the relaxations produced by NANC inhibitory nerve stimulation and ATP (BURNSTOCK and WONG 1978). Since the responses to UV light were unaffected by tetrodotoxin (TTX) and were not associated with ATP release, it was concluded that UV light was probably acting on some part of the purinergic receptor complex.

SU et al. (1971) showed that stimulation of NANC inhibitory nerves in taenia coli preincubated in [^{3}H]adenosine, which was taken up by the nerves and was converted and retained primarily as [^{3}H]ATP, led to release of tritium, while SATCHELL and colleagues showed that high concentrations of 2-substituted imidazolines reduced the inhibitory responses both to ATP and NANC inhibitory nerve stimulation, although they recognised that the blockade was unlikely to be specific. In more recent studies, suramin has been shown to block P2 receptor responses and IJPs in the guinea-pig taenia coli (DEN HERTOG et al. 1989a,b; HOYLE et al. 1990; PIPER and HOLLINGSWORTH 1995). Other papers showed that dipyridamole and hexobendine potentiated the responses to both ATP and nerve stimulation (SATCHELL et al. 1972; SATCHELL and BURNSTOCK 1975), although the result was later contested (JAGER 1976).

Release of ATP from perfused taenia coli during stimulation of NANC inhibitory nerves was demonstrated using the highly sensitive firefly luciferin-luciferase assay technique; no ATP release was detected during comparable relaxations of taenia coli produced by NA (BURNSTOCK et al. 1978). Release of ATP during stimulation of intramural nerves supplying the taenia coli has also been demonstrated more recently using high performance liquid chromatography separation and fluorometric detection (MCCONALOGUE et al. 1996). When taenia coli preparations were incubated in [^{3}H]-adenosine, both neuronal and non-neuronal tritium release was demonstrated (RUTHERFORD and BURNSTOCK 1978). 4-Aminopyridine facilitated purine release from nerve terminals in taenia coli (DEN HERTOG et al. 1985a).

The prostaglandin synthesis inhibitor, indomethacin blocked the "rebound contractions" which characteristically follow the NANC inhibitory responses in the guinea-pig taenia coli (BURNSTOCK et al. 1975). Since adenine nucleotides are known to induce prostaglandin synthesis it was suggested that purinergic signalling may form a link with prostaglandins in the physiological regulation.

SATCHELL and MAGUIRE (1975) carried out the first structure-activity studies of analogues of adenine nucleotides in taenia coli and showed that di- or triphosphate groupings were of prime importance in binding adenine nucleotides to the putative smooth muscle receptor and that hydrolysis of the terminal phosphates was not a requirement for inhibitory activity. Later studies extended these findings (MAGUIRE and SATCHELL 1979; SATCHELL and MAGUIRE 1982) and the actions of enantiomers of 2-azido analogues on taenia were also examined (CUSACK and PLANKER 1979).

In early studies, Vladimirova and Shuba (VLADIMIROVA and SHUBA 1978; SHUBA and VLADIMIROVA 1980) tentatively concluded that apamin was a specific postjunctional blocking agent for purinergic NANC inhibitory transmission. Later studies confirmed that nanomolar concentrations of apamin reduced both inhibition by ATP and NANC nerve stimulation in taenia coli, but also raised the probability that apamin was a selective K^+ channel blocker that is effective against ATP, but probably also against some other agents (BANKS et al. 1979; MAAS and DEN HERTOG 1979; MAAS et al. 1980; MULLER and BAER 1980; CROSSLEY and GILLESPIE 1983; DEN HERTOG et al. 1985a). Gonsopora toxin was shown to facilitate NANC inhibitory responses in taenia coli and it was concluded this was due to augmentation of release of transmitter, possibly ATP (FUJIWARA et al. 1982).

Following the hypothesis that there were separate receptors for adenosine (P1) and ATP/ADP (P2) (BURNSTOCK 1978), studies of the effects of adenosine and ATP in the guinea-pig taenia coli supported this hypothesis (Fig. 11) (BROWN and BURNSTOCK 1981; SATCHELL and MAGUIRE 1982; FERRERO and FRISCHKNECHT 1983) and demonstrated that theophylline blocked relaxations produced by adenosine, but not by ATP (BROWN and BURNSTOCK 1981). The taenia coli preparation was used to study the stereoselectivity of P2 and P1 receptors, and it was shown that there was partial stereoselectivity for P2 receptors on smooth muscle towards enantiomers of ATP, in contrast to the absolute stereospecificity of the receptor for ADP on platelets and for P1 receptors on smooth muscle and autonomic nerve terminals (BURNSTOCK et al.

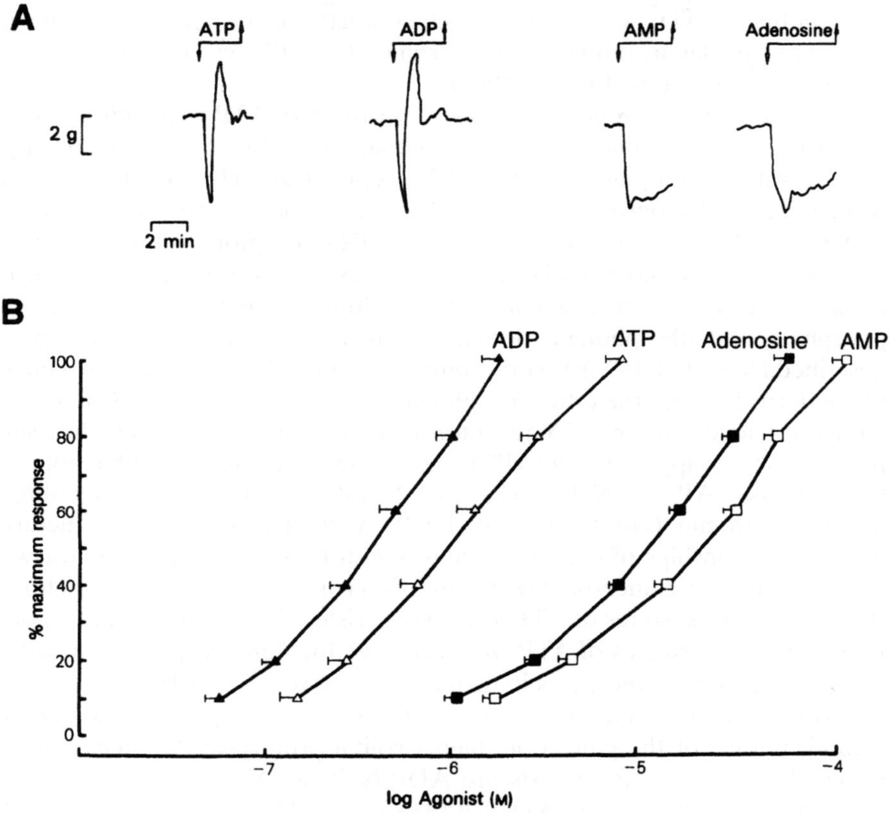

Fig. 11A,B. Relaxation of isolated, carbachol-contracted, guinea-pig taenia coli preparation by purines. **A** Traces of relaxation to ATP (3 μmol/l), ADP (1 μmol/l), AMP (100 μmol/l) and adenosine (100 μmol/l); *horizontal brackets* indicate the period of exposure to the agonist. **B** Log dose-response curves for the relaxation induced by ADP (▲), ATP (△), adenosine (■) and AMP (□). Each point is the mean of at least ten values obtained from at least five experiments; *horizontal lines* show standard error of the mean (from Brown and Burnstock 1981)

1983). In a later study using phosphorothioate analogues of ATP, ADP and AMP it was shown that while P2 receptors mediating inhibitory responses in taenia coli showed marked stereoselectivity, those mediating excitatory responses in guinea-pig bladder showed little stereoselectivity (Burnstock et al. 1984).

The P1 receptor subtype in the guinea-pig taenia coli was identified as A_2 (Burnstock et al. 1984). In a more recent study, adenosine analogues have been shown to relax guinea-pig taenia coli via P1 (A_{2B}) receptors and at a xanthine-resistant site that is unlikely to be an A_3 receptors (Prentice and Hourani 1997). The actions of adenine dinucleotides were examined in taenia coli (Stone 1981; Burnstock and Hoyle 1985). β-Nicotinamide adenine dinucleotide (NAD) was shown to act indirectly on P1 receptors, while β-nicotinamide adenine dinucleotide phosphate (NADP) acts like a P2 receptor agonist (BURNSTOCK and HOYLE 1985).

Structure activity relationships for derivatives of ATP as *agonists* at P2Y receptors were examined in various preparations. In the taenia coli, $\alpha\beta$-meATP was a potent agonist on the P2 receptor (DEN HERTOG et al. 1985b), while the potent agonist N^6-methylATP and the somewhat less potent agonist 2′-deoxyATP were shown to be selective for P2Y receptors, but were inactive at P2X receptors and vascular P2Y receptors (BURNSTOCK et al. 1994). In a later study, structure-activity relationships of pyridoxal-6-arylazo-5′-phosphate and phosphonate derivatives as P2 receptor antagonists were described (KIM et al. 1998); several compounds, including the phenylazo phosphate derivative and the ethyl phosphonate analogue of isoPPADS, had good antagonist actions on the P2Y receptor in guinea-pig taenia coli. Diadenosine polyphosphates appear to act as P2Y agonists in the taenia coli with a potency order $AP_3A = AP_4A > ATP > AP_4 = AP_5A$ and these relaxations are antagonised by suramin (HOURANI et al. 1998). A comparison of the structure-activity relationships of ectonucleotidases with those of the P2 receptor was examined on the guinea-pig taenia coli (WELFORD et al. 1986); it was shown that methylene isosteres of ATP and ADP resisted dephosphylation. Isopolar phosphonate analogues of ATP were selective for P2X receptors in bladder, but inactive on P2Y receptors in taenia coli (CUSACK et al. 1987)

Two relaxation-mediating P2 receptors have been proposed to exist in the smooth muscle of the guinea-pig taenia coli: a prototypic P2Y receptor for which the most potent agonists are ADPβS, 2-meSATP and a separate P2 receptor activated by $\alpha\beta$-meATP (DUDECK et al. 1995; WINDSCHEIF et al. 1995; BÜLTMANN et al. 1996).

Substances other than ATP have been claimed as NANC transmitters in the taenia coli. For example, following the immunohistochemical demonstration of peptides in enteric neurones, the responses of the guinea-pig taenia coli to substance P(SP), NK, SMS, enkephalin (ENK) and VIP were examined; however, unlike ATP, these peptides did not mimic the rapid responses to NANC inhibitory nerve stimulation (COCKS and BURNSTOCK 1979; MACKENZIE and BURNSTOCK 1980). Bradykinin was also shown to be less likely than ATP as a transmitter in NANC inhibitory nerves in taenia coli (FERRARO et al. 1980).

There is immunohistochemical evidence that VIP-containing fibres project from myenteric neurones to the taenia coli (FURNESS et al. 1981) and VIP antiserum was claimed to inhibit selectively both VIP- and neurally-induced relaxations (GRIDER et al. 1985). In a later paper, this latter group claimed that VIP and PACAP together account for neurally mediated relaxation of taenia coli (HILLS et al. 1983; JIN et al. 1994). PACAP has also been claimed to participate in NANC inhibitory transmission in taenia coli (MCCONALOGUE et al. 1995a,b). Another group showed that nicotine evoked a calcium-dependent release of VIP in the taenia coli that was blocked by TTX

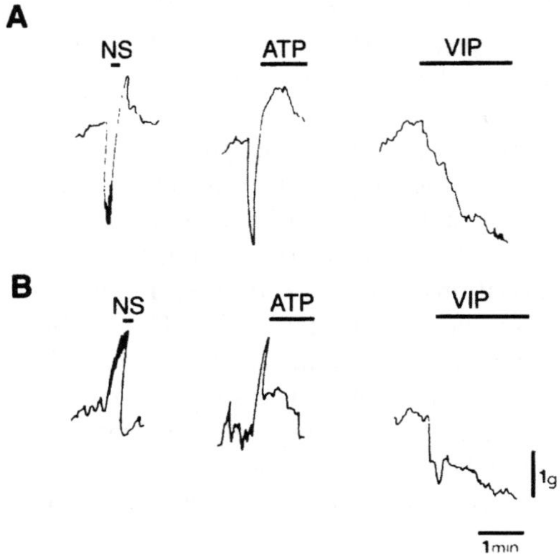

Fig. 12. **A** The responses of the isolated taenia coli to intramural nerve stimulation (NS, pulse width of 0.3 ms, supramaximal voltage and frequency of 0.4 Hz for 10 s), ATP (0.7 μmol/l) and VIP (0.6 μmol/l). **B** After addition of apamin (1 μmol/l) to the organ bath, contractile responses to NS and ATP were obtained, although VIP produced a slightly smaller inhibitory response. The contractile response obtained with ATP was, in other experiments, generally of longer latency. The tone of the preparation was raised by carbachol. Guanethidine (3.4 μmol/l) was present throughout (from MacKenzie and Burnstock 1980)

(Iselin et al. 1988). However, the time course of relaxations produced by VIP is very much slower than that produced by NANC nerve stimulation and ATP (Fig. 12) (MacKenzie and Burnstock 1980; Hills et al. 1983). Further, in a developmental study, NANC inhibitory neuromuscular transmissions were shown to mature earlier than cholinergic neuromuscular transmission in the guinea-pig taenia coli and the responses to VIP and ATP during the same developmental stages did not support VIP as a transmitter responsible for IJPs (Zagorodnyuk et al. 1993).

On the basis that both the nitric oxide synthase (NOS) inhibitor N^G-nitro-L-aginine (L-NNA *or* L-NOARG) and PPADS produce partial inhibition of NANC relaxing responses, NO and ATP have been proposed to co-mediate the NANC response in the guinea-pig taenia coli (Barthó et al. 1998). Other groups support the view that NO is partially or wholly responsible for NANC inhibitory transmission in both guinea-pig (Shuttleworth et al. 1991; Knudsen and Tøttrup 1992; Alomar et al. 1999; Shuttleworth et al. 1999) and human (Tam and Hillier 1992). Two more recent studies, however, argue against NO as an inhibitory transmitter in the taenia coli; one shows that, while NO is released during NANC stimulation, postjunctional effects are not apparent when atropine is present (Ward et al. 1996; Selemidis et al. 1997). In addition, Selemidis et al. (1997), contrary to a previous study, failed to demonstrate any block of NANC relaxations by L-NOARG in the absence of atropine. The existence of two distinct IJPs were described in the guinea-pig taenia coli, but experiments ruled out the involvement of nitrergic transmission (BRIDGEWATER et al. 1995)

The circular smooth muscle of the guinea-pig caecum underlying the taenia coli has also been the subject of study. ATP and adenosine relaxed this preparation and NANC nerve stimulation produced IJPs as for the taenia coli (HOYLE et al. 1988a). In the presence of apamin, excitatory junction potentials (EJPs) and depolarisations were evoked by ATP, which were not blocked by indomethacin (SHUBA and VLADIMIROVA 1980). The inhibitory responses of the caecum of the rat (which has no taenia coli) have been described (MEHTA and KULKARNI 1983) and Reactive blue 2 was shown to reduce the IJPs and hyperpolarisations to $\alpha\beta$-meATP in this preparation (MANZINI et al. 1986), although this was contested in a later paper (SERIO et al. 1996). Inhibitory transmission to the longitudinal muscle of the mouse caecum was claimed to be mediated largely by NO (YOUNG et al. 1996).

V. Colon

A number of papers claim ATP and NO as cotransmitters in NANC inhibitory nerves in the colon (KEEF et al. 1993; BOECKXSTAENS et al. 1993; ZAGORODNYUK and MAGGI 1994; BÖRJESSON et al. 1997; PLUJA et al. 1999). Others claim VIP, NO or VIP together with NO, to be colonic NANC inhibitor transmitters (SUTHANATPONG et al. 1993; GRIDER 1993; KEEF et al. 1993, 1994; KERANEN et al. 1995; QIAN and JONES 1995; SERIO et al. 1995; KISHI et al. 1996). In a thorough study of NANC inhibitory transmission in the circular muscle of guinea-pig colon, however, the authors identified three distinct mechanisms: a fast relaxation in response to low frequency stimulation, probably involving ATP which mobilises intracellular Ca^{2+} leading to the activation of apamin-sensitive K^+ channels; a second fast relaxation at higher frequencies of stimulation involving production and release of NO; a slowly developing relaxation at higher frequencies of stimulation that is apamin- and l-NOARG-resistant probably utilising VIP or a VIP-related peptide (MAGGI and GIULIANI 1996). Other evidence for ATP and NO as cotransmitters includes: the NANC inhibitory nerve responses to ATP are significantly antagonised by apamin or Reactive blue 2 (a P2 receptor blocker) and the NOS inhibitor, L-NOARG (BÖRJESSON et al. 1997); L-NOARG abolished the NANC EJPs to unmask an apamin-sensitive IJP which was inhibited by suramin, while the apamin-resistant response was reduced by L-NOARG (Fig. 13) (ZAGORODNYUK and MAGGI 1994). In further studies of the circular muscle of guinea-pig colon, evidence was presented that ATP, but not PACAP, mediates the apamin-sensitive NANC IJPs (ZAGORODNYUK et al. 1996). Most workers find that apamin (a K^+ channel blocker) selectively blocks the purinergic component in colon (COSTA et al. 1986b; KEEF et al. 1993), but does not block the nitrergic component (MAGGI and GIULIANI 1993). Suramin and apamin reduced the amplitude of

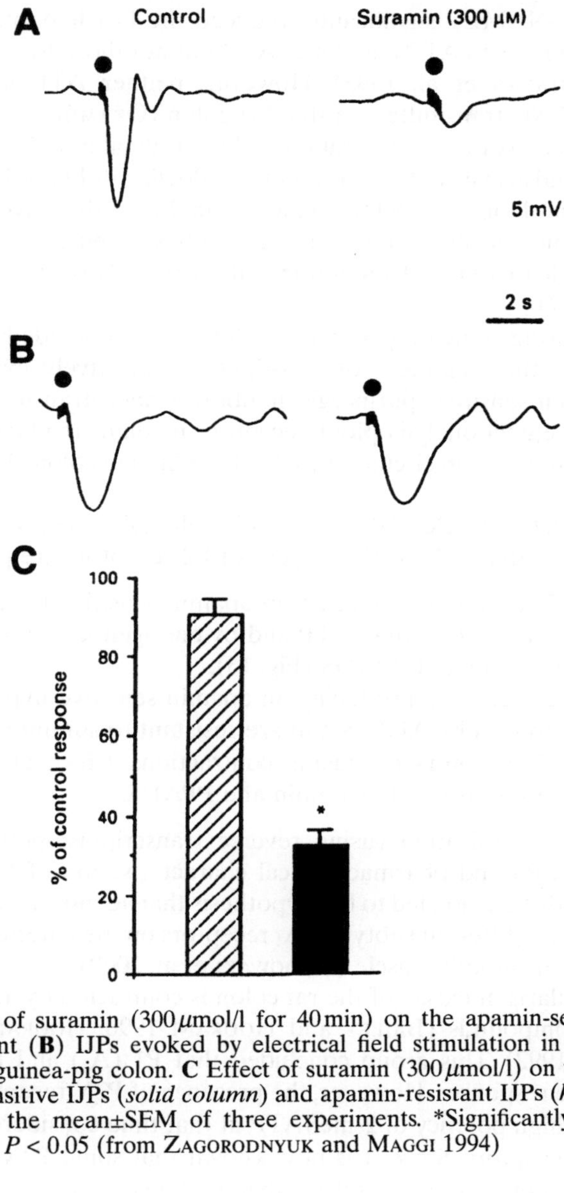

Fig. 13. Effect of suramin (300 μmol/l for 40 min) on the apamin-sensitive (**A**) and apamin-resistant (**B**) IJPs evoked by electrical field stimulation in circular smooth muscle of the guinea-pig colon. **C** Effect of suramin (300 μmol/l) on the amplitude of the apamin-sensitive IJPs (*solid column*) and apamin-resistant IJPs (*hatched column*). Each value is the mean±SEM of three experiments. *Significantly different from control values, $P < 0.05$ (from ZAGORODNYUK and MAGGI 1994)

IJPs recorded in the rat colonic circular muscle without affecting their duration, while L-NOARG reduced the duration, but not the amplitude of the IJP (PLUJA et al. 1999). Apamin-sensitive and -insensitive components of the IJP have been reported in the circular muscle layer of the mouse colon (SPENCER et al. 1998). Both NO and ATP have been implicated in NANC inhibitory regulation of human colonic circular muscle; apamin, but not N^G-nitro-L-aginine methyl ester (L-NAME), significantly reduced the fast hyperpolarisation, as well as the responses to ATP and 2-meSATP, but not those to NO (KEEF et al. 1993; BOECKXSTAENS et al. 1993). However, neither ATP nor VIP were regarded as NANC transmitters in the dog colon (OKAMURA et al. 1998).

ATP elicits relaxation of the rat colon longitudinal muscle probably via a P2Y receptor and also by a P1 (A_2) receptor following its breakdown to adenosine (BAILEY and HOURANI 1992). Single channel recording from cell patches of mouse colonic and ileal smooth muscle cells showed that ATP mediated P2Y-induced release of Ca^{2+} from intracellular stores (KOH et al. 1997; VOGALIS and GOYAL 1997).

In a study using guinea-pig proximal colon, it was concluded that cisapride, a drug used in the treatment of constipation, selectively enhances 5-HT-induced suramin-sensitive purinergic inhibitory neurotransmitters (BRIEJER et al. 1995). Prejunctional δ-opioid receptors mediate inhibition of NANC inhibitory transmission in circular muscle of the human colon (Fig. 14) (HOYLE et al. 1990).

For the circular muscle of the guinea-pig colon, ZAGORODNYUK and MAGGI (1998) concluded that at least three types of P2 receptors are present:

1. Inhibitory P2 receptors, producing an apamin-sensitive hyperpolarisation, which are activated by $\alpha\beta$-meATP and by endogenously released purines, sensitive to suramin and PPADS (Fig. 15).
2. Inhibitory P2 receptors, producing an apamin-sensitive hyperpolarisation, which are activated by ADPβS and are resistant to suramin and PPADS.
3. Excitatory P2 receptors producing contractions, which are activated by ADPβS and are sensitive to suramin and PPADS.

Molecular identification (using reverse transcriptase-polymerase chain reaction, RT-PCR) and pharmacological characterisation of P1 receptors in the guinea-pig distal colon led to the hypothesis that adenosine mediates relaxation through two different subtypes: A_1 receptors on the enteric neurones and A_{2B} receptors on smooth muscle (KADOWAKI et al. 2000).

The muscularis mucosae of the rat colon is contracted by adenosine, ATP and related compounds (BAILEY and HOURANI 1990; HOURANI et al. 1991; BAILEY et al. 1992). This group concluded that P1 (A_1) and P2Y receptors mediated these responses. However, the presence of P2Y receptors was based largely on the high potency of 2-meSATP, at that time considered to be selective for P2Y receptors. Since it is now known that some P2X receptor subtypes are also potently activated by this analogue, it is possible that P2X receptors are present in this gut muscle, which is supported by immunohistochemical expression of $P2X_1$ receptors in the smooth muscle of the muscularis mucosae, but not the muscularis externa (G. BURNSTOCK, unpublished data). A later study from another laboratory confirmed the presence of P1 (A_1) receptors in rat colon muscularis mucosae, mediating contraction (REEVES et al. 1993), although part of the response was due to products of the cyclo-oxygenase pathway (REEVES et al. 1995).

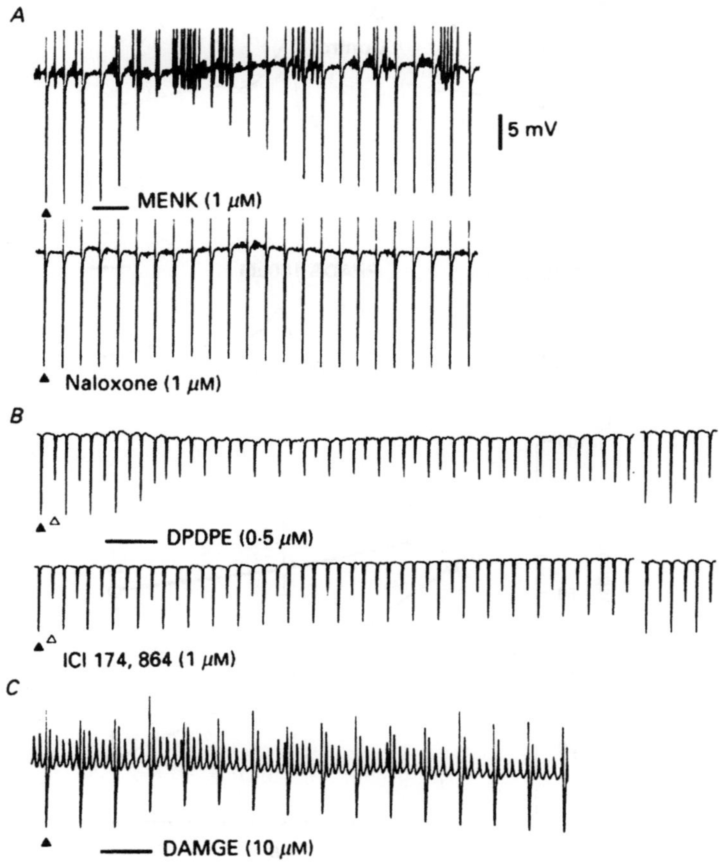

Fig. 14A–C. Antagonism of the enkephalin-induced inhibition of non-adrenergic, non-cholinergic inhibitory junction potentials (IJPs) in isolated preparations of the circular muscle layer of the human distal colon by naloxone, and actions of analogues of enkephalin. **A** Methionine-enkephalin (MENK, 1 μmol/l) caused a large inhibition of the IJP and this action was blocked following incubation with naloxone (1 μmol/l). The application of MENK is indicated by the *bar*, which applies to both traces. **B** The δ-selective agonist [D-Pen2,D-Pen^5]-enkephalin (DPDPE, 0.5 μmol/l) caused an inhibition of the IJP, with little effect on the smooth muscle membrane, and this action was blocked by the δ-selective antagonist ICI 174864 (1 μmol/l). The application of DPDPE is indicated by the *bar* which applies to both traces. **C** The μ-selective agonist ([D-Ala^2, Me Phe, Gly-ol^5]enkephalin) (DAMGE, 10 μmol/l) caused approximately a 50% reduction in IJP amplitude. The application of DAMGE is indicated by the *bar*. In each preparation IJPs were evoked every 30 s (▲) and in **B** membrane conductance was monitored by passing a hyperpolarising current also every 30 s (△). Also in **B**, the break in the traces is for a 10 min period (from HOYLE et al. 1990)

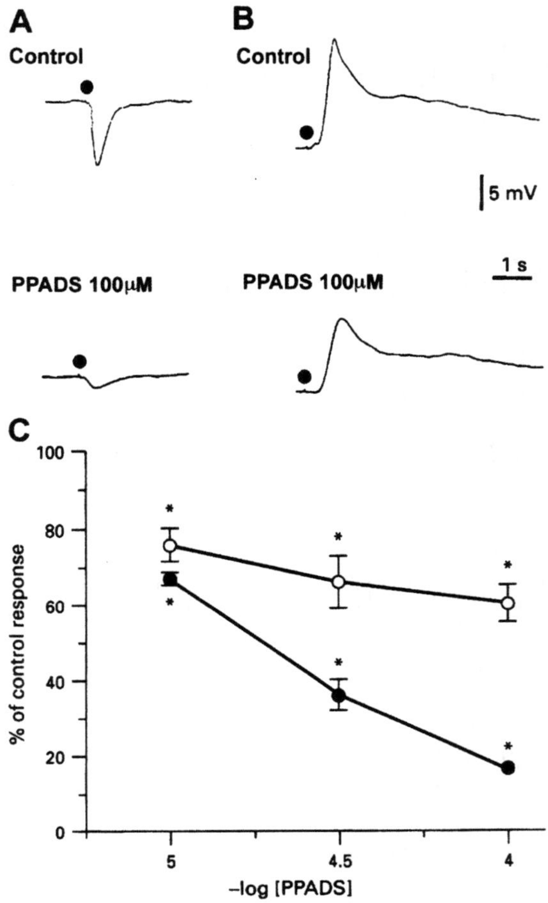

Fig. 15A–C. Effect of pyridoxalphosphate-6-azophenyl-2′,4-disulphonic acid (PPADS) on the NANC inhibitory junction potential (IJP) and excitatory junction potential (EJP) evoked by electrical field stimulation (EFS) in the circular muscle of the guinea-pig colon. **A** EFS (applied at *dots*) evoked a monophasic NANC IJP which was markedly inhibited by PPADS (100 μmol/l for 20 min). **B** In the presence of apamin (0.1 μmol/l) and tetraethylammonium (10 mmol/l), EFS evoked a NANC EJP which was partly inhibited by PPADS (100 μmol/l for 20 min). **C** Concentration-dependent inhibition by PPADS of EFS-evoked NANC IJP (●) and EJP (○), in the presence of apamin and tetraethylammonium; $n = 3 - 12$ experiments. $*P < 0.05$ (from Zagorodnyuk et al. 1996)

Stimulation of lumbar sympathetic nerves evokes contractions of cat colon circular muscle mediated by ATP and NA (Venkova and Krier 1993). Contractions of the circular muscle of the cat colon to ATP and analogues mimic these responses and were claimed to be mediated by P2X receptors, while P2Y and P1 receptors appear to mediate relaxations in this preparation (Venkova et al. 1994). Stimulation of parasympathetic (pelvic) nerves in the presence of atropine produced NANC contractions of the colon and rectum of anaesthetised cats, which were blocked by desensitisation of P2 receptors with repeated injections of $\alpha\beta$-meATP (HEDLUND et al. 1986).

ATP has been shown to inhibit swelling-activated Cl^- currents in canine colonic smooth muscle (DICK et al. 1998). The authors speculate that this may be related to the regulation of myogenic activation in response to distension.

Postnatal development of receptors to purines and pyrimidines in rat colon has been reported (HOURANI 1999; PEACHEY et al. 1999; see BURNSTOCK, Chap. 5, first volume). The contractile responses of the muscularis mucosae to adenosine (via A_1 receptors) and to ATP and UTP (perhaps via $P2Y_1$ and $P2Y_2$ receptors) were seen one day after birth. However, the relaxant responses of the longitudinal muscle coat to P1 and P2 agonists were not established until about two weeks after birth.

VI. Sphincters

NO is clearly established as an important transmitter in all the sphincters in the gastrointestinal tract (WANG et al. 1996). It appears to be dominant in the lower oesophageal and ileo-colonic sphincters, but there is good evidence for the involvement of ATP in the pyloric and internal anal sphincters (SOEDIONO and BURNSTOCK 1994; RAE and MUIR 1996). The role of VIP is more controversial; while VIP immunofluorescent nerve fibres are present in most sphincters, VIP does not always produce relaxations (ALUMETS et al. 1979; MCGREGOR et al. 1984; SOEDIONO and BURNSTOCK 1994). In a recent study of the opossum oesophogastric, pyloric and ileocolonic sphincters, ATP produced relaxations, but these were blocked by TTX, suggesting a purinergic action on myenteric nerves supplying these sphincters (MATSUDA et al. 1997b).

1. Lower Oesophageal Sphincter (LES)

As discussed earlier (Sect. B.I.), most evidence points to a dominant role for NO in NANC transmission to sphincter (KNUDSEN et al. 1992; MATSUDA et al. 1997a). Various neuropeptides have also been implicated, including VIP and SMS in baboon LES (BYBEE et al. 1979), neuropeptide Y and VIP in cat LES (PARKMAN et al. 1989; KORTEZOVA et al. 1996). There is limited evidence for ATP involvement in the NANC inhibitory responses of the LES, although a more recent paper claimed that the IJP in the guinea-pig LES is probably produced by both ATP and NO (IMAEDA and SUZUKI 1997). Also, IJPs recorded in the LES of the guinea-pig were reduced 60% by L-NOARG and the remainder blocked by apamin (YUAN et al. 1998), which is widely regarded as an indicator of purine involvement (RAE and MUIR 1996; OHNO et al. 1996).

2. Pyloric Sphincter

Evidence has been published to support several putative NANC mediators in the pylorus, including immunohistochemical localisation of SP (LINDH et al.

1983; LIDBERG et al. 1983) (although perhaps released from sensory-motor nerves during axon reflex activity); 5-HT (LIDBERG et al. 1984), neuropeptide Y and SMS (perhaps located in sympathetic nerves supplying the pylorus) (LINDH et al. 1986; HOLLE et al. 1991); ENK for NANC vagal excitation and VIP for NANC relaxation of cat and human pylorus (AHLMAN and DAHLSTRÖM 1983; AKKERMANS et al. 1989). VIP innervation of the cat and rabbit pylorus has also been claimed (EDIN et al. 1979; DELOOF et al. 1988) and, more recently, NO (ALLESCHER et al. 1992; ALTDORFER et al. 1996); NO and VIP may be cotransmitters to dog pylorus (WARD et al. 1994). The circular muscle layer of the human pylorus is richly supplied by nerve fibres containing VIP, SP, ENK and CGRP (DOMOTO et al. 1992). Receptors for cholecystokinin were described in the circular smooth muscle layer of the rat pyloric sphincter (SMITH et al. 1984). However, although pyloric muscles of dog are richly innervated by nerves containing opioids, only high frequency nerve stimulation releases sufficient concentrations of opioids to modulate junction potentials (BAYGUINOV and SANDERS 1993).

Studies of NANC inhibitory responses of the rat pyloric sphincter (SOEDIONO and BURNSTOCK 1994; ISHIGUCHI et al. 1999) provides evidence for components mediated by both NO and ATP. L-NAME and P2X antagonists each reduced NANC inhibitory responses by about 50% and PPADS blocked relaxation responses to $\alpha\beta$-meATP. While VIP-containing nerve fibres were abundant in the pyloric sphincter, relaxations to VIP were not observed.

3. Sphincter of "Oddi" (or the Choledocho-Duodenal Junction, CDJ)

This sphincter is located at the junction of the extrahepatic biliary tree and the duodenum; it comprises both circular and longitudinal muscle layers that develop independently from the duodenum and play a major role in regulating the delivery of bile into the duodenum. As usual, most putative NANC transmitters have been implicated, including NO (PAULETZKI et al. 1993; KAUFMAN et al. 1993) and VIP (DAHLSTRAND et al. 1990). However, in a recent study of the rabbit sphincter of Oddi, both fast, apamin-sensitive and slow sustained L-NAME components were observed and a substantial part of fast IJP was suppressed by desensitisation with $\alpha\beta$-meATP (IMOTO et al. 1998). The authors concluded that NO, PACAP and ATP were all involved in NANC relaxation of this muscle. ATP and ADP were earlier shown to have inhibitory actions on the cat sphincter of Oddi (PERSSON 1976).

The identity of NANC (atropine-resistant) excitatory transmitter to the sphincter of Oddi is not clear, but SP-immunoreactive fibres are prominent (CAI et al. 1983; VONGALIS et al. 1989).

4. Ileo-Caecal Sphincter (ICS)

The functional role of the ileo-caecal sphincter is not entirely clear, although the ICS separates ileum from caecum and impedes entry of caecal fauna (bacteria) into the ileum. The pressure in the caecum can exceed that in the terminal ileum, so that the ICS prevents retrograde movement of contents. Neural control appears to be dominated by vagal excitatory control mediated largely by NA and possibly SP (PAHLIN and KEWENTER 1976; RUBIN et al. 1980; ROTHSTEIN et al. 1989). Evidence for NO as a mediator of NANC relaxation of the cat ileo-caecal sphincter has been presented (MIZHORKOVA et al. 1994).

5. Ileo-Colonic Junction

NANC inhibitory transmission has been demonstrated in this sphincter-like junction (PELCKMANS et al. 1989) and follow-up experiments from this group concluded that VIP, SMS, SP, opioids or 5-HT were all unlikely candidates for the transmitter involved (BOECKXSTAENS et al. 1990a), but that NO is the likely transmitter (BOECKXSTAENS et al. 1990b, 1991, 1993). A subsequent publication from another group supported this conclusion (WARD et al. 1992). Evidence has been presented that GABA stimulates $GABA_A$ receptors located on NANC inhibitory nerves in the dog ileo-colonic junction (BOECKXSTAENS et al. 1990c). No evidence to support the view that ATP might be a NANC cotransmitter at this sphincter has been presented.

6. Internal Anal Sphincter

This sphincter has received most attention by a number of different groups and NANC inhibitory transmission demonstrated in rat (NISSAN et al. 1984), cat (BOUVIER and GONELLA 1981; BOUVIER et al. 1986), rabbit (BIANCANI et al. 1985; BITAR et al. 1990; TØTTRUP et al. 1995; KNUDSEN et al. 1995), opossum (NURKO and RATTAN 1988; RATTAN and CHAKDER 1993; RATTAN and SHAH 1988; RATTAN et al. 1995), guinea-pig (COSTA and FURNESS 1973; CREMA et al. 1983; BAIRD and MUIR 1990; WANG et al. 1996; RAE and MUIR 1996), as well as humans (PARKS et al. 1969; FRENCKNER and IHRE 1976; BURLEIGH et al. 1979; BURLEIGH 1983; LIM and MUIR 1986; MCKIRDY 1992; O'KELLY et al. 1993).

In an early paper, BURLEIGH et al. (1979) concluded that some evidence allowed VIP and ATP to be considered as possible NANC inhibitory transmitters on the human internal anal sphincter, with the reservation that this could not be confirmed because selective antagonists were not yet available. Both ATP and adenosine were shown to produce concentration-dependent and TTX-insensitive relaxations of the guinea-pig (CREMA et al. 1983) and rat (NISSAN et al. 1984) anal sphincter, although it was concluded that these receptors were unlikely to be involved in NANC nerve-mediated relaxations (CREMA et al. 1983; RATTAN and SHAH 1988; BOECKXSTAENS et al. 1990d). In a paper published by another group about the rabbit internal anal sphincter, while both ATP and VIP produced relaxations, they favoured VIP as the NANC transmitter (BIANCANI et al. 1985).

Clear evidence in support of ATP as transmitter in the guinea-pig internal anal sphincter came from the laboratory of Muir (LIM and MUIR 1986; BAIRD and MUIR 1990; RAE and MUIR 1996). The first paper showed that

ATP was the most effective agent examined and produced a concentration-dependent membrane potential change comparable in amplitude to that produced by field stimulation of NANC nerves; later they showed that apamin, but not α-chymotrypsin, which antagonised VIP hyperpolarisations, blocked IJPs and ATP hyperpolarisations (LIM and MUIR 1986). Finally they showed that IJPs in the guinea-pig internal anal sphincter consisted of two TTX-sensitive components, one was abolished by apamin and suramin, the other was smaller in amplitude and sensitive to inhibitors of NOS such as L-NAME, and the NO scavenger oxyhaemoglobin (Fig. 16) (RAE and MUIR 1996). They also showed in this study that, while ATP, VIP and PACAP each hyperpolarised and relaxed the preparations, only ATP resembled the evoked IJPs in the time course. Thus, ATP and NO both appeared to act as NANC inhibitory transmitters in the internal anal sphincter. A similar conclusion was reached recently by RAND and colleagues working with rat internal anal sphincter (DE LUCA et al. 1999).

In the opossum, there is good evidence for a nitrergic component (CHAKDER and RATTAN 1993; RATTAN and THATIKUNTA 1993; RATTAN et al. 1995) and the non-nitrergic component was suggested to be VIP by this group (CHAKDER and RATTAN 1992) and also as the inhibitory transmitter responsible for the rectoanal reflex in the opossum (NURKO and RATTAN 1988). In the rabbit internal anal sphincter the additional involvement of carbon monoxide has been considered (TØTTRUP et al. 1995; RATTAN and CHAKDER 1993; CHAKDER et al. 1996). Involvement of NO in NANC inhibitory neurotransmission has also been claimed for the internal anal sphincter of humans (MCKIRDY 1992; O'KELLY et al. 1993) and rabbits (TØTTRUP et al. 1995).

Another substance proposed as an NANC inhibitory transmitter of the internal anal sphincter is ENK in the cat (BOUVIER et al. 1986). SP and bombesin were proposed as NANC *excitatory* transmitters at this sphincter (BITAR et al. 1990). However, the consensus view, typified by a study of rabbit internal anal sphincter (KNUDSEN et al. 1995), appears to be that the innervation is by cholinergic excitatory nerves and by NANC inhibitory nerves involving at least three different transmitters: NO, an apamin-sensitive mediator (probably ATP) and a non-L-NNA, non-apamin-sensitive agent, possibly VIP.

Fig. 16A,B. Effects of apamin and suramin on inhibitory junction potentials (IJPs) from the guinea-pig isolated internal anal sphincter evoked by electrical field stimulation (EFS). **A** IJPs evoked in response to EFS as a single stimulus (s.s.) and as 5 stimuli at 10, 20 and 40 Hz (0.1 ms, supramaximal voltage). Spontaneous electrical depolarisations are a feature of this tissue. Apamin (0.3 μmol/l) initially depolarised the membrane and reduced IJP amplitude at each frequency. **B** Suramin (100 μmol/l) alone reduced but did not abolish IJP amplitude at any frequency; suramin (100 μmol/l) with L-NAME (100 μmol/l) abolished electrical responses, leaving only stimulus artefacts (from RAE and MUIR 1996)

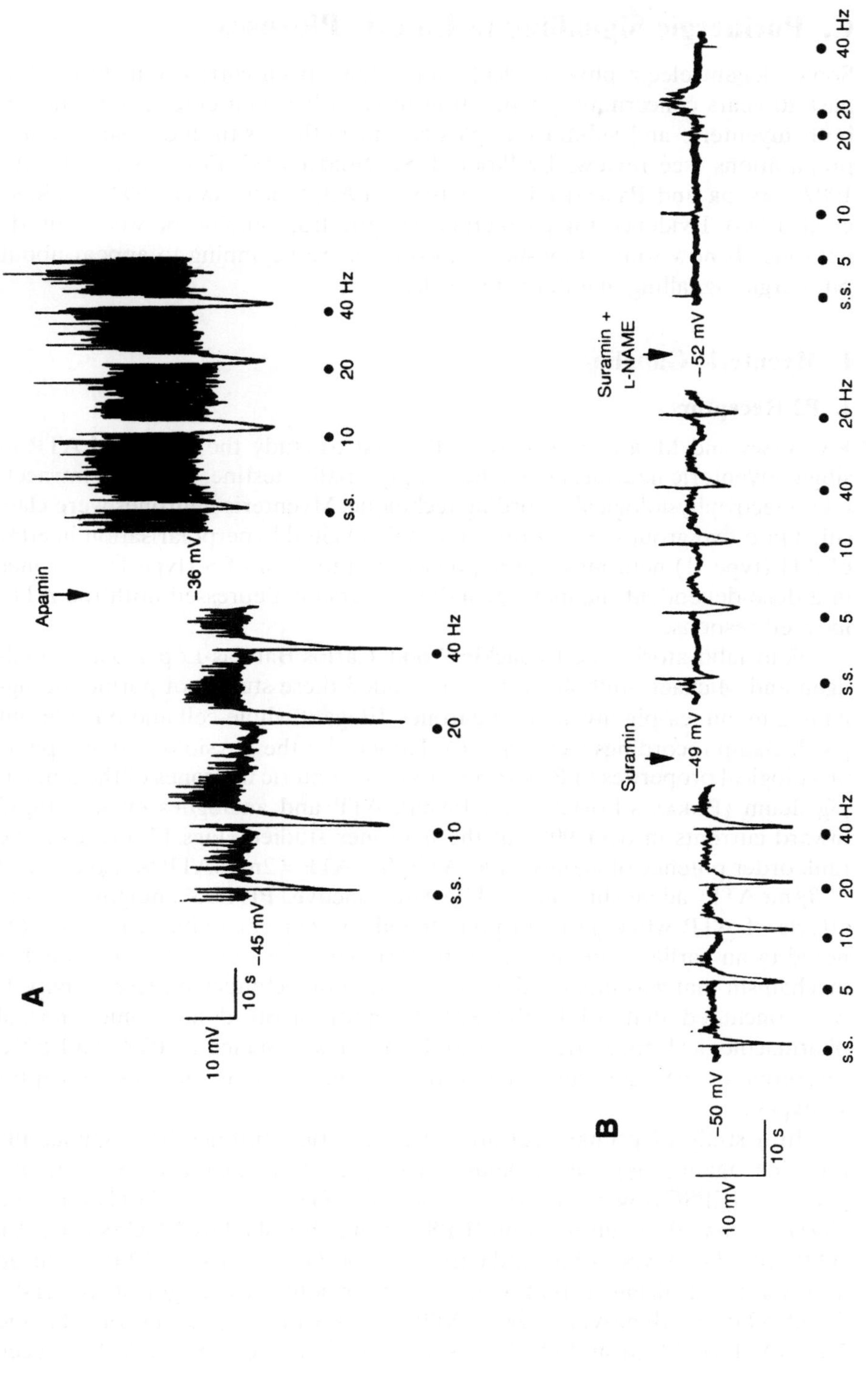
A
10 mV
10 s
−45 mV
s.s.
10
20
40 Hz
Apamin
−36 mV
s.s.
10
20
40 Hz
B
10 mV
10 s
−50 mV
s.s.
5
10
20
40 Hz
Suramin
−49 mV
s.s.
5
10
40
20 Hz
Suramin +
L-NAME
−52 mV
s.s.
5
10
20
20
40 Hz

C. Purinergic Signalling in Enteric Plexuses

Some elegant electrophysiological studies have been carried out during the past 20 years concerning synaptic transmission between enteric neurones in both myenteric and submucous plexuses in both in situ and tissue culture preparations (see reviews by Wood 1981; North 1982; Furness and Costa 1987; Taylor and Bywater 1989; Mihara 1993; Surprenant 1994; Furness et al. 1998). Evidence for purinergic synaptic transmission between enteric neurones is now well established and studies are beginning to appear about purinergic signalling in enteric glial cells.

I. Myenteric Ganglia

1. P2 Receptors

Katayama and Morita (1989) were the first to study the effects of ATP in single myenteric neurones from guinea-pig small intestine, using the intracellular electrophysiological recording technique. Myenteric neurones were classified into two groups at that time and ATP elicited hyperpolarisation in 80% of AH (type II) neurones and depolarisation in 90% of S (type I) neurones in a dose-dependent manner. Quinidine reversibly depressed both the ATP-induced responses.

Four laboratories led by Jackie Wood, Carlos Barajas-López, James Galligan and Michael Mulholland have extended these studies of purinergic signalling in guinea-pig myenteric neurones. Elegant whole-cell and outside-out patch clamp recordings were used to characterise the physiological and pharmacological properties of P2X receptors on myenteric neurones of the guinea-pig ileum (Barajas-López et al. 1996a). ATP and analogues evoked rapid inward currents in over 90% of the neurones studied (Figs. 17 and 18). The rank order potency of agonists was ATPγS = ATP = 2meSATP >> $\alpha\beta$-meATP = $\beta\gamma$-meATP; adenosine and UTP were inactive. PPADS antagonised the effects of ATP, while suramin potentiated the current induced by ATP (as noted in an earlier communication by Barajas-López et al. 1993) through a mechanism that was independent of its action on ectonucleotidase activity. It was concluded that, while the P2X receptor involved has some unusual pharmacological properties, it showed some resemblance to $P2X_1$ and $P2X_6$ receptors or perhaps some heteromeric combinations of these receptor subtypes.

In a study of primary cultures of myenteric neurones from guinea-pig intestine again using patch-clamp techniques, fast excitatory postsynaptic currents (fEPSCs) were recorded (Zhou and Galligan 1996; LePard et al. 1997). Hexamethonium-resistant fEPSCs were abolished by PPADS (Fig. 19). ATP caused two types of inward currents: in 92% of neurones, ATP produced a slowly desensitising current with rank order potencies of agonists of ATP > 2-meSATP >> $\alpha\beta$-meATP > $\beta\gamma$-meATP > ADP with EC_{50} values for ATP and 2-meSATP of 40 μM and 65 μM, respectively; in 8% of nerves, ATP-induced

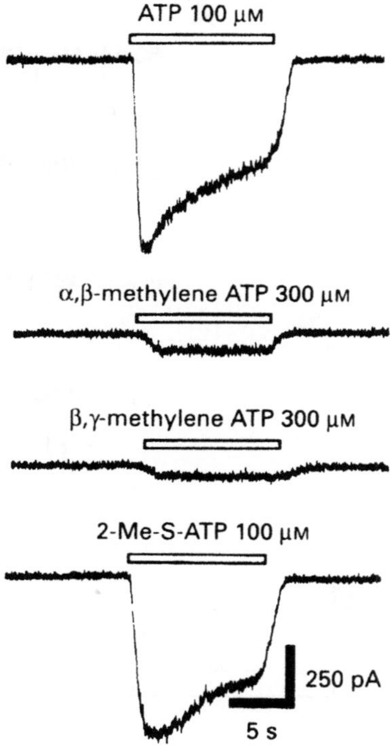

Fig. 17. Potency of several ATP analogues to activate the inward current of guinea-pig myenteric neurones. Traces are the currents induced by the ATP analogues (as indicated) in the same myenteric neurone, *bars* representing the period of agonist application (from BARAJAS-LÓPEZ et al. 1996a)

current desensitised rapidly and $\alpha\beta$-meATP produced similar responses. It was concluded that the slowly desensitising receptors that were $\alpha\beta$-meATP-insensitive were likely to be $P2X_1$ receptors (as was suggested by BARAJAS-LÓPEZ et al. 1996a), whereas the minority of rapidly desensitising receptors might be $P2X_2$ receptors. Later papers from this group claimed that fast excitatory postsynaptic potentials (EPSPs) mediated in part through P2X receptors were prominent in myenteric neurones along the small and large intestine, but were rare in the gastric corpus (LEPARD et al. 1997) and that P2X receptors and nicotinic cholinergic receptors are linked in a mutually inhibitory manner in guinea-pig myenteric neurones (ZHOU and GALLIGAN 1998).

The group of MULHOLLAND carried out studies of purinergic signalling in dispersed primary cultures of guinea-pig myenteric plexus. Extracellular ATP was shown to mediate Ca^{2+}signalling via a PLC-dependent mechanism (KIMBALL et al. 1996). The order of responsiveness of myenteric neurones to purinergic agonists was 2-chloro-ATP = ATP = 2-meSATP > ADP > $\alpha\beta$-meATP = $\beta\gamma$-meATP > AMP > adenosine. ATP-evoked Ca^{2+} transients were inhibited dose-dependently by suramin and Reactive blue 2, and phorbol ester pretreatment caused 76% inhibition, but were not affected by pertussis toxin or nifedipine. Enteric neurones differed from one another in their ability

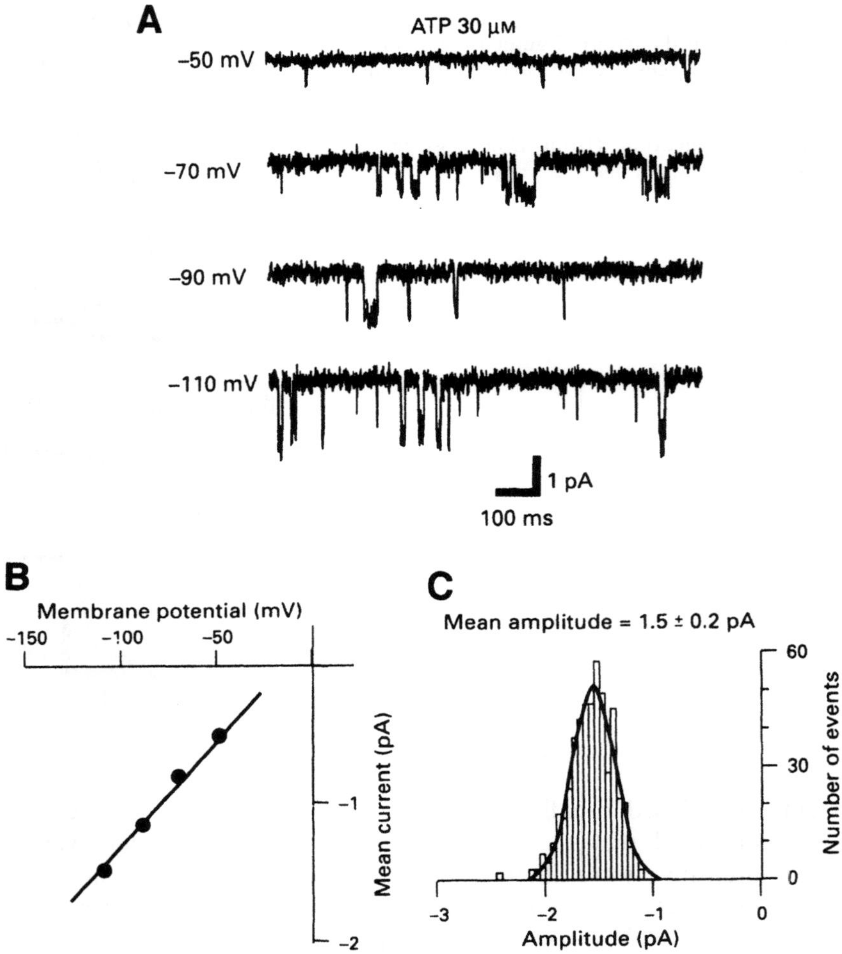

Fig. 18A–C. ATP activates single channel activity in membrane patches of myenteric neurones from guinea-pig ileum. **A** Amplitude of single channel currents was increased by hyperpolarising (mV) the membrane patches. **B** Current-voltage relationship of average single-channel activity shown in **A**. **C** Amplitude histogram of the same channel at a holding potential of –110 mV. Histogram was obtained from a trace sampled at a frequency of 1 kHz (from BARAJAS-LÓPEZ et al. 1996a)

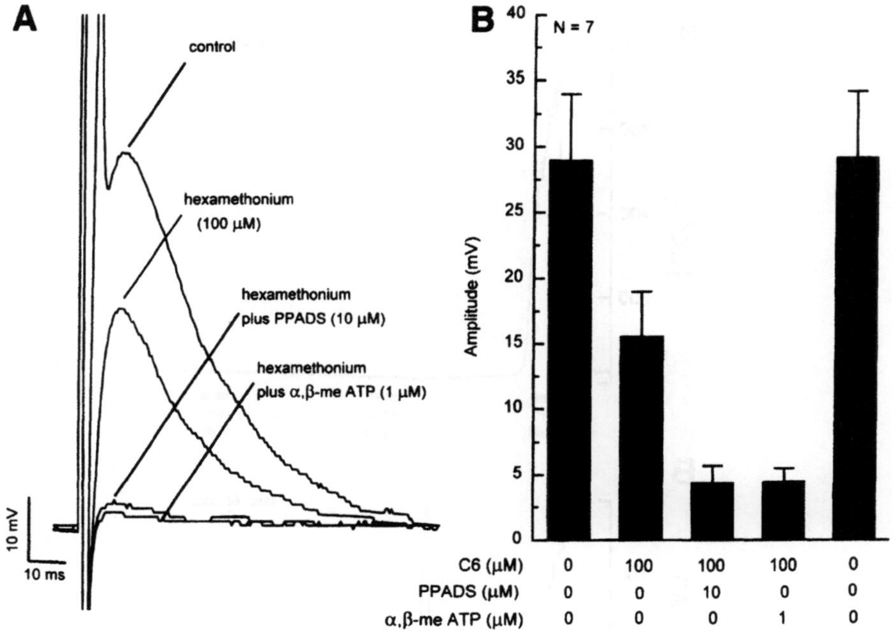

Fig. 19A,B. Effect of purinergic P2X agonist α,β-methyleneATP (α,β-me ATP, 1 μmol/l) on the noncholinergic component of the fast excitatory postsynaptic potential (fEPSP) of guinea-pig myenteric neurones. **A** Representative experiment: the noncholinergic component was totally blocked by PPADS (10 μmol/l). The neurone was allowed to recover in the presence of hexamethonium (not shown). Superfusion with α,β-meATP also totally blocked the noncholinergic component of the fEPSP. **B** Summary: *bar graph* represents the pooled data of all experiments. *First column* is control. *Far right column* denotes recovery (from LePard et al. 1997)

to respond to combinations of ATP with ACh, ATP with SP, ATP with ACh, ATP with ACh and SP, ATP with bombesin or ATP with ACh and bombesin (Fig. 20, Table 1) (Kimball and Mulholland 1995). In their most recent paper, these authors have shown that extracellular ATP also acts on cultured guinea-pig enteric *glia* via an inositol trisphosphate (IP_3) mechanism (probably involving a P2Y receptor) to mobilise intracellular Ca^{2+} (Sarosi et al. 1998). The presence of P2Y receptors on glial cells in the CNS, where astrocytes have been shown to resemble closely enteric glial cells in many respects, has been shown previously (Kastritsis et al. 1992; King et al. 1996; Idestrup and Salter 1998).

Evidence has been presented that two distinct types of P2 receptors are linked to rise in $[Ca^{2+}]_i$ in guinea-pig intestinal myenteric neurones of both AH and S neuronal phenotypes and is not restricted to calbindin-immunoreactive neurones (Christofi et al. 1996, 1997).

ATP regulates synaptic transmission by pre- as well as post-synaptic mechanisms in guinea-pig myenteric neurones, i.e. ATP augments nicotinic fast

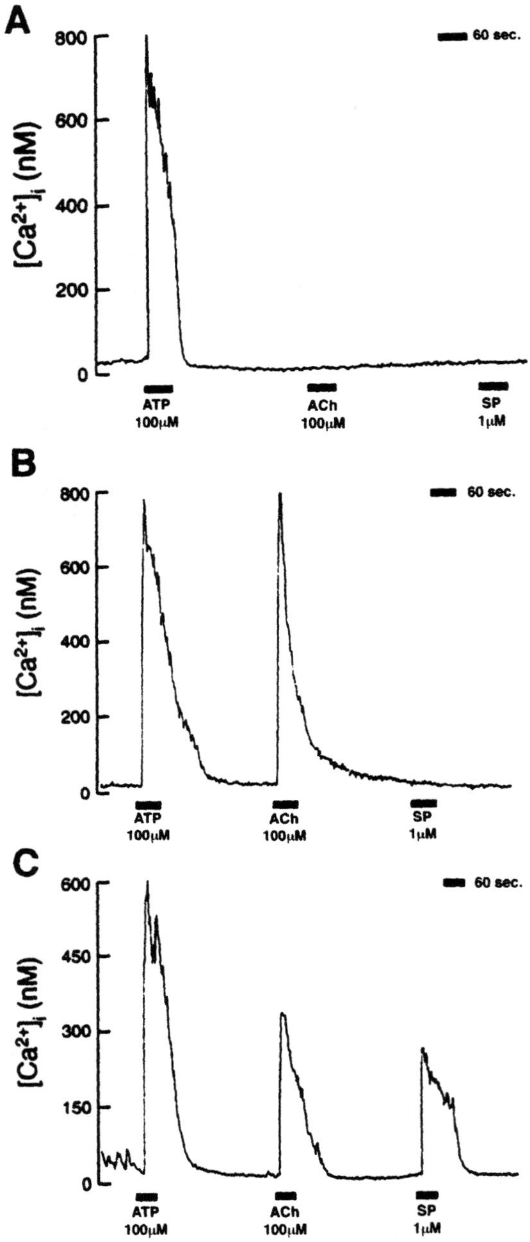

Fig. 20A–C. Representative traces from three cultured myenteric neurones of guinea-pig illustrate the heterogeneity of responses to agonist combinations. The x-axis is time in seconds and lines on this axis represent exposure to agonists. The y-axis is $[Ca^{2+}]_i$ in nmoles/l. Buffer is superfused during the periods between agonist exposure. Neurones were from the same coverslip and the same experiment but differ in responses of $[Ca^{2+}]_i$ to superfusion with ATP, acetylcholine (ACh) and substance P (SP). Each agonist was superfused for 60 s followed by buffer perfusion for 300 s (from KIMBALL and MULLHOLLAND 1995)

Table 1. Percentage of cultured myenteric neurones of guinea-pig responding to combinations of acetylcholine, ATP, substance P (SP) and bombesin

Ligand combinations	% Neurones responding
Acetylcholine + ATP	41
Acetylcholine + bombesin	27
Acetylcholine + SP	62
ATP + bombesin	0
ATP + SP	36
ATP + acetylcholine + SP	50
ATP + acetylcholine + bombesin	0

Agonists were applied individually in random order for 60 s separated by 300 s of buffer superfusion. Cultures were superfused at the end of each experiment with KCl (55 mmol/l) to confirm the cells of interest were viable neurones. Only three ligands could be tested per experiment because of limitations of the multiwell superfusion chamber. Total neurones examined were 72. Acetylcholine 100 μmol/l; ATP 100 μmol/l; SP 1 μmol/l; bombesin 500 nmol/l. (KIMBALL and MULLHOLLAND 1995)

depolarisation produced by ACh, but inhibits muscarinic and SP-mediated depolarisations in both AH and S neurones (KAMIJI et al. 1994).

Exogenous and endogenous ATP, released during increase in intraluminal pressure, inhibit intestinal peristalsis in guinea-pig via different apamin-sensitive purine receptor mechanisms. Exogenous ATP depresses peristalsis mostly via suramin- and PPADS-insensitive P2 receptors, whereas endogenous purines act via P2 receptors sensitive to both suramin and PPADS (HEINEMANN et al. 1999). A preliminary report claims that purinergic transmission is involved in a descending excitatory reflex in guinea-pig small intestine (CLARK et al. 1996). The authors present evidence that excitatory motorneurones have P2X receptors that are excited by anally-directed purinergic interneurons. The possible role(s) of ATP in slow synaptic transmission in enteric ganglia remains to be confirmed. Evidence has been presented recently that ATP plays a major role in excitatory neuro-neuronal transmission in both ascending and descending reflex pathways to the longitudinal and circular muscles of the guinea-pig ileum triggered by mucosal stimulation (SPENCER et al. 2000).

2. P1 Receptors

The first hint that there may be P1 (adenosine) receptors in myenteric neurones was the demonstration that methylxanthines (P1 receptor blockers) antagonised the dipyridamole (adenosine-uptake inhibitor)-induced inhibition of peristaltic activity in the guinea-pig ileum (OKWUASABA et al. 1977; VAN NEUTEN et al. 1977; SHINOZUKA et al. 1985b).

Adenosine was shown in the laboratory of WOOD to inhibit forskolin-induced excitation of myenteric nerves and this suggested that adenosine acts to prevent activation of adenylate cyclase by substances mediating slow EPSPs

(ZAFIROV et al. 1985). Application of adenosine to AH (type II) neurones, but not to S (type I) neurones, results in membrane hyperpolarisation and decrease in input resistance following opening of K^+ channels (PALMER et al. 1987). Adenosine was later shown by this group to suppress nicotinic synaptic transmission in myenteric ganglia of the guinea-pig gastric antrum and small intestine, probably by interacting with presynaptic P1 adenosine receptors located at cholinergic release sites (CHRISTOFI et al. 1992; CHRISTOFI and WOOD 1993a). Further studies showed that the *inhibitory* P1 receptors on AH (type II) neurones were of the high affinity A_1 receptor subtype, linked to a cyclic AMP-independent pathway (CHRISTOFI and WOOD 1933b, 1994), although they showed soon after that there is a minority subset of AH neurones that also express A_2 subtype receptors coupled to adenylate cyclase mediating *excitation* of these neurones (CHRISTOFI et al. 1994). Endogenous adenosine acts at A_1 presynaptic receptors to suppress slow EPSPs and consequently uncovers and amplifies slow inhibitory postsynaptic potentials (IPSPs) in myenteric neurones (CHRISTOFI and WOOD 1993b). In a later study by Kamiji and colleagues, it was shown that ATP inhibited both fast and slow EPSPs recorded in S neurones and slow EPSPs in AH neurones. These actions of ATP were prevented by pretreatment with the P1 receptor antagonists caffeine, theophylline and 8-phenyltheophylline, but the slowly degradable analogues adenosine 5′-*O*-3-thiotriphosphate (ATPγS) and $\alpha\beta$-meATP also depressed EPSPs in both types of neurones; they concluded that ATP regulates synaptic transmission at both pre- and postjunctional sites (KAMIJI et al. 1994).

In whole-cell patch-clamp studies of cultured myenteric neurones from guinea-pig small intestine, 2-CADO was shown to express high voltage activated Ca^{2+} current dose-dependency by reducing both transient and sustained components of the current (BAIDAN et al. 1995) which was supported later by a microelectrode study showing adenosine suppression of cAMP formation in myenteric ganglia *in vitro* (XIA et al. 1997).

In an investigation by another group of the cellular mechanisms underlying adenosine actions on cholinergic transmission in myenteric and submucosal neurones from young guinea-pigs, it was concluded that reduction of cholinergic synaptic transmission via prejunctional A_1 receptors involved the activation of pertussis toxin-insensitive G proteins (BARAJAS-LÓPEZ et al. 1996b). Endogenous adenosine, some of which arises from extracellular ATP metabolism, probably acting through presynaptic A_1 receptors, inhibits SP release from perfused networks of myenteric ganglia from guinea-pig ileum and is postulated to be an important contributor to the overall inhibitory tone present in myenteric ganglia networks (BROAD et al. 1993; MONETA et al. 1997). A neuroprotective role for adenosine in ischaemia has been postulated, consistent with a demonstrated relationship between interstitial adenosine in the myenteric neural network and prevailing O_2 tension (DESHPANDE et al. 1999).

The P1 agonist, 5′-*N*-ethylcarboxamidoadenosine (NECA) is a potent inhibitor of morphine withdrawal induced diarrhoea in rats acting by inhibiting secretion as well as inhibiting intestinal peristalsis and in studies of the P1 receptor subtypes involved, evidence has been presented to suggest that, while A_1 receptors are involved in regulation of peristalsis (TOMARU et al. 1994), A_{2B} receptors mediate inhibition of secretion and it was proposed that A_{2B} adenosine agonists could be of clinical value in the management of some types of diarrhoea (HANCOCK and COUPAR 1995).

3. Storage and Release of ATP from Synaptosomes

The fluorescent antimalarial acridine, quinacrine, is know to bind to high levels of ATP especially when it is associated with peptides contained in large granular vesicles (IRVIN and IRVIN 1954; OLSON et al. 1976; DA PRADA et al. 1978). Subpopulations of enteric neurones in the myenteric plexus of the stomach and intestine of adult guinea-pigs, rabbits and rats and of perinatal rabbits fluoresce strongly following exposure to quinacrine, indicating the presence of high levels of ATP and that they are associated with NANC inhibitory neurones (CROWE and BURNSTOCK 1981a,b; SHINOZUKA et al. 1985a), some being colocalised with NOS (Fig. 21) (BELAI and BURNSTOCK 1994).

Synaptosomal preparations from the guinea-pig ileum myenteric plexus were first described by DOWE et al. (1980) and BRIGGS and COOPER (1981). ATP and adenosine were equipotent in their ability to inhibit the nicotinically-induced release of [^{3}H]ACh; the inhibition by both ATP and adenosine was reversed by theophylline, indicating that a P1 receptor was involved (REESE and COOPER 1982). High concentrations of ATP caused marked increase in the release of [^{3}H]ACh; the authors concluded that a P2 receptor was unlikely to be involved, but in view of the evidence for a P2X receptor mediating release of ACh from cholinergic nerves in the intestine (MOODY and BURNSTOCK 1982) this conclusion may not be justified. Confirmatory support for adenosine modulation of ACh release from synaptosomes prepared from guinea-pig ileum myenteric plexus came from another laboratory (SHINOZUKA et al. 1985a). Affinity studies of various purine nucleosides for P1 receptors on purified myenteric varicosities compared to their efficacy as presynaptic inhibitors of ACh release, led to the conclusion that more than one adenosine receptor is present but that the receptors involved could not be clearly defined as either A_1 or A_2 subtype (CHRISTOFI and COOK 1986, 1987).

A study of evoked release of [^{3}H]NA and ATP from nerve varicosities isolated from the myenteric plexus of guinea-pig ileum led to the conclusion that ATP and [^{3}H]NA may not be released from the same population of secretory vesicles (HAMMOND et al. 1988).

4. Modulation of NANC Purinergic Transmission

Several enteric neurotransmitters have been claimed to modulate release of ATP by acting on NANC neuronal cell bodies in the myenteric plexus.

Morphine or ENK inhibition of NANC-evoked relaxation (reversed by nalaxone), but not relaxations to sympathetic perivascular stimulation

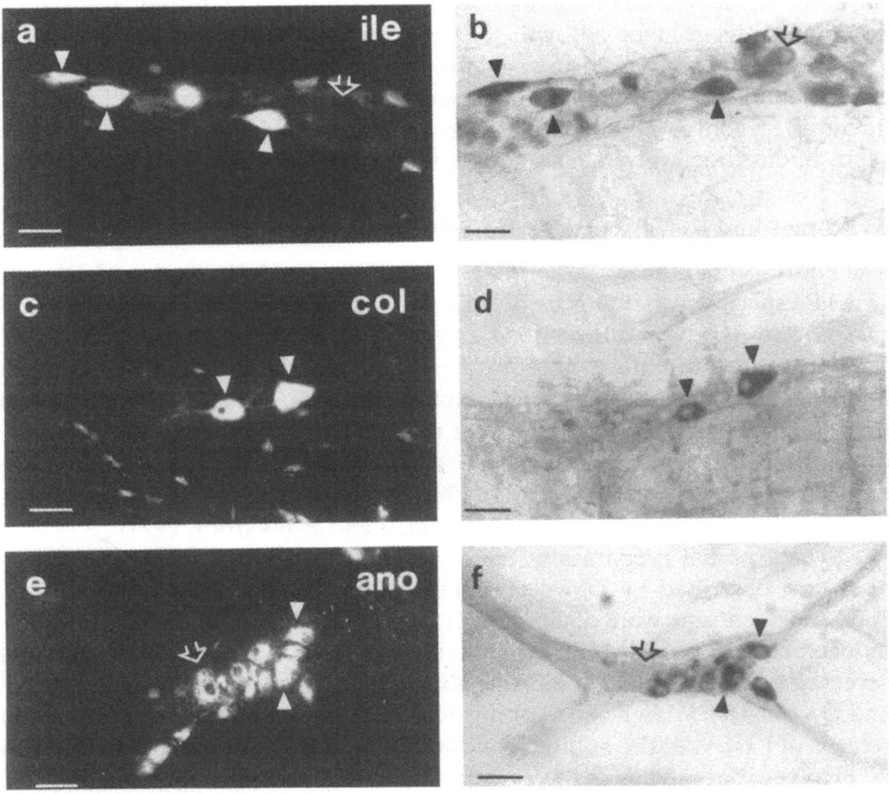

Fig. 21a–f. Micrographs showing colocalisation of ATP and NADPH-diaphorase in myenteric ganglion neurones of ileum and proximal colon and ganglion structure in the anococcygeus muscle of the rat. **a** Quinacrine-fluorescent myenteric neurones of ileum (ile). **b** NADPH-diaphorase-positive myenteric neurones of the same preparation. Most of the fluorescent neurones in a also contain NADPH-diaphorase (*arrowheads*), but there are NADPH-diaphorase-positive but quinacrine-negative neurones (*open arrows*). **c** Quinacrine-fluorescent neurones in the myenteric plexus of rat proximal colon (*col*). **d** NADPH-diaphorase-positive myenteric neurones of the same preparation – note that all quinacrine-fluorescent neurones also contain NADPH-diaphorase (*arrowheads*). **e** Quinacrine-fluorescent ganglion neurones of the anococcygeus muscle (*ano*). **f** In the same ganglion, some of the fluorescent neurones are also shown to contain NADPH-diaphorase (*arrowheads*) – note those neurones that are quinacrine-fluorescent but not NADPH-diaphorase-positive (*open arrows*). *Calibration bars*: 30 μm (from Belai and Burnstock 1994)

in the guinea-pig taenia coli (Shimo and Ishii 1978; Ishii and Shimo 1983), although a later study with ENK did not appear to support this view (Huizinga and Den Hertog 1979). ENK has been shown, however, to be very effective in inhibiting NANC IJPs recorded in human colon (Hoyle et al. 1990).

An early report suggested that excitatory 5-HT receptors were present on purinergic inhibitory neurones in the guinea-pig caecum (DRAKONTIDES and GERSHON 1972); another study claimed that this was not the case (VERMILLION et al. 1979). Later, however, strong evidence was presented that 5-HT released ATP from nerve varicosities isolated from the myenteric plexus of the guinea-pig ileum (AL-HUMAYYD and WHITE 1985). A recent paper showed that enteric neurones express multiple receptors for 5-HT, three excitatory (5-$HT_{1\beta}$, 5-HT_3 and 5-HT_4) and one inhibitory (5-HT_{1A}) that mediates hyperpolarisation of AH neurones and presynaptic inhibition of both fast and slow excitatory neurotransmission (GALLIGAN 1996).

Evidence has been presented that GABA receptors mediate relaxation of rat duodenum by activating intramural NANC neurones in rat duodenum (MAGGI et al. 1984), dog ileocolonic junction (BOECKXSTAENS et al. 1990c) and guinea-pig distal colon (MINOCHA and GALLIGAN 1993). In their study, MAGGI et al. (1984) provided evidence indicating that ATP might be the endogenous substance released by GABA.

II. Submucosal Ganglia

Correlated electrophysiological and histochemical studies of submucous nerves showed that the balance of absorption and secretion of water and electrolytes is activated by sympathetic inhibitory inputs to secretory-motor neurones in the submucous plexus (See BORNSTEIN and FURNESS 1988). MIHARA et al. (1985) were the first to record slow postsynaptic inhibitory and excitatory potentials in S type I neurones of the submucous plexus of the guinea-pig caecum and show mimicry by various transmitters. The non-reversing type of slow *excitatory* postsynaptic potential was mimicked only by ATP.

Adenosine was shown to excite directly, depolarising submucosal neurones by acting at P1 (A_2-like) receptors and also to act presynaptically via P1 (A_1) receptors to inhibit the release of ACh from intramural nerves and of NA from sympathetic nerves in the submucosal plexus (BARAJAS-LÓPEZ et al. 1991; BARAJAS-LÓPEZ 1993). In a later paper, this group concluded that in addition to A_1 presynaptic receptors mediating inhibition of ACh release, ATP acting through "P_3 receptors" might also be involved (BARAJAS-LÓPEZ et al. 1995).

Intracellular recordings from submucosal neurones in guinea-pig small intestine showed that ATP induced fast transient depolarisation of most AH-type neurones and fast transient depolarisation followed by slower onset, longer lasting depolarisation of S-type neurones (BARAJAS-LÓPEZ et al. 1994). When whole-cell patch recordings were employed, superfusion of ATP and analogues evoked rapidly desensitising inward current and ATP-induced single channel currents were also recorded (Fig. 22). In a whole-cell patch-clamp study of ATP-induced membrane currents in guinea-pig small intestinal submucous neurones by another group (GLUSHAKOV et al. 1996), the currents activated by ATP were not blocked by suramin and were often enhanced by Reactive blue 2. This could indicate the involvement of $P2X_4$ or

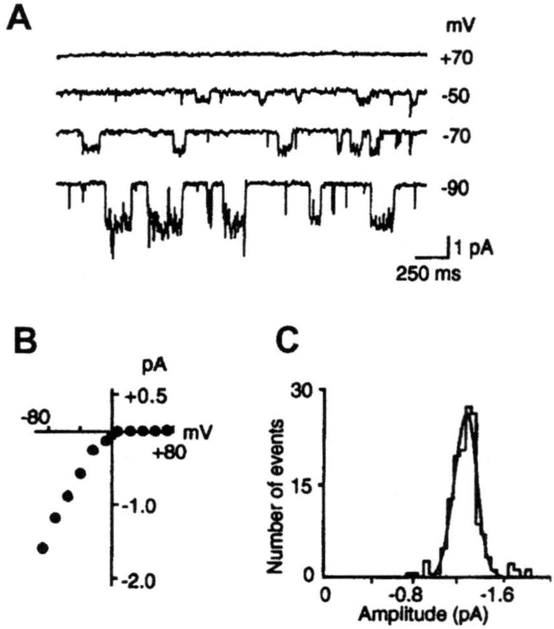

Fig. 22A–C. Voltage dependency of ATP-induced single channel currents from neurones of guinea pig submucous plexus resembles that of whole-cell recordings. **A** ATP-induced single channel currents at different membrane potential (mV). Hyperpolarisation increase open probability and mean open time. **B** The current-voltage relationship of average single-channel activity showed strong inward rectification with no outward currents detected at positive holding potentials. **C** Amplitude histogram of the channel activated by ATP at a holding potential of –70 mV – this histogram was obtained from a 1-min duration trace sampled at a frequency of 1 kHz (from BARAJAS-LÓPEZ et al. 1994)

$P2X_6$ receptors (see BURNSTOCK 1997). The functional interactions between nicotinic and P2X receptors have been investigated in freshly dissociated guinea-pig submucosal neurones in primary culture; whole-cell currents induced by ATP were blocked by PPADS and showed some interdependence on ACh-induced nicotinic currents blocked by hexamethonium (GLUSHAKOV et al. 1996; BARAJAS-LÓPEZ et al. 1998; ZHOU and GALLIGAN 1998). In a recent publication, evidence was presented for two subtypes of P2X receptors in neurones of guinea-pig ileal submucosal plexuses (GLUSHAKOV et al. 1998).

III. Enteric Glial Cells

Enteric glial cells show many similarities to CNS astrocytes, where there is abundant evidence for purinergic signalling (JESSEN and BURNSTOCK 1982; JESSEN and MIRSKY 1983; NEARY et al. 1996; ABBRACCHIO and BURNSTOCK 1998). A few studies provide evidence for P2 receptors on enteric glial cells. The responses of glial cells isolated from guinea-pig myenteric plexus to purines and pyrimidines using calcium signalling in fura-2-acetoxymethylester-loaded cells suggested that responses to ATP and UTP are mediated by P2U receptors coupled to activation of phospholipase C and release of intracellular calcium stores (KIMBALL and MULHOLLAND 1996). In a follow-up paper by this group, they showed that depletion of internal Ca^{2+} stores in enteric glia by extracellular ATP or thapsigargin induces capacitance Ca^{2+} entry which is mediated by a Ni^{2+}- and La^{3+}-sensitive channel that is not inhibited by L- or N-type channel blockers; capacitance Ca^{2+} entry in glial cells is regulated by phosphokinase C and NO (SAROSI et al. 1998).

D. Epithelial Secretion

There is growing recognition of a wide variety of signalling roles for purines and pyrimidines in mucosal and glandular epithelial cells in most regions of the gastrointestinal tract (CHRISTOFI and COOK 1997; ROMAN and FITZ 1999).

I. Salivary Secretion

Salivary acinar and ductal epithelial cells are responsible for the controlled secretion of fluid and electrolytes and of specific proteins and growth factors (COOK and YOUNG 1989). There are several Ca^{2+}-mobilising receptors involved in these activities, including muscarinic receptors, α-adrenoceptors and SP receptors, but it is now well recognised that there are also Ca^{2+}-mobilising receptors for extracellular ATP in rat and mouse parotid acini (GALLACHER 1982; MCMILLIAN et al. 1987; SOLTOFF et al. 1990; FUKUSHI et al. 1997) and in rat, mouse and human submandibular acinar and duct cells (YU and TURNER 1991; GIBB et al. 1994; HURLEY et al. 1996; KURIHARA et al. 1997; TURNER et al. 1997, 1998).

Potassium-evoked release of purines from rat submaxillary gland has been demonstrated, although it was not possible in the experiments described to discriminate between neuronal and non-neuronal elements as the source of purines released by depolarisation (FILINGER et al. 1989).

Intraarterial administration of various nucleotides, including ATP and ADP to the cat submandibular salivary gland led to increase in blood flow, possible mimicking neurally released ATP as a cotransmitter in parasympathetic and/or sympathetic nerves (JONES et al. 1980).

1. Parotid Gland

In a seminal paper, GALLACHER (1982) showed that in acinar cells of the parotid gland, ATP evoked a marked increase in membrane conductance, K^+ efflux and amylase secretion (Fig. 23); a P2 receptor was implicated, since adenosine had no effect and the responses could be blocked by quinidine, but not by theophylline. Extracellular ATP was later shown to elevate intracellu-

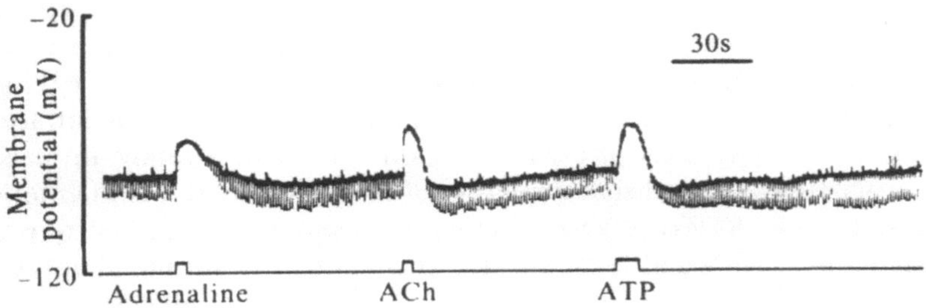

Fig. 23. ATP-evoked changes in membrane potential from acinar cells of mouse parotid gland. The potential changes superimposed on the record of resting membrane potential are due to repetitive injection of hyperpolarising current pulses through the recording electrode (2 nA, 100 ms). The amplitude of these electronic potentials corresponds to the input resistance of the acinus. Responses are seen to the ionophoretic application of the agonists adrenaline, acetylcholine (ACh) and ATP. The initial potential change is due to an increase in passive membrane permeability primarily to K^+ and Na^+. The resulting efflux of K^+ and influx of Na^+ stimulate an electrogenic Na/K pump which gives rise to the delayed hyperpolarisation seen when the input resistance has returned to prestimulus levels. The ATP micropipette contained 0.5–1.0 mol/l ATP in distilled water. Ejection current was 300 nA applied for 8 s. Responses were recorded with hyperpolarising ejection charges of ≥50 nC. Depolarising ejection charges were ineffective, as was ionophoresis of a control solution of distilled water titrated to pH 3.0. Adrenaline and ACh ionophoresis were achieved by depolarising ejection currents of 300 and 200 nA respectively for 4 and 3 s. Note the similar nature of the membrane response to the three agonists (from GALLACHER 1982)

lar free calcium in rat parotid acinar cells and the possibility that ATP plays a neurotransmitter role in the parotid gland raised (MCMILLIAN et al. 1987). In fact, ATP was found to be more effective than muscarinic and α-adrenergic agonists and SP as a stimulus for elevating $[Ca^{2+}]_i$ levels (MCMILLIAN et al. 1988). Coomassie Brilliant blue G was a more potent antagonist of P2 receptor-mediated responses of rat parotid acinar cells than Reactive blue 2 (Cibacron blue 3GA) (SOLTOFF et al. 1989). Further studies by this group led them to suggest that ATP may function as a neurotransmitter to modulate salivary fluid secretion by stimulating Ca^{2+}-sensitive Cl^- and K^+ channels and multiple Na^+ uptake pathways in the rat parotid acinar cell (SOLTOFF et al. 1990). They showed that some of these pathways were similar to those activated by carbachol while others were unique to ATP.

Two distinct $[Ca^{2+}]_i$ responses to ATP were distinguished in rat parotid acinar cells raising the possibility that both $P2X_7$ and P2Y receptors were implicated (SOLTOFF et al. 1992; MCMILLIAN et al. 1993). To learn more about the ATP binding site of the $P2X_7/P_{2Z}$ receptor in these acinar cells, the isothiocyanate compound DIDS was examined and found to be an effective antagonist at the parotid $P2X_7/P_{2Z}$ receptor (SOLTOFF et al. 1993). A study by another group showed an inhibitory effect of ATP^{4-} on the ACh-mediated response of rat parotid acini and presented evidence to suggest that this was due to interactions of the activated $P2X_7/P_{2Z}$ receptor with the phospholipase C-coupled processes underlying the muscarinic cholinergic response (JORGENSEN et al. 1995). More recently it has been shown that ATP, acting through $P2X_7/P_{2Z}$ receptors, causes Na^+ entry by opening cation-permeable channels and thereafter the increase in $[Na^+]_i$ triggers Ca^{2+} release from intracellular ryanodine-sensitive stores, while UTP acting through P2U (= $P2Y_2$ and/or $P2Y_4$)-type receptors caused Ca^{2+} release independent of external Na^+ (FUKUSHI et al. 1997). In a somewhat conflicting paper, activation of $P2X_7/P_{2Z}$ receptors on rat parotid acinar cells was claimed to cause a large entry of Ca^{2+} into the cells (TOJYO et al. 1997). In a recent abstract, a $P2X_7$ receptor was identified in rat parotid salivary glands which was modulated by Mn^{2+} and Ni^{2+}, but not by Cd^{2+} (ARKLEET al. 1998). Ectonucleotidase and 5′-nucleotidase levels in parotid acini have been reported (DOWD et al. 1996).

Using a rat parotid secretory granule preparation, ATP was shown to activate Ca^{2+}-independent membrane-associated phospholipase A_2 (PLA_2) (MIZUNO-KAMIYA et al. 1998). ATPγS was active to a lesser extent, while UTP, CTP and GTPγS showed little activation. It was suggested that the PLA_2 located in the granular membranes may participate in the liberation of arachidonic acid in parietal cells that is regulated through a mechanism mediated by ATP.

The results of a recent study using RT-PCR showed strong expression of $P2X_4$ and $P2X_7$ mRNA in parotid glands which correlated well with the responses of the parotid acinar cells to extracellular ATP (TENNETI et al. 1998). It was further shown that parasympathetic denervation of the parotid gland increased the number of cells with $P2X_4$ responses and the levels of $P2X_4$ mRNA, opening up important general issues about transsynaptic regulation of P2X receptor expression. Another recent study describes how ATP, acting through $P2X_7$ receptor-mediated phospholipase D, may produce a Ca^{2+}-independent protein kinase C to account for the finding that ATP shortened the duration and decreased the magnitude of ACh-induced Ca^{2+} release from rat parotid acinar cells (FUKUSHI 1999).

2. Submandibular Gland

The effect of ATP on various types of preparations from submandibular salivary glands has also been reported, including the human submandibular duct cell line, HSG-PA (YU and TURNER 1991; KIM et al. 1996; KURIHARA et al. 1997), rat submandibular gland acini (HURLEY et al. 1993, 1994,1996), crude cell suspensions of whole rat submandibular glands (DEHAYE 1993; LACHISH et al. 1996; MÉTIOUI et al. 1996; TURNER et al. 1997; ZENG et al. 1997) or ductal cells (AMSALLEM et al. 1996) and a mouse submandibular epithelial salivary cell line, ST_{885} (GIBB et al. 1994).

The studies of HSG-PA cells indicated the presence of a P_{2U}-like receptor where the agonist profile was UTP = ATP > ATPγS > ADP > ADPβS with both $\alpha\beta$-meATP and 2-meSATP having little or no effect, but no evidence for the presence of the $P2X_7/P_{2Z}$ receptor identified in parotid acinar cells (YU and

TURNER 1991). The cells challenged by UTP hyperpolarised which provided the driving force for net Cl^- efflux (KIM et al. 1996). In the most recent paper on HSG cells, in addition to a P2U receptor mediating IP_3 formation to nucleotides, the authors suggested that Ca^{2+} influx might be mediated by a second, perhaps P2X receptor (KURIHARA et al. 1997).

Studies of rat submandibular glial acini identified, as for parotid acini, a $P2X_7/P_{2Z}$ receptor activated by ATP^{4-} which promoted Ca^{2+} influx, but not release, from intracellular stores (HURLEY et al. 1993, 1994) and in later papers these authors claimed that ATP promotes an increase in both $[Ca^{2+}]_i$ and $[Na^+]_i$ via a common influx pathway (HURLEY et al. 1996).

For mixed duct and acinar cell suspensions, again a $P2X_7/P_{2Z}$ receptor coupled to a non-selective cation channel was described, occupation of which by ATP potentiates the responses to both carbachol and SP (MÉTIOUI et al. 1996; LACHISH et al. 1996). In a subsequent paper from this group, using suspensions of submandibular ductal cells only, two purinergic receptors were identified, a metabotropic, probably $P2Y_1$ receptor and a P2X ionotropic receptor coupled to a manganese-permeant calcium channel and to kallikrein secretion (Fig. 24) (AMSALLEM et al. 1996). In the most recent study by another

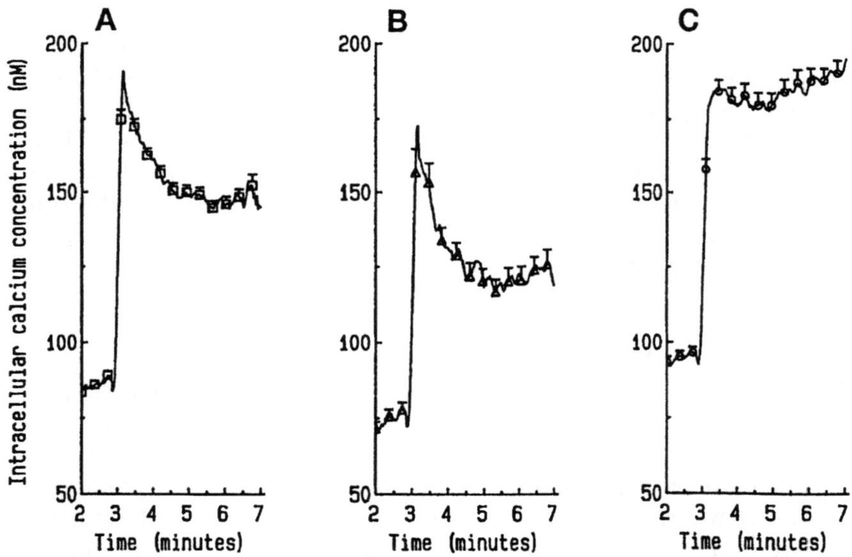

Fig. 24. Effect of: **A** carbachol 100 μmol/l, **B** adrenaline 100 μmol/l, **C** ATP 1 mmol/l on intracellular Ca^{2+} concentration ($[Ca^{2+}]_i$) in rat submandibular ductal cells. Rat submandibular ductal cells were loaded with fura 2-acetoxymethyl ester and after washing were resuspended in 2 ml of a medium without Mg^{2+} or Ca^{2+}. At 1 min, 1 mmol/l Ca^{2+} was added, and at 3 min the cells were exposed to one of the agonists. Results are expressed in $[Ca^{2+}]_i$ and are the means±SEM of 6 (**A**), 6 (**B**) and 15 (**C**) experiments (from AMSALLEM et al. 1996)

group, coordinated actions of $P2X_7/P_{2Z}$ (luminal) and P_{2U}-like (basolateral) receptors were proposed that mediate part of the transcellular cystic fibrosis transmembrane regulator (CFTR)-like Cl^- transport by acinar and duct cells to determine the final electrolyte composition of salivary fluid (ZENG et al. 1997). P2Y receptors (= old P2U receptor) identified in both acinar and ductal cells of rat submandibular gland increased with time in culture and it was speculated that changes in expression of the $P2Y_2$ receptor on salivary gland cells may be related to pathological challenges to the gland in vivo (TURNER et al. 1997). This group included a review of the field in an experimental paper, their main observations being:

1. $P2Y_1$ receptor activity is present in submandibular glands, although it tends to decline with age.
2. $P2Y_2$ receptors are present in cell lines and are up-regulated during short-term culture of normal glands and following ligation of the main secretory duct of submandibular gland.
3. The P2X subtypes, $P2X_4$ and $P2X_7$, and the P2Y subtypes, $P2Y_1$ and $P2Y_2$, are co-expressed in salivary glands and salivary cell lines, and exhibit distinct basolateral, as opposed to apical, localisation in polarised cell monolayers as well as having discrete patterns of intracellular signalling (TURNER et al. 1998).

The conclusion from a study of nucleotide actions of the mouse submandibular salivary cell line, ST_{885}, was that two P2 receptor subtypes were probably present, one where ATP and UTP were equipotent (probably $P2Y_2$) and another where 2-meSATP was active (possibly a $P2Y_1$ receptor) (GIBB et al. 1994).

There is convincing evidence for the expression of Na^+-H^+ exchanger (NHE) isoforms in the basolateral membrane of rat submandibular gland duct and acinar cells (NHE1) and NHE2 and NHE3 in the luminal membrane of these cells which shows that the activities of the basolateral and luminal NHEs are regulated by P2 receptors (P2U receptors in the isolated membrane and $P2X_7/P_{2Z}$ receptors in the luminal membrane) in a membrane-specific manner which may play an important role in co-ordinating the overall process of Na^+ absorption (LEE et al. 1998). Another recent study (KABRÉ et al. 1999) has shown that extracellular ATP and benzoyl-ATP, a potent agonist for $P2X_7$ receptors, substantially increased the release of arachidonic acid from rat submandibular gland ductal cells; these effects involved activation of PLA_2 by the purinergic agonists.

An RT-PCR and pharmacological study of postnatal development of purinergic signalling in salivary glands, using dispersed cell aggregate preparations from the submandibular-sublingual gland complex of 1-day old and 1-, 2-, 3- and 4-week old rats, showed that functional $P2Y_1$ receptors were expressed in immature (1 day postnatal) salivary glands and that receptor activity decreased as the glands matured, suggesting that $P2Y_1$ receptors may have an important role during salivary gland development (PARK et al. 1997).

P2Y$_1$ receptors have been shown to play important roles in embryonic chick development (Meyer et al. 1999).

II. Ciliated Epithelium of Oesophagus

Extracellular ATP has long been recognised as a stimulant of ciliary activity in water and mucus-transporting epithelia (see Levin et al. 1997). Studies have been carried out on monolayer tissue cultures of epithelial cells grown from frog oesophagus (Ovadyahu et al. 1988; Weiss et al. 1992). ATP enhances ciliary beat frequency 2–3-fold and induces pronounced changes in the metachronal wave parameters (Gheber and Preil 1994). In addition, membrane fluidisation was induced and that increases in cytosolic Ca^{2+}, principally from internal stores, coupled to membrane hyperpolarisation were necessary to activate all these cellular effects (Tarasiuk et al. 1995; Alfahel et al. 1996). Studies from this group have also established the existence of two P2 receptors, one of which is probably a P2Y receptor (Gheber et al. 1995) and that the actions of ATP depends on protein kinase C producing sustained enhancement of ciliary beat frequency via activation of calcium influx through non-voltage operated Ca^{2+} channels (Levin et al. 1997).

III. Gastric Acid Secretion

The first report of purinergic modulation of gastric acid secretion was by Kidder (1973), who showed that ATP or the ATP analogue 5′-adenylyl methylene diphosphonate (AMP-PCP) added to the serosal bathing solution of the bullfrog gastric mucosa inhibited gastric acid secretion, although they were unaware at that time of purinergic receptors and did not explain their findings in these terms. Another study at this time of the effect of vagal nerve stimulation on gastric acid secretion in anaesthetised dogs led to the conclusion that, in addition to cholinergic nerves, an unsuspected second neural pathway existed which was capable of influencing gastric acid secretion (Tansey et al. 1975). Gastric hypersecretion of pylorus-ligated rats was inhibited dose-dependently by ADP and AMP (Mózsik et al. 1978a). The question of whether ATPases were involved in regulation of the gastric acid secretory process was raised in early papers (Mózsik et al. 1978b; Forte and Lee 1977; Nandi et al. 1981; Sachs et al. 1982).

There were also early suggestions that extracellular receptors to adenosine were responsible for modulation of acid secretions to the secretagogues, histamine and methacholine (Skoglund et al. 1982; Gerber et al. 1984, 1985) using a dog gastric fundus preparation; theophylline was shown to block the adenosine actions. In a study of basal acid secretion in whole rat stomach, it was shown that, while adenosine caused significant reduction in basal acid secretion, ATP and ADP significantly increased basal acid secretion (Gandarias et al. 1985). Vagally mediated stimulation of gastric acid secretion by intravenously administered adenosine derivations was demonstrated in anaesthetised rats (PUURUNEN et al. 1986). The authors took this to indicate that adenosine can stimulate gastric acid secretion by activating the vagus nerves via adenosine receptors in afferent pathways. The potent effects of the adenosine analogue R-PIA on inhibitory gastric acid secretion in the rat was taken to indicate that the P1 receptor involved was of the A_1 subtype (SCARPIGNATO et al. 1987). Gastric acid secretion was measured in conscious rats with an indwelling gastric cannula (GLAVIN et al. 1987). The potent P1 receptor antagonist 8-phenyltheophylline augmented gastric acid output, supporting a role for adenosine as a regulator of gastric acid secretion. Data was presented to suggest that endogenous adenosine generated by gastric cells interact with parietal cell adenosine receptors to mediate acid secretion to histamine (GERBER et al. 1985; GERBER and PAYNE 1988).

Following up their earlier study of the effect of intravenous adenosine in anaesthetised rat, PUURUNEN and HUTTUNEN (1988) presented evidence to indicate that adenosine inhibits gastric acid secretion by a decrease in stimulation of vagal impulses to the stomach and that it acts in the brain via P1 receptors insensitive to xanthine. Another study, using unanaesthetised rat with indwelling gastric cannulas, showed a rank order for P1 receptor agonists in decreasing gastric acid output of NECA = R-PIA > 2-CADO > N^6-*S*-isopropyladenosine (S-PIA) (WESTERBERG and GEIGER 1989). NECA decreased the volume of gastric secretion, whereas R-PIA had no effects on volume, but significantly increased the pH of the secretions. In an attempt to characterise the effects of adenosine, ATP and ADP on acid secretion in isolated rabbit gastric cells, it was claimed that there were stimulating receptors to adenosine that were inhibited by methylxanthines, perhaps mediated via P1 receptors, and inhibitory receptors to ATP, $\alpha\beta$-meATP and ADP, which were reduced by indomethacin, perhaps mediated via P2 receptors (AINZ et al. 1989, 1993; GIL-RODRIGO et al. 1990). In later papers, it was claimed that there are P1 (A_2 subtype) receptors on rabbit parietal cells which mediate the stimulatory effects of adenosine and analogues on gastric acid production (OTA et al. 1989; AINZ et al. 1993). In the most recent paper from this group, data was presented to suggest that ATP selectively inhibits histamine-stimulated gastric acid secretion by acting directly on parietal cells, perhaps mediated by P2Y receptors with some part due to prostaglandin production (GIL-RODRIGO et al. 1996).

Adenosine has been shown to decrease or increase production of gastrin, a known stimulant of gastric secretions, perhaps via A_1 or A_2 subtypes, thereby indirectly modulating gastric acid secretion (WESTERBERG and GEIGER 1988; SCHEPP et al. 1990).

It has been suggested that muscularis mucosae may augment gastric acid secretion and in a study designed to test this hypothesis it was shown that there are contractions of the muscularis mucosae to ATP and ADP, but it was concluded that muscularis mucosae relaxation, rather than contraction, might be allied to acid secretion (PERCY et al. 1999).

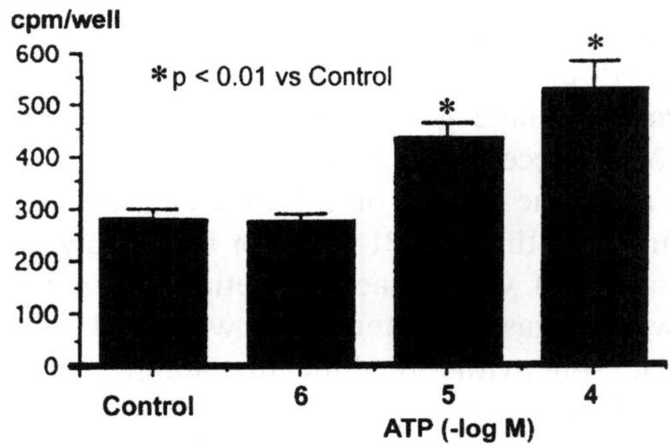

Fig. 25. Dose-dependent effect of ATP on mucus secretion from rabbit gastric mucous cells in primary culture. Cells prelabelled with [^{3}H]glucosamine were incubated with ATP for 30 min. Secreted proteins (solubilised in with NaOH and neutralised with 1 N HCl) were measured using a scintillation counter and expressed as counts per minute per well. Values are means ± SEM for four determinations (from OTA et al. 1994)

IV. Gastric Mucus Secretion

ATP stimulated mucous glycoprotein secretion by rabbit gastric mucous cells in primary culture (Fig. 25) (OTA et al. 1994). The order of potency of ATP analogues was $\alpha\beta$-meATP > ATP > 2meSATP; the efficacy of ATP analogues to increase $[Ca^{2+}]_i$ was similar. A study of mucin secretion in the goblet cell line, HT29-Cl.16E, suggested that both ATP and carbachol produce exocytotic release of mucin by acting on the same granular pool (BERTRAND et al. 1999). P1 receptor agonists had no effect. P2 receptor-mediated stimulation of mucus secretion appeared to be mediated by intracellular calcium, not by endogenous PGE_2. An autoradiographic study of sections of rabbit fundus with [^{35}S]2′-deoxyATPαS, regarded as a radioligand for P2Y receptors, shows a selective distribution over the mucosa, but not muscle layer, and was paralleled by high density binding on gastric gland plasma membranes (VALLEJO et al. 1996).

V. Intestinal Secretion

The first hint that extracellular ATP might be involved in electrolyte secretion in the intestine was the observation that ATP, either in the mucosal or serosal fluid, caused a transient increase in the potential difference and short-circuit current across the wall of rat small intestine or colon (Fig. 26) (KOHN et al. 1970). Later, ATP, ADP and AMP, but not adenosine, were shown to increase cAMP-mediated stimulation of active ion transport in dispersed enterocytes prepared from the guinea-pig small intestine (KORMAN et al. 1982) and later, ATP was shown to stimulate Ca^{2+}-uptake in isolated rat intestinal epithelial cells (RICHARDS et al. 1987).

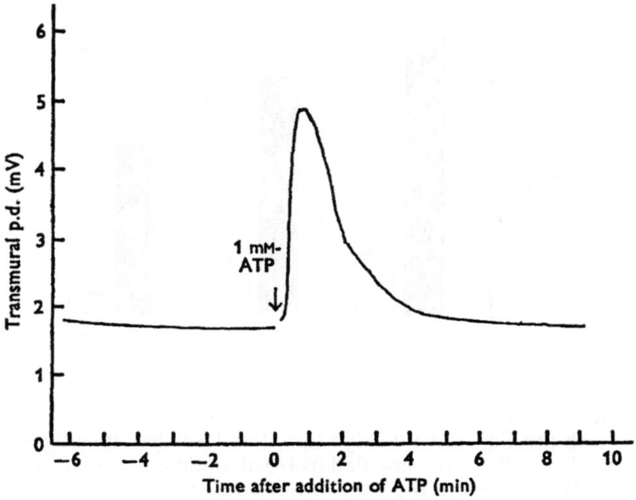

Fig. 26. Time course of the change in potential difference across the wall of the rat jejunum following the addition of 1 mmol/l ATP to the mucosal bathing medium. Positive values for potential indicate that the serosal side is positive with respect to the mucosal side (from KOHN et al. 1970)

Rabbit ileal mucosa, when mounted in a flux chamber and subjected to electrical field stimulation, secreted Cl^-, a change reflected in an increase in short-circuit current and it was suggested that the mediator was likely to be a combination of ACh and NANC neurotransmitters released from nerves lying close to the secretory epithelium (HUBEL 1984). While the identity of the NANC mediator was not clear, it was noted that one of the early antagonists proposed for ATP, namely $ANAPP_3$, did not significantly reduce the response to electrical field stimulation.

Experiments carried out by CUTHBERT and HICKMAN (1985) confirmed the earlier reports about the effects of ATP on transepithelial ion transport but, since they found that TTX virtually abolished the effects of ATP on electrogenic chloride secretion, they suggested that the effects of ATP were indirect, via neural elements in the intramural plexus.

Under resting conditions the mammalian distal colon has a NaCl-absorptive epithelium, the absorption occurring at surface cells in colonic crypts and intracellular Ca^{2+} or cAMP are important second messengers that activate NaCl secretion (BINDER and SANDLE 1994). A study of ATP actions in isolated crypts of rat distal colon, using the fura-2 technique to measure $[Ca^{2+}]_i$ (Fig. 27) (LEIPZIGER et al. 1997) led to the following conclusions:

1. Basolateral ATP induces $[Ca^{2+}]_i$ in isolated crypts and acts as a secretagogue in the distal rat colon.
2. A basolateral P2Y receptor is responsible for this ATP-induced NaCl secretion.

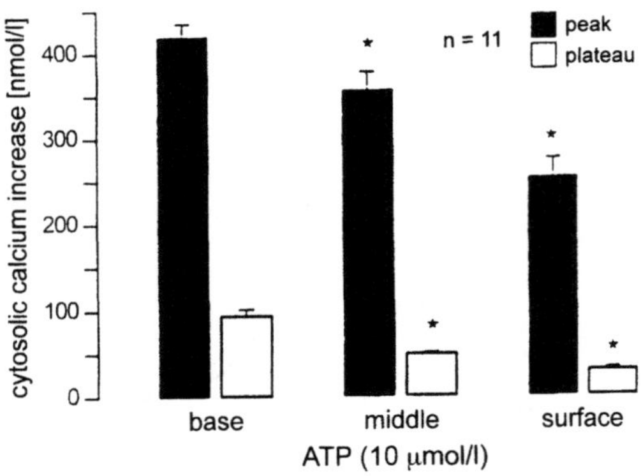

Fig. 27. ATP-induced (10 μmol/l) $[Ca^{2+}]_i$ increase in the rat isolated, intact colonic crypt was measured with the Ca^{2+}-sensitive dye fura-2 and a video imaging set-up along the axis of the crypt. The cells were classified as basal, middle and surface cells according to their location. Base refers to the very base, middle refers to 50% of total length in the middle of the crypt and surface are only cells at the very surface. Both the ATP-induced $[Ca^{2+}]_i$ peak and plateau decrease along the crypt axis. *Asterisks* indicate significant differences of peak and plateau $[Ca^{2+}]_i$ increases compared between basal and middle and middle and surface cells (from LEIPZIGER et al. 1997)

3. The ATP action is not mediated by adenosine.
4. The ATP-induced $[Ca^{2+}]_i$ signals are mostly located at the crypt base, which is the secretory part of the colonic crypt.

The rank order of potencies for these actions was 2-meSATP > ADP > ATP >> UTP, suggesting that a $P2Y_1$ receptor might be involved. In a later abstract, this group reported that luminal ATP induces K^+ secretion via a $P2Y_2$ receptor in rat distal colonic mucosa (KERSTAN et al. 1997). It is interesting that 2′,3′-*O*-(2,4,6-trinitrocyclohexadienylidene)ATP (TNP-ATP), which has since been identified as a potent antagonist at $P2X_1$ and $P2X_3$ receptors (VIRGINIO et al. 1998), was shown to block colonic Cl^- channels (VENGLARIK et al. 1993), although there do not appear to be reports of the effects of TNP-ATP on P2Y receptors. In situ hybridisation studies have shown that the mRNA for $P2X_4$ receptors is localised in rat intestinal crypts (TANAKA et al. 1996).

ATP regulation of Cl^- secretion has also been demonstrated in a human intestinal epithelial cell line, Caco-2, grown on permeable membrane supports and assayed for Cl^- secretion by measuring short circuit current (Fig. 28) (INOUE et al. 1997). The potency order on the apical side was UTP > ATP > UDP > 2-meSATP = ADP and on the basolateral side UTP = 2-meSATP = ATP > ADP >> UDP, suggesting that two different P2Y receptor subtypes are involved.

The strong presence of ecto-diphosphohydrolase (apyrase) in rat small intestinal brush-border membranes has been demonstrated (SCHWEICKHARDT

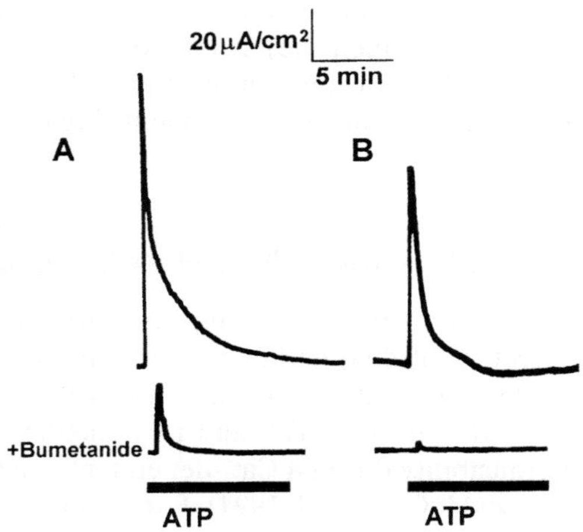

Fig. 28. Representative short circuit current (I_{sc}) recordings from Caco-2 cells (a human epithelial cell line), illustrating stimulation by extracellular ATP (10 µmol/l) applied to the **A** apical and **B** basolateral surface of the cells. *Lower traces* show pretreatment with bumetanide (100 µmol/l in basolateral bathing solution) which substantially inhibited responses to subsequent ATP (from INOUE et al. 1997)

et al. 1995), consistent with the view that nucleotides have potent actions in mucosal epithelial cells.

Adenosine was claimed to stimulate electrolytic secretion in isolated epithelia of rabbit colon (GRASL and TURNHEIM 1984) and the P1 receptor antagonist, theophylline, caused an increase in short circuit current and reversed the direction of net Cl^- movement in rabbit ileum (AHSAN et al. 1987). An examination of the effects of various analogues of adenosine led to the conclusion that the A_1 receptor subtype is present in rat jejunal mucosal epithelial cells (REYMANN and GNIESS 1988). Earlier studies showed that increases in short-circuit current in both small and large intestine were preferential to ATP, with adenosine having significantly less effect (CUTHBERT and HICKMAN 1985; KOHN et al. 1970).

A study of goblet cell-like clone derived from colonic HT-29 cells led to the conclusion that ATP-stimulated increase in Cl^- current does not require an increase in $[Ca^{2+}]_i$ suggesting the involvement of either another signalling pathway or direct activation of Cl^- channels via purinergic receptors (GUO et al. 1995). ATP-stimulated electrolyte and mucin secretion by this human intestinal goblet cell line has also been reported (MERLIN et al. 1994).

Adenosine has been shown to inhibit intestinal fluid secretion and a study of the relative actions of various adenosine agonists and antagonists led to the conclusion that the P1 (A_{2B}) receptor subtype is involved (HANCOCK and COUPAR 1995).

VI. Purinergic Regulation of Epithelial Cell Transport

Exogenous ATP added to the medium bathing the mucosal surface of the intestine inhibits calcium transport to reduce the unidirectional flux of Ca^{2+} from the mucosal to serosal side (WRÓBEL and MICHALSKA 1977).

Inhibition of uptake of amino acids (including leucine, lysine, alanine, valine and isoleucine) from isolated intestinal epithelial cells by extracellular ATP has been demonstrated (REISER and CHRISTIANSEN 1971) as well as regulation of Na^+-dependent sugar transport (KIMMICH and RANDLES 1980).

VII. Biliary Epithelial Cell (Cholangiocyte) Secretion

The biliary epithelium contributes to bile formation through absorption and secretion of fluid and electrolytes. In an early study of the actions of purine nucleotides and nucleosides on isolated bile duct epithelial cells, it was considered that both ATP and its breakdown product adenosine enhanced Cl^- permeability through Ca^{2+}-dependent and Ca^{2+}-independent pathways respectively (MCGILL et al. 1994). Extracellular nucleotides were claimed to modulate secretory and absorptive function of cholangiocytes by activating Na^+/H^+ exchange mechanisms (ELSING et al. 1996).

Since ATP and UTP were equipotent in increasingly short circuit currents, in voltage-clamped cultured rat cholangiocytes it was concluded that the exposure of polarised cholangioytes to ATP results in luminal Cl^- secretion through activation of P2U (now regarded as $P2Y_2$ or $P2Y_4$) receptors in the apical membrane (Fig. 29) and that release of ATP into bile may serve as an autocrine or paracrine signal regulating cholangiocyte secretory function (SCHLENKER et al. 1997). ATP released into the bile by hepatocytes and cholangiocytes, in addition to activating Cl^-conductance, also promotes Na^+/H^+ exchange through apical $P2Y_2$ receptors (ZSEMBERY et al. 1998).

The source of the extracellular ATP producing increases in membrane Cl^- permeability of biliary epithelial cells was examined recently with human and rat cultures (see Fig. 30). ATP release from these cells increased rapidly during cell swelling induced by hypotonic stress (ROMAN et al. 1999). Removal of extracellular ATP with apyrase or P2 receptor blockade with suramin reversibly inhibited whole cell Cl^- current activation and prevented cell volume recovery during hypotonic stress; furthermore, apyrase induced cell swelling under isotonic conditions.

Electrical field stimulation of the isolated guinea-pig common bile duct produced a slowly developing contraction, followed by a quick phasic "off-peak" contraction; PPADS selectively blocked the "off-peak" contraction and the contractions to ATP (PATACCHINI et al. 1998). It was concluded that ATP is involved as a NANC excitatory transmitter, as well as tachykinins.

A valuable review of the emerging role of purinergic signalling in gastrointestinal epithelial secretion and hepatobiliary function has been published recently (ROMAN and FITZ 1999).

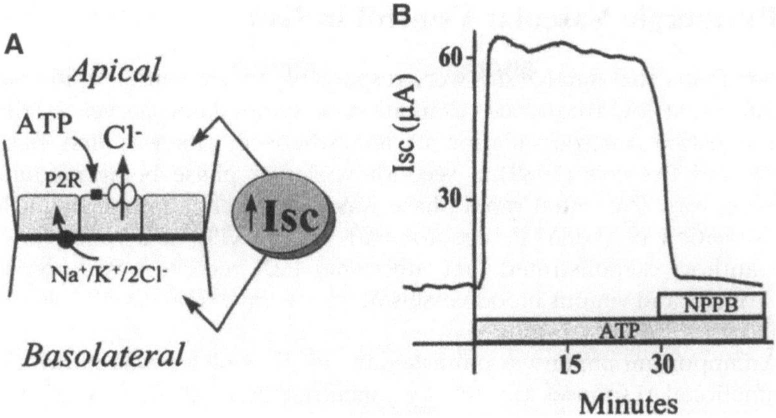

Fig. 29A,B. ATP stimulates transepithelial secretion across rat cholangiocyte monolayers in culture. In this example, polarised cell monolayers are mounted in an Ussing chamber, and the transepithelial movement of Cl^- ions from basolateral to apical space is measured as short-circuit current (I_{sc}, **A**). Under basal conditions, I_{sc} is low (**B**). However, addition of ATP (*bar*) activates $P2Y_2$ receptors (P2R) in the apical membrane and elicits a brisk Cl^- secretory response as demonstrated by a marked increase in I_{sc}. Apical Cl^- secretion occurs via opening of apical membrane Cl^- channels and is inhibited by application of the selective anion channel blocker 5-nitro-2-(3-phenylpropylamino) benzoic acid (NPPB, *bar*) (from Roman and Fitz 1999, adapted from Schlenker et al. 1997)

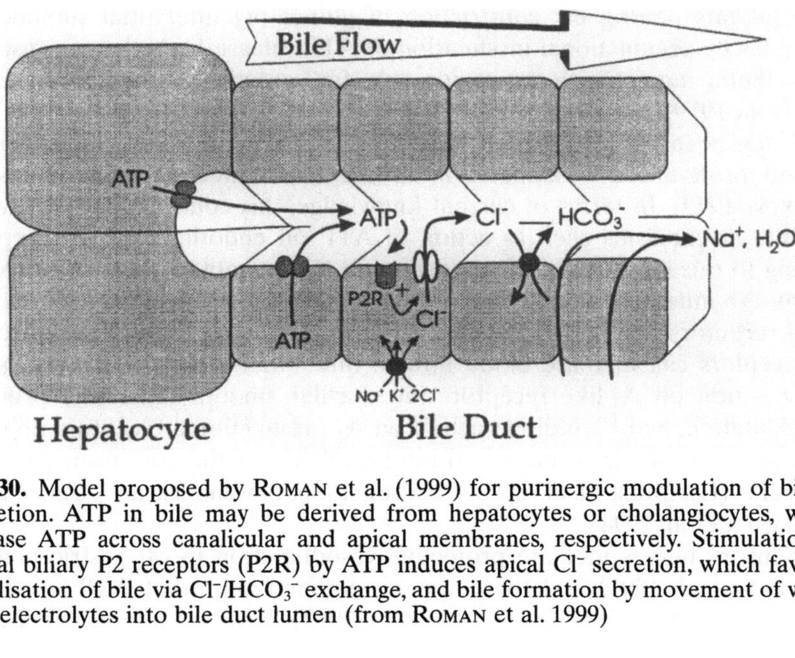

Fig. 30. Model proposed by Roman et al. (1999) for purinergic modulation of biliary secretion. ATP in bile may be derived from hepatocytes or cholangiocytes, which release ATP across canalicular and apical membranes, respectively. Stimulation of apical biliary P2 receptors (P2R) by ATP induces apical Cl^- secretion, which favours alkalisation of bile via Cl^-/HCO_3^- exchange, and bile formation by movement of water and electrolytes into bile duct lumen (from Roman et al. 1999)

E. Purinergic Vascular Control in Gut

The possibility that nucleotides were responsible for the initial rapid response, particularly at low frequency stimulation of sympathetic nerves, of the perfused intestinal microcirculation of anaesthetised cats was first raised by TAYLOR and PARSONS (1989), a secondary, slower phase being mediated by adrenoceptors. The initial rapid phase was completely abolished by selective desensitisation of the ATP receptor with $\alpha\beta$-meATP. In a follow-up study, these authors demonstrated that functional P2X receptors were present in both arterial and venous blood vessels of the cat intestinal circulation (TAYLOR and PARSONS 1991).

An important paper was published in 1992 in which it was shown that the postjunctional responses (excitatory junction potentials as well as constrictions) of guinea-pig submucosal arterioles to sympathetic nerve stimulation were mediated solely through the activation of P2X receptors by ATP or a related purine nucleotide (Fig. 31); the function of neurally released NA was to act through prejunctional α_2-adrenoceptors to depress transmitter release (EVANS and SURPRENANT 1992). The finding was confirmed and extended by another laboratory, which examined the relative potencies of a number of purinergic agonists on guinea-pig submucosal arterioles and showed that the constrictions were antagonised by suramin and PPADS (GALLIGAN et al. 1995). Surprenant and her colleagues speculated in review articles (SURPRENANT 1994; VANNER and SURPRENANT 1996) that sympathetic purinergic control of arterioles is involved in extrinsically-mediated mucosal reflex activity, particularly during inflammatory conditions. In a recent paper it was considered that ACh inhibits neurogenic constriction of guinea-pig interstitial submucosal arterioles by prejunctional modulation of ATP release from the perivascular sympathetic nerves with no major role for endothelial paracrine factors (KOTECHA 1999).

It has been known for some time that ATP, when injected interarterially, elicited pronounced vasodilatation in the denervated cat small intestine (FOLKOW 1949). In terms of current knowledge, this could be due to several possible mechanisms such as action of ATP on endothelial P2Y receptors leading to release of NO, a direct action on P2Y receptors located on vascular smooth muscle or breakdown of ATP by ectoenzymes to adenosine to act on P1 receptors (see BURNSTOCK and RALEVIC 1994). Adenosine acting through P1 receptors can increase blood flow in interstitial vessels in two ways: by direct action on A_2-like receptors on vascular smooth muscle to produce vasodilatation; and by indirect action on A_1 prejunctional receptors on sympathetic vasoconstrictor nerves to inhibit release of the cotransmitters' NA and ATP (GRANGER et al. 1980; PROCTOR 1986; SAWMILLER and CHOU 1991; PENNANEN et al. 1994).

Sensory nerves mediate protective vasodilatation in rat gastric mucosa (HOLZER et al. 1991) and there was an earlier report that ATP causes an increase in gastric blood flow (YANO et al. 1983). This is of interest since ATP

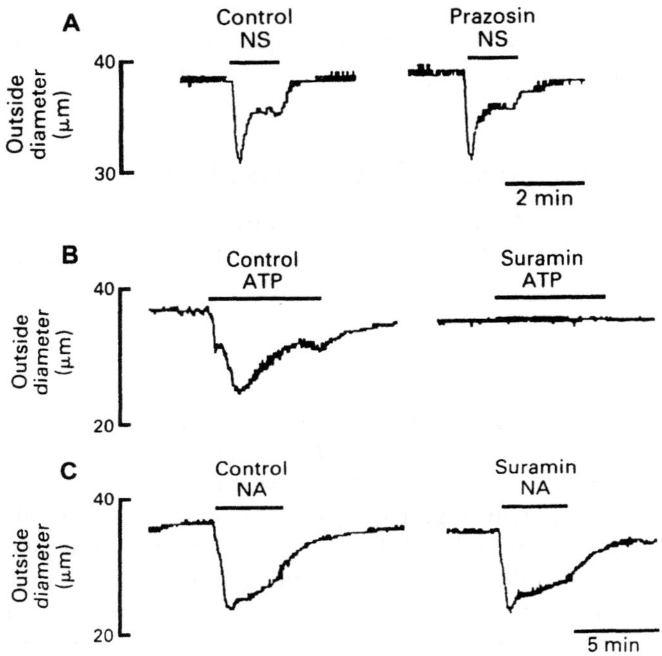

Fig. 31A–C. Constrictions of submucosal arteries of guinea-pig in response to nerve stimulation are not mediated by noradrenaline but through the activation of P2 receptors. **A** Nerve-evoked constrictions (NS, 100 pulses at 10 Hz) were unaffected by the α1-adrenoceptor antagonist, prazocin (0.1 μmol/l). **B** Constrictions to exogenously applied ATP (3 μmol/l) were abolished by the P2 receptor antagonist, suramin (100 μmol/l). **C** Suramin (100 μmol/l) had no effect on the contraction evoked by the exogenous application of noradrenaline (NA, 3 μmol/l). Vessel diameter of isolated superfused submucosal arteries was measured using an on-line computer analysis of TV images with an Imaging Technology system, sampling data at 10–20 Hz with a resolution of less than 1 μm (from Evans and Surprenant 1992)

is a cotransmitter in sensory-motor nerves and upon release acts on P2Y receptors present in the vascular smooth muscle (Burnstock and Warland 1987).

Mesenteric arteries supplying the gastrointestinal tract of rat, guinea-pig, rabbit and dog have been shown to be innervated by sympathetic nerves in which ATP is a major cotransmitter with NA (Krishnamurty and Kadowitz 1983; Ishikawa 1985; von Kügelgen and Starke 1985; Muramatsu 1986; Machaly et al. 1988; Bulloch and Starke 1989; Sjöblom-Widfeld et al. 1990; Bültmann et al. 1991; Donosoet al. 1997; Onaka et al. 1997). For the sympathetic nerves in the jejunal branches of the rabbit mesenteric artery, like those supplying submucosal arterioles, ATP appears to be the sole transmitter, while NA acts prejunctionally (Ramme et al. 1987). In a later paper (Evans and Cunnane 1992) it was concluded that, while contraction of the rabbit jejunal artery to short trains of stimuli is predominantly purinergic, a noradrenergic component can be revealed at higher frequencies of stimulation or during longer trains of stimuli.

F. Purinergic Signalling in Diseased Gut

There are only a limited number of studies to date on changes in purinergic signalling in the gut in disease conditions, but this is undoubtedly a growth area.

I. Gastric Ulcers

Gastric hyperacidity is generally regarded as a causative factor in the development of gastric ulceration as well as *Helicobacter pylori* infection (GOLDMAN and ROSOFF 1968; MCCOLL 1997). There was an early study of the role of ATP in the development of gastric hypersecretion and ulceration in pylorus-ligated rats (MÓZSIK et al. 1979). It was concluded that in the pylorus-ligated rats gastric acid secretion was an ATP-dependent process and that the cAMP system (modulated via adenosine acting on P1 receptors) had an inhibitory effect on the development of ulceration. Methylxanthines, perhaps acting by blocking the actions of adenosine, stimulate the volume and acid content of gastric secretions (JOHANNESSON et al. 1985) and promote gastric ulceration (HENRY and STEPHENS 1980; ERNSTER 1984). Adenosine has been found to decrease gastric ulceration induced by exposing animals to cold, stressful environments (WESTERBERG et al. 1986; DALY 1982). Dipyridamole, probably by increasing extracellular adenosine, significantly reduced the extent of gastric bleeding and ulcer formation (PARET et al. 1982). Some possible mechanisms involved in the actions of adenosine as a protective agent against gastric ulceration are discussed by WESTERBERG and GEIGER (1988). It should be mentioned that there is one early report that adenosine, administered intracerebrally or subcutaneously, increased stress-induced gastric lesions (USHIJIMA et al. 1985).

The involvement of central P1 receptors in the development of gastric ulcer formation has been raised by several groups. The general consensus is that adenosine receptor activation in the brain reduces stress-induced ulcer formation (GEIGER and GLAVIN 1985). Intracerebroventricularly administered adenosine and analogues reduced formation of gastric ulcers in stressed rats and this protective effect was reversed by the P1 antagonist, 8-sulphophenyltheophylline, given centrally or peripherally (WESTERBERG and GEIGER 1987). This finding was supported by another group who concluded that central adenosine inhibits gastric secretion by decreasing the stimulatory vagal impulses to the stomach (PUURUNEN and HUTTUNEN 1988).

Adenosine deaminase activity was studied in the gastric mucosa of patients with peptic ulcer (NAMIOT et al. 1991). Enzyme activities were higher at a distance of over 2 cm from the ulcer margin than that measured close to the ulcer, and there was a significant decrease in adenosine deaminase activity after treatment with ranitidine (NAMIOT et al. 1993).

II. Chagas' Disease

Chagas' disease (American trypanosomiasis) is due to the parasitic haemaflagellate protozoan *Trypanosoma cruzi*, which is transmitted from animals to man by the Reduviidae bug. *T. cruzi* is unable to synthesise its own adenine (FERNANDES and CASTELLANI 1958). The disease is characterised by inflammatory lesions on various organs and nerves, including the gut. The gut disorders include megaoesophagus, loss of propulsive activity in the oesophagus and achalasia of the lower oesophageal sphincter in advanced cases, as well as megacolon where dilatation extends to the rectum in 80% of the cases; motility is lower than normal and constipation is common (SAID et al. 1985). Evidence for loss of distal ganglia in Auerbach's and Meissner's plexuses, that results in loss of postganglionic intrinsic nerves, has been reported and hyperexcitability of intestinal muscle to various agents as well as disturbances of absorption. Profuse salivary secretion is also common, due to salivary gland denervation, especially in patients with megaoesophagus and megacolon. Pyloric dysfunction is relatively common too (see MARTINS CAMPOS and TAFURI 1973; LONG 1983).

While little is known about the neurotransmitters most affected in Chagas' disease, there are some hints that purinergic signalling might be impaired. For example, both low affinity Mg^{2+}-activated ATPase and high affinity (Ca^{2+}-Mg^{2+}) ATPase (FRASCH et al. 1978; CATALDI DE FLOMBAUM and STOPPONI 1992) as well as adenosine kinase (KIDDER 1982) have been shown to be present in *T. cruzi*, which rapidly break down extracellular nucleotides. Furthermore, in a recent abstract, enhancement of $P2X_7/P_{2Z}$ receptor-associated cell permeabilisation during the acute phase of Chagas' disease was reported (COUTINHO et al. 1998). Undoubtedly, new experiments examining changes in purinergic signalling in the gut of Chagas' disease patients would be rewarding.

III. Hirschsprung's Disease

Hirschsprung's disease is a congenital abnormality of the enteric nervous system and is characterised by the absence of ganglion cells in the submucosal and myenteric plexuses of variable lengths of the hind gut from the rectum forward and by chronic constriction of the aganglionic region. It is known that there is hyperinnervation by extrinsic sympathetic nerves containing NA and preganglionic parasympathetic nerves containing ACh (see GANNON et al. 1969; IKAWA 1981), but enteric nerves arising from intrinsic neurones containing various neuropeptides, 5-HT, NO and presumably ATP, show marked reductions in density and may be absent from aganglionic segments (ROGAWSKI et al. 1978; LARSSON 1994; LARSSON et al. 1995). IJPs could not be evoked in aganglionic segments of human colon (FRIGO et al. 1973; OKASORA and OKAMOTO 1986), nor in aganglionic segments of piebald-lethal mouse colon (RICHARDSON 1975; BYWATER and TAYLOR 1982; OKASORA and OKAMOTO 1986). Evidence has been presented earlier (see Sect. B.V) that IJPs in the colon are partly mediated by ATP. However, only one paper appears to have addressed directly the involvement of purinergic signalling in Hirschsprung's disease (ZAGORODNYUK et al. 1989). These authors reported that no IJPs could be evoked by intramural nerve stimulation of the longitudinal muscle excised from the rectosigmoidal part of the large intestine of Hirschsprung's patients and that excitatory purine receptor-mediated contractions of the longitudinal muscle were produced in the majority of preparations from the constricted region of the large intestine from patients with Hirschsprung's disease.

IV. Diabetes

Alterations in various activities of the gut have been described in diabetic patients, including gastric motility, gastroparesis manifested by nausea, vomiting, early satiety, postprandial bloating and diffuse epigastric pain, as well as episodic diarrhoea and more prominently sustained constipation (LOO et al. 1984; LOW 1996). Dysphagia of the oesophagus and impairment of gastric emptying have also been reported (see PFEIFER et al. 1993; SPALLONE et al. 1995). A complexity of progressive changes in adrenergic, serotoninergic and peptidergic nerves has been described in the gastrointestinal tract of streptozotocin diabetic rats (BELAI et al. 1988). Abnormalities in cholinergic (NOWAK et al. 1986) and nitrergic activities (MARTINEZ-CUESTA et al. 1995; JENKINSON and REID 1995;TAKAHASHI et al. 1997) have also been reported.

Electrical recording from gastric smooth muscle from streptozotocin-induced diabetic rats during transmural nerve stimulation showed IJPs of reduced amplitude and no EJPs (XUE and SUZUKI 1997). Some studies of the responses to purines in streptozotocin rats have been carried out. Electrophysiological studies showed that the IJPs and the hyperpolarisations in response to ATP were similar in the circular muscle of the caecum of streptozotocin diabetic (8 week) and untreated control rats, although the rate of hyperpolarisation of single IJPs was slower in the diabetic tissues (HOYLE et al. 1988b). BELAI et al. (1991) showed that where ATP-induced relaxations of longitudinal strips from the gastric fundus were not significantly different in control and diabetic rats, the stimulation-induced release of ATP *increased* threefold in diabetic compared to control gastric fundus. Desensitisation of receptors to ATP with $\alpha\beta$-meATP reduced the relaxant responses to both ATP and electrical field stimulation, suggesting a role for ATP in NANC neurotransmission in rat gastric fundus and this reduction was greater in diabetic tissues. In view of this data, the authors suggested that the purinergic component of the vagal NANC responses of the stomach may be increased in diabetes, a finding reminiscent of an increased purinergic component in parasympathetic control of the bladder in interstitial cystitis (PALEA et al. 1993) and in sympathetic nerves supplying blood vessels in spontaneously hypertensive rats (VIDAL et al. 1986; BULLOCH and McGRATH 1992; BROCK and VAN HELDEN 1995). The effects of ATP and adenosine on the urinary bladder and colon muscle from streptozotocin diabetic rats have been examined (GÜR and KARAHAN 1997). It was shown that, while maximum relaxant responses and sensitivity of the colon to ATP were unchanged in 8-week diabetic rats, the responses to adenosine were reduced.

V. Ischaemia

Autopsies suggest that death resulting from ischaemic damage to the gastrointestinal tract is more common than often realised. Damage takes different forms, but essentially necrosis of the colonic wall rapidly become gangrenous because of its contact with intraluminal microbial flora; in this way, the vascular origin of the lesion is often hidden (see SAEGESSER et al. 1979).

Mg-ATP has been recommended as an adjunct for the treatment of ischaemia and shock (CHAUDRY 1983, 1991; CHAUDRY et al. 1986), although other groups have contested this (e.g. SCHLOERB et al. 1981) including its use for the treatment of intestinal ischaemic shock (VAN DER MEER et al. 1982). Shock is considered to be the result of an inadequate circulation with diminished blood flow to tissues, resulting in hypoxic conditions and significant reduction in the levels of ATP (CHAUDRY 1991)

Intestinal ischaemia is characterised by rapid early inhibition of salt and nutrient absorptive function and the appearance of net secretion, resulting in significant intraluminal fluid sequestration often leading to early diarrhoea. In a study of the crypt-like epithelial cell line, T84, a large Cl^- secretory current was observed during the initial phases of cellular ATP depletion, an event highly reminiscent of the early diarrhoeal response of mature intestine to ischaemia. This secretory response was shown to be due to endogenous release of adenosine from the epithelial monolayer (I would suggest, arising partly by direct release from hypoxic cells and partly following ectoenzymic breakdown of released ATP), in quantities sufficient to interact locally with cell surface P1 receptors in autocrine fashion (MATTHEWS et al. 1995). ATP-$MgCl_2$ has been claimed to restore gut absorptive capacity early after trauma-haemorrhagic shock (SINGH et al. 1993).

It is not clear whether the ATP administered is effective in ischaemia by replenishing depleted ATP stores (CHAUDRY 1991) or because it acts on P2Y receptors on endothelial cells to release NO with subsequent vasodilatation and increased blood flow (see BURNSTOCK 1988, 1990). A more recent paper indicated that ATP-$MgCl_2$ reduces intestinal permeability during mesenteric ischaemia, without alteration in the ATP tissue level or increases in intestinal blood flow (KREIENBERG et al. 1996).

Adenosine has been implicated as a potential anti-inflammatory or cytoprotective agent in postischaemic intestine (KAMINSKI and PROCTOR 1992). Experiments have been carried out in dogs to test the hypothesis that adenosine exerts beneficial effects on intestinal motility after experimental mesenteric ischaemia (FEIT and ROCHE 1988).

VI. Food Intake

Peripherally administered purines suppress food intake in rats (CAPOGROSSI et al. 1979) and intracerebroventricular adenosine has also been shown to inhibit food intake (LEVINE and MORLEY 1983). Caffeine blocked the suppressive effect of adenosine on feeding and opioid-induced feeding (WAGER-SRDAR et al. 1984). Administration of the adenosine ligand R-PIA increased food intake, an effect not blocked by caffeine, while 2-CADO and NECA failed to alter feeding behaviour, leaving it unclear as to which P1 receptor subtypes were involved (LEVINE et al. 1989).

Methylxanthine adenosine antagonists enhance taste (SCHIFFMAN et al. 1985). This potentiation is thought to be due to local modulation of endogenously released adenosine at the level of the taste receptors themselves or the sensory nerve endings of the chorda tympani associated with taste cells of the anterior tongue (BARRACO 1991). Human subjects taking theophylline have an increased sensitivity to taste that is reversed when the drug is discontinued and increased taste sensitivity with caffeine (SCHIFFMAN et al. 1985), another adenosine antagonist, which may be one of the factors that lead the elderly to consume greater amounts of tea and coffee. $P2X_2$ and $P2X_3$ ATP ion-gated receptors have recently been localised on taste buds of the rat tongue and are thought to play a nociceptive role (BO et al. 1999).

VII. Colonic Tumours

ATP was claimed to exhibit significant anticancer activities against $CT6_Z$ colon adenocarcinoma in mice, while adenosine was ineffective under identical conditions (RAPAPORT and FONTAINE 1989). Daily intraperitoneal injections of adenine nucleotides in large volumes of saline, starting after the tumours become palpable, resulted in inhibition of tumour growth. Intraperitoneal injections of ATP led to expansion of erythrocyte ATP pools over a period of hours, slow release of ATP from these pools leading to several-fold increases in the blood plasma (extracellular) ATP levels at concentrations sufficient to account for the anticancer activities demonstrated in in vitro systems (RAPAPORT 1983; RAPAPORT et al. 1983).

The human colonic adenocarcinoma cell line, HT-29 has been used for a number of studies of the antitumour activity of nucleotides. Purine and pyrimidine nucleotides depolarised these cells dose-dependently with the potency sequence UTP > ATP > ITP > GTP > TTP > CTP; ATP had greater effects than ADP and adenosine (LOHRMANN et al. 1992). In another study it was shown that ATP-stimulated increase of Cl^- current did not require an increase in $[Ca^{2+}]_i$, suggesting the involvement of either another signalling pathway or direct activation of Cl^- channels by purinergic receptors (GUO et al. 1995). Human colonic HT-29 cells express endogenous P_{2U}-like receptors (PARR et al. 1994; ZHANG and ROOMANS 1997) and ATP increased extracellular acidification rates via these receptors (RICHARDS et al. 1997). Subclones of the HT-29 colonic cancer cell line, a mucin-secretory cell line HT29-Cl.16E and the sister cell line HT29-Cl.19A which lacks mucin granules, both responded to ATP with transient increases in Cl^- conductance (GUO et al. 1997). $P1(A_1)$ adenosine receptors are also expressed in HT-29 colonic adenocarcinoma cells and their block leads to reduction of proliferation (LELIÈVE et al. 1998a). Examination of the effects of adenosine on proliferation on groups of adenocarcinoma cells at different stages of enterocyte differentiation showed that proliferation was decreased by adenosine deaminase, NECA and 8-phenyltheophylline in HT-29, DLD-1, Caco-2 and SW403 cells (LELIÈVE et al. 1998b).

In a recent publication, HÖPFNER et al. (1998) demonstrated expression of functional P2 purinergic receptors in primary cultures of human colorectal carcinoma cells with a rank order of potency ATP > UTP > ATPγS > ADP > adenosine, which is characteristic of the P_{2U}-like ($P2Y_2$ and/or $P2Y_4$) receptor that they suggested may play a role in regulation of cell proliferation and apoptosis.

RT-PCR studies showed that genes encoding $P2Y_2$, $P2Y_4$ and $P2Y_6$ receptors were expressed in Caco-2 human colonic adenocarcinoma cells (MCALROY et al. 1999), while Cl^- secretion was shown to be regulated by ATP in T84 cells, another human colonic adenocarcinoma cell line (DHO et al. 1992). In this latter cell line, adenosine also apparently caused Cl^- secretion, which was reversed by 8-phenyltheophylline (BARRETT et al. 1990).

G. Future Directions

The field of purinergic signalling in general is exploding in many directions and gut physiology and pathophysiology is no exception. From this review it is clear that there is by now abundant evidence for purinergic signalling in neuromuscular and synaptic transmission in the gut as well as in neural control of various epithelial cell functions and in control of vascular tone, but few studies to date concerning purinergic signalling in enteric glial cells (see Sect. C.III) or interstitial cells of Cajal. A few areas ripe for development with hints of things to come have been selected for discussion in this final section.

I. Pain and Inflammation

The recent discovery of the $P2X_3$ receptor (a ligand-gated ion channel) that is selectively expressed on small diameter nociceptive sensory nerves (CHEN et al. 1995; LEWIS et al. 1995), has led to the exploration of the involvement of purinergic signalling in pain (see BURNSTOCK and WOOD 1996) and to the possible sources of ATP triggering $P2X_3$ receptors in the skin and visceral organs (BURNSTOCK 1996, 1999). It is well known that distension of the gut is often associated with pain and a new hypothesis has recently been put forward for purinergic mechanosensory transduction in tubes, including ureter, salivary duct, cervix, bile duct and gut, and sacs such as urinary bladder, gall bladder and lung, where distension leads to release of ATP from epithelial cells lining the lumen that then acts on $P2X_3$ receptors located on the dense subepithelial sensory nerve plexus sending information to pain centres in the CNS via no dose, trigeminal and dorsal root ganglia (Fig. 32) (BURNSTOCK 1999). This hypothesis is yet to be tested in the gut, but there are hints that are encouraging. For example, there was early recognition that slow distension of the stomach evoked a non-painful sensation like fullness, but at a greater inten-

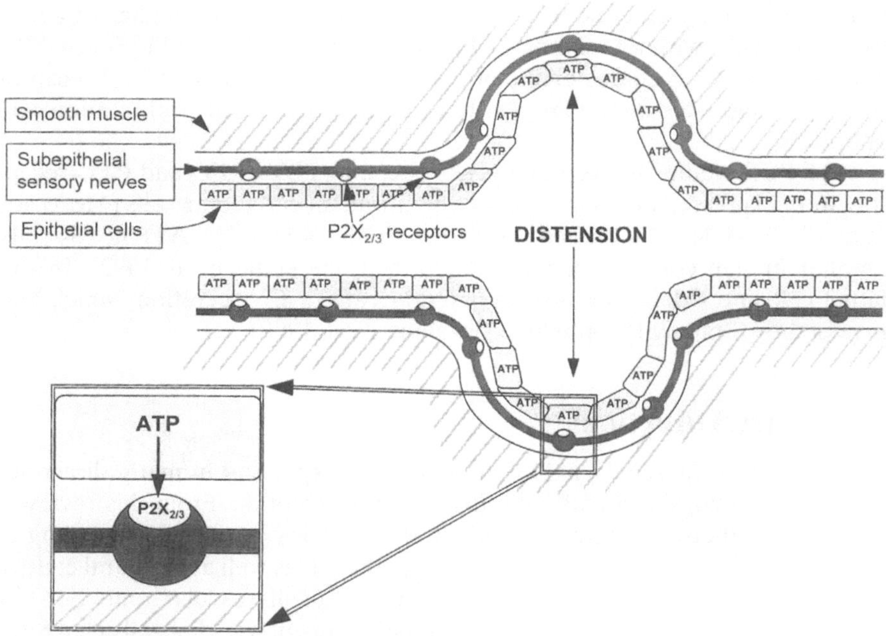

Fig. 32. Schematic representation of hypothesis for purinergic mechanosensory transduction in tubes (e.g. gastrointestinal tract, salivary and bile ducts, ureter, vagina) and sacs (e.g. urinary and gall bladders, and lung). It is proposed that distension leads to release of ATP from epithelium lining the tube or sac, which then acts on $P2X_{2/3}$ receptors on subepithelial sensory nerves to convey sensory/nociceptive information to the CNS (from BURNSTOCK 1999)

sity of distending pressure the sense of fullness was replaced by pain (Hertz 1911). The receptive endings of the mucosal afferents are generally located in or below the mucosal epithelium and are sensitive to the neurotoxic action of capsaicin (Holzer and Barthó 1996). These afferents are generally mechanosensitive and exhibit a rapidly adaptive response to mechanical probing of the mucosa (Blumberg et al. 1983; Jänig and Koltzenberg 1991). Thirty percent of pelvic nerve colonic afferents have high thresholds for responses to distending pressures greater than 30 mmHg (see Sengupta and Gebhart 1994, 1998) and a behavioural study has demonstrated that colonic distension greater than or equal to 30 mmHg is noxious in the rat (Ness and Gebhart 1988). There is also a recent study showing mechanosensitivity of visceral afferents in stomach and colon (Raybould et al. 1999). On the basis of their response to noxious intensity of distension, it has been proposed that high threshold afferents are probably nociceptive-specific visceral neurones (Cervero 1994). In a recent study, $\alpha\beta$-meATP, administered intraarterially, induced dose-dependent increases in mesenteric afferent nerve discharge in nerves supplying the jejunum of anaesthetised rats; PPADS and suramin antagonised this discharge (Kirkup et al. 1999).

Irritable bowel syndrome (IBS) is associated with abnormal sensory perceptions of fullness, discomfort and pain and it has been suggested that this may be due to hypersensitivity of the primary afferents or sensitisation of the central neurones upon which they terminate (Mayer and Gebhart 1994; Bueno et al. 1997). Inflammation of the gastrointestinal tract is another painful condition, where the release of endogenous inflammatory substances is triggered. There are indications that purines are involved in inflammatory conditions in joints and in the brain (see Abbracchio and Burnstock 1998; Brambilla et al. 1999). It will be of interest in future studies to see if purinergic nociceptive mechanisms are involved in these and other pathological conditions. For example, intravenous boluses of adenosine produced transient epigastric discomfort indistinguishable from spontaneous pain in five of six patients with endoscopically-confirmed duodenal ulcer (Watt et al. 1987). On the other hand, intravenous infusion of adenosine has been reported to induce antinociception in humans in experimental and surgical pain (Segerdahl et al. 1994, 1995; Forsberg et al. 1999). ATP released from sympathetic nerves has been claimed to modulate immune cells in the intestinal mucosa and may thereby be of functional relevance in neuroimmunomodulation during inflammatory bowel disease (Straub et al. 1999).

Recent developments implicate $P2Y_6$ receptors in human T cells, suggesting a role in the pathogenesis of inflammatory bowel disease-mediated intestinal damage (Somers et al. 1998).

II. Long-Term (Trophic) Actions of Purines

Most of the examples of purinergic signalling included in this review concern short-term events such as neuromuscular and synaptic transmission and secretion, but there is growing recognition of long-term (trophic) roles of purines and pyrimidines in such activities and embryonic development, cell proliferation, growth and differentiation, neuritogenesis, wound healing and programmed cell death (see ABBRACCHIO 1996; NEARY et al. 1996; ABBRACCHIO and BURNSTOCK 1998; NEARY and ABBRACCHIO, Chap. 11, first volume). However, investigators of purinergic trophic events in the gut are only just beginning.

For example, it has been shown that $P2X_5$ receptors are strongly represented in proliferating and differentiating cells, while $P2X_7$ receptors label keratinised epithelial cells of various stratified squamous epithelial tissue, including footpad, skin, tongue, oesophagus, vagina, soft palate and cornea (GRÖSCHEL-STEWART et al. 1999a). This finding prompted an extension of these studies in our laboratory (GRÖSCHEL-STEWART et al. 1999b) to examine the turnover of mucosal epithelium, known to occur over 3- to 4-day cycles in the small intestine (MADARA 1991). We showed that antibodies to the $P2X_5$ receptor immunostained the membranes of the narrow "stem" of villus goblet cells where the nucleus and cell organelles reside, while $P2X_7$ receptor immunoreactivity was seen only on the membranes of enterocytes and goblet cells at the tip of the villus, where cells are undergoing apoptosis and are exfoliated into the lumen.

Another indication of trophic purinergic signalling in gut comes from studies of explants of myenteric plexus into the brain (striatum) where it was claimed that neurite outgrowth of striatal neurones was evoked by combined, perhaps synaptic, actions of purines released from NANC inhibitory nerves and growth factors, probably released from enteric glial cells (HÖPKER et al. 1996). Trophic actions of 2-CADO acting synergistically with basic fibroblast growth factor (bFGF) to increase neurite outgrowth of myenteric neurones in dissociated cell cultures has also been reported (SCHÄFER et al. 1995).

The mechanisms involved in the plasticity of changes in structure, neurochemical coding and survival of enteric neurones and their effectors that occur in pathological conditions such as obstruction leading to hypertrophy or after surgery or denervation of extrinsic nerves or in ageing or even related to diet (GIACOBINI ROBECCHI et al. 1985; DAHL et al. 1987; SANTER and BAKER 1988; GABELLA 1984, 1989, 1990; SANTER and CONBOY 1990; LUCK et al. 1993; YUNKER and GALLIGAN 1994; JOHNSON et al. 1998) are not known. It will be interesting to see if long-term purinergic signalling is involved in these events.

References

Abbracchio MP (1996) P1 and P2 receptors in cell growth and differentiation. Drug Dev Res 39:393–406

Abbracchio MP, Burnstock G (1998) Purinergic signalling: pathophysiological roles. Jpn J Pharmacol 78:113–145

Ahlman H, Dahlstrom A (1983) Vagal mechanisms controlling serotonin release from the gastrointestinal tract and pyloric motor function. J Auton Nerv Syst 9:119–140

Ahsan MA, Ilundain A, Naftalin RJ, Sandhu BK, Smith PM (1987) Effects of theophylline, choleragen and loperamide on rabbit ileal fluid and electrolyte transport *in vitro*. Br J Pharmacol 92:743–754

Ainz LF, Gil-Rodrigo CE, Gómez R, Malillos M, Requejo D, Gandarias JM (1989) Effects of various physiologic adenine derivatives on the secretion of acid in isolated gastric glands in rabbits. Rev Esp Fisiol 45:281–286

Ainz LF, Salgado C, Gandarias JM, Gómez R, Vallejo A, Gil-Rodrigo CE (1993) $P_1(A_2/Ra)$-purinoceptors may mediate the stimulatory effect of adenosine and adenosine analogs on acid formation in isolated rabbit parietal cells. Pharmacol Res 27:319–334

Akbarali HI, Bieger D, Triggle CR (1986) Tetrodotoxin-sensitive and -insensitive relaxations in the rat oesophageal tunica muscularis mucosae. J Physiol 381:49–63

Akkermans LMA, Houghton LA, Brown NJ (1989) Neural and hormonal control of pyloric sphincter function. Scand J Gastroenterol 24:27–31

Al Humayyd M, White TD (1985) 5-Hydroxytryptamine releases adenosine 5′-triphosphate from nerve varicosities isolated from the myenteric plexus of guinea-pig ileum. Br J Pharmacol 84:27–34

Alfahel E, Korngreen A, Parola AH, Priel Z (1996) Purinergically induced membrane fluidization in ciliary cells: characterization and control by calcium and membrane potential. Biophys J 70:1045–1053

Allescher HD, Tougas G, Vergara P, Lu S, Daniel EE (1992) Nitricoxide as a putative nonadrenergic noncholinergic inhibitory transmitter in the canine pylorus in vivo. Am J Physiol 262:G695–G702

Alomar CC, Santos C, Puig-Parellada P (1999) Evidence that inhibitory neurotransmission differs between the proximal and distal segments of guinea-pig taenia caeci. Eur J Pharmacol 369:215–219

Altdorfer K, Feher E, Donath T, Feher J (1996) Nitric oxide synthase-containing nerve elements in the pylorus of the cat. Neurosci Lett 212:195–198

Alumets J, Schaffalitzky de Muckadell O, Fahrenkrug J, Sundler F, Hakanson R, Uddman R (1979) A rich VIP nerve supply is characteristic of sphincters. Nature 280:155–156

Amsallem H, Métioui M, Vanden Abeele A, Elyamani A, Moran A, Dehaye JP (1996) Presence of a metabotropic and an ionotropic purinergic receptor on rat submandibular ductal cells. Am J Physiol 271:C1546–C1555

Andrews PLR, Lawes INC (1985) Characteristics of the vagally driven non-adrenergic, non-cholinergic inhibitory innervation of ferret gastric corpus. J Physiol 363:1–20

Arkle S, Hanahoe A, Shum CMC (1998) Effects of KN-62, RO-31–8220, Mn^{2+}, Ni^{2+} and Co^{2+} on ATP^{4-}-stimulated responses in rat parotid salivary glands in vitro. Br J Pharmacol 125:83P [Abstract]

Baccari MC, Calamai F, Staderini G (1990a) Effects of arterial infusions of adenosine 5′-triphosphate (ATP) and vasoactive intestinal polypeptide (VIP) on vagal excitatory motor responses in the rabbit stomach in vivo. J Auton Nerv Syst 30 (Suppl): S15–S18

Baccari MC, Calamai F, Staderini G (1990b) Depression by fenoprofen of the "rebound contractions" elicited by vagal stimulation and arterial infusion of ATP in the rabbit stomach in vivo. Exp Physiol 75:415–418

Baccari MC, Calamai F, Staderini G (1991) Vagally-induced non-adrenergic, non-cholinergic inhibitory motility in the rabbit stomach in vivo. Funct Neurol 6:239–242

Baccari MC, Calamai F, Staderini G (1992) The influence of the vagally induced rebound contractions on the non-adrenergic, non-cholinergic (NANC) inhibitory motility of the rabbit stomach and the role of prostaglandins. J Auton Nerv Syst 37:125–135

Baccari MC, Calamai F, Staderini G (1994) Modulation of cholinergic transmission by nitric oxide, VIP and ATP in the gastric muscle. Neuroreport 5:905–908

Baccari MC, Calamai F, Staderini G (1996) Prostaglandin E_2 modulates neurally induced nonadrenergic, noncholinergic gastric relaxations in the rabbit in vivo. Gastroenterology 110:129–138

Baer HP, Frew R (1979) Relaxation of guinea-pig fundic strip by adenosine, adenosine triphosphate and electrical stimulation: lack of antagonism by theophylline or ATP treatment. Br J Pharmacol 67:293–299

Baidan LV, Zholos AV, Wood JD (1995) Modulation of calcium currents by G-proteins and adenosine receptors in myenteric neurones cultured from adult guinea-pig small intestine. Br J Pharmacol 116:1882–1886

Bailey SJ, Hourani SMO (1990) A study of the purinoceptors mediating contraction in the rat colon. Br J Pharmacol 100:753–756

Bailey SJ, Hourani SMO (1992) Effects of purines on the longitudinal muscle of the rat colon. Br J Pharmacol 105:885–892

Bailey SJ, Hickman D, Hourani SMO (1992) Characterization of the P1-purinoceptors mediating contraction of the rat colon muscularis mucosae. Br J Pharmacol 105:400–404

Baird AA, Muir TC (1990) Membrane hyperpolarization, cyclic nucleotide levels and relaxation in the guinea-pig internal anal sphincter. Br J Pharmacol 100:329–335

Banks BEC, Brown C, Burgess GM, Burnstock G, Claret M, Cocks TM, Jenkinson DH (1979) Apamin blocks certain neurotransmitter-induced increases in potassium permeability. Nature 282:415–417

Barajas-López C (1993) Adenosine reduces the potassium conductance of guinea pig submucosal plexus neurons by activating protein kinase A. Pflugers Arch 424:410–415

Barajas-López C, Surprenant A, North RA (1991) Adenosine A_1 and A_2 receptors mediate presynaptic inhibition and postsynaptic excitation in guinea pig submucosal neurons. J Pharmacol Exp Ther 258:490–495

Barajas-López C, Barrientos M, Espinosa-Luna R (1993) Suramin increases the efficacy of ATP to activate an inward current in myenteric neurons from guinea-pig ileum. Eur J Pharmacol 250:141–145

Barajas-López C, Espinosa-Luna R, Gerzanich V (1994) ATP closes a potassium and opens a cationic conductance through different receptors in neurons of guinea pig submucous plexus. J Pharmacol Exp Ther 268:1397–1402

Barajas-López C, Muller MJ, Prieto-Gómez B, Espinosa-Luna R (1995) ATP inhibits the synaptic release of acetylcholine in submucosal neurons. J Pharmacol Exp Ther 274:1238–1245

Barajas-López C, Huizinga JD, Collins SM, Gerzanich V, Espinosa-Luna R, Peres AL (1996a) P_{2x}-Purinoceptors of myenteric neurones from the guinea-pig ileum and their unusual pharmacological properties. Br J Pharmacol 119:1541–1548

Barajas-López C, Peres AL, Espinosa-Luna R (1996b) Cellular mechanisms underlying adenosine actions on cholinergic transmission in enteric neurons. Am J Physiol 271:C264–C275

Barajas-López C, Espinosa-Luna R, Zhu Y (1998) Functional interactions between nicotinic and P2X channels in short-term cultures of guinea-pig submucosal neurons. J Physiol Lond 513:671–683

Barraco RA (1991) Behavioral actions of adenosine and related substances. In: Phillis JW (ed) Adenosine and adenine nucleotides as regulators of cellular function. CRC Press, Boca Raton, FL, pp 339–366.

Barrett KE, Cohn JA, Huott PA, Wasserman SI, Dharmsathaphorn K (1990) Immune-related intestinal chloride secretion. II. Effect of adenosine on T84 cell line. Am J Physiol 258:C902–C912

Barthó L, Petho G, Ronai Z (1985) Theophylline-sensitive modulation of non-cholinergic excitatory neurotransmission in the guinea-pig ileum. Br J Pharmacol 86:315–317

Barthó L, Lénárd LJ, Maggi CA (1997) Evidence for the involvement of P2-purinoceptors in the cholinergic contraction of the guinea-pig ileum. Br J Pharmacol 121:1507–1508
Barthó L, Lénárd LJ, Szigeti R (1998) Nitric oxide and ATP co-mediate the NANC relaxant response in the guinea-pig taenia coli. Naunyn Schmiedeberg's Arch Pharmacol. 358:496–499
Bartlett V, Stewart RR, Nakatsu K (1979) Evidence for two adenine derivative receptors in rat ileum which are not involved in the nonadrenergic, noncholinergic response. Can J Physiol Pharmacol 57:1130–1137
Bauer V (1993) NANC transmission in intestines and its pharmacological modulation. Acta Neurobiol Exp Warsz 53:65–77
Bauer V, Kuriyama H (1982) The nature of non-cholinergic, non-adrenergic transmission in longitudinal and circular muscles of the guinea-pig ileum. J Physiol 332:375–391
Bayguinov O, Sanders KM (1993) Regulation of neural responses in the canine pyloric sphincter by opioids. Br J Pharmacol 108:1024–1030
Beck K, Calamai F, Staderini G, Susini T (1988) Gastric motor responses elicited by vagal stimulation and purine compounds in the atropine-treated rabbit. Br J Pharmacol 94:1157–1166
Behar J, Guenard V, Walsh JH, Biancani P (1989) VIP and acetylcholine: neurotransmitters in esophageal circular smooth muscle. Am J Physiol 257:G380–G385
Belai A, Burnstock G (1994) Evidence for coexistence of ATP and nitric oxide in non-adrenergic, non-cholinergic (NANC) inhibitory neurones in the rat ileum, colon and anococcygeus muscle. Cell Tissue Res 278:197–200
Belai A, Lincoln J, Milner P, Burnstock G (1988) Progressive changes in adrenergic, serotonergic, and peptidergic nerves in proximal colon of streptozotocin-diabetic rats. Gastroenterology 95:1234–1241
Belai A, Lefebvre RA, Burnstock G (1991) Motor activity and neurotransmitter release in the gastric fundus of streptozotocin-diabetic rats. Eur J Pharmacol 194:225–234
Bennett M, Burnstock G, Holman ME (1966) Transmission from intramural inhibitory nerves to the smooth muscle of the guinea-pig taenia coli. J Physiol 182:541–558
Bertrand CA, Laboisse CL, Hopfer U (1999) Purinergic and cholinergic agonists induce exocytosis from the same granule pool in HT29-Cl.16 emonolayers. Am J Physiol 276:C907–C914
Biancani P, Walsh JH, Behar J (1984) Vasoactive intestinal polypeptide: a neurotransmitter for lower esophageal sphincter relaxation. J Clin Invest 73:963–967
Biancani P, Walsh J, Behar J (1985) Vasoactive intestinal peptide: a neurotransmitter for relaxation of the rabbit internal anal sphincter. Gastroenterology 89:867–874
Binder HJ, Sandle GI (1994) Electrolyte transport in the mammalian colon. In: Johnson LR (ed) Physiology of the gastrointestinal tract. Raven Press, New York, pp 2133–2171
Bitar KN, Hillemeier C, Biancani P (1990) Differential regulation of smooth muscle contraction in rabbit internal anal sphincter by substance P and bombesin. Life Sci 47:2429–2434
Blottiére HM, Loirand G, Pacaud P (1996) Rise in cytosolicCa^{2+} concentration induced by P2-purinoceptor activation in isolated myocytes from the rat gastrointestinal tract. Br J Pharmacol 117:775–780
Blumberg H, Haupt P, Jänig W, Kohler W (1983) Encoding of visceral noxious stimuli in the discharge patterns of visceral afferent fibres from the colon. Pflugers Arch 398:33–40
Bo X, Alavi A, Xiang Z, Oglesby IB, Ford APDW, Burnstock G (1999) Localization of $P2X_2$ and $P2X_3$ receptor immunoreactive nerves in rat taste buds. Neuroreport 10:1107–1111
Boeckxstaens GE, Pelckmans PA (1997) Nitric oxide and the non-adrenergic non-cholinergic neurotransmission. Comp Biochem Physiol A Physiol 118:925–937

Boeckxstaens GE, Pelckmans PA, Rampart M, Verbeuren TJ, Herman AG, Van Maercke YM (1990a) Nonadrenergic noncholinergic mechanisms in the ileocolonic junction. Arch Int Pharmacodyn Ther 303:270–281

Boeckxstaens GE, Pelckmans PA, Bult H, De Man JG, Herman AG, Van Maercke YM (1990b) Non-adrenergic non-cholinergic relaxation mediated by nitric oxide in the canine ileocolonic junction. Eur J Pharmacol 190:239–246

Boeckxstaens GE, Pelckmans PA, Rampart M, Ruytjens IF, Verbeuren TJ, Herman AG, Van Maercke YM (1990c) $GABA_A$receptor-mediated stimulation of non-adrenergic non-cholinergic neurones in the dog ileocolonic junction. Br J Pharmacol 101:460–464

Boeckxstaens GE, Pelckmans PA, Rampart M, Verbeuren TJ, Herman AG, Van Maercke YM (1990d) Evidence against ATP being the inhibitory transmitter released by nonadrenergic noncholinergic nerves in the canine ileocolonic junction. J Pharmacol Exp Ther 254:659–663

Boeckxstaens GE, Pelckmans PA, Ruytjens IF, Bult H, De Man JG, HermanAG, Van Maercke YM (1991) Bioassay of nitric oxide released upon stimulation of non-adrenergic non-cholinergic nerves in the canine ileocolonic junction. Br J Pharmacol 103:1085–1091

Boeckxstaens GE, De Man JG, Pelckmans PA, Cromheeke KM, Herman AG, Van Maercke YM (1993) Ca^{2+} dependency of the release of nitric oxide from non-adrenergic non-cholinergic nerves. Br J Pharmacol 110:1329–1334

Börjesson L, Nordgren S, Delbro DS (1997) DMPP causes relaxation of rat distal colon by a purinergic and a nitrergic mechanism. Eur J Pharmacol 334:223–231

Börjesson L, Nordgren S, Delbro DS (1999) K^+-induced neurogenic relaxation of rat distal colon. J Pharmacol Exp Ther 291:717–724

Bornstein JC, Furness JB (1988) Correlated electrophysiological and histochemical studies of submucous neurons and their contribution to understanding enteric neural circuits. J Auton Nerv Syst 25:1–13

Bouvier M, Gonella J (1981) Nervous control of the internal anal sphincter of the cat. J Physiol 310:457–469

Bouvier M, Kirschner G, Gonella J (1986) Actions of morphine and enkephalins on the internal anal sphincter of the cat: relevance for the physiological role of opiates. J Auton Nerv Syst 16:219–232

Brambilla R, Burnstock G, Bonazzi A, Ceruti S, Cattabeni F, Abbracchio MP (1999) Cyclooxygenase-2 mediates P2Y receptor-induced reactive astrogliosis. Special Report. Br J Pharmacol 126:563–567

Bridgewater M, Cunnane TC, Brading AF (1995) Characteristic features of inhibitory junction potentials evoked by single stimuli in the guinea-pig isolated taenia caeci. J Physiol 485:145–155

Briejer MR, Veen GJ, Akkermans LM, Lefebvre RA, Schuurkes JA (1995) Cisapride and structural analogs selectively enhance 5-hydroxytryptamine (5-HT)-induced purinergic neurotransmission in the guinea pig proximal colon. J Pharmacol Exp Ther 274:641–648

Briggs CA, Cooper JR (1981) A synaptosomal preparation from the guinea pig ileum myenteric plexus. J Neurochem 36:1097–1108

Brizzi E, Calamai F, Staderini G, Susini T, Viligiardi R (1985) Inhibition by ATP of vagal relaxatory motility in the rabbit stomach in vivo. IRCS Med Sci 13:376–377

Broad RM, McDonald TJ, Brodin E, Cook MA (1992) Adenosine A_1 receptors mediate inhibition of tachykinin release from perifused enteric nerve endings. Am J Physiol 262:G525–G531

Broad RM, McDonald TJ, Cook MA (1993) Adenosine and 5-HT inhibit substance P release from nerve endings in myenteric ganglia by distinct mechanisms. Am J Physiol 264:G454–G461

Brock JA, Van Helden DF (1995) Enhanced excitatory junction potentials in mesenteric arteries from spontaneously hypertensive rats. Pflugers Arch – Eur J Physiol 430:901–908

Brown C, Burnstock G (1981) Evidence in support of the P_1/P_2 purinoceptor hypothesis in the guinea-pig taenia coli. Br J Pharmacol 73:617–624
Brownhill VR, Hourani SMO, Kitchen I (1997) Ontogeny of P2-purinoceptors in the longitudinal muscle and muscularis mucosae of the rat isolated duodenum. Br J Pharmacol 122:225–232
Bueno L, Fioramonti J, Delvaux M, Frexinos J (1997) Mediators and pharmacology of visceral sensitivity: from basic to clinical investigations. Gastroenterology 112:1714–1743
Bulloch JM, McGrath JC (1992) Evidence for increased purinergic contribution in hypertensive blood vessels exhibiting co-transmission. Br J Pharmacol (Suppl) 107:145P
Bulloch JM, Starke K (1989) Presynaptic α_2-autoinhibition in a vascular neuroeffector junction where ATP and noradrenaline act as co-transmitters. Br J Pharmacol 99:279–284
Bültmann R, von Kügelgen I, Starke K (1991) Adrenergic and purinergic cotransmission in nicotine-evoked vasoconstriction in rabbit ileocolic arteries. Naunyn Schmiedebergs Arch Pharmacol 344:174–182
Bültmann R, Dudeck O, Starke K (1996) Evaluation of P2-purinoceptor antagonists at two relaxation-mediating P2-purinoceptors in guinea-pig taenia coli. Naunyn Schmiedebergs Arch Pharmacol 353:445–451
Burleigh DE (1983) Non-cholinergic, non-adrenergic inhibitory neurons in human internal anal sphincter muscle. J Pharm Pharmacol 35:258–260
Burleigh DE, D'Mello A, Parks AG (1979) Responses of isolated human internal anal sphincter to drugs and electrical field stimulation. Gastroenterology 77:484–490
Burnstock G (1969) Evolution of the autonomic innervation of visceral and cardiovascular systems in vertebrates. Pharmacol Rev 21:247–324
Burnstock G (1972) Purinergic nerves. Pharmacol Rev 24:509–581
Burnstock G (1978) A basis for distinguishing two types of purinergicreceptor. In: Straub RW, Bolis L (eds) Cell membrane receptors for drugs and hormones: a multidisciplinary approach. Raven Press, New York, pp 107–118
Burnstock G (1988) Local purinergic regulation of blood pressure. (The First John T. Shepherd Lecture). In: Vanhoutte PM (ed) Vasodilatation: vascular smooth muscle, peptides, autonomic nerves, and endothelium. Raven Press, New York, pp 1–14
Burnstock G (1990) Local mechanisms of blood flow control by perivascular nerves and endothelium. J Hypertens 8 (Suppl 7): S95–S106
Burnstock G (1992) Neuromuscular transmission and neuromodulation in the gastrointestinal tract. In: Heading RC, Wood JD (eds) Gastrointestinal dysmotility: focus on cisapride. Proceedings of the 2nd International Cisapride Investigators Meeting, Nice, 3–4 December 1990. Raven Press, New York, pp 41–60
Burnstock G (1996) Purinoceptors: ontogeny and phylogeny. Drug Dev Res 39:204–242
Burnstock G (1997) Commentary on paper by G. Burnstock, G. Campbell, D. Satchell and A. Smythe (1970) entitled "Evidence that adenosine triphosphate or a related nucleotide is the transmitter substance released by non-adrenergic inhibitory nerves in the gut". In: Birmingham AT, Brown DA (eds) Landmarks in pharmacology. Br J Pharmacol (Golden Jubilee 1946–1996), pp 334–357
Burnstock G (1999) Release of vasoactive substances from endothelial cells by shear stress and purinergic mechano-sensory transduction. J Anat 194:335–342
Burnstock G, Hoyle CHV (1985) Actions of adenine dinucleotides in the guinea-pig taenia coli: NAD acts indirectly on P_1-purinoceptors; NADP acts like a P_2-purinoceptor agonist. Br J Pharmacol 84:825–831
Burnstock G, Ralevic V (1994) New insights into the local regulation of blood flow by perivascular nerves and endothelium. Br J Plast Surg 47:527–543
Burnstock G, Warland JJI (1987) P_2-purinoceptors of two subtypes in the rabbit mesenteric artery: reactive blue 2 selectively inhibits responses mediated via the P_{2y}- but not the P_{2x}-purinoceptor. Br J Pharmacol 90:383–391

Burnstock G, Wong H (1978) Comparison of the effects of ultraviolet light and purinergic nerve stimulation on the guinea-pig taenia coli. Br J Pharmacol 62:293–302
Burnstock G, Wood JN (1996) Purinergic receptors: their role in nociception and primary afferent neurotransmission. Curr Opin Neurobiol 6:526–532
Burnstock G, Campbell G, Bennett M, Holman ME (1963) Inhibition of the smooth muscle of the taenia coli. Nature 200:581–582
Burnstock G, Campbell G, Bennett M, Holman ME (1964) Innervation of the guinea-pig taenia coli: are there intrinsic inhibitory nerves which are distinct from sympathetic nerves? Int J Neuropharmacol 3:163–166
Burnstock G, Campbell G, Rand MJ (1966) The inhibitory innervation of the taenia of the guinea-pig caecum. J Physiol 182:504–526
Burnstock G, Campbell G, Satchell D, Smythe A (1970) Evidence that adenosine triphosphate or a related nucleotide is the transmitter substance released by non-adrenergic inhibitory nerves in the gut. Br J Pharmacol 40:668–688
Burnstock G, Satchell DG, Smythe A (1972) A comparison of the excitatory and inhibitory effects of non-adrenergic, non-cholinergic nerve stimulation and exogenously applied ATP on a variety of smooth muscle preparations from different vertebrate species. Br J Pharmacol 46:234–242
Burnstock G, Cocks T, Paddle B, Staszewska-Barczak J (1975) Evidence that prostaglandin is responsible for the "rebound contraction" following stimulation of non-adrenergic, non-cholinergic ("purinergic') inhibitory nerves. Eur J Pharmacol 31:360–362
Burnstock G, Cocks T, Kasakov L, Wong HK (1978) Direct evidence for ATP release from non-adrenergic, non-cholinergic ("purinergic") nerves in the guinea-pig taenia coli and bladder. Eur J Pharmacol 49:145–149
Burnstock G, Cusack NJ, Hills JM, Mackenzie I, Meghji P (1983) Studies on the stereoselectivity of the P_2-purinoceptor. Br J Pharmacol 79:907–913
Burnstock G, Cusack NJ, Meldrum LA (1984) Effects of phosphorothioate analogues of ATP, ADP and AMP on guinea-pig taenia coli and urinary bladder. Br J Pharmacol 82:369–374
Burnstock G, Fischer B, Hoyle CHV, Maillard M, Ziganshin AU, Brizzolara AL, von Isakovics A, Boyer JL, Harden TK, Jacobson KA (1994) Structure activity relationships for derivatives of adenosine 5′-triphosphate as agonists at P_2 purinoceptors: heterogeneity within P_{2X} and P_{2Y} subtypes. Drug Dev Res 31:206–219
Bybee DE, Brown FC, Georges LP, Castell DO, McGuigan JE (1979) Somatostatin effects on lower esophageal sphincter function. Am J Physiol 237:E77–E81
Bywater RAR, Taylor GS (1982) Electrophysiological studies on the colon of the piebald-lethal mouse. Proc Aust Physiol Pharmacol Soc 13:250P [Abstract]
Cai W, Gu J, Huang W, McGregor GP, Ghatei MA, Bloom SR, Polak JM (1983) Peptide immunoreactive nerves and cells of the guinea-pig gall bladder and biliary pathways. Gut 24:1186–1193
Capogrossi MC, Francendese A, DiGirolamo M (1979) Suppression of food intake by adenosine and inosine. Am J Clin Nutr 32:1762–1768
Castell DO (1975) The lower esophageal sphincter. Physiologic and clinical aspects. Ann Intern Med 83:390–401
Cataldi de Flombaum MA, Stoppani AO (1992) High-affinity calcium-stimulated, magnesium-dependent adenosine triphosphatase in *Trypanosoma cruzi*. Comp Biochem Physiol B 103:933–937
Cervero F (1994) Sensory innervation of the viscera: peripheral basis of visceral pain. Physiol Rev 74:95–138
Chakder S, Rattan S (1992) Neurally mediated relaxation of opossum internal anal sphincter: influence of superoxide anion generator and the scavenger. J Pharmacol Exp Ther 260:1113–1118
Chakder S, Rattan S (1993) Release of nitric oxide by activation of nonadrenergic noncholinergic neurons of internal anal sphincter. Am J Physiol 264:G7–G12

Chakder S, Rathi S, Ma XL, Rattan S (1996) Heme oxygenase inhibitor zinc protoporphyrin IX causes an activation of nitric oxide synthase in the rabbit internal anal sphincter. J Pharmacol Exp Ther 277:1376–1382
Chaudry IH (1983) Cellular mechanisms in shock and ischemia and their correction. Am J Physiol 245:R117–R134
Chaudry IH (1991) Use of ATP following shock and ischemia. Ann NYAcad Sci 603:130–141
Chaudry IH, Clemens MG, Baue AE (1986) The role of ATP-magnesium in ischemia and shock. Magnesium 5:211–220
Chen C-C, Akopian AN, Sivilotti L, Colquhoun D, Burnstock G, Wood JN (1995) A P2X purinoceptor expressed by a subset of sensory neurons. Nature 377:428–431
Christofi FL, Cook MA (1986) Affinity of various purine nucleosides for adenosine receptors on purified myenteric varicosities compared to their efficacy as presynaptic inhibitors of acetylcholine release. J Pharmacol Exp Ther 237:305–311
Christofi FL, Cook MA (1987) Possible heterogeneity of adenosine receptors present on myenteric nerve endings. J Pharmacol Exp Ther 243:302–309
Christofi FL, Cook MA (1997) Purinergic modulation of gastrointestinal function. In: Jacobson KA, Jarvis MF (eds) Purinergic approaches in experimental therapeutics. Wiley-Liss, New York, pp 261–282
Christofi FL, Wood JD (1993a) Presynaptic inhibition by adenosine A_1 receptors on guinea pig small intestinal myenteric neurons. Gastroenterology 104:1420–1429
Christofi FL, Wood JD (1993b) Endogenously released adenosine acts at A_1 receptors to suppress slow excitatory transmission (slow EPSP) and enhance slow inhibitory transmission (slow IPSP) in the myenteric plexus of guinea-pig small intestine. Gastroenterology 104:A490 [Abstract]
Christofi FL, Wood JD (1994) Electrophysiological subtypes of inhibitory P1 purinoceptors on myenteric neurones of guinea-pig small bowel. Br J Pharmacol 113:703–710
Christofi FL, McDonald TJ, Cook MA (1990) Adenosine receptors are coupled negatively to release of tachykinin(s) from enteric nerve endings. J Pharmacol Exp Ther 253:290–295
Christofi FL, Tack J, Wood JD (1992) Suppression of nicotinic synaptic transmission by adenosine in myenteric ganglia of the guinea-pig gastric antrum. Eur J Pharmacol 216:17–22
Christofi FL, Baidan LV, Fertel RH, Wood JD (1994) Adenosine A_2 receptor-mediated excitation of a subset of AH/type 2 neurons and elevation of cAMP levels in myenteric ganglia of guinea-pig ileum. Neurogastroenterol Motility 6:67–78
Christofi FL, Guan Z, Lucas JH, Rosenberg Schaffer LJ, Stokes BT (1996) Responsiveness to ATP with an increase in intracellular free Ca^{2+} is not a distinctive feature of calbindin-D28 immunoreactive neurons in myenteric ganglia. Brain Res 725:241–246
Christofi FL, Guan Z, Wood JD, Baidan LV, Stokes BT (1997) Purinergic Ca^{2+} signaling in myenteric neurons via P2 purinoceptors. Am J Physiol 272:G463–G473
Clark SR, Costa M, Tonini M, Brookes SJ (1996) Purinergic transmission is involved in a descending excitatory reflex in the guinea-pig small intestine. Proc Aust Neurosci Soc 7:176
Cocks T, Burnstock G (1979) Effects of neuronal polypeptides on intestinal smooth muscle; a comparison with non-adrenergic, non-cholinergic nerve stimulation and ATP. Eur J Pharmacol 54:251–259
Collier HO, Tucker JF (1983) Novel form of drug-dependence – on adenosine in guinea pig ileum. Nature 302:618–621
Communi D, Parmentier M, Boeynaems JM (1996) Cloning, functional expression and tissue distribution of the human $P2Y_6$ receptor. Biochem Biophys Res Commun 222:303–308
Cook DI, Young JA (1989) Fluid and electrolyte secretion by salivary glands. In: Forte JG (ed) Handbook of physiology, the gastrointestinal system. salivary, pancreatic,

gastric and hepatobiliary secretion, section 6, vol III. American Physiological Society, Bethesda, Md, pp 1–23
Costa M, Furness JB (1973) The origins of the adrenergic fibres which innervate the internal anal sphincter, the rectum, and other tissues of the pelvic region in the guinea-pig. Z Anat Entw Gesch 140:129–142
Costa M, Furness JB, Gibbins IL (1986a) Chemical coding of enteric neurones. In: Hökfelt T, Fuxe K, Pernow B (eds) Coexistence of neuronal messengers: a new principle in chemical transmission. Elsevier, Amsterdam, pp 217–239
Costa M, Furness JB, Humphreys CM (1986b) Apamin distinguishes two types of relaxation mediated by enteric nerves in the guinea-pig gastrointestinal tract. Naunyn Schmiedebergs Arch Pharmacol 332:79–88
Coupar JM (1999) Characterization and tissue location of the neural adenosine receptor in the rat ileum. Br J Pharmacol 126:1269–1275
Coutinho CMLM, Pons AH, Araujo-Jorge TC, Persechini PM, Coutinho-Silva R (1998) Enhancement of P2Z-associated cell permeabilization during acute phase of Chagas' disease. Drug Dev Res 43:38 [Abstract]
Crema A, Frigo GM, Lecchini S, Manzo L, Onori L, Tonini M (1983) Purine receptors in the guinea-pig internal anal sphincter. Br J Pharmacol 78:599–603
Crist JR, He XD, Goyal RK (1992) Both ATP and the peptide VIP are inhibitory neurotransmitters in guinea-pig ileum circular muscle. J Physiol 447:119–131
Crossley AW, Gillespie JS (1983) The effect of an inhibitory factor from the bovine retractor penis on the gastro-intestinal tract and gallbladder of the guinea-pig. Br J Pharmacol 78:213–220
Crowe R, Burnstock G (1981a) Perinatal development of quinacrine-positive neurons in the rabbit gastrointestinal tract. J Auton Nerv Syst 4:217–230
Crowe R, Burnstock G (1981b) Comparative studies of quinacrine-positive neurones in the myenteric plexus of stomach and intestine of guinea-pig, rabbit and rat. Cell Tissue Res 221:93–107
Currò D, Preziosi P (1998) Non-cholinergic non-adrenergic relaxation of the rat stomach. Gen Pharmacol 31:697–703
Cusack NJ, Planker M (1979) Relaxation of isolated taenia coli of guinea-pig by enantiomers of 2-azido analogues of adenosine and adenine nucleotides. Br J Pharmacol 67:153–158
Cusack NJ, Hourani SM, Loizou GD, Welford LA (1987) Pharmacological effects of isopolar phosphonate analogues of ATP on P2-purinoceptors in guinea-pig taenia coli and urinary bladder. Br J Pharmacol 90:791–795
Cuthbert AW, Hickman ME (1985) Indirect effects of adenosine triphosphate on chloride secretion in mammalian colon. J Membr Biol 86:157–166
Da Prada M, Richards JG, Lorez HP (1978) Blood platelets and biogenic monoamines: biochemical, pharmacological and morphological studies. In: de Gaetano G, Grattini S (eds) Platelets: a multidisciplinary approach. Raven Press, New York, pp 331–353
Dahl JL, Bloom DD, Epstein ML, Fox DA, Bass P (1987) Effect of chemical ablation of myenteric neurons on neurotransmitter levels in the rat jejunum. Gastroenterology 92:338–344
Dahlstrand C, Dahlstrom A, Theodorsson E, Rehfeld J, Ahlman H (1990) Is the CCK-8 induced relaxation of the feline sphincter of Oddi mediated by VIP neurons? J Auton Nerv Syst 31:75–84
Daly JW (1982) Adenosine receptors: targets for future drugs. J Med Chem 25:197–207
Daniel EE, Helmy-Elkholy A, Jager LP, Kannan MS (1983) Neither apurine nor VIP is the mediator of inhibitory nerves of opossum oesophageal smooth muscle. J Physiol 336:243–260
Daniel EE, Jager LP, Jury J, Helmy-Elkholy A, Kannan MS, Posey-Daniel V (1984) The mediators and mechanisms causing the non-adrenergic, non-cholinergic nerve responses in opossum esophagus: Role of interstitial cells of Cajal. Biomed Res 5:67–84

Daniel EE, Jager LP, Jury J (1989) Vasoactive intestinal polypeptide and non-adrenergic, non-cholinergic inhibition in lower oesophageal sphincter of opossum. Br J Pharmacol 96:746–752
De Carle DJ, Christensen J (1976) Letter: "Purinergic" neurotransmitters in the esophagus. Ann Intern Med 84:220–221
De Luca A, Li CG, Rand MJ (1999) Nitrergic and purinergic mechanisms and their interactions for relaxation of the rat internal anal sphincter. J Auton Pharmacol 19:29–37
Dehaye JP (1993) ATP^{4-} increases the intracellular calcium concentration in rat submandibular glands. Gen Pharmacol 24:1097–1100
Delbro D, Fändriks L (1982) ATP induces non-cholinergic, non-adrenergic gastric relaxation in vivo. Acta Physiol Scand Suppl 508:67 [Abstract]
Delbro D, Fändriks L (1984) Inhibition of vagally induced non-adrenergic, non-cholinergic gastric relaxation by P2-purinoceptor desensitization. Acta Physiol Scand 120:12A [Abstract]
Deloof S, Croix D, Tramu G (1988) The role of vasoactive intestinal polypeptide in the inhibition of antral and pyloric electrical activity in rabbits. J Auton Nerv Syst 22:167–173
Den Hertog A, Pielkenrood J, Van den Akker J (1985a) Effector mechanisms for α,β-methylene ATP and ATP derivatives in guinea-pig taenia caeci. Eur J Pharmacol 110:95–101
Den Hertog A, Pielkenrood J, Van den Akker J (1985b) Responses evoked by electrical stimulation, adenosine triphosphate, adenosine and 4-aminopyridine in taenia caeci of the guinea-pig. Eur J Pharmacol 109:373–380
Den Hertog A, Nelemans A, Van den Akker J (1989a) The inhibitory action of suramin on the P_2-purinoceptor response in smooth muscle cells of the guinea-pig taenia caeci. Eur J Pharmacol 166:531–534
Den Hertog A, Van den Akker J, Nelemans A (1989b) Suramin and the inhibitory junction potential in taenia caeci of the guinea-pig. Eur J Pharmacol 173:207–209
Deshpande NA, McDonald TJ, Cook MA (1999) Endogenous interstitial adenosine in isolated myenteric neural networks varies inversely with prevailing Po_2. Am J Physiol 276:G875–G885
Dho S, Stewart K, Foskett JK (1992) Purinergic receptor activation of Cl^- secretion in T84 cells. Am J Physiol 262:C67–C74
Dick GM, Bradley KK, Horowitz B, Hume JR, Sanders KM (1998) Functional and molecular identification of a novel chloride conductance in canine colonic smooth muscle. Am J Physiol 275:C940–C950
DiMarino AJ Jr (1974) Characteristics of lower esophageal sphincter function in symptomatic diffuse esophageal spasm. Gastroenterology 66:1–6
Domoto T, Oki M, Kotoh T, Nakamura T (1992) Heterogeneous distribution of peptide-containing nerve fibres within the circular muscle layer of the human pylorus. Clin Auton Res 2:403–407
Donoso MV, Steiner M, Huidobro Toro JP (1997) BIBP 3226, suramin and prazosin identify neuropeptide Y, adenosine 5′-triphosphate and noradrenaline as sympathetic cotransmitters in the rat arterial mesenteric bed. J Pharmacol Exp Ther 282:691–698
Dowd FJ, Murphy HC, Li L (1996) Metabolism of extracellular ATP by rat parotid cells. Arch Oral Biol 41:855–862
Dowdle EB, Maske R (1980) The effects of calcium concentration on the inhibition of cholinergic neurotransmission in the myenteric plexus of guinea-pig ileum by adenine nucleotides. Br J Pharmacol 71:245–252
Dowe GH, Kilbinger H, Whittaker VP (1980) Isolation of cholinergic synaptic vesicles from the myenteric plexus of guinea-pig small intestine. J Neurochem 35:993–1003
Drakontides AB, Gershon MD (1972) Studies of the interaction of 5-hydroxytryptamine and the perivascular innervation of the guinea-pig caecum. Br J Pharmacol 45:417–434

Dudeck O, Bultmann R, Starke K (1995) Two relaxation-mediating P2-purinoceptors in guinea-pig taenia caeci. Naunyn Schmiedebergs Arch Pharmacol 351:107–110
Edin R, Lundberg JM, Ahlman H, Dahlstrom A, Fahrenkrug J, Hokfelt T, Kewenter J (1979) On the VIP-ergic innervation of the feline pylorus. Acta Physiol Scand 107:185–187
Elsing C, Kassner A, Stremmel W (1996) Sodium, hydrogen antiporter activation by extracellular adenosine triphosphate in biliary epithelial cells. Gastroenterology 111:1321–1332
Ernster VL (1984) Epidemiologic studies of caffeine and human health. Prog Clin Biol Res 158:377–400
EvansRJ, Cunnane TC (1992) Relative contributions of ATP and noradrenaline to the nerve evoked contraction of the rabbit jejunal artery. Dependence on stimulation parameters. Naunyn Schmiedebergs Arch Pharmacol 345:424–430
Evans RJ, Surprenant A (1992) Vasoconstriction of guinea-pig submucosal arterioles following sympathetic nerve stimulation is mediated by the release of ATP. Br J Pharmacol 106:242–249
Feit C, Roche M (1988) Action of adenosine on intestinal motility after experimental mesenteric ischaemia in the dog. Gastroenterol Clin Biol12:803–809
Fernandes JF, Castellani O (1958) Nucleotide and polynucleotide synthesis in *Trypanosoma cruzi*. I Precursors of purine compounds. Exp Parasitol 7:224–235
Fernández E, Guo X, Vergara P, Jimenez M (1998) Evidence supporting a role for ATP as non-adrenergic non-cholinergic inhibitory transmitter in the porcine ileum. Life Sci 62:1303–1315
Ferrero JD, Frischknecht R (1983) Different effector mechanisms for ATP and adenosine hyperpolarization in the guinea-pig taenia coli. Eur J Pharmacol 87:151–154
Ferrero JD, Cocks T, Burnstock G (1980) A comparison between ATP and bradykinin as possible mediators of the responses of smooth muscle to non-adrenergic non-cholinergic nerves. Eur J Pharmacol 63:295–302
Filinger EJ, Perec CJ, Stefano FJ (1989) Potassium-evoked efflux of [^{3}H]purines from the rat submaxillary gland. Gen Pharmacol 20:285–288
Fisher RS, Cohen S (1975) Disorders of the lower esophageal sphincter. Annu Rev Med 26:373–390
Folkow B (1949) The vasodilator action of adenosine of adenosine triphosphate. Acta Physiol Scand 17:311–317
Forsberg C, Sollevi A, Thorn SE, Segerdahl M (1999) Effects of adenosine infusion on gastric emptying in healthy volunteers. Acta Anaesthesiol Scand 43:87–90
Forte JG, Lee HC (1977) Gastric adenosine triphosphatases: a review of their possible role in HCl secretion. Gastroenterology 73:921–926
Foster HE, Hooper M, Spedding M, Sweetman AJ, Weetman DF (1978) Antagonism of the inhibitory effects of adenosine 5′-triphosphate on the isolated taenia of the guinea-pig caecum: structure-activity relationships within a series of isatogen derivatives. Br J Pharmacol 63:309–314
Foster HE, Hooper M, Imam SH, Lovett GS, Nicholson J, Swain CJ, Sweetman AJ, Weetman DF (1983) Increased inhibitory action against adenosine 5′-triphosphate in the isolated taenia of the guinea-pig caecum by substitution in the A-ring of 2-phenylisatogen. Br J Pharmacol 79:273–278
Fox DA, Herman JR, Bass P (1986) Differentiation between myenteric plexus and longitudinal muscle of the rat jejunum as the site of action of putative enteric neurotransmitters. Eur J Pharmacol 131:39–47
Frasch AC, Cazzulo JJ, Stoppani AO (1978) Solubilization and some properties of the Mg^{2+}-activated adenosine triphosphatase from *Trypanosoma cruzi*. Comp Biochem Physiol B 61:207–212
Frenckner B, Ihre T (1976) Influence of autonomic nerves on the internal anal sphincter in man. Gut 17:306–312
Frew R, Lundy PM (1982) Evidence against ATP being the nonadrenergic, noncholinergic inhibitory transmitter in guinea pig stomach. Eur J Pharmacol 81:333–336

Frew R, Lundy PM (1986) Arylazido aminopropionyl ATP ($ANAPP_3$): interaction with adenosine receptors in longitudinal smooth muscle of the guinea-pig ileum. Eur J Pharmacol 123:395–400

Frigo GM, Del Tacca M, Lecchini S, Crema A (1973) Some observations on the intrinsic nervous mechanism in Hirschsprung's disease. Gut 14:35–40

Fujiwara M, Hong SC, Muramatsu I (1982) Effects of goniopora toxin on non-adrenergic, non-cholinergic response and purine nucleotide release in guinea-pig taenia coli. J Physiol 326:515–526

Fukushi Y (1999) Heterologous desensitization of muscarinic receptors by P_{2Z} purinoceptors in rat parotid acinar cells. Eur J Pharmacol 364:55–64

Fukushi Y, Ozawa T, Kanno T, Wakui M (1997) Na^+-dependent release of intracellular Ca^{2+} induced by purinoceptors in parotid acinar cells of the rat. Eur J Pharmacol 336:89–97

Furness JB, Costa M (1987) Identification of transmitters of functionally defined enteric neurons. In: Handbook of Physiology – the Gastrointestinal System I. American Physiological Society, Bethesda, Md, pp 387–402

Furness JB, Costa M, Walsh JH (1981) Evidence for and significance of the projection of VIP neurons from the myenteric plexus to the taenia coli in the guinea pig. Gastroenterology 80:1557–1561

Furness JB, Young HM, Pompolo S, Bornstein JC, Kunze WA, McConalogue K (1995) Plurichemical transmission and chemical coding of neurons in the digestive tract. Gastroenterology 108:554–563

Furness JB, Kunze WA, Bertrand PP, Clerc N, Bornstein JC (1998) Intrinsic primary afferent neurons of the intestine. Prog Neurobiol 54:1–18

Gabella G (1984) Size of neurons and glial cells in the intramural ganglia of the hypertrophic intestine of the guinea-pig. J Neurocytol 13:73–84

Gabella G (1989) Fall in the number of myenteric neurons in aging guinea pigs. Gastroenterology 96:1487–1493

Gabella G (1990) On the plasticity of form and structure of enteric ganglia. J Auton Nerv Syst 30:S59–66

Gaion RM, Dorigo P, Trolese B, Borin E, Adami R, Gambarotto L (1988) Involvement of P1-purinoreceptors in the relaxing effect of adenosine in rat duodenum. J Auton Pharmacol 8:135–140

Gallacher DV (1982) Are there purinergic receptors on parotid acinar cells? Nature 296:83–86

Galligan JJ (1996) Electrophysiological studies of 5-hydroxytryptamine receptors on enteric neurons. Behav Brain Res 73:199–201

Galligan JJ, Herring A, Harpstead T (1995) Pharmacological characterization of purinoceptor-mediated constriction of submucosal arterioles in guinea pig ileum. J Pharmacol Exp Ther 274:1425–1430

Gandarias JM, Ainz LF, Gil-Rodrigo CE, Goiriena JJ, Gómez R, Martinez I (1985) Effect of various adenine derivatives on gastric acid secretion in the isolated rat stomach. Rev Esp Fisiol 41:83–87

Gannon BJ, Burnstock G, Noblett HR, Campbell PE (1969) Histochemical diagnosis of Hirschsprung's disease. Lancet i: 894–895

Geiger JD, Glavin GB (1985) Adenosine receptor activation in brain reduces stress-induced ulcer formation. Eur J Pharmacol 115:185–190

Gerber JG, Payne NA (1988) Endogenous adenosine modulates gastric acid secretion to histamine in canine parietal cells. J Pharmacol Exp Ther 244:190–194

Gerber JG, Fadul S, Payne NA, Nies AS (1984) Adenosine: a modulator of gastric acid secretion in vivo. J Pharmacol Exp Ther 231:109–113

Gerber JG, Nies AS, Payne NA (1985) Adenosine receptors on canine parietal cells modulate gastric acid secretion to histamine. J Pharmacol Exp Ther 233:623–627

Gershon MD, Wade PR (1994) New developments in the enteric nervous system. Curr Opin Gastroenterol 10:183–192

Gheber L, Priel Z (1994) Metachronal activity of cultured mucociliary epithelium under normal and stimulated conditions. Cell Motil Cytoskeleton 28:333–345
Gheber L, Priel Z, Aflalo C, Shoshan Barmatz V (1995) Extracellular ATP binding proteins as potential receptors in mucociliary epithelium: characterization using [^{32}P]-3′-*O*-(4-benzoyl)benzoyl ATP, a photoaffinity label. J Membr Biol 147:83–93
Giacobini Robecchi MG, Cannas M, Filogamo G (1985) Increase in the number and volume of myenteric neurons in the adult rat. Int J Dev Neurosci 3:673–675
Gibb CA, Singh S, Cook DI, Poronnik P, Conigrave AD (1994) A nucleotide receptor that mobilizes Ca^{2+} in the mouse submandibular salivary cell line ST885. Br J Pharmacol 111:1135–1139
Gil-Rodrigo CE, Galdiz B, Gandarias JM, Gomez R, Ainz LF (1990) Characterization of the effects of adenosine, adenosine 5′-triphosphate and related purines on acid secretion in isolated rabbit gastric glands. Pharmacol Res 22:103–113
Gil-Rodrigo CE, Bergaretxe I, Carou M, Galdiz B, Salgado C, Ainz LF (1996) Inhibitory action of extracellular adenosine 5′-triphosphate on parietal cells isolated from rabbit gastric mucosa. Gen Physiol Biophys 15:251–264
Gintzler AR, Musacchio JM (1975) Interactions of morphine, adenosine, adenosine triphosphate and phosphodiesterase inhibitors on the field-stimulated guinea-pig ileum. J Pharmacol Exp Ther 194:575–582
Glasgow I, Mattar K, Krantis A (1998) Rat gastroduodenal motility in vivo: involvement of NO and ATP in spontaneous motor activity. Am J Physiol 275:G889–G896
Glavin GB, Westerberg VS, Geiger JD (1987) Modulation of gastric acid secretion by adenosine in conscious rats. Can J Physiol Pharmacol 65:1182–1185
Glushakov AV, Melishchuk AI, Skok VI (1996) ATP-induced currents in submucous plexus neurons of the guinea-pig small intestine. Neurophysiology (Moscow) 28:77–85
Glushakov AV, Glushakova HY, Skok VI (1998) Two types of P_{2X}-purinoceptors in neurons of the guinea pig ileum submucous plexus. Neurophysiology (Moscow) 30:242–245
Goldman H, Rosoff CB (1968) Pathogenesis of acute gastric stress ulcers. Am J Pathol 52:227–244
Goyal RK, Rattan S, Said SI (1980) VIP as a possible neurotransmitter of non-cholinergic non-adrenergic inhibitory neurones. Nature 288:378–380
Granger DN, Richardson PD, Kvietys PR, Mortillaro NA (1980) Intestinal blood flow. Gastroenterology 78:837–863
Grasl M, Turnheim K (1984) Stimulation of electrolyte secretion in rabbit colon by adenosine. J Physiol 346:93–110
Grider JR (1993) Interplay of VIP and nitric oxide in regulation of the descending relaxation phase of peristalsis. Am J Physiol 264:G334–G340
Grider JR, Makhlouf GM (1986) Colonic peristaltic reflex: identification of vasoactive intestinal peptide as mediator of descending relaxation. Am J Physiol 251:40–45
Grider JR, Said SI, Makhlouf GM (1982) VIP and ATP as mediators of non-adrenergic, non-cholinergic gastric relaxation. Gastroenterology 82:1075 [Abstract]
Grider JR, Cable MB, Bitar KN, Said SI, Makhlouf GM (1985) Vasoactive intestinal peptide. Relaxant neurotransmitter in tenia coli of the guinea pig. Gastroenterology 89:36–42
Gröschel-Stewart U, Bardini M, Robson T, Burnstock G (1999a) Localisation of $P2X_5$ and $P2X_7$ receptors by immunohistochemistry in rat stratified squamous epithelia. Cell Tissue Res 296:599–605
Gröschel-Stewart U, Bardini M, Robson T, Burnstock G (1999b) P2X receptors in the rat duodenal villus. Cell Tissue Res 297:111–117
Guimaraes CR, Rodrigues LA, Vettore O, Antonio A (1988) The relaxing response of the isolated rat duodenum to nicotine. Gen Pharmacol 19:655–659

Guo X, Merlin D, Harvey RD, Laboisse C, Hopfer U (1995) Stimulation of Cl^- secretion by extracellular ATP does not depend on increased cytosolic Ca^{2+} in HT-29.cl16E. Am J Physiol 269:C1457–C1463
Guo XW, Merlin D, Laboisse C, Hopfer U (1997) Purinergic agonists, but not cAMP, stimulate coupled granule fusion and Cl^- conductance in HT29-Cl.16E. Am J Physiol 273:C804–C809
Gür S, Karahan ST (1997) Effects of adenosine 5′-triphosphate, adenosine and acetylcholine in urinary bladder and colon muscles from streptozotocin diabetic rats. Arzneimittelforschung 47:1226–1229
Gustafsson LE (1984) Adenosine antagonism and related effects of theophylline derivatives in guinea pig ileum longitudinal muscle. Acta Physiol Scand 122:191–198
Gustafsson L, Hedqvist P, Fredholm BB, Lundgren G (1978) Inhibition of acetylcholine release in guinea pig ileum by adenosine. Acta Physiol Scand 104:469–478
Gustafsson L, Fredholm BB, Hedqvist P (1981) Theophylline interferes with the modulatory role of endogenous adenosine on cholinergic neurotransmission in guinea pig ileum. Acta Physiol Scand 111:269–280
Gustafsson LE, Wiklund NP, Lundin J, Hedqvist P (1985) Characterization of pre- and post-junctional adenosine receptors in guinea-pig ileum. Acta Physiol Scand 123:195–203
Hammond JR, MacDonald WF, White TD (1988) Evoked secretion of [^{3}H]noradrenaline and ATP from nerve varicosities isolated from the myenteric plexus of the guinea pig ileum. Can J Physiol Pharmacol 66:369–375
Hancock DL, Coupar IM (1995) Functional characterization of the adenosine receptor mediating inhibition of peristalsis in the rat jejunum. Br J Pharmacol 115:739–744
Hayashi E, Mori M, Yamada S, Kumitomo M (1978) Effects of purine compounds on cholinergic nerves. Specificity of adenosine and related compounds on acetylcholine release in electrically stimulated guinea pig ileum. Eur J Pharmacol 48:297–307
Hayashi E, Maeda T, Shinozuka K (1982) Sites of actions of adenosine in intrinsic cholinergic nerves of ileal longitudinal muscle from guinea pig. Eur J Pharmacol 84:99–102
Hayashi E, Maeda T, Shinozuka K (1985) Adenosine and dipyridamole: actions and interactions on the contractile response of guinea-pig ileum to high frequency electrical field stimulation. Br J Pharmacol 84:765–771
Heazell MA (1975) Is ATP an inhibitory neurotransmitter in the rat stomach? Br J Pharmacol 55:285P-286P [Abstract]
Hedlund H, Fändriks L, Delbro D, Fasth S (1986) Effect of α,β-methylene ATP on distal colonic and rectal motility – a possible involvement of P_2-purinoceptors in pelvic nerve mediated non-adrenergic, non-cholinergic contraction. Acta Physiol Scand 127:425–432
Heinemann A, Shahbazian A, Barthó L, Holzer P (1999) Different receptors mediating the inhibitory action of exogenous ATP and endogenously released purines on guinea-pig intestinal peristalsis. Br J Pharmacol 128:313–320
Henry JP, Stephens PM (1980) Caffeine as an intensifier of stress-induced hormonal and pathophysiologic changes in mice. Pharmacol Biochem Behav 13:719–727
Hertz AF (1911) The sensibility of the alimentary canal in health and disease. Lancet i:1051–1056
Hills JM, Collis CS, Burnstock G (1983) The effects of vasoactive intestinal polypeptide on the electrical activity of guinea-pig intestinal smooth muscle. Eur J Pharmacol 88:371–376
Holle GE, Milenov K, Forth W (1991) Adrenergic control of interdigestive and digestive motility via the pyloric region. J Gastroint Motil 3:131–137
Holzer P, Barthó L (1996) Sensory neurons in the intestine. In: Geppeti P, Holzer P (eds) Neurogenic inflammation. CRC Press, New York, pp 153–167

Holzer P, Livingston EH, Saria A, Guth PH (1991) Sensory neurons mediate protective vasodilatation in rat gastric mucosa. Am J Physiol 260:G363–G370

Hooper M, Spedding M, Sweetman AJ, Weetman DF (1974) Proceedings: 2–2′ pyridylisatogen tosylate: an antagonist of the inhibitory effects of ATP on smooth muscle. Br J Pharmacol 50:458P–459P

Höpfner M, Lemmer K, Jansen A, Hanski C, Riecken EO, Gavish M, Mann B, Buhr H, Glassmeier G, Scherübl H (1998) Expression of functional P2-purinergic receptors in primary cultures of human colorectal carcinoma cells. Biochem Biophys Res Commun 251:811–817

Höpker V, Saffrey MJ, Burnstock G (1996) Neurite outgrowth of striatal neurons in vitro: involvement of purines in the growth promoting effect of myenteric plexus explants. Int J Dev Neurosci 14:439–451

Hourani SMO (1999) Postnatal development of purinoceptors in rat visceral smooth muscle preparations. Gen Pharmacol 32:3–7

Hourani SMO, Bailey SJ, Nicholls J, Kitchen I (1991) Direct effects of adenylyl 5′-(β,γ-methylene)diphosphonate, a stable ATP analogue, on relaxant P_1-purinoceptors in smooth muscle. Br J Pharmacol 104:685–690

Hourani SMO, Bailey SJ, Johnson CR, Tennant JP (1998) Effects of adenosine 5′-triphosphate, uridine 5′-triphosphate, adenosine 5′-tetraphosphate and diadenosine polyphosphates in guinea-pig taenia caeci and rat colon muscularis mucosae. Naunyn Schmiedebergs Arch Pharmacol 358:464–473

Hoyle CHV (1992) Transmission: purines. In: Burnstock G, Hoyle CHV (eds) The autonomic nervous system. Autonomic neuroeffector mechanisms. Harwood Academic Publishers, Chur, Switzerland, pp 367–407

Hoyle CHV (1996) Purinergigic cotransmission: parasympathetic and enteric nerves. Semin Neurosci 8:207–215

Hoyle CHV, Burnstock G (1989) Neuromuscular transmission in the gastrointestinal tract. In: Wood JD (ed) Handbook of physiology, sect 6: The gastrointestinal system, vol I: Motility and circulation. American Physiological Society, Bethesda, Md, pp 435–464

Hoyle CHV, Vladimirova IA, Burnstock G (1988a) Pre- and postjunctional actions of purine and xanthine compounds in the guinea-pig caecum circular muscle. Br J Pharmacol 95:653–663

Hoyle CHV, Reilly WM, Lincoln J, Burnstock G (1988b) Adrenergic, but not cholinergic or purinergic, responses are potentiated in the cecum of diabetic rats. Gastroenterology 94:1357–1367

Hoyle CHV, Kamm MA, Burnstock G, Lennard-Jones JE (1990) Enkephalins modulate inhibitory neuromuscular transmission in circular muscle of human colon via d-opioid receptors. J Physiol 431:465–478

Hubel KA (1984) Electrical stimulus-secretion coupling in rabbit ileal mucosa. J Pharmacol Exp Ther 231:577–582

Huizinga JD, Den Hertog A (1979) The effect of enkephalins on the intramural inhibitory non-adrenergic nerve responses of smooth muscle. Eur J Pharmacol 54:389–391

Huizinga JD, Den Hertog A (1980) Inhibition of fundic strips from guinea-pig stomach: the effect of theophylline on responses to adenosine, ATP and intramural nerve stimulation. Eur J Pharmacol 63:259–265

Huizinga JD, Pielkenrood JM, Den Hertog A (1981) Dual action of high energy adenine nucleotides in comparison with responses evoked by other adenine derivatives and intramural nerve stimulation on smooth muscle. Eur J Pharmacol 74:175–180

Hunt WB, Parsons DG, Wahid A, Wilkinson J (1978) Influence of 2-2′-pyridylisatogen tosylate on responses produced by ATP and by neural stimulation on the rat gastric corpus. Br J Pharmacol 63:378P-379P [Abstract]

Hurley TW, Shoemaker DD, Ryan MP (1993) Extracellular ATP prevents the release of stored Ca^{2+} by autonomic agonists in rat submandibular gland acini. Am J Physiol 265:C1472–C1478

Hurley TW, Ryan MP, Shoemaker DD (1994) Mobilization of Ca^{2+} influx, but not of stored Ca^{2+}, by extracellular ATP in rat submandibular gland acini. Arch Oral Biol 39:205–212

Hurley TW, Ryan MP, Moore WC (1996) Regulation of changes in cytosolic Ca^{2+} and Na^{+}concentrations in rat submandibular gland acini exposed to carbachol and ATP. J Cell Physiol 168:229–238

Idestrup CP, Salter MW (1998) P2Y and P2U receptors differentially release intracellular Ca^{2+} via the phospholipasec/inositol 1,4,5-triphosphate pathway in astrocytes from the dorsal spinal cord. Neuroscience 86:913–923

Ikawa H (1981) Study of acetylcholine and muscarinic receptors in Hirschsprung's disease. Jpn J Pediatr Surg 17:237–247

Imaeda K, Suzuki H (1997) Properties of inhibitory transmission in smooth muscle of the guinea pig lower esophageal sphincter. J Auton Nerv Syst 65:132 [Abstract]

Imoto A, Inoue R, Tanaka M, Ito Y (1998) Inhibitory NANC neurotransmission in choledocho-duodenal junction of rabbits – a possible role of PACAP. J Auton Nerv Syst 70:189–199

Inoue CN, Woo JS, Schwiebert EM, Morita T, Hanaoka K, Guggino SE, Guggino WB (1997) Role of purinergic receptors in chloride secretion in Caco-2 cells. Am J Physiol 272:C1862–C1870

Irie K, Furukawa K, Nomoto T, Fujii E, Muraki T (1994) Developmental changes in the response of rat isolated duodenum to nicotine. Eur J Pharmacol 251:75–81

Irvin JL, Irvin EM (1954) The interaction of quinacrine with adenine nucleotides. J Biol Chem 210:45–56

Iselin CE, Martin JL, Magistretti PJ, Ferrero JD (1988) Stimulation by nicotine of enteric inhibitory nerves and release of vasoactive intestinal peptide in the taenia of the guinea-pig caecum. Eur J Pharmacol 148:179–186

Ishiguchi T, Takahashi T, Owyang C (1999) Nitric oxide and adenosine 5′-triphosphate (ATP) regulate pyloric relaxation in rats. Gastroenterology 116:G4590 [Abstract]

Ishii T, Shimo Y (1983) Nerve-mediated non-adrenergic inhibitory responses of guinea-pig taenia caeci: further evidence of depression by morphine. J Pharm Pharmacol 35:828–830

Ishikawa S (1985) Actions of ATP and α,β-methylene ATP on neuromuscular transmission and smooth muscle membrane of the rabbit and guinea-pig mesenteric arteries. Br J Pharmacol 86:777–787

Itoh H, Sakai J, Imoto A, Creed KE (1995) The control of smooth muscle tissues by nonadrenergic, noncholinergic (NANC) nerve fibres in the autonomic nervous system. J Smooth Musc Res 31:67–78

Jager LP (1976) Effects of dipyridamole on the smooth muscle cells of the guinea-pig's taenia coli. Arch Int Pharmacodyn Ther 221:40–53

Jager LP, Schevers JA (1980) A comparison of effects evoked in guinea-pig taenia caecum by purine nucleotides and by "purinergic" nerve stimulation. J Physiol 299:75–83

Jänig W, Koltzenburg M (1991) Receptive properties of sacral primary afferent neurons supplying the colon. J Neurophysiol 65:1067–1077

Jenkinson KM, Reid JJ (1995) Effect of diabetes on relaxations to non-adrenergic, non-cholinergic nerve stimulation in longitudinal muscle of the rat gastric fundus. Br J Pharmacol 116:1551–1556

Jessen KR, Burnstock G (1982) The enteric nervous system in tissue culture: a new mammalian model for the study of complex nervous networks. In: Kalsner S (ed) Trends in autonomic pharmacology, vol II. Urban and Schwartzenberg, Baltimore/Munich, pp 95–115

Jessen KR, Mirsky R (1983) Astrocyte-like glia in the peripheral nervous system: an immunohistochemical study of enteric glia. J Neurosci 3:2206–2218

Jin J-G, Katsoulis S, Schmidt WE, Grider JR (1994) Inhibitory transmission in tenia coli mediated by distinct vasoactive intestinal peptide and apamin-sensitive

pituitary adenylate cyclase activating peptide receptors. J Pharmacol Exp Ther 270:433–439
Johannesson N, Andersson KE, Joelsson B, Persson CG (1985) Relaxation of lower esophageal sphincter and stimulation of gastric secretion and diuresis by antiasthmatic xanthines. Role of adenosine antagonism. Am Rev Respir Dis 131:26–30
Johnson CR, Hourani SM (1994) Contractile effects of uridine 5′-triphosphate in the rat duodenum. Br J Pharmacol 113:1191–1196
Johnson CR, Charlton SJ, Hourani SMO (1996) Responses of the longitudinal muscle and the muscularis mucosae of the rat duodenum to adenine and uracil nucleotides. Br J Pharmacol. 117:823–830
Johnson RJ, Schemann M, Santer RM, Cowen T (1998) The effects of age on the overall population and on sub-populations of myenteric neurons in the rat small intestine. J Anat 192:479–488
Jones CJ, Mann GE, Smaje LH (1980) The role of cyclic nucleotides and related compounds in nerve-mediated vasodilatation in the cat submandibular gland. Br J Pharmacol 68:485–497
Jorgensen TD, Gromada J, Tritsaris K, Nauntofte B, Dissing S (1995) Activation of P_{2z} purinoceptors diminishes the muscarinic cholinergic-induced release of inositol 1,4,5-trisphosphate and stored calcium in rat parotid acini. ATP as a co-transmitter in the stimulus-secretion coupling. Biochem J312:457–464
Kabré E, Chaïb N, Boussard P, Merino G, Devleeschouwer M, Dehaye JP (1999) Study on the activation of phospholipases A_2 by purinergic agonists in rat submandibular ductal cells. Biochim Biophys Acta 1436:616–627
Kadowaki M, Takeda M, Tokita K, Hanaoka K, Tomoi M (2000) Molecular identification and pharmacological characterization of adenosine receptors in the guinea-pig colon. Br J Pharmacol 129:871–876
Kamiji T, Morita K, Katayama Y (1994) ATP regulates synaptic transmission by pre- and postsynaptic mechanisms in guinea-pig myenteric neurons. Neuroscience 59:165–174
Kamikawa Y, Shimo Y (1982) Modulating effects of opioids, purine compounds, 5-hydroxytryptamine and prostaglandin E_2 on cholinergic neurotransmission in a guinea-pig oesophagus preparation. J Pharm Pharmacol 34:794–797
Kamikawa Y, Serizawa K, Shimo Y (1977) Some possibilities for prostaglandin mediation in the contractile response to ATP of the guinea-pig digestive tract. Eur J Pharmacol 45:199–203
Kaminski PM, Proctor KG (1992) Extracellular and intracellular actions of adenosine and related compounds in the reperfused rat intestine. Circ Res 71:720–731
Kasakov L, Milenov K (1979) Myoelectrical patterns of "purinergic" gastric relaxation. Abstracts of Symposium on Physiology and Pharmacology of Smooth Muscle, Sofia, Bulgaria, October 9–11, 1979. p 65
Kastritsis CH, Salm AK, McCarthy K (1992) Stimulation of the P2Y purinergic receptor on type 1 astroglia results in inositol phosphate formation and calcium mobilization. J Neurochem 58:1277–1284
Katayama Y, Morita K (1989) Adenosine 5′-triphosphate modulates membrane potassium conductance in guinea-pig myenteric neurones. J Physiol 408:373–390
Katsuragi T, Furuta K, Harada T, Furukawa T (1985) Cholinergic neuromodulation by ATP, adenosine and its N^6-substituted analogues in guinea-pig ileum. Clin Exp Pharmacol Physiol 12:73–78
Katsuragi T, Shiakabe K, Ogawa S, Soejima O, Furukawa T (1990) Involvement of dihydropyridine-sensitive Ca^{2+}channels in adenosine-evoked inhibition of acetylcholine release from guinea pig ileal preparation. J Neurochem 55:363–369
Kaufman HS, Shermak MA, May CA, Pitt HA, Lillemoe KD (1993) Nitric oxide inhibits resting sphincter of Oddi activity. Am J Surg 165:74–80

Kazic T, Milosavljevic D (1977) Influence of pyridylisatogen tosylate on contractions produced by ATP and by purinergic stimulation in the terminal ileum of the guinea-pig. J Pharm Pharmacol 29:542–545

Keef KD, Du C, Ward SM, McGregor B, Sanders KM (1993) Enteric inhibitory neural regulation of human colonic circular muscle: role of nitric oxide. Gastroenterology 105:1009–1016

Keef KD, Shuttleworth CW, Xue C, Bayguinov O, Publicover NG, Sanders KM (1994) Relationship between nitric oxide and vasoactive intestinal polypeptide in enteric inhibitory neurotransmission. Neuropharmacology 33:1303–1314

Kennedy I, Humphrey PP (1994) Evidence for the presence of two types of P2 purinoceptor in the guinea-pig ileal longitudinal smooth muscle preparation. Eur J Pharmacol 261:273–280

Keranen U, Vanhatalo S, Kiviluoto T, Kivilaakso E, Soinila S (1995) Co-localization of NADPH diaphorase reactivity and vasoactive intestinal polypeptide in human colon. J Auton Nerv Syst 54:177–183

Kerstan D, Leipziger J, Gordjani N, Nitschke R, Greger R (1997) Luminal addition of ATP induces K^+ secretion via a $P2Y_2$ receptor in rat distal colonic mucosa. Pflugers Arch 433:R128

Kidder GW (1973) Effects of the ATP analog 5′-adenylylmethylene diphosphonate on acid secretion in frog gastric mucosa. Biochim Biophys Acta 298:732–742

Kidder GW (1982) Adenosine kinase from *Trypanosoma cruzi*. Biochem Biophys Res Commun 107:381–388

Kim HD, Bowen JW, James-Kracke MR, Landon LA, Camden JM, Burnett JE, Turner JT (1996) Potentiation of regulatory volume decrease by P_{2U} purinoceptors in HSG-PA cells. Am J Physiol 270:C86–C97

Kim Y-C, Camaioni E, Ziganshin AU, Ji X-D, King BF, Wildman SS, Rychkov A, Yoburn J, Kim H, Mohanram A, Harden TK, Boyer JL, Burnstock G, Jacobson, KA (1998) Synthesis and structure-activity relationships of pyridoxal-6-azoaryl-5′-phosphate and phosphonate derivatives as P2 receptor antagonists. Drug Dev Res 45:52–66

Kimball BC, Mulholland MW (1995) Neuroligands evoke calcium signaling in cultured myenteric neurons. Surgery 118:162–169

Kimball BC, Mulholland MW (1996) Enteric glia exhibit P_{2U} receptors that increase cytosolic calcium by a phospholipase C-dependent mechanism. J Neurochem 66:604–612

Kimball BC, Yule DI, Mulholland MW (1996) Extracellular ATP mediates Ca^{2+} signaling in cultured myenteric neurons via a PLC-dependent mechanism. Am J Physiol 270:G587–G593

Kimmich G, Randles J (1980) Regulation of Na^+-dependent sugar transport in intestinal epithelial cells by exogenous ATP. Am J Physiol 238:C177–C183

King BF, Neary JT, Zhu Q, Wang S, Norenberg MD, Burnstock G (1996) P2 purinoceptors in rat cortical astrocytes: expression, calcium-imaging and signalling studies. Neuroscience 74:1187–1196

Kirkup AJ, Booth CE, Chessell IP, Humphrey PPA, Grundy D (1999) Excitatory effect of P2X receptor activation on mesenteric afferent nerves in the anaesthetised rat. J Physiol 520:551–563

Kishi M, Takeuchi T, Suthamnatpong N, Ishii T, Nishio H, Hata F, Takewaki T (1996) VIP- and PACAP-mediated nonadrenergic, noncholinergic inhibition in longitudinal muscle of rat distal colon: involvement of activation of charybdotoxin- and apamin-sensitive K^+channels. Br J Pharmacol 119:623–630

Knudsen MA, Tøttrup A (1992) A possible role of the l-arginine-nitric oxide pathway in the modulation of cholinergic transmission in the guinea-pig taenia coli. Br J Pharmacol 107:837–841

Knudsen MA, Svane D, Tøttrup A (1992) Action profiles of nitricoxide, S-nitroso-L-cysteine, SNP, and NANC responses in opossum lower esophageal sphincter. Am J Physiol 262:G840–G846

Knudsen MA, Glavind EB, Tøttrup A (1995) Transmitter interactions in rabbit internal anal sphincter. Am J Physiol 269:G232–G239

Koh SD, Dick GM, Sanders KM (1997) Small-conductance Ca^{2+}-dependent K^+ channels activated by ATP in murine colonic smooth muscle. Am J Physiol 273:C2010–C2021

Kohn PG, Newey H, Smyth DH (1970) The effect of adenosine triphosphate on the transmural potential in rat small intestine. J Physiol 208:203–220

Korman LY, Lemp GF, Jackson MJ, Gardner JD (1982) Mechanism of action of ATP on intestinal epithelial cells. Cyclic AMP-mediated stimulation of active ion transport. Biochim Biophys Acta 721:47–54

Kortezova N, Mizhorkova Z, Milusheva E, Varga G, Vizi ES, Papasova M (1996) Non-adrenergic non-cholinergic neuron stimulation in the cat lower esophageal sphincter. Eur J Pharmacol 304:109–115

Kotecha N (1999) Mechanisms underlying ACh induced modulation of neurogenic and applied ATP constrictions in the submucosal arterioles of the guinea-pig small intestine. Br J Pharmacol 126:1625–1633

Kreienberg PB, Darling RC, Shah DM, Vincent PA, Blumenstock FA (1996) ATP-$MgCl_2$ reduces intestinal permeability during mesenteric ischemia. J Surg Res 66:69–74

Krishnamurty VS, Kadowitz PJ (1983) Influence of adenosine triphosphate on the isolated perfused mesenteric artery of the rabbit. Can J Physiol Pharmacol 61:1409–1417

Kuchii M, Miyahara JT, Shibata S (1973) [^{3}H]-Adenine nucleotide and [^{3}H]-noradrenaline release evoked by electrical field stimulation, perivascular nerve stimulation and nicotine from the taenia of the guinea-pig caecum. Br J Pharmacol 49:258–266

Kurihara K, Nakanishi N, Ueha T (1997) A calcium channel in human submandibular duct cell line, HSG cells, not regulated by P_{2U} purinergic receptor-mediated intracellular calcium mobilization. Arch Oral Biol 42:547–557

Kwok YN, McIntosh C, Brown J (1990) Augmentation of release of gastric somatostatin-like immunoreactivity by adenosine, adenosine triphosphate and their analogs. J Pharmacol Exp Ther 255:781–788

Lachish M, Alzola E, Chaib N, Métioui M, Grosfils K, Kabré E, Moran A, Marino A, Dehaye JP (1996) Study of nonspecific cation channel coupled to P_{2z} purinergic receptors using an acid load technique. Am J Physiol 271:C1920–C1926

Larsson LT (1994) Hirschsprung's disease – immunohistochemical findings. Histol Histopathol 9:615–629

Larsson LT, Shen Z, Ekblad E, Sundler F, Alm P, Andersson KE (1995) Lack of neuronal nitric oxide synthase in nerve fibers of aganglionic intestine: a clue to Hirschsprung's disease. J Pediatr Gastroenterol Nutr 20:49–53

Lee MG, Schultheis PJ, Yan M, Shull GE, Bookstein C, Chang E, Tse M, Donowitz M, Park K, Muallem S (1998) Membrane-limited expression and regulation of Na^+-H^+ exchanger isoforms by P_2 receptors in the rat submandibular gland duct. J Physiol 513:341–357

Lefebvre RA (1986) Study on the possible neurotransmitter of the non-adrenergic non-cholinergic innervation of the rat gastric fundus. Arch Int Pharmacodyn Ther 280:110–136

Lefebvre RA, Burnstock G (1990) Effect of adenosine triphosphate and related purines in the rat gastric fundus. Arch Int Pharmacodyn Ther 303:199–215

Lefebvre RA, Willems JL (1979) Gastric relaxation by apomorphine and ATP in the conscious dog. J Pharm Pharmacol 31:561–563

Lefebvre RA, de Beurme FA, Sas S (1991) Effect of apamin on the responses to VIP, ATP and NANC neurone stimulation in the rat and cat gastric fundus. J Auton Pharmacol 11:73–83

Leipziger J, Kerstan D, Nitschke R, Greger R (1997) ATP increases $[Ca^{2+}]_i$ and ion secretion via abasolateral P2Y-receptor in rat distal colonic mucosa. Pflugers Arch 434:77–83
Lelièvre V, Muller JM, Falcòn J (1998a) Adenosine modulates cell proliferation in human colonic adenocarcinoma. I. Possible involvement of adenosine A_1 receptor subtypes in HT29 cells. Eur J Pharmacol 341:289–297
Lelièvre V, Muller JM, Falcòn J (1998b) Adenosine modulates cell proliferation in human colonic carcinoma. II. Differential behavior of HT29, DLD-1, Caco-2 and SW403 cell lines. Eur J Pharmacol 341:299–308
LePard KJ, Messori E, Galligan JJ (1997) Purinergic fast excitatory postsynaptic potentials in myenteric neurons of guinea pig: distribution and pharmacology. Gastroenterology 113:1522–1534
Levin R, Braiman A, Priel Z (1997) Protein kinase C induced calcium influx and sustained enhancement of ciliary beating by extracellular ATP. Cell Calcium 21: 103–113
Levine AS, Morley JE (1983) Effect of intraventricular adenosine on food intake in rats. Pharmacol Biochem Behav 19:23–26
Levine AS, Grace M, Krahn DD, Billington CJ (1989) The adenosine agonist N^6-R-phenylisopropyladenosine (R-PIA) stimulates feeding in rats. Brain Res 477:280–285
Lewis C, Neidhart S, Holy C, North RA, Surprenant A (1995) Coexpression of $P2X_2$ and $P2X_3$receptor subunits can account for ATP-gated currents in sensory neurons. Nature 377:432–435
Lidberg P, Dahlström A, Lundberg JM, Ahlman H (1983) Different modes of action of substance P in the motor control of the feline stomach and pylorus. Regul Pept 7:41–52
Lidberg P, Dahlström A, Ahlman H (1984) Is 5-HT a mediator in the motor control of the feline pylorus? Scand J Gastroenterol 19:321–328
Lim SP, Muir TC (1986) Neuroeffector transmission in the guinea-pig internal anal sphincter: an electrical and mechanical study. Eur J Pharmacol 128:17–24
Lincoln J, Hoyle CHV, Burnstock G (1997) Nitric oxide in health and disease. Cambridge University Press, Cambridge
Lindh B, Dalsgaard CJ, Elfvin LG, Hokfelt T, Cuello AC (1983) Evidence of substance P immunoreactive neurons in dorsal root ganglia and vagal ganglia projecting to the guinea pig pylorus. Brain Res 269:365–369
Lindh B, Hokfelt T, Elfvin LG, Terenius L, Fahrenkrug J, Elde R, Goldstein M (1986) Topography of NPY-, somatostatin-, and VIP-immunoreactive, neuronal subpopulations in the guinea pig celiac-superior mesenteric ganglion and their projection to the pylorus. J Neurosci 6:2371–2383
Lohrmann E, Cabantchik ZI, Greger R (1992) Transmitter-induced changes of the membrane voltage of HT_{29} cells. Pflugers Arch 421:224–229
Long RG (1983) Chagas' disease. In: Bannister R (ed) Autonomic failure: a textbook of clincal disorders of the autonomic nervous system. Oxford University Press, Oxford, New York, Toronto, pp 596–614
Loo FD, Palmer DW, Soergel KH, Kalbfleisch JH, Wood CM (1984) Gastric emptying in patients with diabetes mellitus. Gastroenterology 86:485–494
Low PA (1996) Diabetic autonomic neuropathy. Semin Neurol 16:143–151
Luck MS, Dahl JL, Boyeson MG, Bass P (1993) Neuroplasticity in the smooth muscle of the myenterically and extrinsically denervated rat jejunum. Cell Tissue Res 271:363–374
Lundgren O (1983) Vagal control of the motor functions of the lower esophageal sphincter and the stomach. J Auton Nerv Syst 9:185–197
Maas AJ, Den Hertog A (1979) The effect of apamin on the smooth muscle cells of the guinea-pig taenia coli. Eur J Pharmacol 58:151–156
Maas AJ, Den Hertog A, Ras R, Van den Akker J (1980) The action of apamin on guinea-pig taenia caeci. Eur J Pharmacol 67:265–274

Machaly M, Dalziel HH, Sneddon P (1988) Evidence for ATP as a cotransmitter in dog mesenteric artery. Eur J Pharmacol 147:83–91

Mackenzie I, Burnstock G (1980) Evidence against vasoactive intestinal polypeptide being the non-adrenergic, non-cholinergic inhibitory transmitter released from nerves supplying the smooth muscle of the guinea-pig taenia coli. Eur J Pharmacol 67:255–264

Madara JL (1991) Functional morphology of epithelium of the small intestine. In: Fields H, Frizzell RA, Schultz SG (eds) Handbook of physiology, sect 6: The gastrointestinal system, vol iv: Intestinal absorption and secretion. American Physiological Society, Bethesda, Md, pp 83–120

Maggi CA, Giuliani S (1993) Multiple inhibitory mechanisms mediate non-adrenergic non-cholinergic relaxation in the circular muscle of the guinea-pig colon. Naunyn Schmiedebergs Arch Pharmacol 347:630–634

Maggi CA, Giuliani S (1996) Characterization of the apamin- and L-nitroarginine-resistant NANC inhibitory transmission to the circular muscle of guinea-pig colon. J Auton Pharmacol 16:131–145

Maggi CA, Manzini S, Meli A (1984) Evidence that $GABA_A$ receptors mediate relaxation of rat duodenum by activating intramural nonadrenergic-noncholinergic neurones. J Auton Pharmacol 4:77–85

Maggi CA, Manzini S, Giuliani S, Santicioli P, Meli A (1986) Extrinsic origin of the capsaicin-sensitive innervation of rat duodenum: possible involvement of calcitonin gene-related peptide (CGRP) in the capsaicin-induced activation of intramural non-adrenergic non-cholinergic neurons. Naunyn Schmiedebergs Arch Pharmacol 334:172–180

Maguire MH, Satchell DG (1979) The contribution of adenosine to the inhibitory actions of adenine nucleotides on the guinea-pig taenia coli: studies with phosphate-modified adenine nucleotide analogs and dipyridamole. J Pharmacol Exp Ther 211:626–631

Mahmod SM, Huddart H (1993) Purinergic modulation of spontaneous activity and of responses to high potassium and acetylcholine in rat ileal smooth muscle. Comp Biochem Physiol C 106:79–85

Manzini S, Maggi CA, Meli A (1985) Further evidence for involvement of adenosine-5′-triphosphate in non-adrenergic non-cholinergic relaxation of the isolated rat duodenum. Eur J Pharmacol 113:399–408

Manzini S, Hoyle CHV, Burnstock G (1986) An electrophysiological analysis of the effect of reactive blue 2, a putative P_2-purinoceptor antagonist, on inhibitory junction potentials of rat caecum. Eur J Pharmacol 127:197–204

Martin MT, Gonalons E, Fernandez E (1998) Contribution of inhibitory neurotransmitters to the CCK induced relaxation of the circular muscle of avian ileum. Life Sci 62:937–946

Martinez-Cuesta MA, Massuda H, Whittle BJ, Moncada S (1995) Impairment of nitrergic-mediated relaxation of rat isolated duodenum by experimental diabetes. Br J Pharmacol 114:919–924

Martins Campos JV, Tafuri WL (1973) Chagas enteropathy. Gut 14:910–919

Mashimo H, He XD, Huang PL, Fishman MC, Goyal RK (1996) Neuronal constitutive nitric oxide synthase is involved in murine enteric inhibitory neurotransmission. J Clin Invest 98:8–13

Matharu MS, Hollingsworth M (1992) Purinoceptors mediating relaxation and spasm in the rat gastric fundus. Br J Pharmacol 106:395–403

Matsuda NM, Lemos MC, Feitosa Junior RL, de Oliveira RB, Ballejo G (1997a) Nonadrenergic-noncholinergic relaxations of isolated circular muscle from South American opossum esophagogastric junction: is nitric oxide the inhibitory mediator? J Auton NervSyst 66:119–125

Matsuda NM, Oliveira RB, Ballejo G (1997b) Characterization of nerve-induced relaxation of gastrointestinal sphincteric smooth muscle in a South American opossum (*Didelphis albiventris*). Braz J Med Biol Res 30:793–799

Matsuo K, Katsuragi T, Fujiki S, Sato C, Furukawa T (1997) ATP release and contraction mediated by different P2-receptor subtypes in guinea-pig ileal smooth muscle. Br J Pharmacol 121:1744–1749

Matthews JB, Tally KJ, Smith JA, Zeind AJ, Hrnjez BJ (1995) Activation of Cl secretion during chemical hypoxia by endogenous release of adenosine in intestinal epithelial monolayers. J Clin Invest 96:117–125

Matusák O, Bauer V (1986) Effect of desensitization induced by adenosine 5′-triphosphate, substance P, bradykinin, serotonin, γ-aminobutyric acid and endogenous noncholinergic-nonadrenergic transmitter in the guinea-pig ileum. Eur J Pharmacol 126:199–209

Mayer EA, Gebhart GF (1994) Basic and clinical aspects of visceral hyperalgesia. Gastroenterology 107:271–293

McAlroy HL, Collett A, Ahmed S, Ko WH, Baines DL, Wilson SM (1999) Expression of multiple P2Y receptor subtypes by Caco-2 human colonic adenocarcinoma cells. J Physiol 517:97P [Abstract]

McColl KE (1997) Pathophysiology of duodenal ulcer disease. Eur J Gastroenterol Hepatol 9 (Suppl 1):S9–S12

McConalogue K, Furness JB, Vremec MA, Holst JJ, Tornoe K, Marley PD (1995a) Histochemical, pharmacological, biochemical and chromatographic evidence that pituitary adenylyl cyclase activating peptide is involved in inhibitory neurotransmission in the taenia of the guinea-pig caecum. J Auton Nerv Syst 50:311–322

McConalogue K, Lyster DJ, Furness JB (1995b) Electrophysiological analysis of the actions of pituitary adenylyl cyclase activating peptide in the taenia of the guinea-pig caecum. Naunyn Schmiedebergs Arch Pharmacol 352:538–544

McConalogue K, Todorov L, Furness JB, Westfall DP (1996) Direct measurement of the release of ATP and its major metabolites from the nerve fibres of the guinea-pig taenia coli. Clin Exp Pharmacol Physiol 23:807–812

McGill JM, Basavappa S, Mangel AW, Shimokura GH, Middleton JP, Fitz JG (1994) Adenosine triphosphate activates ion permeabilities in biliary epithelial cells. Gastroenterology 107:236–243

McGregor GP, Bishop AE, Blank MA, Christofides ND, Yiangou Y, Polak JM, Bloom SR (1984) Comparative distribution of vasoactive intestinal polypeptide (VIP), substance P and PHI in the enteric sphincters of the cat. Experientia 40:469–471

McKirdy HC (1992) Innervation of internal anal sphincter – in vitro studies. Int J Colorectal Dis 7:43–44

McKirdy HC, McKirdy ML, Lewis MJ, Marshall RW (1992) Evidence for involvement of nitric oxide in the non-adrenergic non-cholinergic (NANC) relaxation of human lower oesophageal sphincter muscle strips. Exp Physiol 77:509–511

McMillian MK, Soltoff SP, Cantley LC, Talamo BR (1987) Extracellular ATP elevates intracellular free calcium in rat parotid acinar cells. Biochem Biophys Res Commun 149:523–530

McMillian MK, Soltoff SP, Lechleiter JD, Cantley LC, Talamo BR (1988) Extracellular ATP increases free cytosolic calcium in rat parotid acinar cells. Differences from phospholipase C-linked receptor agonists. Biochem J 255:291–300

McMillian MK, Soltoff SP, Cantley LC, Rudel R, Talamo BR (1993) Two distinct cytosolic calcium responses to extracellular ATP in rat parotid acinar cells. Br J Pharmacol 108:453–461

Mehta AK, Kulkarni SK (1983) Evidence for the presence of P_1- and P_2-purinoceptors in rat caecum. Arch Int Pharmacodyn Ther 264:187–194

Merlin D, Augeron C, Tien X-Y, Guo X, Laboisse CL, Hopfer U (1994) ATP-stimulated electrolyte and mucin secretion in the human intestinal goblet cell line HT29-C1.16E. J Membr Biol 137:137–149

Métioui M, Amsallem H, Alzola E, Chaib N, Elyamani A, Moran A, Marino A, Dehaye JP (1996) Low affinity purinergic receptor modulates the response of rat submandibular glands to carbachol and substance P. J Cell Physiol 168:462–475

Meyer MP, Clarke JDW, Patel K, Townsend-Nicholson A, Burnstock G (1999) Selective expression of purinoceptor $cP2Y_1$ suggests a role for nucleotide signalling in development of the chick embryo. Dev Dyn 214:152–158

Mihara S (1993) Intracellular recordings from neurones of the submucous plexus. Prog Neurobiol 40:529–572

Mihara S, Katayama Y, Nishi S (1985) Slow postsynaptic potentials in neurones of the submucous plexus of guinea pig caecum and their mimickry by noradrenaline and various peptides. Neuroscience 16:1057–1066

Minker E, Matejka Z (1981) Purinergic reflex activated by cathartics in the rat. Acta Physiol Acad Sci Hung 57:99–107

Minocha A, Galligan JJ (1993) Excitatory and inhibitory responses mediated by $GABA_A$ and $GABA_B$ receptors in guinea pig distal colon. Eur J Pharmacol 230:187–193

Mishra SK, Raviprakash V (1988) Excitatory P1-purinoceptors on pre- and postganglionic cholinergic nerve terminals in the chick oesophagus. J Pharm Pharmacol 40:441–442

Mizhorkova Z, Kortezova N, Bredy Dobreva G, Papasova M (1994) Role of nitric oxide in mediating non-adrenergic non-cholinergic relaxation of the cat ileocecal sphincter. Eur J Pharmacol 265:77–82

Mizuno-Kamiya M, Kameyama Y, Yashiro K, Fujita A (1998) ATP-mediated activation of Ca^{2+}-independent phospholipase A_2 in secretory granular membranes from rat parotid gland. J Biochem Tokyo 123:205–212

Moneta NA, McDonald TJ, Cook MA (1997) Endogenous adenosine inhibits evoked substance P release from perifused networks of myenteric ganglia. Am J Physiol 272:G38–G45

Moody CJ, Burnstock G (1982) Evidence for the presence of P_1-purinoceptors on cholinergic nerve terminals in the guinea-pig ileum. Eur J Pharmacol 77:1–9

Moody CJ, Meghji P, Burnstock G (1984) Stimulation of P_1-purinoceptors by ATP depends partly on its conversion to AMP and adenosine and partly on direct action. Eur J Pharmacol 97:47–54

Moritoki H, Kanbe T, Maruoka M, Ohara M, Ishida Y (1978) Potentiation by dipyridamole of the inhibition of guinea-pig ileum twitch response caused by adenine derivatives. J Pharmacol Exp Ther 204:343–350

Mózsik G, Beck Z, Füzesi Z, Kiss J, Nagy L, Palotai Z, Szilágyi A, Tárnok F, Tóth E, Vizi F (1978a) Cellular mechanisms of gastric hypersecretion in pylorus-ligated rats. Acta Physiol Scand Special Suppl 187–198

Mózsik G, Kutas J, Nagy L, Tárnok F, Vizi F (1978b) Interrelationships between the cholinergic influences, gastric mucosa Na^+–K^+-dependent ATPase, ATP, ADP, ions of gastric juice and basal secretion inpatients. Acta Physiol Scand Special Suppl 199–208

Mózsik G, Fiegler M, Lörincz P, Nagy L, Tárnok F (1979) The role of the ATP-adenylate cyclase-cAMP system and its pharmacological regulation in the development of gastric hypersecretion and ulceration. Acta Med Acad Sci Hung 36:427–448

Muller MJ, Baer HP (1980) Apamin, a nonspecific antagonist of smooth muscle relaxants. Naunyn Schmiedebergs Arch Pharmacol 311:105–107

Muramatsu I (1986) Evidence for sympathetic, purinergic transmission in the mesenteric artery of the dog. Br J Pharmacol 87:478–480

Murr MM, Balsiger BM, Farrugia G, Sarr MG (1999) Role of nitric oxide, vasoactive intestinal polypeptide, and ATP in inhibitory neurotransmission in human jejunum. J Surg Res 84:8–12

Murray J, Du C, Ledlow A, Bates JN, Conklin JL (1991) Nitric oxide: Mediator of nonadrenergic noncholinergic responses of opossum esophageal muscle. Am J Physiol 261:G401–G406

Murthy KS, Makhlouf GM (1998) Coexpression of purinergic ligand-gated P2X and G protein-coupled P2Y receptors in smooth muscle. Preferential activation of P2Y

receptors coupled to phospholipase C(PLC)-$\beta 1$ via $G\alpha_{q/11}$ and to PLC-$\beta 3$ via $G\beta\gamma_{i3}$. J Biol Chem 273:4695–4704
Murthy KS, McHenry L, Grider JR, Makhlouf GM (1995) Adenosine A_1 and A_{2b} receptors coupled to distinct interactive signaling pathways in intestinal muscle cells. J Pharmacol Exp Ther 274:300–306
Nagata K, Saito H, Matsuki N (1993) Adenosine induces contractions in suncus ileum. Jpn J Pharmacol 63:415–421
Namiot Z, Rutkiewicz J, Stasiewicz J, Baranczuk E, Marcinkiewicz M (1991) Adenosine deaminase activity in the gastric mucosa in patients with gastric ulcer. Effects of ranitidine and sucralfate. Eur J Pharmacol 205:101–103
Namiot Z, Marcinkiewicz M, Jaroszewicz W, Stasiewicz J, Gorski J (1993) Mucosal adenosine deaminase activity and gastric ulcer healing. Eur J Pharmacol 243:301–303
Nandi J, Ray TK, Sen PC (1981) Studies of gastric Ca^{2+}-stimulated adenosine triphosphatase. I. Characterization and general properties. Biochim Biophys Acta 646:457–464
Neary J, Rathbone M, Cattabeni F, Abbracchio MP, Burnstock G (1996) Trophic actions of extracellular nucleotides and nucleosides on glial and neuronal cells. Trends Neurosci 19:13–18
Ness TJ, Gebhart GF (1988) Colorectal distension as a noxious visceral stimulus: physiologic and pharmacologic characterization of pseudaffective reflexes in the rat. Brain Res 450:153–169
Nicholls J, Hourani SMO (1997) Characterization of adenosine receptors on rat ileum, ileal longitudinal muscle and muscularis mucosae. Eur J Pharmacol 338:143–150
Nicholls J, Brownhill VR, Hourani SMO (1996) Characterization of P1-purinoceptors on rat isolated duodenum longitudinal muscle and muscularis mucosae. Br J Pharmacol 117:170–174
Nissan S, Vinograd Y, Hadari A, Merguerian P, Zamir O, Lernau O, Hanani M (1984) Physiological and pharmacological studies of the internal anal sphincter in the rat. J Pediatr Surg 19:12–14
Nitahara K, Kittel A, Liang SD, Vizi ES (1995) A_1-receptor-mediated effect of adenosine on the release of acetylcholine from the myenteric plexus: role and localization of ecto-ATPase and 5′-nucleotidase. Neuroscience 67:159–168
North RA (1982) Electrophysiology of the enteric nervous system. Neuroscience 7:315–325
Northway MG, Burks TF (1980) Stimulation of cholinergic nerves in dog intestine by adenine nucleotides. Eur J Pharmacol 65:11–19
Nowak TV, Harrington B, Kalbfleisch JH, Amatruda JM (1986) Evidence for abnormal cholinergic neuromuscular transmission in diabetic rat small intestine. Gastroenterology 91:124–132
Nurko S, Rattan S (1988) Role of vasoactive intestinal polypeptide in the internal anal sphincter relaxation of the opossum. J Clin Invest 81:1146–1153
Ny L, Alm P, Larsson B, Ekstrom P, Andersson KE (1995) Nitric oxide pathway in cat esophagus: localization of nitric oxide synthase and functional effects. Am J Physiol 268:G59–G70
O'Kelly T, Brading A, Mortensen N (1993) Nerve mediated relaxation of the human internal anal sphincter: the role of nitric oxide. Gut 34:689–693
Ohga A, Taneike T (1977) Dissimilarity between the responses to adenosine triphosphate or its related compounds and non-adrenergic inhibitory nerve stimulation in the longitudinal smooth muscle of pig stomach. Br J Pharmacol 60:221–231
Ohkawa H (1974) An analysis of the mechanical responses of the isolated ileum to single transmural stimulation and to drugs. Bull Yamaguchi Med School 21:31–45
Ohkawa H (1984) Effects of ATP, its related nucleotides and VIP on the inhibitory potentials in the smooth muscle cells of the guinea-pig duodenum. Tohoku J Exp Med 142:409–422

Ohkawa H, Watanabe M (1976) Non-adrenergic inhibition of the electrical activity of the antrum muscle fibers of the guinea-pig stomach. Tohoku J Exp Med 120:11–18
Ohno N, Ito KM, Yamamoto Y, Suzuki H (1993) Suramin selectively inhibits the non-adrenergic non-cholinergic inhibitory junction potential in the guinea-pig stomach. Eur J Pharmacol 249:121–123
Ohno N, Xue L, Yamamoto Y, Suzuki H (1996) Properties of the inhibitory junction potential in smooth muscle of the guinea-pig gastric fundus. Br J Pharmacol 117:974–978
Okamura T, Tanobe Y, Fujioka H, Ayajiki K, Toda N (1998) Mechanism of neurogenic relaxation and modification of the response by enteric substances in isolated dog colon. Eur J Pharmacol 358:245–252
Okasora T, Okamoto E (1986) Electrophysiological and pharmacological study on innervation of the aganglionic colon in Hirschsprung's disease of human and murine model. Z Kinderchir 41:93–96
Okwuasaba FK, Hamilton JT, Cook MA (1977) Relaxations of guinea-pig fundic strip by adenosine, adenine nucleotides and electrical stimulation: antagonism by theophylline and desensitization to adenosine and its derivatives. Eur J Pharmacol 46:181–198
Okwuasaba FK, Hamilton JT, Cook MA (1978) Evidence for the cell surface locus of presynaptic purine nucleotide receptors in the guinea-pig ileum. J Pharmacol Exp Ther 207:779–786
Olson L, Ålund M, Nordberg K-A (1976) Fluorescence-microscopical demonstration of a population of gastro-intestinal nerve fibres with a selective affinity for quinacrine. Cell Tissue Res 171:407–423
Onaka U, Fujii K, Abe I, Fujishima M (1997) Enhancement by exogenous and locally generated angiotensin II of purinergic neurotransmission via angiotensin type 1 receptor in the guinea-pig isolated mesenteric artery. Br J Pharmacol 122:942–948
Ota S, Hiraishi H, Terano A, Mutoh H, Kurachi Y, Shimada T, Ivey KJ, Sugimoto T (1989) Effect of adenosine and adenosine analogs on [^{14}C]aminopyrine accumulation by rabbit parietal cells. Dig Dis Sci 34:1882–1889
Ota S, Yoshiura K, Takahashi M, Hata Y, Kohmoto O, Kawabe T, Shimada T, Hiraishi H, Mutoh H, Terano A, Sugimoto T, Omata M (1994) P2 purinergic receptor regulation of mucus glycoprotein secretion by rabbit gastric mucous cells in a primary culture. Gastroenterology 106:1485–1492
Otsuguro K-I, Ito S, Ohta T, Nakazato Y (1996) Influence of purines and pyrimidines on circular muscle of the rat proximal stomach. Eur J Pharmacol 317:97–105
Otsuguro K-I, Ohta T, Ito S, Nakazato Y (1998) Two types of relaxation-mediating P_2 receptors in rat gastric circular muscle. Jpn J Pharmacol 78:209–215
Ovadyahu D, Eshel D, Priel Z (1988) Intensification of ciliary motility by extracellular ATP. Biorheology 25:489–501
Pacaud P, Feolde E, Frelin C, Loirand G (1996) Characterization of the P_{2Y}-purinoceptor involved in the ATP-induced rise in cytosolic Ca^{2+} concentration in rat ileal myocytes. Br J Pharmacol 118:2213–2219
Pahlin PE, Kewenter J (1976) The vagal control of the ileo-cecal sphincter in the cat. Acta Physiol Scand 96:433–442
Palea S, Artibani W, Ostardo E, Trist DG, Pietra C (1993) Evidence for purinergic neurotransmission in human urinary bladder affected by interstitial cystitis. J Urol 150:2007–2012
Palmer JM, Wood JD, Zafirov DH (1987) Purinergic inhibition in the small intestinal myenteric plexus of the guinea-pig. J Physiol 387:357–369
Paret RS, Kumashiro R, Kodama Y, Matsumoto T (1982) The effect of dipyridamole on experimentally induced stress ulcers. Am Surg 48:594–598
Park MK, Garrad RC, Weisman GA, Turner JT (1997) Changes in $P2Y_1$ nucleotide receptor activity during the development of rat salivary glands. Am J Physiol 272:C1388–C1393

Parkman HP, Reynolds JC, Ogorek CP, Kicsak KM (1989) Neuropeptide Y augments adrenergic contractions at feline lower esophageal sphincter. Am J Physiol 256:G589–G597

Parks AG, Fishlock DJ, Cameron JDH, May H (1969) Preliminary investigation of the pharmacology of the human internal anal sphincter. Gut 10:674–677

Parr CE, Sullivan DM, Paradiso AM, Lazarowski ER, Burch LH, Olsen JC, Erb L, Weisman GA, Boucher RC, Turner JT (1994) Cloning and expression of a human P_{2U} nucleotide receptor, a target for cystic fibrosis pharmacotherapy. Proc Natl Acad Sci USA 91:3275–3279

Patacchini R, De Giorgio R, Barthó L, Barbara G, Corinaldesi R, Maggi CA (1998) Evidence that tachykinins are the main NANC excitatory neurotransmitters in the guinea-pig common bile duct. Br J Pharmacol 124:1703–1711

Paton DM, Olsson RA (1991) Potency of N^6 secondary and tertiary alkyladenosine analogues at presynaptic A_1 adenosine receptors in guinea-pig ileum. J Auton Pharmacol 11:85–91

Paul ML, Miles DL, Cook MA (1982) The influence of glucosidic conformation and charge distribution on activity of adenine nucleosides as presynaptic inhibitors of acetylcholine release. J Pharmacol Exp Ther 222:241–245

Pauletzki JG, Sharkey KA, Davison JS, Bomzon A, Shaffer EA (1993) Involvement of L-arginine-nitric oxide pathways in neural relaxation of the sphincter of Oddi. Eur J Pharmacol 232:263–270

Peachey JA, Hourani SMO, Kitchen I (1999) Ontogeny of adenosine receptors in the longitudinal muscle and muscularis mucosae of the rat distal colon. Naunyn Schmiedebergs Arch Pharmacol 359:140–146

Pelckmans PA, Boeckxstaens GE, Van Maercke YM, Herman AG, Verbeuren TJ (1989) Acetylcholine is an indirect inhibitory transmitter in the canine ileocolonic junction. Eur J Pharmacol 170:235–242

Pencheva N (1997) Dependence of gamma-aminobutyric acid modulation of cholinergic transmission on nitric oxide and purines in cat terminal ileum. Eur J Pharmacol 339:193–200

Pennanen MF, Bass BL, Dziki AJ, Harmon JW (1994) Adenosine: differential effect on blood flow to subregions of the upper gastrointestinal tract. J Surg Res 56:461–465

Percy WH, Miller AJ, Brunz JT (1997) Pharmacologic characteristics of rabbit esophageal muscularis mucosae in vitro. Dig Dis Sci 42:2537–2546

Percy WH, Warren JM, Brunz JT (1999) Characteristics of the muscularis mucosae in the acid-secreting region of the rabbit stomach. Am J Physiol 276:G1213–G1220

Persson CG (1976) Inhibitory innervation of cat sphincter of Oddi. Br J Pharmacol 58:479–482

Pfeifer MA, Jung S, Crain G, Schumer M (1993) Autonomic neuropathy. Diabet Med 10:70S-73S

Piper AS, Hollingsworth M (1995) The purinoceptors of the guinea-pig isolated taenia caeci. Eur J Pharmacol 280:125–134

Pluja L, Fernandez E, Jimenez M (1999) Neural modulation of the cyclic electrical and mechanical activity in the rat colonic circular muscle: putative role of ATP and NO. Br J Pharmacol 126:883–892

Postorino A, Serio R, Mule F (1990) On the purinergic system in rat duodenum: existence of P1 and P2 receptors on the smooth muscle. Arch Int Physiol Biochim 98:53–58

Prentice DJ, Hourani SMO (1997) Adenosine analogues relax guinea-pig taenia caeci via an adenosine A_{2B} receptor and a xanthine-resistant site. Eur J Pharmacol 323:103–106

Proctor KG (1986) Possible role for adenosine in local regulation of absorptive hyperemia in rat intestine. Circ Res 59:474–481

Puurunen J, Huttunen P (1988) Central gastric antisecretory action of adenosine in the rat. Eur J Pharmacol 147:59–66

Puurunen J, Aittakumpu R, Tanskanen T (1986) Vagally mediated stimulation of gastric acid secretion by intravenously administered adenosine derivatives in anaesthetized rats. Acta Pharmacol Toxicol Copenh 58:265–271
Qian YM, Jones RL (1995) Inhibition of rat colon contractility by prostacyclin (IP-) receptor agonists: involvement of NANC neurotransmission. Br J Pharmacol 115:163–171
Rae MG, Muir TC (1996) Neuronal mediators of inhibitory junction potentials and relaxation in the guinea-pig internal anal sphincter. J Physiol 493:517–527
Ramme D, Regenold JT, Starke K, Busse R, Illes P (1987) Identification of the neuroeffector transmitter in jejunal branches of the rabbit mesenteric artery. Naunyn Schmiedebergs Arch Pharmacol 336:267–273
Rand MJ (1992) Nitrergic transmission: nitric oxide as a mediator of non-adrenergic, non-cholinergic neuro-effector transmission. Clin Exp Pharmacol Physiol 19:147–169
Rapaport E (1983) Treatment of human tumor cells with ADP or ATP yields arrest of growth in the S phase of the cell cycle. J Cell Physiol 114:279–283
Rapaport E, Fontaine J (1989) Anticancer activities of adenine nucleotides in mice are mediated through expansion of erythrocyte ATP pools. Proc Natl Acad Sci USA 86:1662–1666
Rapaport E, Fishman RF, Gercel C (1983) Growth inhibition of human tumor cells in soft-agar cultures by treatment with low levels of adenosine 5′-triphosphate. Cancer Res 43:4402–4406
Rattan S, Chakder S (1993) Inhibitory effect of CO on internal anal sphincter: heme oxygenase inhibitor inhibits NANC relaxation. Am J Physiol 265:G799–G804
Rattan S, Goyal RK (1980) Evidence against purinergic inhibitory nerves in the vagal pathway to the opossum lower esophageal sphincter. Gastroenterology 78:898–904
Rattan S, Shah R (1988) Influence of purinoceptors' agonists and antagonists on opossum internal anal sphincter. Am J Physiol 255:G389–G394
Rattan S, Thatikunta P (1993) Role of nitric oxide in sympathetic neurotransmission in opossum internal anal sphincter. Gastroenterology 105:827–836
Rattan S, Rosenthal GJ, Chakder S (1995) Human recombinant hemoglobin (rHb1.1) inhibits nonadrenergic noncholinergic (NANC) nerve-mediated relaxation of internal anal sphincter. J Pharmacol Exp Ther 272:1211–1216
Raybould HE, Gschossman JM, Ennes H, Lembo T, Mayer EA (1999) Involvement of stretch-sensitive calcium flux in mechanical transduction in visceral afferents. J Auton Nerv Syst 75:1–6
Reese JH, Cooper JR (1982) Modulation of the release of acetylcholine from ileal synaptosomes by adenosine and adenosine 5′-triphosphate. J Pharmacol Exp Ther 223:612–616
Reeves JJ, Coates J, Jarvis JE, Sheehan MJ, Strong P (1993) Characterization of the adenosine receptor mediating contraction in rat colonic muscularis mucosae. Br J Pharmacol 110:1255–1259
Reeves JJ, Jarvis JE, Sheehan MJ, Strong P (1995) Further investigations into adenosine A_1 receptor-mediated contraction in rat colonic muscularis mucosae and its augmentation by certain alkylxanthine antagonists. Br J Pharmacol 114:999–1004
Reiser S, Christiansen PA (1971) Inhibition of amino acid uptake by ATP in isolated intestinal epithelial cells. Biochim Biophys Acta 233:480–484
Reymann A, Gniess A (1988) Evidence for adenosine A_1 receptor action in rat jejunal mucosa. Eur J Pharmacol 149:155–158
Rhee PL, Koh KC, Paik SW, Rhee JW, Xu W, Kim KW, Kang TM. (1996) ATP as an excitatory NANC neurotransmitter in guinea-pig gastric antral circular muscle. Gastroenterology 110:A744 [Abstract]
Richards NW, Allbee WE, Gaginella TS, Wallace LJ (1987) Exogenous ATP-stimulated calcium uptake in isolated rat intestinal epithelial cells. Life Sci 40:1665–1672

Richards M, van Giersbergen P, Zimmermann A, Lesur B, Hoflack J (1997) Activation of neurotensin receptors and purinoceptors in human colonic adenocarcinoma cells detected with the microphysiometer. Biochem Pharmacol 54:825–832
Richardson J (1975) Pharmacologic studies of Hirschsprung's disease on a murine model. J Pediatr Surg 10:875–884
Rikimaru A, Fukushi Y, Suzuki T (1971) Effects of imidazole and phentolamine on the relaxant responses of guinea-pig taenia coli to transmural stimulation and to adenosine triphosphate. Tohoku J Exp Med 105:199–200
Rodrigo J, Uttenthal LO, Peinado MA, Esteban FJ, Fernández AP, Serrano J, Martínez de Velasco J, Santacana M, Bentura ML, Martínez-Murillo R, Pedrosa JA (1998) Distribution of nitric oxide synthase in the esophagus of the cat and monkey. J Auton Nerv Syst 70:164–179
Rogawski MA, Goodrich JT, Gershon MD, Touloukian RJ (1978) Hirschsprung's disease: absence of serotonergic neurons in the aganglionic colon. J Pediatr Surg 13:608–615
Roman RM, Fitz JG (1999) Emerging roles of purinergic signaling in gastrointestinal epithelial secretion and hepatobiliary function. Gastroenterology 116:964–979
Roman RM, Feranchak AP, Salter KD, Wang Y, Fitz JG (1999) Endogenous ATP release regulates Cl^- secretion in cultures of human and rat biliary epithelial cells. Am J Physiol 276:G1391–G1400
Rothstein RD, Johnson E, Ouyang A (1989) Substance P: mechanism of action and receptor distribution at the feline ileocecal sphincter region. Am J Physiol 257:G447–G453
Rubin MR, Fournet J, Snape WJ Jr, Cohen S (1980) Adrenergic regulation of ileocecal sphincter function in the cat. Gastroenterology 78:15–21
Rutherford A, Burnstock G (1978) Neuronal and non-neuronal components in the overflow of labelled adenyl compounds from guinea-pig taenia coli. Eur J Pharmacol 48:195–202
Sachs G, Wallmark B, Saccomani G, Rabon E, Stewart HB, DiBona DR, Berglindh T (1982) The ATP-dependent component of gastric acid secretion. Current Topics in Membrane and Transport. Academic Press, San Diego, pp 135–159
Saegesser F, Roenspies U, Robinson JW (1979) Ischemic diseases of the large intestine. Pathobiol Annu 9:303–337
Saffrey MJ, Hassall CJS, Hoyle CHV, Belai A, Moss J, Schmidt HHHW, Förstermann U, Murad F, Burnstock G (1992) Colocalisation of nitric oxide synthase and NADPH-diaphorase activity in cultured myenteric neurones. Neuroreport 3:333–336
Said G, Joskowicz M, Barreira AA, Eisen H (1985) Neuropathy associated with experimental Chagas' disease. Ann Neurol 18:676–683
Sakai K, Akima M, Matsushita H (1979) Analysis of the contractile responses of the ileal segment of the isolated blood-perfused small intestine of rats to adenosine triphosphate and related compounds. Eur J Pharmacol 58:157–162
Sakamoto Y, Morishita H, Imanaga I (1987) The long-lasting contraction induced by transmural stimulation in the longitudinal muscle of the guinea-pig gastric corpus. Jpn J Physiol 37:621–630
Sanders KM (1998) G protein-coupled receptors in gastrointestinal physiology. IV. Neural regulation of gastrointestinal smooth muscle. Am J Physiol 275:G1–G7
Sanders KM, Ward SM (1992) Nitric oxide as a mediator of nonadrenergic noncholinergic neurotransmission. Am J Physiol 262:G379–G392
Santer RM, Baker DM (1988) Enteric neuron numbers and sizes in Auerbach's plexus in the small and large intestine of adult and aged rats. J Auton Nerv Syst 25:59–67
Santer RM, Conboy VB (1990) Prenatal under nutrition permanently decreases enteric neuron number and sympathetic innervation of Auerbach's plexus in the rat. J Anat 168:57–62

Sarosi GA, Barnhart DC, Turner DJ, Mulholland MW (1998) Capacitative Ca^{2+} entry in enteric glia induced by thapsigargin and extracellular ATP. Am J Physiol 275:G550–G555

Satchell DG (1981) Nucleotide pyrophosphatase antagonises responses to adenosine 5′-triphosphate and non-adrenergic, non-cholinergic inhibitory nerve stimulation in the guinea-pig isolated taenia coli. Br J Pharmacol 74:319–321

Satchell DG, Burnstock G (1971) Quantitative studies of the release of purine compounds following stimulation of non-adrenergic inhibitory nerves in the stomach. Biochem Pharmacol 20:1694–1697

Satchell DG, Burnstock G (1975) Comparison of the inhibitory effects on the guinea-pig taenia coli of adenine nucleotides and adenosine in the presence and absence of dipyridamole. Eur J Pharmacol 32:324–328

Satchell DG, Maguire MH (1975) Inhibitory effects of adenine nucleotide analogs on the isolated guinea-pig taenia coli. J Pharmacol Exp Ther 195:540–548

Satchell DG, Maguire MH (1982) Evidence for separate receptors for ATP and adenosine in the guinea-pig taenia coli. Eur J Pharmacol 81:669–672

Satchell DG, Lynch A, Bourke PM, Burnstock G (1972) Potentiation of the effects of exogenously applied ATP and purinergic nerve stimulation on the guinea-pig taenia coli by dipyridamole and hexobendine. Eur J Pharmacol 19:343–350

Satchell D, Burnstock G, Dann P (1973) Antagonism of the effects of purinergic nerve stimulation and exogenously applied ATP on the guinea-pig taenia coli by 2-substituted imidazolines and related compounds. Eur J Pharmacol 23:264–269

Sato C, Tsujioka Y, Katsuragi T (1999) Cross desensitizations on contractions by P2-agonists of guinea pig ileum. Jpn J Pharmacol 80:311–317

Sawmiller DR, Chou CC (1991) Adenosine is a vasodilator in the intestinal mucosa. Am J Physiol 261:G9–15

Sawynok J, Jhamandas KH (1976) Inhibition of acetylcholine release from cholinergic nerves by adenosine, adenine nucleotides and morphine: antagonism by theophylline. J Pharmacol Exp Ther 197:379–390

Scarpignato C, Tramacere R, Zappia L, Del Soldato P (1987) Inhibition of gastric acid secretion by adenosine receptor stimulation in the rat. Pharmacology 34:264–268

Schäfer K, Saffrey MJ, Burnstock G (1995) Trophic actions of 2-chloroadenosine and bFGF on cultured myenteric neurones. Neuroreport 6:937–941

Schepp W, Soll AH, Walsh JH (1990) Dual modulation by adenosine of gastrin release from canine G-cells in primary culture. Am J Physiol 259:G556–G563

Schiffman SS, Gill JM, Diaz C (1985) Methylxanthines enhance taste: evidence for modulation of taste by adenosine receptor. Pharmacol Biochem Behav 22:195–203

Schlenker T, Romac JM, Sharara AI, Roman RM, Kim SJ, LaRusso N, Liddle RA, Fitz JG (1997) Regulation of biliary secretion through apical purinergic receptors in cultured rat cholangiocytes. Am J Physiol273:G1108–G1117

Schloerb PR, Sieracki L, Botwin AJ, Winblad JM, Maguire MH (1981) Intravenous adenosine triphosphate (ATP) in hemorrhagic shock in rats. Am J Physiol 240:R52–R60

Schweickhardt C, Sabolic I, Brown D, Burckhardt G (1995) Ecto-adenosine triphosphatase in rat small intestinal brush-border membranes. Am J Physiol 268:G663–G672

Segerdahl M, Ekblom A, Sollevi A (1994) The influence of adenosine, ketamine, and morphine on experimentally induced ischemic pain in healthy volunteers. Anesth Analg 79:787–791

Segerdahl M, Ekblom A, Sandelin K, Wickman M, Sollevi A (1995) Peroperative adenosine infusion reduces the requirements for isoflurane and postoperative analgesics. Anesth Analg 80:1145–1149

Selemidis S, Satchell DG, Cocks TM (1997) Evidence that NO acts as a redundant NANC inhibitory neurotransmitter in the guinea-pig isolated taenia coli. Br J Pharmacol 121:604–611

Sengupta JN, Gebhart GF (1994) Characterisation of mechanosensitive pelvic nerve afferent fibers innervating the colon of the rat. J Neurophysiol 71:2046–2060

Sengupta JN, Gebhart GF (1998) The sensory innervation of the colon and its modulation. Curr Opin Gastroenterol 14:15–20

Serio R, Mule F, Adamo EB, Postorino A (1990) Evidence against purines being neurotransmitters of non-adrenergic, non-cholinergic nerves in rat duodenum. Eur J Pharmacol 182:487–495

Serio R, Mule F, Postorino A (1995) Nonadrenergic, noncholinergic inhibitory junction potentials in rat proximal colon: role of nitric oxide. Can J Physiol Pharmacol 73:79–84

Serio M, Mulè F, Postorino A, Vetri T, Bonvissuto F (1996) Apamin-sensitive and -insensitive components of inhibitory junction potentials in rat caecum: role of nitric oxide. J Auton Pharmacol 16:183–189

Shimo Y, Ishii T (1978) Effects of morphine on non-adrenergic inhibitory responses of the guinea-pig taenia coli. J Pharm Pharmacol 30:596–597

Shinozuka K, Maeda T, Hayashi E (1985a) Effects of adenosine on ^{45}Ca uptake and [^{3}H]acetylcholine release in synaptosomal preparation from guinea-pig ileum myenteric plexus. Eur J Pharmacol 113:417–424

Shinozuka K, Maeda T, Hayashi E (1985b) Possibilities for adenosine modulation of peristaltic reflex in guinea pig isolated ileum. J Pharmacobiodyn 8:877–884

Shuba MF, Vladimirova IA (1980) Effect of apamin on the electrical responses of smooth muscle to adenosine 5′-triphosphate and to non-adrenergic, non-cholinergic nerve stimulation. Neuroscience 5:853–859

Shuttleworth CW, Murphy R, Furness JB (1991) Evidence that nitric oxide participates in non-adrenergic inhibitory transmission to intestinal muscle in the guinea-pig. Neurosci Lett 130:77–80

Shuttleworth CWR, Sweeney KM, Sanders KM (1999) Evidence that nitric oxide acts as an inhibitory neurotransmitter supplying taenia from guinea-pig caecum. Br J Pharmacol 127:1495–1501

Singh G, Chaudry KI, Chaudry IH (1993) ATP-$MgCl_2$ restores gut absorptive capacity early after trauma-hemorrhagic shock. Am J Physiol 264:R977–R983

Sjöblom-Widfeldt N, Gustafsson H, Nilsson H (1990) Transmitter characteristics of small mesenteric arteries from the rat. Acta Physiol Scand 138:203–212

Sjöqvist A, Fahrenkrug J, Hemlin M, Jodal M, Lundgren O (1985) Effects of intra-arterially infused adenosine triphosphate (ATP) on release of vasoactive intestinal polypeptide (VIP) from the gastrointestinal tract of the cat. Acta Physiol Scand 125:693–698

Skoglund ML, Vinik AI, Feller MR (1982) Inhibition of acid secretion by adenosine. The Physiologist 25:219

Small RC, Weston AH (1979) Theophylline antagonizes some effects of purines in the intestine but not those of intramural inhibitory nerve stimulation. Br J Pharmacol 67:301–308

Smith GT, Moran TH, Coyle JT, Kuhar MJ, O'Donahue TL, McHugh PR (1984) Anatomic localization of cholecystokinin receptors to the pyloric sphincter. Am J Physiol 246:R127–R130

Smits GJ, Lefebvre RA (1996) ATP and nitric oxide: inhibitory NANC neurotransmitters in the longitudinal muscle-myenteric plexus preparation of the rat ileum. Br J Pharmacol 118:695–703

Sneddon JD, Smythe A, Satchell D, Burnstock G (1973) An investigation of the identity of the transmitter substance released by non-adrenergic, non-cholinergic excitatory nerves supplying the small intestine of some lower vertebrates. Comp Gen Pharmacol 4:53–60

Soediono P, Burnstock G (1994) Contribution of ATP and nitric oxide to NANC inhibitory transmission in rat pyloric sphincter. Br J Pharmacol 113:681–686

Soltoff SP, McMillian MK, Talamo BR (1989) Coomassie Brilliant Blue G is a more potent antagonist of P_2 purinergic responses than Reactive Blue 2 (Cibacron Blue 3GA) in rat parotid acinar cells. Biochem Biophys Res Commun 165:1279–1285

Soltoff SP, McMillian MK, Lechleiter JD, Cantley LC, Talamo BR (1990) Elevation of $[Ca^{2+}]_i$ and the activation of ion channels and fluxes by extracellular ATP and phospholipase C-linked agonists in rat parotid acinar cells. Ann NY Acad Sci 603:76–90

Soltoff SP, McMillian MK, Talamo BR (1992) ATP activates a cation-permeable pathway in rat parotid acinar cells. Am J Physiol 262:C934–C940

Soltoff SP, McMillian MK, Talamo BR, Cantley LC (1993) Blockade of ATP binding site of P2 purinoceptors in rat parotid acinar cells by isothiocyanate compounds. Biochem Pharmacol 45:1936–1940

Somers GR, Hammet FM, Trute L, Southey MC, Venter DJ (1998) Expression of the $P2Y_6$ purinergic receptor in human T cells infiltrating inflammatory bowel disease. Lab Invest 78:1375–1383

Somogyi GT, Vizi ES (1988) Evidence that cholinergic axon terminals are equipped with both muscarinic and adenosine receptors. Brain Res Bull 21:575–579

Spallone V, Uccioli L, Menzinger G (1995) Diabetic autonomic neuropathy. Diabetes Metab Rev 11:227–257

Spedding M, Weetman DF (1976) Identification of separate receptors for adenosine and adenosine 5′-triphosphate in causing relaxations of the isolated taenia of the guinea-pig caecum. Br J Pharmacol 57:305–310

Spedding M, Sweetman AJ, Weetman DF (1975) Antagonism of adenosine 5′-triphosphate-induced relaxation by 2-2′-pyridylisatogen in the taenia of guinea-pig caecum. Br J Pharmacol 53:575–583

Spencer NJ, Bywater RA, Holman ME, Taylor GS (1998) Spontaneous and evoked inhibitory junction potentials in the circular muscle layer of mouse colon. J Auton Nerv Syst 69:115–121

Spencer NJ, Walsh M, Smith TK (2000) Purinergic and cholinergic neuro-neuronal transmission underlying reflexes activated by mucosal stimulation in the isolated guinea-pig ileum. J Physiol 522:321–331

Sperlágh B, Vizi ES (1990) Stimulation of presynaptic P1 and P2 receptors at ATP in Auerbach's plexus. Eur J Pharmacol 183:1680

Sperlágh B, Vizi ES (1991) Effect of presynaptic P2 receptor stimulation on transmitter release. J Neurochem 56:1466–1470

Stone TW (1981) Actions of adenine dinucleotides on the vas deferens, guinea-pig taenia caeci and bladder. Eur J Pharmacol 75:93–102

Straub RH, Mayer M, Falk W, Scholmerich J (1999) Neurotransmitters of the sympathetic nerve terminal are powerful chemoattractants for monocytes/macrophages. Gastroenterology 116:G3582 [Abstract]

Su C, Bevan JA, Burnstock G (1971) [^{3}H]Adenosine triphosphate: release during stimulation of enteric nerves. Science 173:337–339

Surprenant A (1994) Control of the gastrointestinal tract by enteric neurons. Annu Rev Physiol 56:117–140

Suthamnatpong N, Hata F, Kanada A, Takeuchi T, Yagasaki O (1993) Mediators of non-adrenergic, noncholinergic inhibition in the proximal, middle and distal regions of rat colon. Br J Pharmacol 108:348–355

Szewczak SM, Behar J, Billett G, Hillemeier C, Rhim BY, Biancani P (1990) VIP-induced alterations in cAMP and inositol phosphates in the lower esophageal sphincter. Am J Physiol 259:G239–G244

Takahashi T, Nakamura K, Itoh H, Sima AA, Owyang C (1997) Impaired expression of nitric oxide synthase in the gastric myenteric plexus of spontaneously diabetic rats. Gastroenterology 113:1535–1544

Tam FS, Hillier K (1992) The role of nitric oxide in mediating non-adrenergic non-cholinergic relaxation in longitudinal muscle of human taenia coli. Life Sci 51:1277–1284

Tanaka J, Murate M, Wang CZ, Seino S, Iwanaga T (1996) Cellular distribution of the $P2X_4$ ATP receptor mRNA in the brain and non-neuronal organs of rats. Arch Histol Cytol 59:485–490
Tansey MF, Probst SJ, Martin JS (1975) Evidence of nonvagal neural stimulation of canine gastric acid secretion. Surg Gynecol Obstet 140:861–867
Tarasiuk A, Bar Shimon M, Gheber L, Korngreen A, Grossman Y, Priel Z (1995) Extracellular ATP induces hyperpolarization and motility stimulation of ciliary cells. Biophys J 68:1163–1169
Taylor GS, Bywater RAR (1989) Novel autonomic neurotransmitters and intestinal function. Pharmacol Ther 40:404–438
Taylor EM, Parsons ME (1989) Adrenergic and purinergic neurotransmission in arterial resistance vessels of the cat intestinal circulation. Eur J Pharmacol 164:23–33
Taylor EM, Parsons ME (1991) Effects of α,β-methylene ATP on resistance and capacitance blood vessels of the cat intestinal circulation; a comparison with other vasoconstrictor agents and sympathetic nerve stimulation. Eur J Pharmacol 205:35–41
Tenneti L, Gibbons SJ, Talamo BR (1998) Expression and trans-synaptic regulation of P_{2x4} and P_{2z} receptors for extracellular ATP in parotid acinar cells. Effects of parasympathetic denervation. J Biol Chem 273:26799–26808
Tojyo Y, Tanimura A, Matsui S, Matsumoto Y (1997) Effects of extracellular ATP on cytosolic Ca^{2+} concentration and secretory responses in rat parotid acinar cells. Arch Oral Biol 42:393–399
Tomaru A, Ishii A, Kishibayashi N, Shimada J, Suzuki F, Karasawa A (1994) Possible physiological role of endogenous adenosine in defecation in rats. Eur J Pharmacol 264:91–94
Tomita T, Watanabe H (1973) A comparison of the effects of adenosine triphosphate with noradrenaline and with the inhibitory potential of the guinea-pig taenia coli. J Physiol 231:167–177
Tøttrup A, Knudsen M, Gregerson H (1991a) The role of the L-arginine-nitric oxide pathway for lower oesophageal sphincter. Br J Pharmacol 104:113–116
Tøttrup A, Svane D, Forman A (1991b) Nitric oxide mediating NANC inhibition in opossum lower esophageal sphincter. Am J Physiol 260:G385–G389
Tøttrup A, Ny L, Alm P, Larsson B, Forman A, Andersson KE (1993) The role of the l-arginine/nitric oxide pathway for relaxation of the human lower oesophageal sphincter. Acta Physiol Scand 149:451–459
Tøttrup A, Knudsen MA, Hanberg Sorensen F, Glavind EB (1995) Pharmacological identification of different inhibitory mediators involved in the innervation of the internal anal sphincter. Br J Pharmacol 115:158–162
Turner JT, Weisman GA, Camden JM (1997) Upregulation of $P2Y_2$ nucleotide receptors in rat salivary gland cells during short-term culture. Am J Physiol 273:C1100–C1107
Turner JT, Weisman GA, Landon LA, Park M, Camden JM (1998) Salivary gland nucleotide receptors: evidence for functional expression of both P2X and P2Y subtypes. Eur J Morphol 36:170–175
Ushijima I, Mizuki Y, Yamada M (1985) Development of stress-induced gastric lesions involves central adenosine A_1-receptor stimulation. Brain Res 339:351–355
Vallejo AI, Bo X, Burnstock G (1996) P2Y-purinoceptors in gastric gland plasma membranes. Eur J Pharmacol 312:209–214
Van der Meer C, Snijders PM, Valkenburg PW (1982) The effect of ATP on survival in intestinal ischemia shock, hemorrhagic shock, and endotoxin shock in rats. Circ Shock 9:619–628
Van Nueten JM, Fontaine J, Helsen L, Janssen PA (1977) Inhibition by purines of peristaltic activity in the guinea-pig ileum. Arch Int Pharmacodyn Ther 227:168–170
Vanner S, Surprenant A (1996) Neural reflexes controlling intestinal microcirculation. Am J Physiol 271:G223–G230

Venglarik CJ, Singh AK, Wang R, Bridges RJ (1993) Trinitrophenyl-ATP blocks colonic Cl^- channels in planar phospholipid bilayers. Evidence for two nucleotide binding sites. J Gen Physiol 101:545–569
Venkova K, Krier J (1993) Stimulation of lumbar sympathetic nerves evokes contractions of cat colon circular muscle mediated by ATP and noradrenaline. Br J Pharmacol 110:1260–1270
Venkova K, Milne A, Krier J (1994) Contractions mediated by α_1-adrenoceptors and P_2-purinoceptors in a cat colon circular muscle. Br J Pharmacol 112:1237–1243
Vermillion DL, Gillespie JP, Cooke AR, Wood JD (1979) Does 5-hydroxytryptamine influence "purinergic" inhibitory neurons in the intestine? Am J Physiol 237:E198–E202
Vidal M, Hicks PE, Langer SZ (1986) Differential effects of α,β-methylene ATP on responses to nerve stimulation in SHR and WKY tail arteries. Naunyn Schmiedebergs Arch Pharmacol 332:384–390
Vigne P, Pacaud P, Loirand G, Breittmayer JP (1998) PPADS inhibits $P2Y_1$ purinoceptors in rat brain capillary endothelial cells and in rat ileal myocytes by an indirect mechanism. Biochem Biophys Res Commun 244:332–335
Virginio C, Robertson G, Surprenant A, North RA (1998) Trinitrophenyl-substituted nucleotides are potent antagonists selective for $P2X_1$, $P2X_3$ and heteromeric $P2X_{2/3}$ receptors. Mol Pharmacol 53:969–973
Vizi ES, Knoll J (1976) The inhibitory effect of adenosine and related nucleotides on the release of acetylcholine. Neuroscience 1:391–398
Vizi ES, Somogyi GT, Magyar K (1981) Evidence that morphine and opioid peptides do not share a common pathway with adenosine in inhibiting acetylcholine release from isolated intestine. J Auton Pharmacol 1:413–419
Vladimirova IA, Shuba MF (1978) Effect of strychnine, hydrastine and apamin on synaptic transmission in smooth muscle cells [in Russian]. Neirofiziologiia 10:295–299
Vogalis F, Goyal RK (1997) Activation of small conductance Ca^{2+}-dependent K^+ channels by purinergic agonists in smooth muscle cells of the mouse ileum. J Physiol 502:497–508
von Kügelgen I, Starke K (1985) Noradrenaline and adenosine triphosphate as cotransmitters of neurogenic vasoconstriction in rabbit mesenteric artery. J Physiol 367:435–455
Vongalis F, Bywater RA, Taylor GS (1989) Nerve-mediated contractile and electrical activity of the guinea-pig choledocho-duodenal junction. J Auton Nerv Syst 29:19–28
Wager-Srdar S, Levine AS, Morley JE (1984) Food intake: opioid/purine interactions. Pharmacol Biochem Behav 21:33–38
Wang XY, Wong WC, Ling EA (1996) Localization of nicotinamide adenine dinucleotide phosphate diaphorase (NADPH-d) activity in the gastrointestinal sphincters in the guinea pig. J Auton Nerv Syst 58:51–55
Ward SM, Dalziel HH, Bradley ME, Buxton IL, Keef K, Westfall DP, Sanders KM (1992) Involvement of cyclic GMP in non-adrenergic, non-cholinergic inhibitory neurotransmission in dog proximal colon. Br J Pharmacol 107:1075–1082
Ward SM, Xue C, Sanders KM (1994) Localization of nitric oxide synthase in canine ileocolonic and pyloric sphincters. Cell Tissue Res 275:513–527
Ward SM, Dalziel HH, Khoyi MA, Westfall AS, Sanders KM, Westfall DP (1996) Hyperpolarization and inhibition of contraction mediated by nitric oxide released from enteric inhibitory neurones in guinea-pig taenia coli. Br J Pharmacol 118:49–56
Watt AH, Lewis DJ, Horne JJ, Smith PM (1987) Reproduction of epigastric pain of duodenal ulceration by adenosine. Br Med J 294:10–12
Weiss T, Gheber L, Shoshan Barmatz V, Priel Z (1992) Possible mechanism of ciliary stimulation by extracellular ATP: involvement of calcium-dependent potassium channels and exogenous Ca^{2+}. J Membr Biol 127:185–193

Welford LA, Cusack NJ, Hourani SMO (1986) ATP analogues and the guinea-pig taenia coli: a comparison of the structure-activity relationships of ectonucleotidases with those of the P_2-purinoceptor. Eur J Pharmacol 129:217–224
Westerberg VS, Geiger JD (1987) Central effects of adenosine analogs on stress-induced gastric ulcer formation. Life Sci 41:2201–2205
Westerberg VS, Geiger JD (1988) Adenosine and gastric function. Trends Pharmacol Sci 9:345–347
Westerberg VS, Geiger JD (1989) Adenosine analogs inhibit gastric acid secretion. Eur J Pharmacol 160:275–281
Westerberg VS, Glavin GB, Geiger JD (1986) Intracerebroventricular administration of (–)-phenylisopropyladenosine protects rats against stress induced ulcer formation. Proc West Pharmacol Soc 29:425–427
Weston AH (1973) The effect of desensitization to adenosine triphosphate on the peristaltic reflex in guinea-pig ileum. Br J Pharmacol 47:606–608
Wiklund NP, Gustafsson LE (1987) On the nature of endogenous purines modulating cholinergic neurotransmission in the guinea-pig ileum. Acta Physiol Scand 131:11–18
Wiklund NP, Gustafsson LE (1988a) Agonist and antagonist characterization of the P_2-purinoceptors in the guinea pig ileum. Acta Physiol Scand 132:15–22
Wiklund NP, Gustafsson LE (1988b) Indications for P_2-purinoceptor subtypes in guinea pig smooth muscle. Eur J Pharmacol 148:361–370
Wiklund NP, Gustafsson LE, Lundin J (1985) Pre- and postjunctional modulation of cholinergic neuroeffector transmission by adenine nucleotides. Experiments with agonist and antagonist. Acta Physiol Scand 125:681–691
Will S, Triggle CR, Bieger D (1990) Mastocyte and smooth muscle purinoceptors of the rat oesophagus. In: Abstracts of IUPHAR Satellite Symposium, Noordwijk, July 6–8, 1990, p 33
Windscheif U, Pfaff O, Ziganshin AU, Hoyle CHV, Bäumert HG, Mutschler E, Burnstock G, Lambrecht G (1995) Inhibitory action of PPADS on the relaxant responses to adenine nucleotides or electrical field stimulation in guinea-pig taenia-coli and rat duodenum. Br J Pharmacol 115:1509–1517
Wood JD (1981) Physiology of the enteric nervous system. In: Johnson LR (ed) Physiology of the gastrointestinal tract. Raven Press, New York, pp 1–37
Wörl J, Mayer B, Neuhuber WL (1994) Nitrergic innervation of the rat esophagus: focus on motor endplates. J Auton Nerv Syst 49:227–233
Wróbel J, Michalska L (1977) The effect of exogenous ATP on intestinal calcium transport. Comp Biochem Physiol 58A:421–425
Xia Y, Fertel RH, Wood JD (1997) Suppression of cAMP formation by adenosine in myenteric ganglia from guinea-pig small intestine. Eur J Pharmacol 320:95–101
Xue L, Suzuki H (1997) Electrical responses of gastric smooth muscles in streptozotocin-induced diabetic rats. Am J Physiol 272:G77–G83
Xue L, Fukuta H, Yamamoto Y, Suzuki H (1996) Properties of junction potentials in gastric smooth muscle of the rat. Jpn J Physiol 46:123–130
Xue L, Imaeda Y, Suzuki H (1998) Effects of suramin on electrical and mechanical activities in antrum smooth muscle of guinea-pig stomach. J Auton Pharmacol 18:325–331
Xue L, Farrugia G, Sarr MG, Szurszewski JH (1999) ATP is a mediator of the fast inhibitory junction in human jejunal circular smooth muscle. Am J Physiol 276:G1373–G1379
Yagasaki O, Nabata H, Yanagiya I (1983) Effects of desensitization to adenosine 5′-triphosphate and vasoactive intestinal polypeptide on non-adrenergic inhibitory responses of longitudinal and circular muscles in the rat ileum. J Pharm Pharmacol 35:818–820
Yano S, Fujiwara A, Ozaki Y, Harada M (1983) Gastric blood flow responses to autonomic nerve stimulation and related pharmacological studies in rats. J Pharm Pharmacol 35:641–646

Young HM, Ciampoli D, Johnson PJ, Stebbing MJ (1996) Inhibitory transmission to the longitudinal muscle of the mouse caecum is mediated largely by nitric oxide acting via soluble guanylyl cyclase. J Auton Nerv Syst 61:103–108
Yu HX, Turner JT (1991) Functional studies in the human submandibular duct cell line, HSG-PA, suggest a second salivary gland receptor subtype for nucleotides. J Pharmacol Exp Ther 259:1344–1350
Yuan S, Costa M, Brookes SJ (1998) Neuronal pathways and transmission to the lower esophageal sphincter of the guinea pig. Gastroenterology 115:661–671
Yunker AM, Galligan JJ (1994) Extrinsic denervation increases NADPH diaphorase staining in myenteric nerves of guinea pig ileum. Neurosci Lett 167:51–54
Zafirov DH, Palmer JM, Wood JD (1985) Adenosine inhibits forskolin-induced excitation in myenteric neurons. Eur J Pharmacol 113:143–144
Zagorodnyuk V, Maggi CA (1994) Electrophysiological evidence for different release mechanism of ATP and NO as inhibitory NANC transmitters in guinea-pig colon. Br J Pharmacol 112:1077–1082
Zagorodnyuk V, Maggi CA (1998) Pharmacological evidence for the existence of multiple P2 receptors in the circular muscle of guinea-pig colon. Br J Pharmacol 123:122–128
Zagorodniuk VP, Shuba MF (1986) Nature of non-adrenergic inhibition in the smooth muscles of the human intestine [Russian]. Neirofiziologiia 18:373–381
Zagorodnyuk VP, Vladimirova IA, Vovk EV, Shuba MF (1989) Studies of the inhibitory non-adrenergic neuromuscular transmission in the smooth muscle of the normal human intestine and from a case of Hirschsprung's disease. J Auton Nerv Syst 26:51–60
Zagorodnyuk V, Hoyle CHV, Burnstock G (1993) An electrophysiological study of developmental changes in the innervation of the guinea-pig taeniacoli. Pflugers Arch 423:427–433
Zagorodnyuk V, Santicioli P, Maggi CA, Giachetti A (1996) The possible role of ATP and PACAP as mediators of apamin-sensitive NANC inhibitory junction potentials in circular muscle of guinea-pig colon. Br J Pharmacol 119:779–786
Zeng W, Lee MG, Muallem S (1997) Membrane-specific regulation of Cl^- channels by purinergic receptors in rat submandibular gland acinar and duct cells. J Biol Chem 272:32956–32965
Zhang W, Roomans GM (1997) Regulation of ion transport by P_{2U} purinoceptors and α_{2A} adrenoceptors in HT29 cells. Cell Biol Int 4:195–200
Zhou X, Galligan JJ (1996) P2X purinoceptors in cultured myenteric neurons of guinea-pig small intestine. J Physiol 496:719–729
Zhou X, Galligan JJ (1998) Non-additive interaction between nicotinic cholinergic and P2X purine receptors in guinea-pig enteric neurons in culture. J Physiol 513:685–697
Ziganshin AU, Hoyle CHV, Ziganshina LE, Burnstock G (1995) Prejunctional inhibition of cholinergic neuromuscular transmission in the guinea-pig ileum by diadenosine polyphosphates. Br J Pharmacol 114:440P
Zsembery A, Spirlì C, Granato A, LaRusso NF, Okolicsanyi L, Crepaldi G, Strazzabosco M (1998) Purinergic regulation of acid/base transport in human and rat biliary epithelial cell lines. Hepatology 28:914–920

CHAPTER 22

P1 Receptors in the Respiratory System

I. BIAGGIONI and I. FEOKTISTOV

A. Introduction

In this chapter the role of adenosine receptors in the modulation respiratory system is reviewed. The chapter is divided into sections based on the effects of adenosine receptors on the different aspects of pulmonary physiology. The potential role of P1 purinergic receptors in the regulation of epithelial and endothelial function, ventilatory drive, pulmonary vascular tone and airway tone, and in the pathophysiology of inflammatory processes is discussed (Table 1). P2 receptor mediated effects on the respiratory system are reviewed elsewhere by WEGNER (chap. 23, this volume).

Because adenosine actions are mediated via cell membrane receptors, its importance as a regulatory modulator is directly proportional to its interstitial concentrations. Interstitial accumulation of adenosine depends on a balance between its release and its metabolism.

I. Sources of Adenosine within the Lung

Adenosine is an intermediate product in the metabolism of ATP. As such, it is present in every cell and released when metabolic demands exceed oxygen supply. In theory, therefore, any cell located within the lung is a potential source of adenosine. In particular, adenosine release has been demonstrated during the activation of platelets, mast cells, and neutrophils. Episodes of bronchoconstriction in asthmatics are associated with increased levels of adenosine in bronchoalveolar lavage fluid (DRIVER et al. 1993) and in plasma (MANN et al. 1986). Angiotensin II also specifically induces adenosine release from rat lung (BOTTIGLIERI et al. 1990) with noradrenaline, bradykinin, and vasopressin having no effect on adenosine release. The mechanisms by which angiotensin II releases adenosine, and the cellular source of adenosine have not been defined.

II. Metabolism of Adenosine by the Lung

Several studies have explored the clearance of circulating adenosine by the lung. In isolated porcine lungs perfused with a buffer solution, approximately

Table 1. Summary of the effects of adenosine receptors on respiratory function

Receptor	Cells/Structure	Effect	Representative Reference
Pulmonary secretion			
A_1	Epithelium	Phosphotidylcholine secretion	Gobran and Rooney (1990)
A_1	Epithelium	Cl^- transport	McCoy et al. (1995)
A_2	Epithelium	Phosphotidylcholine secretion	Griese et al. (1991)
A_{2B}?	Epithelium	Cl^- transport	Lazarowski et al. (1992)
Modulation of Respiration			
A_1	Ventral respiratory group	Ventilation	Dong and Feldman (1995)
A_{2B}?	Nucleus tractus solitarii	Ventilation	Phillis et al. (1997)
A_1	Pulmonary sensory afferents	Apnea	Kwong et al. (1998)
A_{2A}	Carotid chemoreceptors	Ventilation	Maxwell et al. (1987)
Pulmonary Vascular Tone			
A_1	?	Vasoconstriction	Neely et al. (1991)
A_{2A}? A_{2B}?	Pulmonary vasculature	Vasodilation	McCormack et al. (1989)
Airway Tone			
A_1	Trachea/bronchi	Bronchoconstriction	Nyce and Metzger (1997)
A_2	Trachea	Relaxation of preconstricted airway	Ghai et al. (1987)
Inflammation			
A_2	Basophils	Inhibition of secretion	Hughes et al. (1983)
A_{2A}	Human fetal blood mast cells	Inhibition of secretion	Suzuki et al. (1998)
A_{2B}	Human and dog mast cells	Activation of secretion	Feoktistov and Biaggioni (1995)
A_3	Rat mast cells	Potentiation of secretion	Ramkumar et al. (1993)
A_3	Eosinophils	Inhibition of chemotaxis	Knight et al. (1997)
A_2	Rabbit alveolar macrophages	Inhibition of secretion	Hasday and Sitrin (1987)

half of the adenosine was cleared in a single pass through this organ. Uptake was inhibited by dipyridamole and also by adenine (HELLEWELL and PEARSON 1983). It was assumed that this uptake takes place in the endothelial layer, as has been demonstrated in the coronary circulation. Similar findings were reported in the isolated rat lung (BAKHLE and CHELLIAH 1983a), but dipyridamole was found to be less effective in blocking adenosine uptake in rat lungs compared to guinea-pig lung (BAKHLE and CHELLIAH 1983a). In this regard, it is recognized that adenosine uptake is less sensitive to dipyridamole in rats. Clearance of adenosine by the lung appears to be more efficient in humans than in other species; it is estimated that the extraction of adenosine across the human lung is about 75% (UTTERBACK et al. 1994).

A few studies have explored the impact of disease states in pulmonary metabolism of adenosine. No differences were detected in adenosine uptake, intracellular metabolism, or extracellular metabolism in lung from spontaneously hypertensive rats vs Wistar-Kyoto rats (BOTTIGLIERI et al. 1988). Similarly, in lungs from streptozotocin-treated rats, hydrolysis of AMP to adenosine was decreased compared to control rats, but adenosine metabolism and uptake was not affected in this model of diabetes (BAKHLE and CHELLIAH 1983b).

B. Adenosine Receptors in Epithelial and Endothelial Function

The main function of the lung is gas exchange, which takes place in the alveoli through the blood-gas barrier. This barrier, which separates the blood in the pulmonary capillaries from the alveolar air is composed of capillary endothelial cells, an interstitial layer, and alveolar epithelial cells (LAITINEN and LAITINEN 1994). Epithelial cells are covered by a layer of surfactant the function of which is to minimize surface tension so that the alveoli remain open. This barrier is less than half a micron thick in places. Alveoli are lined by two types of epithelial cell: type I pneumocytes (squamous epithelial cells), and type II pneumocytes (cuboidal epithelial cells). Surfactant is a lipid material composed primarily of dipalmitoyl phosphatidylcholine and secreted by type II pneumocytes.

Adenosine A_2-like receptors are expressed in primary cultures of rat type II alveolar pneumocytes. The stable adenosine analog, 5′-*N*-ethylcarboxaminoadenosine (NECA) was found to be more potent than the selective A_1 agonists (*R*)-N^6-phenylisopropyladenosine (R-PIA) and 2-chloroadenosine (2-CADO) in increasing intracellular cAMP levels and phosphatidylcholine secretion (GILFILLAN and ROONEY 1987). NECA increased phosphatidylcholine secretion with an EC_{50} value of approximately 90 nmol/l. [^{3}H]NECA bound to membranes from these cells with a K_d value of 240 nmol/l (GRIESE et al. 1991). The relatively high affinity of NECA suggests the presence of an A_{2A} receptor subtype. Conversely, it was suggested that A_1 receptors restrain phosphatidylcholine secretion at least in vitro, because addition of A_1 agonists restrains the effects of stimulants that act through activation of adenylate cyclase (GOBRAN and ROONEY 1990).

The possibility that adenosine receptors regulate surfactant secretion was addressed in the newborn rabbit. Treatment with the adenosine receptor antagonist 8-phenyltheophylline (8-PT) resulted in significant inhibition of surfactant secretion immediately after birth (ROONEY and GOBRAN 1988), implying that endogenous adenosine had a tonic effect promoting surfactant secretion. This study did not explore the receptor type that mediates this action.

The volume and composition of the pulmonary secretion by epithelial cells is modulated by Cl^- transport. Adenosine analogs stimulate Cl^- transport in human airway epithelium, an effect accompanied by elevation of cAMP. The order of potency of the adenosine analogues suggested the involvement of the low affinity, A_{2B} adenosine receptor (LAZAROWSKI et al. 1992). It was also suggested that A_1 adenosine receptors may stimulate Cl^- currents in human airway epithelium. This was concluded because 2-CADO and N^6-cyclopentyladenosine (CPA), but not the selective A_{2A} agonist (CGS 21680) at a concentration of 30 nmol/l activated Cl^--selective membrane current. Inhibition with 100 nmol/l 1,3-dipropyl-8-cyclopentylxanthine (DPCPX) was taken as additional evidence of A_1-dependent mechanism (RUGOLO et al. 1993). However, A_{2B} adenosine receptors may fit even better to this pharmacological profile. CGS 21680 binds to A_1 receptors with a K_i of 2.6 μmol/l (JACOBSON and SUZUKI 1996), and therefore should have been effective in this preparation if A_1 receptors were involved. On the other hand, CGS21680 is virtually ineffective on A_{2B} receptors in concentrations up to 1 μmol/l (FEOKTISTOV and BIAGGIONI 1993). Furthermore, the selective A_1 antagonist, DPCPX, has been shown to bind human A_{2B} receptors with a K_i of 40 nmol/l (ROBEVA et al. 1996). Adenosine A_1 receptors indeed have been identified in human airway epithelium using reverse transcriptase-polymerase chain reaction, but they were functionally coupled to inhibition and not activation of Cl^- secretion (Fig. 1) (MCCOY et al. 1995). Functional coupling of adenosine A_{2B} receptors to Cl^- transport was also observed in other epithelial tissues, including the intestine (for review see FEOKTISTOV and BIAGGIONI 1997).

Cl^- secretion from epithelial cells is particularly important during fetal development. Lungs in the fetus remain "inflated" by liquid that is secreted by the pulmonary epithelium. This secretion is driven by active Cl^- transport and Cl^- secretion was modulated by adenosine in fetal tracheal epithelium from 14- and 18-day fetal rats (BARKER and GATZY 1998). The adenosine receptor type that mediates these actions was not determined, but it should be noted that the EC_{50} value for adenosine was in the low micromolar range, which is characteristic for the A_{2B} adenosine receptors.

Cl^- secretion is also of great importance in cystic fibrosis, an autosomal recessive disorder of defective chloride ion channels affecting epithelial cells

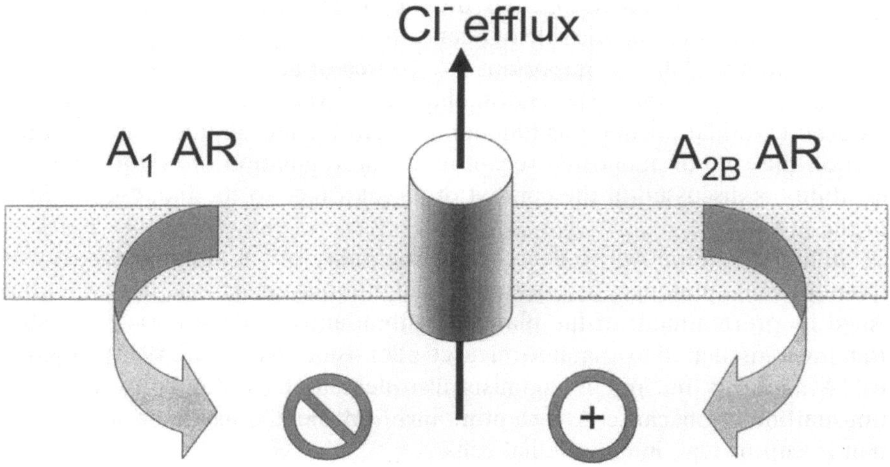

Fig. 1. Regulation of chloride transport in the airway epithelium by adenosine receptors. A_1 adenosine receptors (A_1 AR) inhibit chloride efflux, whereas A_{2B} adenosine receptors (A_{2B} AR) stimulate it

through the body. The disease is caused by mutations in the cystic fibrosis transmembrane regulator gene (CFTR). The CFTR gene product, an apical membrane cAMP- and ATP- activated Cl^- channel (RIORDAN et al. 1989), contributes to chloride secretion by epithelial tissues of the lung, pancreas, and gastrointestinal tract. Patients die of lung complications related to local inflammation and predisposition to infections because of the decreased water secretion and increased mucus viscosity. The potential role of adenosine on Cl^- secretion in cystic fibrosis was explored in cystic fibrosis pancreatic adenocarcinoma (CFPAC) cells. As we mentioned before, stimulation of adenosine A_2-like receptors would facilitate Cl^- secretion, whereas stimulation of adenosine A_1 receptors would inhibit it. Treatment of CFPAC cells with the A_1 antagonists DPCPX and xanthine amine congener (XAC) was found to activate $^{36}Cl^-$ efflux, implying that in this cellular model there was enough production of endogenous adenosine to activate A_1 receptors that inhibit Cl^- efflux. This possibility was reinforced by the fact that the effects of A_1 antagonists could be reversed by the addition of adenosine agonists or by elimination of endogenous adenosine with adenosine deaminase (EIDELMAN et al. 1992). It has been suggested that activation of $^{36}Cl^-$ efflux by A_1 antagonists may be specific for this genetic defect, because it is not observed in cells lacking the cystic fibrosis defect or in CFPAC cells transfected with wild-type CFTR gene (EIDELMAN et al. 1992). However, other group of investigators, using whole cell patch-clamp techniques instead of $^{36}Cl^-$ efflux, found that antagonism of A_1 receptors increased cAMP-activated Cl^- secretion in both normal (9HTEo-) and cystic fibrosis (2CFSMEo-) human airway epithelial cells and in cystic fibrosis epithelial cells in primary culture

(McCoy et al. 1995). Taken together, these findings warrant further exploration of the potential therapeutic role of adenosine A_1 antagonists in cystic fibrosis.

Endothelial cells are important as a source of adenosine and in its clearance by cellular uptake. The endothelium may contribute also to adenosine-induced vasodilation, and this function is explored later in this chapter. In this section the role of adenosine receptors in the regulation of endothelial permeability is discussed in the context of its relevance to the blood-gas barrier of the alveoli.

Platelets are known to decrease permeability of ^{125}I-albumin across confluent bovine pulmonary artery endothelial monolayers. This activity was abolished by pretreatment of the platelets with adenosine deaminase, suggesting that it was mediated by platelet-released adenosine (Paty et al. 1992). Of interest, A_1 agonists, but not A_2 agonists, also decreased permeability. This is an unusual finding because A_1 receptors have not been considered a predominant receptor type in endothelial cells.

An increase in permeability will favor water flux into the alveolar interstitium and eventually the alveolar space (pulmonary edema). Transvascular water flux was measured in an isolated perfused guinea pig lung model using the capillary filtration coefficient (Kf,c). Platelets added to the perfusate reduced Kf,c by 29%. Pretreatment of platelets with adenosine deaminase abolished this response, suggesting that platelet-released adenosine also mediated this phenomenon (Paty et al. 1992).

C. Adenosine Receptors in the Control of Ventilation

A detailed description of the neural pathways involved in the regulation of respiration is outside the scope of this review and can found elsewhere (Strohl et al. 1998). However, those structures and pathways affected by adenosine receptors are shown in Fig. 2. The respiratory system is equipped with sensors at multiple sites that continuously monitor changes in blood chemistry (peripheral and central chemoreceptors that monitor pH, and arterial tension of O_2 and CO_2). Of these, the carotid body chemoreceptors are significantly influenced by adenosine. These chemoreceptors are a collection of cells (type I glomus cells) forming the carotid body (located at the bifurcation of the internal and external carotid arteries) and the aortic body (located in the arch of the thoracic aorta). These structures are richly vascularized and are very sensitive to decreases in arterial oxygen tension, responding with an increase in afferent firing that travels through discrete neural pathways and end in neurons of the dorsal respiratory group associated with the nucleus tractus solitarii (NTS) of the brainstem to mediate an increase in ventilation. There are several other groups of neurons located in the pons and the brainstem that are involved in respiratory control. These centers form a network that provides respiratory pacemaker function. Adenosine also modulates ventilation by acting on these central respiratory centers.

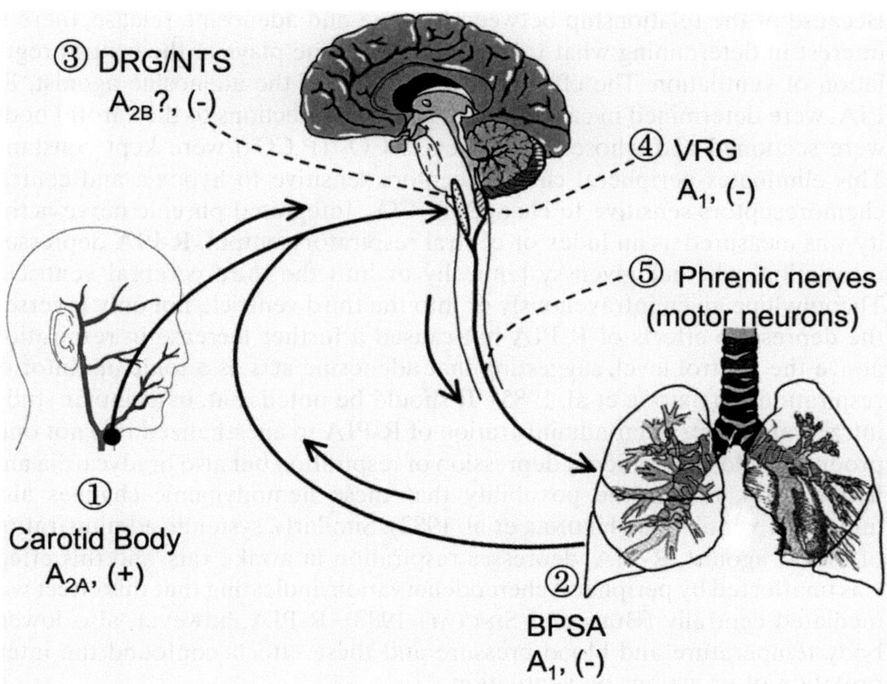

Fig. 2. Simplified scheme of structures involved in respiratory regulation. Only those proposed to interact with adenosine are shown. Adenosine interacts with diverse structures involved in the regulation of respiration, both centrally and peripherally, and produces either activation or inhibition of respiration. *1.* Carotid bodies contain chemoreceptors that primarily sense P_aO_2 levels in blood. Animal studies suggest that A_{2A} receptors mimic hypoxia and trigger chemoreceptor activation. Infusion of adenosine into blood vessels irrigating these areas results in respiratory stimulation in humans. *2.* Bronchopulmonary sensory afferents (BPSA) are also located in the bronchial tree and are activated by A_1 receptors, resulting in inhibition of respiration. *3.* Afferent nerves arising from carotid and aortic body chemoreceptors synapse in neurons associated with the ventrolateral nucleus tractus solitarii (NTS) and the dorsal respiratory group (DRG), one of the brainstem nuclei thought to be important in respiratory "pacemaker" function. Microinjections of adenosine analogs into the NTS results in inhibition of respiration. The adenosine receptor subtype that mediates these actions has not been characterized but the relative agonist potency is consistent with A_{2B} receptors. *4.* Another group of neurons that modulate inspiration and expiration are organized around the ventral respiratory group (VRG). Application of adenosine to there neurons inhibits respiration through A_1 receptors. *5.* Central integration of these brainstem nuclei results in respiratory outflow that controls breathing movements through spinal motor neurons that run with the phrenic nerve

I. Adenosine Receptors and Central Modulation of Ventilation

Mammals respond to severe hypoxia with a biphasic change in respiration, consisting of an initial increase followed by respiratory depression which, under severe and sustained hypoxic conditions, can lead to central apnea.

Because of the relationship between hypoxia and adenosine release, there is interest in determining what role, if any, adenosine plays in the central regulation of ventilation. The effects on respiration of the adenosine agonist, R-PIA, were determined in cats in whom neural connections to the carotid body were sectioned and whose arterial levels CO_2 (P_aCO_2) were kept constant. This eliminates peripheral chemoreceptors sensitive to hypoxia and central chemoreceptors sensitive to changes in CO_2. Integrated phrenic nerve activity was measured as an index or central respiratory output. R-PIA depressed respiration, whether given systemically or into the third cerebral ventricle. Theophylline given intravenously or into the third ventricle not only reversed the depressive effects of R-PIA but caused a further increase in respiration above the control level, suggesting that adenosine acts as a tonic inhibitor of respiration (ELDRIDGE et al. 1985). It should be noted that, in a similar study, intracerebroventricular administration of R-PIA to anesthetized rats not only produced a dose-dependent depression of respiration but also bradycardia and hypotension, raising the possibility that these hemodynamic changes also impact on ventilation (HEDNER et al. 1982). Similarly, systemic administration of the A_1 agonist, R-PIA, depresses respiration in awake rats, and this effect was unaffected by peripheral chemodenervation, indicating that this effect was mediated centrally (BURR and SINCLAIR 1988). R-PIA, however, also lowers body temperature and blood pressure and these effects confound the interpretation of its actions on ventilation.

Adenosine was applied by intracellular iontophoresis to respiratory neurons of the ventral respiratory group, assuming that this approach results in transient increases in extracellular adenosine concentrations in the vicinity of the cell. Adenosine hyperpolarized neurons and decreased spontaneous and stimulus-evoked postsynaptic activities (SCHMIDT et al. 1995), effects that would result in respiratory depression. The selective A_1 antagonist, DPCPX, blocked the effects of exogenous adenosine and also those induced by hypoxia. The A_1 adenosine agonist, CPA, also depressed respiratory neurons in an in vitro brainstem/spinal cord preparation from neonatal rats, an effect that was blocked by the selective A_1 receptor antagonist 8-cyclopentyltheophylline (CPT) (DONG and FELDMAN 1995).

Among the brainstem centers involved in cardiovascular and respiratory modulation, the effects of adenosine have been studied in detail in the nucleus tractus solitarius (NTS). This brainstem center is also important in autonomic control of blood pressure because it harbors the first synapse for afferent impulses arising from the carotid baroreceptors. Several subnuclei of the NTS receive respiratory afferent input and have neurons that discharge in phase with inspiration (STROHL et al. 1998). Activation of glutamatergic nerve terminals in the NTS results in decreases in blood pressure, heart rate, sympathetic tone, and respiratory frequency (REIS et al. 1981). Microinjections of adenosine into the caudal NTS decrease heart rate, arterial blood pressure (TSENG et al. 1988; MOSQUEDA-GARCIA et al. 1991; BARRACO et al. 1991) and respiratory frequency (PHILLIS et al. 1997). Similarly, microinjection of

NECA into the medial region of the caudal NTS at the level of the caudal tip of the area postrema produces a decrease in ventilatory rate but increases tidal volume (BARRACO et al. 1990). Decerebration decreased the cardiovascular, but not the ventilatory effects, of NECA (BARRACO et al. 1990).

Studies with microinjections of selective agonists into the NTS suggest that A_{2A} receptors mediate their cardiovascular actions (PHILLIS et al. 1997). On the other hand, it is not clear which receptor subtype mediates the ventilatory effects of adenosine in the NTS. It has been suggested that this action is coupled to activation of adenylate cyclase, because microinjections of cAMP analogs or forskolin reproduced the decrease in respiratory frequency (BARRACO et al. 1988). Neither the A_1 agonist, CPA, nor the A_{2A} agonist, CGS 21680, had any effect on ventilation parameters (PHILLIS et al. 1997), suggesting that ventilatory depression is mediated by A_{2B} receptors in the NTS (PHILLIS et al. 1997), but this remains speculative.

It is important to note that the cardiovascular effects of adenosine in the NTS imply a neuroexcitatory action because they are identical to those observed by microinjection of glutamate (REIS et al. 1981). Given the inhibitory effects of adenosine elsewhere in the central nervous system, it is remarkable that injections of adenosine into the NTS has the same effects as injections of an excitatory amino acid. This apparent paradox has been resolved by studies showing that adenosine elicits glutamate release in the NTS (MOSQUEDA-GARCIA et al. 1991). It is not known if a similar process is involved in adenosine-induced ventilatory depression in the NTS.

II. Adenosine Receptors and Peripheral Modulation of Respiration

Over a decade ago Ribeiro and colleagues showed that intracarotid injections of adenosine in cats (MCQUEEN and RIBEIRO 1986) and rats (MONTEIRO and RIBEIRO 1987) increase carotid sinus nerve activity, implying carotid body chemoreceptor activation. This effect is also observed in the isolated superfused cat carotid (RUNOLD et al. 1990b) and aortic (RUNOLD et al. 1990a) bodies. Adenosine-induced chemoreceptor activation is thought to be mediated by A_2-like receptors because NECA is more potent in increasing chemoreceptor activation than R-PIA (MCQUEEN and RIBEIRO 1986). Blockade of adenosine receptors with 8-PT not only abolished the effects of these agonists but also reduced the sensitivity of carotid chemoreceptors to hypoxia (MCQUEEN and RIBEIRO 1986).

Arterial chemoreceptor activation appears to be particularly prominent in humans and explains the cardiovascular effects and the increase in minute ventilation produced by intravenous adenosine (WATT and ROUTLEDGE 1985); adenosine increases blood pressure and stimulates respiration if injected into the aortic arch at a site proximal to the origin of the carotid arteries, but decreases blood pressure and has no effect on ventilation when infused into the descending aorta (WATT et al. 1987; BIAGGIONI et al. 1987).

Furthermore, ventilatory stimulation is not observed in patients with impaired carotid bodies secondary to bilateral carotid endarterectomy (GRIFFITHS et al. 1990). Given that the respiratory effects of intravenous adenosine in humans are blocked by theophylline but not by the A_{2B} antagonist enprofylline (MAXWELL et al. 1987), and the agonist profile from animal studies suggesting A_2 receptors, it is possible that this effect is mediated by A_{2A} receptors.

In contrast to the respiratory stimulation due to activation of arterial chemoreceptor afferents, KWONG et al. (1998) have recently reported an A_1-mediated excitatory action of adenosine on bronchopulmonary sensory afferents. Activation of these vagal C type afferent fibers leads to apnea. Such a reflex was elicited by injections of adenosine into the right atria of anesthetized rats, and was accompanied by bradycardia and hypotension. No effect was seen when adenosine was injected into the left ventricle, making it unlikely that it represented an effect mediated in the central nervous system. Furthermore, perineural capsaicin treatment of the cervical vagi blocked the adenosine-induced respiratory inhibition. Pretreatment with the selective A_1 antagonist, DPCPX, attenuated this effect, suggesting the involvement of A_1 receptors (KWONG et al. 1998).

In a follow-up report, these investigators were able to record afferent traffic from single pulmonary C fibers in anesthetized, artificially ventilated rats (HONG et al. 1998). Right atrial injection of adenosine increased afferent traffic several-fold. Of interest, whereas the effect of capsaicin was immediate, that of adenosine had a latency ranging from 3s to 18s. The reason for this latency is not clear, but a similar phenomenon is described in adenosine-induced activation of chemosensitive afferent fibers in the human forearm (COSTA and BIAGGIONI 1993, 1994). C-fiber stimulation was blocked by the non-selective antagonist theophylline and by the A_1 antagonist DPCPX, but not by the A_{2A} antagonist, 3,7-dimethyl-1-propargylxanthine. The conclusion from these studies is that A_1 receptors activate pulmonary afferent C fibers which may induce respiratory depression (HONG et al. 1998). The physiological relevance of this phenomenon is unknown, since intravenous administration of adenosine as bolus injections uniformly produces profound respiratory stimulation in humans, rather than depression (BIAGGIONI et al. 1987).

III. Studies in Intact Animals and Functional Relevance

1. Regulation of Fetal Breathing by Adenosine Receptors

Fetuses have spontaneous breathing movements about 30–50% of the time, and these are thought to contribute to lung development and the establishment of neural pathways regulating respiration. Fetal breathing is almost completely eliminated by hypoxia (KOOS et al. 1994), and this is associated with a 2- to 3-fold increase in interstitial levels of adenosine in the brain stem, estimated with the microdialysis technique (Koos et al. 1994). It should be noted, however, that in this study adenosine uptake inhibitors had to be added to the microdialysis perfusate to reduce cellular clearance of adenosine and demonstrate a hypoxia-induced increase in interstitial levels. Intra-arterial infusion of adenosine also inhibits fetal breathing in sheep (Koos and Matsuda 1990). A similar effect is observed when an adenosine analog is injected into the cisterna magna (Bissonnette et al. 1991). This effect appears to be mediated in brainstem centers because it is eliminated by rostral brainstem transection. In such animals, intra-arterial injections of adenosine produce an opposite effect, and increase the rate and amplitude of respiration, presumably by activating chemoreceptors (Koos and Doany 1991). Systemic administration of the A_1 agonist R-PIA caused fetal apnea that was reduced by theophylline but not enprofylline (Avital et al. 1993), suggesting that this is not due to A_{2B} receptors. In a more direct approach, central administration of R-PIA into the fourth ventricle also reduced fetal breathing in unanesthetized chronically instrumented sheep, and this effect was blocked by theophylline (Bissonnette et al. 1991). Finally, intravascular infusion of theophylline attenuates the hypoxia-induced inhibition of fetal breathing (Koos and Matsuda 1990; Bissonnette et al. 1990).

A constant intra-arterial infusion of the selective A_{2A} agonist, CGS 21680, produced a biphasic respiratory response in fetal sheep, initial stimulation being followed by respiratory depression (Koos and Chau 1998). Denervation of both carotid bodies abolished the initial increases in ventilation. The subsequent ventilatory depression was not blocked by carotid body denervation or by A_1 antagonists. The authors concluded that A_{2A} receptors in fetal sheep activate arterial chemoreceptors to produce respiratory stimulation, and central neurons to produce respiratory depression (Koos and Chau 1998). Similarly, in fetal sheep in which central pathways were eliminated by brain stem section, intraarterial adenosine produced an increase in ventilation that was prevented by carotid body denervation in these animals (Koos et al. 1992). These studies indicate that adenosine activates arterial chemoreceptors even during fetal development, but this effect is often masked by central inhibition of respiration.

Avital et al. (1993) found that theophylline, but not enprofylline, increased the incidence of spontaneous fetal breathing during the first day of administration, but by day two fetal breathing had returned to baseline values. This phenomenon was attributed to adenosine receptor upregulation due to chronic exposure to antagonist. Conversely, Koos and Matsuda (1990) found that during 9 h of continuous infusion of adenosine in fetal lambs, the incidence of fetal breathing was depressed initially but returned to control values by 6 h, suggesting agonist-induced adenosine receptor downregulation. There has been concern that chronic exposure to methylxanthines in utero will result in sudden withdrawal at birth, with increased incidence of apnea in neonates (Horowitz et al. 1982), as endogenous adenosine acts on upregulated receptors.

2. Regulation by Adenosine Receptors of Neonatal Breathing

Whereas hypoxia produces an initial stimulation of ventilation in adult animals, it depresses ventilation in neonates. Given the effects of adenosine on ventilation there has been great interest in exploring its role in the control of ventilation in the neonates and determining how it differs from adults, particularly because caffeine and theophylline have been widely used in the treatment of neonatal apnea despite the lack of controlled clinical trials (for review see ARANDA and TURMEN 1979). This beneficial effect of methylxanthines is believed to be due to stimulation of central respiratory centers because it is present even after section of carotid sinus nerves and the vagi (LEMESSURIER 1936). It should be noted, however, that BLANCHARD et al. (1986) found that chemodenervation abolished the ventilatory response to caffeine in neonatal lambs. This beneficial effect of systemic caffeine administration implies that the central depressant effect of adenosine predominate over its peripheral chemostimulant actions. This appears to be particularly true in neonatal animals, in whom adenosine stimulation of peripheral chemoreceptors is reduced or absent. For example, intravenous bolus injections of adenosine stimulates ventilation in adult rabbits, but depresses ventilation in neonatal animals (WATT et al. 1986).

The adenosine analogue, R-PIA ($5\,\mu$mol/kg) administered intraperitoneally caused marked respiratory depression in urethane-anaesthetized decerebrate rabbit pups, and in unanesthetized intact animals during natural sleep. This effect was more pronounced in younger than in older animals, and it was reversed with theophylline (20mg/kg). R-PIA also caused respiratory depression when administered onto the exposed surface of the fourth ventricle (LAGERCRANTZ et al. 1984). A more direct approach was used by HERLENIUS et al. (1997), to examine the effect of A_1 receptors on modulation of respiratory brainstem neurons in neonatal rats. They used an in vitro brainstem-spinal cord preparation from neonatal (0–4days old) rats. Both dipyridamole and the A_1 agonist R-PIA decreased the activity of respiratory neurons. This effect was more prominent in the younger animals and was reversed by theophylline and the A_1 antagonist DPCPX.

As discussed previously, adenosine acts in the NTS to depress ventilation. It is possible, therefore, that hypoxia increases adenosine levels in the NTS, resulting in respiratory depression. To explore this possibility YAN et al. (1995) measured interstitial levels of adenosine in the NTS using microdialysis in anesthetized spontaneously breathing piglets. Adenosine levels in the NTS increased during hypoxia by 40%.

DARNALL and BRUCE (1987) found that the respiratory depression induced by the A_1 agonist R-PIA or by hypoxia could be blocked by theophylline but not enprofylline, in newborn piglets. Similarly, both intravenous caffeine and intraventricular 8-PT abolished the late ventilatory depression associated with hypoxia in the newborn piglet (LOPES et al. 1994; ELNAZIR et al. 1996).

In summary, there is substantial evidence supporting a role of endogenous adenosine on the central control of ventilation during development. It is possible that adenosine plays a role in the regulation of fetal breathing and in the occurrence of neonatal apnea. These findings provide a rationale for the use of methylxanthines in the treatment of neonatal apnea. Theophylline has also been reported to reduce apneic episodes in infants at risk of sudden-infant death syndrome (SIDS) and in asymptomatic siblings with a positive family history of SIDS (HUNT et al. 1983). The central depressor effects of adenosine appear to predominate early in life, when adenosine-induced stimulation of peripheral arterial chemoreceptors is reduced or even absent. It remains speculative if the lack of ventilatory stimulation by adenosine acting on peripheral chemoreceptors contributes to neonatal apnea. This issue could be explored by systemic administration of adenosine to neonates. There is limited experience about the use of adenosine in infants, but given the extremely short half-life of adenosine it can be given with close monitoring as previously done in infants with pulmonary hypertension (PATOLE et al. 1998) and respiratory failure (FULLERTON et al. 1996).

3. Studies in Adult Animals

In anesthetized, vagotomized, and glomectomized cats, hypoxia depressed phrenic nerve activity, an index of central respiratory output. Phrenic activity remained significantly depressed below the original control level for more than 1 h after return to the hyperoxic state. Pretreatment of animals with theophylline reduced the depression of respiratory activity produced by hypoxia, and respiratory activity resumed to the original control level within 10 min after return to hyperoxia. Adenosine, therefore, appeared to mediate the long-lasting depression of respiration following hypoxia and to be partially responsible for the acute depression of respiration seen initially during hypoxia (MILLHORN et al. 1984).

Several other studies have evaluated the effects of xanthines on hypoxia-induced respiratory depression in intact animals and humans. In most cases, however, the pharmacological tools and methodology used in these studies does not allow a determination of whether these effects are mediated centrally or peripherally, or to determine the receptor type involved. Although it is likely that adenosine plays a role in the modulation of respiration by hypoxia in adult humans, definite proof is lacking.

Despite these limitation, it is of interest that theophylline has been used in the treatment of adult patients with Cheyne-Stokes respiration since 1927 (VOGI 1927). This respiratory pattern is characterized by central apnea with periodic breathing, and is seen in patients with neurological disorders or with severe heart failure. In a double-blind crossover study, theophylline therapy resulting in plasma levels of approximately 60 μmol/l produced a significant decrease in the frequency and severity of episodes of central apnea (JAVAHERI et al. 1996).

D. Adenosine Receptors in the Regulation of Pulmonary Vascular Tone

I. Adenosine Receptors in the Pulmonary Vasculature

Adenosine agonists relaxed pulmonary vessels isolated from juvenile rabbits with the order of potency of NECA > adenosine > N^6-cyclohexyladenosine (CHA), suggesting an A_2-like receptor mediated effect. 8-*p*-Sulfophenyltheophylline (8-SPT) produced a competitive antagonism. Endothelium removal, and inhibition of nitric oxide synthase (NOS) with Nω-nitro-L-arginine attenuated relaxations to NECA. Similar findings were observed in pulmonary veins. These results suggest that adenosine and NECA dilate pulmonary arteries and veins isolated from young rabbits via a mechanism that is partially dependent on endothelium-derived nitric oxide (Steinhorn et al. 1994). A_2 receptors also dilate the preconstricted cat pulmonary vascular bed because the A_{2A} agonist 5′-(*N*-cyclopropyl)-carboxamidoadenosine (CPCA) was ten times more potent than adenosine, and its vasodilation was blocked by the A_{2A} antagonist, KF-17837. Contrary to the previous study, adenosine-induced vasodilation was not altered by the NOS inhibitors Nω-nitro-L-arginine benzyl ester or methylene blue (Cheng et al. 1996).

Isolated human pulmonary arteries are vasodilated by adenosine agonist with a profile consistent with A_2 receptors (McCormack et al. 1989). The half maximal vasodilation produced by NECA was observed at concentrations greater than 10 μmol/l. It is unclear whether this indicates a role of A_{2B} receptors in this response. Mechanical removal of endothelium produced a small but not significant shift to the right of the adenosine-mediated vasodilation, and no differences were observed in the sensitivity to adenosine between large (7–10 mm internal diameter) or small (200–400 μm) arteries (McCormack et al. 1989). A similar profile was observed in the hypoxic constricted isolated blood-perfused rat lung. NECA, 10 μmol/l, produced vasodilation that was attenuated by 67 μmol/l 8-SPT but not by 1 μmol/l DPCPX. Neither the selective A_1 agonist 2-chloro-N^6-cyclopentyladenosine (CCPA) or the A_{2A} agonist CGS 21680 produced vasodilation, implying that this effect is mediated by A_{2B} receptors. NECA-induced vasodilation was not blocked with the K^+ channel blockers tetraethylammonium and glibenclamide, or by the nitric oxide synthase inhibitor Nω nitro-L-arginine methyl ester (Haynes et al. 1995).

In summary, the pulmonary vasodilation produced by adenosine in the lung is similar to that seen in other vascular beds. There is little doubt that it is mediated by A_2 receptors, and the few studies that have examined the receptor subtype suggest the involvement of A_{2B} receptors. It is possible that A_{2A} receptors also mediate pulmonary vasodilation. It is assumed that vasodilation is related to activation of adenylate cyclase, but the intracellular signaling pathways have not been examined. At least in some experimental models generation of nitric oxide contributes, but does not completely accounts for adenosine-induced vasodilation. It should be noted that the contribution of the endothelium and of NO to adenosine-induced vasodilation of other vascular beds is also controversial. Results differ depending on the vascular bed and species studied, and both endothelium-dependent and -independent vasodilation have been reported (COSTA and BIAGGIONI 1998). Furthermore, even in studies showing endothelium-dependency, the vasodilatory effects of adenosine may or may not be mediated by NO (HEADRICK and BERNE 1990). There is currently no evidence for a substantial contribution of activation of K^+ channels in adenosine-induced pulmonary vasodilation, but this issue has not been extensively explored.

II. Therapeutic Potential of Adenosine as a Pulmonary Vasodilator

Adenosine has a very short half-life in humans due to rapid cellular uptake. It is difficult to measure its precise half-life, but it has been estimated to be less than 1s in human blood (MOSER et al. 1989). Given these pharmacological characteristics, adenosine has the theoretical advantages of a selective pulmonary vasodilator. It would be possible, for example, to infuse adenosine intravenously to produce preferential pulmonary vasodilation with significant inactivation by uptake as it crosses the pulmonary circulation before reaching the systemic circulation, thus avoiding systemic hypotension. This premise was examined experimentally in anesthetized patients undergoing coronary artery bypass surgery. Adenosine was infused into the right ventricle in patients with normal pulmonary arterial pressures undergoing coronary artery bypass surgery. Plasma concentrations of adenosine were measured in the right and left atria to calculate its extraction across the pulmonary vascular bed. As expected from animal studies, pulmonary extraction in humans was extensive (74% ± 5%). The mean maximal decrease in pulmonary vascular resistance, 49% ± 10%, occurred at an adenosine infusion rate of 30 μg/kg per min, a dose that produced no change in systemic vascular resistance (UTTERBACK et al. 1994; FULLERTON et al. 1996).

Recent studies have explored the use of adenosine as a pulmonary vasodilator in the treatment of clinical conditions characterized by pulmonary hypertension. Of particular interest has been the study of patients with primary pulmonary hypertension, a condition of unknown cause and poor prognosis. Acute administration of vasodilator drugs is commonly used in these patients to identify those with reversible pulmonary vasoconstriction. MORGAN et al. (1991) infused adenosine (up to 50 μg/kg per min) into the pulmonary artery in seven patients with primary pulmonary hypertension and found a dose-dependent and significant decrease in pulmonary vascular resistance, and an increase in cardiac output. Systemic vascular resistance also decreased, but this effect was less than that seen in the pulmonary circulation. At the doses used by these investigators, adenosine has only marginal effects on systolic blood pressure and heart rate, and very little if any effects secondary to arterial chemoreceptor activation, such as respiratory stimulation and sympathetic activation (BIAGGIONI et al. 1991). These findings have been confirmed in subsequent studies using intravenous adenosine but bigger doses have been required (average of about 250 μg/kg per min), and this has resulted in an increase in cardiac output and systemic vasodilation (SCHRADER et al. 1992).

Elevation of pulmonary pressure is also present in patients with end-stage heart failure. Adenosine, 100 μg/kg per min, was found to be a potent and relatively selective pulmonary vasodilator and preferable to sodium nitroprusside as a test for the reversibility of pulmonary vasoconstriction in patients with severe heart failure. However, adenosine increased left atrial pressure which was thought to be due to diastolic stiffening of the ventricle or to a negative inotropic effect, potentially deleterious actions. Therefore, adenosine was not considered a useful therapeutic agent in patients awaiting heart transplantation (HAYWOOD et al. 1992). More recent studies, however, have shown that this paradoxical rise in left atrial pressure primarily results from changes in vascular loading rather than from direct effects on cardiac diastolic or systolic function (NUSSBACHER et al. 1995).

High pulmonary artery pressure is physiologically seen in the fetus. This is reversed at birth by pulmonary vasodilation occurs in response to the increase in arterial O_2 pressure (P_aO_2). This phenomenon can be studied in fetal animals by exposing them to oxygen, which results in pulmonary vasodilation and increase in P_aO_2. In this model, 8-PT increased baseline pulmonary vascular resistance and inhibited the pulmonary vasodilation caused by oxygen (KONDURI et al. 1993). Pulmonary artery pressure normalizes quickly after birth, but may persist abnormally high in some infants. KONDURI et al. studied term infants with persistent pulmonary hypertension of the newborn. Four of five infants receiving 25–50 μg/kg per min adenosine intravenously over a 24-h period had a significant improvement in oxygenation. In contrast, none of the infants randomly assigned to the placebo group showed improvement. No changes in arterial blood pressure or heart rate were observed in either group (KONDURI et al. 1996).

III. Paradoxical Pulmonary Vasoconstriction Elicited by Adenosine Receptors

Despite the substantial evidence for adenosine-induced pulmonary vasodilation, there are instances where the opposite effect is observed. Perhaps the first evidence of this phenomenon was provided by DRURY and SZENT-GYÖRGYI (1929). In their seminal description of the cardiovascular effects of adenosine, increments in pulmonary pressure were observed when adenosine was administered in a dog heart-lung preparation. Cardiac output was not measured in these experiments and this finding was interpreted as being secondary to an increase in flow, rather than to pulmonary vasoconstriction.

More recently, a similar phenomenon has been described in chronically instrumented awake sheep in vivo. Bolus injections of adenosine into the lung produced a dose-dependent increase in pulmonary vascular resistance

(Biaggioni et al. 1989). This effect was blocked by theophylline and 1,3-dipropyl 8-*p*-sulfophenylxanthine (DPSPX). Vasoconstriction was also completely abolished by cyclooxygenase inhibition with indomethacin or ibuprofen, and by the thromboxane/endoperoxide receptor antagonist, SQ 29,548 (Biaggioni et al. 1989). Thromboxane A_2 contributes to adenosine-induced pulmonary vasoconstriction because adenosine increases the trans-pulmonary gradient of its metabolite thromboxane B_2. The contribution of other vasoconstrictive eicosanoids is also possible. Adenosine-induced vascular contractions are reviewed in detail elsewhere (Biaggioni 1992; Biaggioni and Mosqueda-Garcia 1995).

A similar phenomenon has been described in other animal species. In cats, adenosine vasoconstricts the resting pulmonary vasculature but produces vasodilation if the lung is preconstricted (Neely et al. 1989). The order of potency of adenosine receptor agonists to produce pulmonary vasoconstriction suggests the involvement of A_1 receptors. Likewise, intralobar injections of the A_1 agonist, CPA, increased pulmonary pressure, an effect that was blocked by the A_1 antagonist 1,3-dipropyl-8-noradamantylxanthine (KW-3902). Adenosine-induced pulmonary vasoconstriction is also blocked by the cyclooxygenase inhibitor meclofenamate, and by the thromboxane receptor antagonist SQ 29548 (Neely et al. 1989; 1991).

In guinea pig main pulmonary arteries precontracted with noradrenaline, adenosine exerted an initial phasic contraction followed by a tonic contraction and a slow relaxation. Selective antagonism of A_1 receptors with 10 nmol/l DPCPX blocked only the tonic contraction. Furthermore, the order of potency of the adenosine analogues for purine-induced phasic contraction was CPA > R-PIA > NECA = N^6-2-(4-aminophenyl) ethyl adenosine (APNEA) > CGS 21680, suggesting the involvement of A_1 type adenosine receptors in the contraction phase. Once vasoconstrictor A_1 receptors were blocked with 300 nmol/l DPCPX, the rank order of potency for the purine-induced pulmonary vasodilation was NECA > 2-chloroadenosine > CGS 21680 = R-PIA > CPA, suggesting the involvement of vasodilating A_{2B} receptors. Furthermore, neither NECA- nor adenosine-induced relaxation was influenced by the A_{2A} antagonist, CP 66713 (300 nmol/l). Neither removal of the endothelium nor inhibition of nitric oxide synthase with 100 μM Nω-nitro-L-arginine blocked NECA-induced vasodilation. In summary, in guinea pigs, adenosine induces endothelium- and nitric oxide-independent pulmonary vasodilation via A_{2B} receptors, and pulmonary vasoconstriction via A_1 receptors (Szentmiklosi et al. 1995).

Adenosine can also produce pulmonary vasoconstriction by mechanisms independent of eicosanoid release. Adenosine produces contractile responses in guinea pig pulmonary artery strips precontracted with norepinephrine (Wiklund et al. 1987). The purine also potentiates the vasoconstriction produced by transmural electrical nerve stimulation or by exogenous norepinephrine in guinea pig pulmonary arteries (Wiklund et al. 1989). NECA was more potent than R-PIA in this model, suggesting the involvement of A_2-like receptors. These effects were blocked by 8-SPT. It is suggested that the vasoconstrictive effects were at least partially due to prejunctional stimulation of norepinephrine release by A_2 receptors (WIKLUND et al. 1989).

IV. Adenosine Receptors in the Regulation of Pulmonary Vascular Tone

It has long been recognized that hypoxia produces pulmonary vasoconstriction. In isolated perfused lungs this hypoxic vasoconstriction is transient and is followed by a return of pulmonary artery pressure to baseline values despite maintenance of hypoxic conditions (GOTTLIEB et al. 1984). Given that hypoxia induces the release of adenosine in most vascular beds, and that adenosine is capable of both pulmonary vasoconstriction or dilation, it is tempting to speculate that adenosine contributes to hypoxia-induced changes in pulmonary vascular tone.

Available evidence about the role of endogenous adenosine, however, is contradictory. Early studies showed that theophylline reduced hypoxic vasoconstriction in intact dogs (HALES and KAZEMI 1974). More recently a similar result was obtained in artificially ventilated rats. Systemic hypoxia increased pulmonary artery pressure and this effect was abolished by 8-SPT (THOMAS and MARSHALL 1993). On the other hand, adenosine deaminase did not alter the vascular responses to hypoxia in isolated ferret lungs. The authors concluded, therefore, that, unless endogenous adenosine was somehow protected from adenosine deaminase, neither hypoxic vasoconstriction nor the subsequent vasodilation were mediated by adenosine (GOTTLIEB et al. 1984).

E. Adenosine Receptors in the Regulation of Airway Tone and Their Role in Asthma and Other Inflammatory Processes

I. Adenosine Receptors in the Regulation of Airway Tone

The early indication that adenosine receptors regulate respiratory smooth muscle tone came from observation that adenosine can reverse the relaxation of guinea pig tracheal smooth muscle produced by theophylline (FREDHOLM et al. 1979). The role of adenosine in the regulation of airway smooth muscle tone is complex. Adenosine has been shown to exert diverse effects on airway smooth muscle depending on species studied and the experimental model used. Adenosine and its analogs produced constriction of isolated guinea pig trachea presumably via A_1 adenosine receptors, but at higher doses they relaxed the trachea previously contracted by various spasmogens via A_2 adenosine receptors (GHAI et al. 1987; FARMER et al. 1988; LOSINSKI and ALEXANDER 1995). It is possible that the balance between stimulatory and inhibitory actions of adenosine on airway smooth muscle determines the apparent lack of bronchomotor responses to exogenous adenosine observed in many normal animal models (ALI et al. 1994a; PAUWELS et al. 1995) as well as in normal human subjects (CUSHLEY et al. 1983).

It should also be noted that the effects of adenosine on airway smooth muscle may be indirect, resulting from activation of a neuronal reflex or from a primary action, not on smooth muscle but on other cells present in the airways. The vagus nerves and fibers from the upper four to five thoracic sympathetic ganglia form anterior and posterior plexuses from which the two main nerve networks arise: the peribronchial and periarterial plexuses. The former innervates airways down to the level of respiratory bronchioles and the latter innervates bronchial vessels. Cholinergic nerves are the dominant neural bronchoconstrictor pathway in animal and human airways whereas adrenergic β-receptors stimulation produces bronchodilation. In addition to classical cholinergic and adrenergic mechanisms, there exists a third component of neural control with neuropeptides acting as neurotransmitters. Reflex cholinergic bronchoconstriction may occur due to stimulation of afferent sensory C-fiber endings by inflammatory mediators. The autonomic nervous system may regulate not only airway smooth muscle tone, but also secretion from submucosal glands, transport of fluid across epithelium, permeability and blood flow in the bronchial circulation, and release of mediators from inflammatory cells (BARNES et al. 1998).

The role of neural mechanisms in the regulation of bronchomotor tone by adenosine remains controversial. Acetylcholine antagonists attenuated adenosine-dependent bronchoconstriction in inbred rats (PAUWELS and VAN DER STRAETEN 1987), isolated canine bronchial segments (SAKAI et al. 1989), and in asthmatic patients (CRIMI et al. 1992), implicating a possible involvement of cholinergic neural pathways. On the other hand, A_{2B} receptors have been shown to potentiate neurally-mediated cholinergic bronchoconstriction through an enprofylline-sensitive process (WALDAY and AAS 1991). However, other investigators failed to show any role for cholinergic neurons in the bronchoconstrictor effects of adenosine in humans (MANN et al. 1985) or in the guinea pig trachea (GRUNDSTROM et al. 1981). Activation of sensory C-fibers has been also implicated in bronchoconstrictor responses to adenosine because desensitization of neuropeptide-producing nerves with capsaicin reduced adenosine effects in guinea pigs (MANZINI and BALLATI 1990). Desensitization of airway C-fibers by repetitive challenge with bradykinin, attenuated the bronchoconstrictor responses to adenosine, but not to histamine, in asthmatics (POLOSA et al. 1992). However, inhibition of neutral endopeptidase, the enzyme that cleaves neuropeptides, with phosphoramidone, which should enhance tachykinin-mediated effects, failed to elicit any significant increase in the adenosine-induced bronchospastic response (POLOSA et al. 1997). It should be noted that adenosine is capable of activating sensory C-fibers within the lung (as described in the section dealing with peripheral modulation of ventilation) and elsewhere in the body (for review see BIAGGIONI 1992).

Adenosine seems to be a fairly weak or inactive contractile agent in the isolated normal human bronchi. However, isolated bronchi from asthmatic patients selectively constricted in response to adenosine but not to histamine or leukotriene C_4. Adenosine-induced contractions were antagonized by 1 μmol/l 2-thio-DPCPX and 30 μmol/l 8-SPT. The subtypes of adenosine receptors involved in this mechanism remain unknown, because the antagonists were used at high concentrations and therefore were not selective. The contractile effects of adenosine were blocked by antagonists of histamine and leukotrienes, suggesting that adenosine acts indirectly by liberation of pro-inflammatory mediators, possibly from mast cells (Björck et al. 1992). These findings correlate with clinical observations where pre-medication with anti-histamines (Rafferty et al. 1987; Phillips et al. 1987) and potent inhibitors of cyclooxygenase (Crimi et al. 1995) inhibit the acute bronchoconstrictor response to inhaled AMP in asthmatic subjects.

Adenosine could also modulate bronchial tone through an epithelial-dependent mechanism. The functional coupling of adenosine A_{2A} receptors to nitric oxide production was reported in rabbit bronchial epithelium. The epithelium-dependent relaxation of airway smooth muscles was inhibited by Nω-nitro-L-arginine and the inhibition was reversed by L-arginine, but not D-arginine. The order of potency of the adenosine analogues suggested the involvement of A_{2A} adenosine receptors. Furthermore, the selective A_{2A} agonist CGS 21680 increased nitric oxide production in primary culture of epithelial cells (Ali et al. 1997). Finally, adenosine can also modulate bronchial tone indirectly by acting on mast cells, as reviewed in the following sections of this chapter.

II. Adenosine Receptors in Inflammatory Cells of the Airways

Inflamation in airways is a complex response to tissue injury that involves the activation and release of mediators from many inflammatory and structural cells including mast cells, eosinophils, basophils, T-lymphocytes, macrophages, neutrophils, dendritic cells, platelets, epithelial cells, smooth muscle cells, endothelial cells, and fibroblasts (for review see Barnes et al. 1998). Purinergic receptors are expressed in most of these cells and their role in inflammatory processes is reviewed elsewhere in this book (Dubyak, Chap. 25, this volume; Montesinos and Cronstein, Chap. 24, this volume). The actions of adenosine in mast cells, eosinophils, basophils, and alveolar macrophages that are thought to be associated with asthma and other airway inflammatory diseases are discussed further below.

All these inflammatory cells originate from a common bone marrow totipotent hematopoietic stem cell. Mast cell progenitors leave the bone marrow as progenitors and complete their differentiation after arriving in peripheral tissues. Depending on tissue where they reside, mast cells develop a certain phenotype. There are two major groups of mast cells based on their protease composition: MC_{TC} cells containing tryptase, chymase, cathepsin G, and carboxypeptidase, and MC_T cells containing only tryptase. In pulmonary tissue MC_T cells are the predominant mast cell type (SCHWARTZ 1990). In contrast to mast cells, basophils and eosinophils do not migrate to tissues unless recruited into an inflamed site. Macrophages also originate from a common bone marrow progenitor cell for both granulocytes and monocytes/macrophages. Alveolar macrophages are derived from blood monocytes and from proliferating macrophage precursors in the interstitium of the lung. Migration of monocytes into the lung appears to be a random phenomenon in the absence of localized inflammation.

Most of the mast cells in lung are resident at tissue sites which are exposed to the environment. They are found within epithelium, subepithelial connective tissue and between smooth muscle cells. Mast cells, by virtue of their location and capacity to release a wide variety of pro-inflammatory mediators, are believed to initiate the early bronchoconstrictor response to allergen and probably also the bronchoconstriction induced by exercise, fog, or cold air. This event is followed by an inflammatory response in which leukocytes, especially eosinophils, infiltrate the affected site. About 3–12h after recovery of the airways from the early response, the late phase bronchoconstriction can be observed in approximately 50% of allergic asthmatics and 30% of patients with exercise-induced asthma (PRADALIER 1993). It has been proposed that basophils may play a role in this late phase response. These cells have been identified in the upper airways in allergic rhinitis and during nasal late phase response (LICHTENSTEIN and BOCHNER 1991). However, according to other authors, evidence of the involvement of basophils in the late phase constrictor response in the lower airways remains inconclusive (REDINGTON et al. 1995). The presence of eosinophils in the airways is characteristic of the airway inflammation found in asthma. Eosinophils may cause airway hyperresponsiveness, chemotaxis of inflammatory cell, and denudation of epithelium. The unique feature of alveolar macrophages is that they function both in connective tissue of the septum and in the air space of the alveolus, scavenging the surface to remove inhaled particulate matter. These cells constantly sample their microenvironment and secrete a variety of pro- and anti-inflammatory products in response to different stimuli. The role of alveolar macrophages in the pathogenesis of asthma is unclear, but these cells can be involved in numerous inflammatory events in the lung.

1. Adenosine Receptors in Mast Cells

The immediate response of sensitized airways to allergen occurs through interaction between specific antibodies of the IgE class bound to the high affinity IgE receptors (FcεRI) of mast cells and the relevant allergen. As a result, mast cells release a variety of preformed mediators, consisting mostly of histamine, proteoglycans, and the neutral proteases. Following this event, the stimulated mast cells produce and release a wide variety of newly-generated pro-inflammatory mediators including lipid-derived substances and cytokines (Fig. 3).

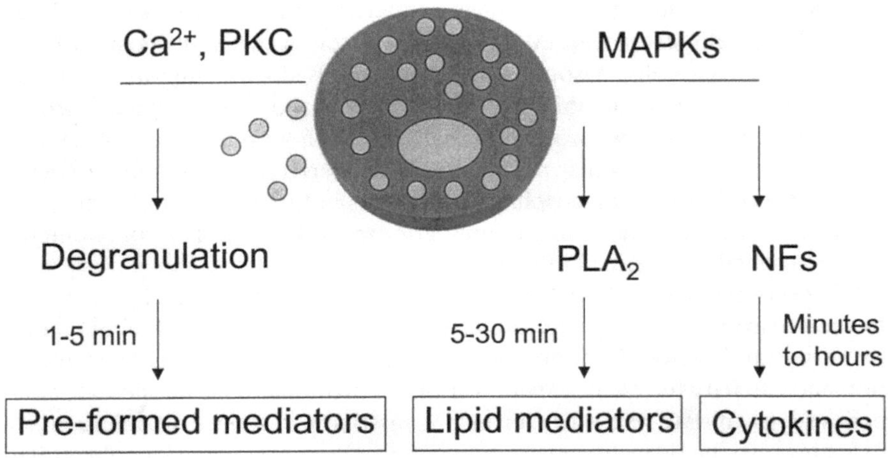

Fig. 3. IgE-mediated release of mediators by mast cells. The principal events and time required for the release of mediators are indicated. MAPKs, mitogen-activated protein kinases; NFs, nuclear factors; PLA_2, phospholipase A_2; PKC, protein kinase C

Most of the recent advances in the understanding of biochemical pathways linked to FcεRI receptor activation were made by using rodent mast cells as a model (Fig. 4). FcεRI is a heterotetrameric receptor, which includes one IgE-binding α subunit, one β subunit, and two disulfide-bonded γ subunits. The multivalent binding of allergen to receptor-bound IgE and the subsequent aggregation of FcεRI receptors triggers activation of β subunit-bound protein kinase Lyn, which phosphorylates tyrosine residues in immunoreceptor tyrosine-based activation motifs of β and γ subunits. Phosphorylated γ and β subunits, in turn, recruit and activate tyrosine kinase Syk and additional Lyn respectively. Activated Lyn phosphorylates and activates Bruton's tyrosine kinase, which together with Syk is believed to be responsible for tyrosine phosphorylation and activation of phospholipase C-γ. Resulting activation of protein kinase C by diacylglycerol and rise in intracellular calcium induced by inositol 1,4,5-trisphosphate are believed to be necessary for release of preformed mast cell mediators. It is now recognized that the synthesis of newly-generated mediators requires activation of mitogen-activated protein kinases (MAPK) cascades. It has been demonstrated that activation of FcεRI receptor-associated protein tyrosine kinases eventually leads to stimulation of all three major MAPK pathways: ERK, p38 MAPK, and JNK (KAWAKAMI et al. 1998; ISHIZUKA et al. 1998). Stimulation of the ERK pathway is considered to be important for activation of phospholipase A_2 and release of arachidonic acid, a precursor for many lipid-derived mediators (HIRASAWA et al. 1995). Stimulation of ERK and other MAPK pathways initiates synthesis of proinflammatory cytokines via activation of various nuclear transcription factors (for a review on FcεRI, see KINET 1999).

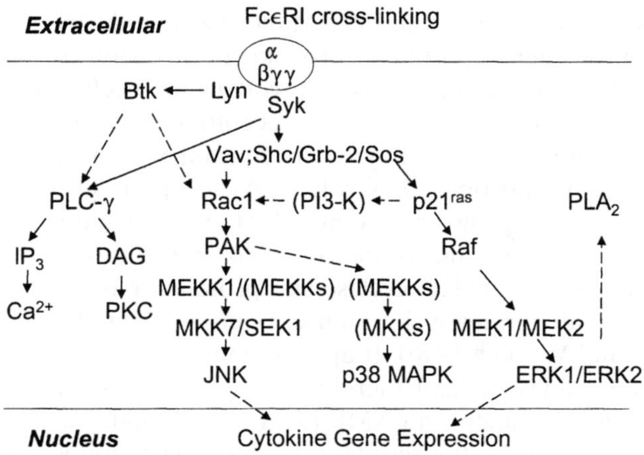

Fig. 4. Schematic representation of intracellular pathways coupled to $Fc_{\varepsilon}RI$ receptors in rodent mast cells. *Broken lines* indicate pathways which may include intermediate steps. See text for details. $Fc_{\varepsilon}RI$, high affinity Fc receptor for IgE; Lyn, non-receptor protein tyrosine kinase *lyn* of *src*-family; Syk, non-receptor protein tyrosine kinase *syk* of Syk/Zap-70family; Btk, Bruton's tyrosine kinase; Vav, hematopoietic cell-specific guanine nucleotide exchange factor protein; Shc, *src* homology/collagen adaptor protein to Grb-2; Grb-2, growth-factor-receptor binding protein 2; Sos, Son of sevenless guanine nucleotide exchange factor protein; Rac 1, small GTP-binding protein of the p21 Ras-related Rho subfamily; $p21^{ras}$, small GTP-binding protein; PI3-K, phosphatidylinositol 3-kinase; PAK, p21 (Cdc42/Rac)-activated kinase; Raf-1, serine/threonine kinase Raf-1; MEKK, MEK kinase; MKK, MAPK kinase; SEK1, SAPK/ERK kinase 1; MEK, MAPK/ERK kinase; JNK, c-Jun N-terminal kinase; MAPK, mitogen-activated protein kinase; ERK, extracellular signal-regulated protein kinase; PLA_2, phospholipase A_2; PLC-γ, phospholipase C-γ; IP_3, inositol 1,4,5-trisphosphate; DAG, diacylglycerol; PKC, protein kinase C

Marquardt et al. (1978) were the first to report that adenosine, while ineffective alone, potentiated histamine release induced by anti-IgE, concanavalin A, compound 48/80, or the calcium ionophore A23187 in isolated rat mast cells. The mechanisms that mediate potentiation of these cells remain unclear. Stimulation of adenylate cyclase by adenosine was blocked by 8-PT, but potentiation of histamine release was not, suggesting that these effects are mediated by different adenosine receptors (Church et al. 1986).

Because potentiation of rat peritoneal mast cells is insensitive to 8-PT, the possibility was raised that this effect is mediated by A_3 receptors, since the rat A_3 receptor has remarkably low affinity for xanthines (Zhou et al. 1992). This possibility was examined in the rat basophilic leukemia cell line RBL-2H3, which has been used as a model for rat mast cells. Adenosine analogs stimulated phospholipase C, increased cytoplasmic calcium, and potentiated mediator release in these cells with a pharmacological profile consistent with A_3 receptors (Ramkumar et al. 1993). Expression of A_3 receptors in RBL-2H3 cells, along with A_{2A} and A_{2B}, was confirmed by radioligand binding and detection of mRNA (RAMKUMAR et al. 1993; MARQUARDT et al. 1994). Furthermore, A_3 agonists have been reported to provoke bronchoconstriction in vivo in BDE strain rats through a process likely involving mast cell activation (MEADE et al. 1996). Of interest, rat mast cell A_3 receptors are also implicated in the mediation of adenosine-induced hypotension (HANNON et al. 1995; FOZARD et al. 1996), further supporting a key role of A_3 receptors in mast cells of this species. There is also growing evidence that A_{2B} receptors modulate mast cell function (FEOKTISTOV and BIAGGIONI 1996, 1997; FEOKTISTOV et al. 1998). Adenosine activates adenylate cyclase and protein kinase C, and potentiates stimulated mediator release in mouse bone marrow-derived mast cells (MARQUARDT and WALKER 1990). It appears that the abilities of adenosine to activate protein kinase C and thereby to augment mast cell degranulation are independent of changes in cAMP (MARQUARDT and WALKER 1994). Both A_{2A} and A_{2B}, but not A_1, transcripts were detected in mouse bone marrow-derived mast cells (MARQUARDT et al. 1994). The failure of the A_{2A}-selective agonist CGS 21680 to enhance mediator release suggests that A_{2B} is the receptor type that modulates degranulation of these mast cells (MARQUARDT et al. 1994).

A_{2B} receptors have also been shown to activate the human mast cell line HMC-1 (FEOKTISTOV and BIAGGIONI 1995). HMC-1 cells were derived from a patient with mast cell leukemia and have a neutral proteases content similar to that of human lung mast cells. These cells co-express A_{2A} and A_{2B} receptors, and both are coupled to adenylate cyclase through G_s-proteins (Fig. 5).

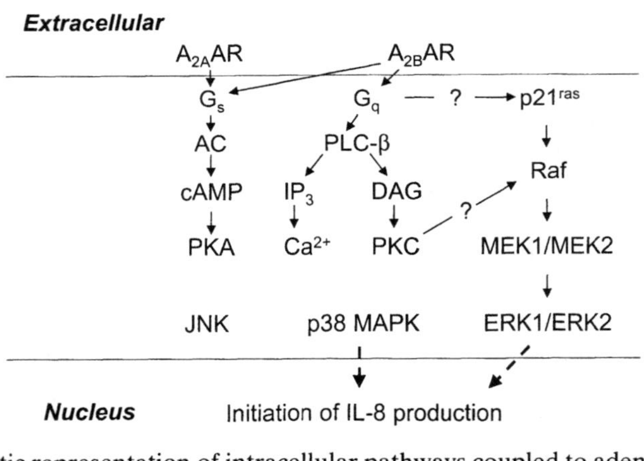

Fig. 5. Schematic representation of intracellular pathways coupled to adenosine receptors in human mast cells. *Broken lines* indicate pathways which may include intermediate steps. See text for details. AR, adenosine receptors; AC, adenylate cyclase; PKA, protein kinase A; PLC-β, phospholipase C-β. For other abbreviations see legend for Fig. 4. Reproduced from FEOKTISTOV et al. (1999) with permission from the American Society for Pharmacology and Experimental Therapeutics

However, only A_{2B} receptors activate HMC-1 cells, as indicated by stimulation of IL-8 secretion with the non-selective A_2 agonist NECA, but not with the selective A_{2A} agonist CGS 21680. Cyclic AMP does not appear to modulate this process because neither forskolin nor 8-Br-cAMP influenced IL-8 secretion. On the other hand, A_{2B} receptors appear to be coupled to phospholipase C-β, as evidenced by increase in inositol phosphate production with consequent mobilization of intracellular calcium. These A_{2B} receptor-dependent pathways were stimulated through a cholera toxin- and pertussis toxin-insensitive G-protein, presumably of the G_q family (FEOKTISTOV and BIAGGIONI 1995). Stimulation of A_{2B} receptors activates also the small GTP-binding protein $p21^{ras}$. This event triggers ERK signaling pathway with sequential stimulation of Raf, MEK1/2, and ERK1/2 protein kinase activities. Adenosine receptors also couple to JNK and p38 MAPK signaling pathways. Furthermore, blockade of ERK pathway with the selective MEK inhibitor PD 098058 or p38 MAPK pathway with the selective p38 MAPK inhibitors SB 202190 and SB 203580 revealed that these MAP kinase pathways are essential for adenosine A_{2B} receptor-mediated production of IL-8 (FEOKTISTOV et al. 1999). The exact mechanism of coupling of A_{2B} receptors with MAPK signaling pathways is currently not known. Mechanisms whereby G-protein-coupled receptors regulate tyrosine protein kinase activity have been proposed (for review see LUTTRELL et al. 1999). Further study of convergence of adenosine receptors and FcεRI signaling pathways in mast cells would help to explain the role of adenosine in mast cell activation.

An A_{2B}-mediated mechanism of secretion was also reported in a canine BR mastocytoma cell line. Adenosine analogs stimulated β-hexaminidase release, inositol phosphate production, and intracellular calcium mobilization through a pertussis toxin insensitive G-protein with a pharmacological profile consistent with activation of A_{2B} receptors (AUCHAMPACH et al. 1997).

In contrast, adenosine inhibited tryptase release and ERK activity in mast cells derived from human umbilical cord blood. This inhibition was attributed to A_{2A} adenosine receptor activation because the selective A_{2A} agonist CGS 21680 was more potent than the A_1 agonist CPA or A_3 agonist N^6-benzyl-NECA, and was blocked by the selective A_{2A} antagonist ZM 241385 (SUZUKI et al. 1998).

Mast cells from different species, and even from different anatomical sites within the same species, can vary substantially in their morphological and biochemical characteristics. It is not surprising, therefore, that more than one adenosine receptor is involved in activation of mast cells obtained from different species and tissues. Therefore, it is important to determine the characteristics of adenosine receptors present in human mast cells which are relevant to asthma. In parenchymal human lung mast cells, obtained from normal sections of excised lung, adenosine analogs do not directly evoke release of histamine and leukotriene C_4, but potentiate mediator release from immunologically activated cells at low micromolar concentrations (PEACHELL et al.

1991; MARONE et al. 1989). The order of potency of adenosine analogs and the low affinity of this process suggested that the response of human lung mast cells to adenosine is mediated by A_{2B} receptors. To further emphasize heterogeneity of mast cells, the same group of investigators found that adenosine analogs did not potentiate but rather inhibited both histamine and PGD_2 release in immunologically activated mast cells isolated from human skin tissue (MARONE et al. 1989).

2. Adenosine Receptors in Eosinophils

Recruitment of eosinophils, in preference to neutrophils, is one of the prominent features of inflammation of human asthmatic airways. Blood eosinophilia accompanies the late phase, but not the early asthmatic response. The accumulation of eosinophils and eosinophil products have been documented in bronchoalveolar lavage fluid (BAL) during allergen-induced late-phase reactions. The eosinophil products "major basic protein" and "eosinophil cationic protein" are both cytotoxic to the respiratory epithelium and may contribute to the denudation of the epithelium seen in asthma (for a review on the role of eosinophils in asthma, see STREK and LEFF 1997). Whereas adenosine increases intracellular calcium and potentiates PAF-induced superoxide production in human eosinophils, it suppresses human and guinea pig eosinophils activated with opsonized zymosan (YUKAWA et al. 1989). Radioligand binding studies, in situ hybridization, and RNA protection assay identified the adenosine receptors in human eosinophils as belonging to the A_3 subtype (WALKER et al. 1997; KOHNO et al. 1996). Furthermore, the selective A_3 receptor agonist 2-chloro-N^6-(3-iodobenzyl)-adenosine-5′-*N*-methyluronamide (Cl-IB-MECA) but not CPA and CGS 21680, selective for A_1 and A_{2A} receptors respectively, induced release of calcium from internal stores (KOHNO et al. 1996). The selective A_3 receptor agonist N^6-(4-amino-3-iodobenzyl)adenosine (I-ABA) inhibited PAF-induced eosinophil chemotaxis and this effect was blocked by the selective A_3 receptor antagonist, 3-(3-iodo-4-aminobenzyl)-8-(4-oxyacetate)phenyl-1-propylxanthine (I-ABOPX) (WALKER et al. 1997). Similar effects of A_3 adenosine receptor activation were observed when eosinophil chemotaxis was induced by leukotriene B_4 or RANTES (KNIGHT et al. 1997). The relevance of adenosine's effects on eosinophils to asthma and other inflammatory lung diseases remains to be determined.

3. Adenosine Receptors in Basophils

Adenosine and its analogs stimulate adenylate cyclase, and inhibit histamine and leukotriene C_4 release in human basophils. The order of potency of adenosine analogs suggested that the response of human basophils to adenosine is mediated by A_2-like receptors (HUGHES et al. 1983; MARONE et al. 1985, 1989; PEACHELL et al. 1988). Whereas basophils are implicated in the late phase allergic reactions in the upper airways (LICHTENSTEIN and BOCHNER 1991), their role in the human lung is less clear (REDINGTON et al. 1995).

4. Adenosine Receptors in Alveolar Macrophages

The vast majority of cells recovered from bronchoalveolar lavage (BAL) in normal and asthmatic subjects are macrophages. However, the precise role of macrophages in asthma remains unknown. Adenosine can modulate secretion of pro-inflammatory cytokines in macrophages from various sources. Adenosine A_{2B} receptors inhibit IFN-γ-induced expression of MHC class II genes and nitric oxide synthase in murine bone marrow-derived macrophages (XAUS et al. 1999). In the human U-937 and in murine J774.1 macrophage cell lines the A_3 adenosine receptors inhibit tumor necrosis-alpha (TNF-α) formation (JACOBSON et al. 1997; BOWLIN et al. 1997). Less is known about the role of adenosine receptors in alveolar macrophages. The presence of adenosine A_2 receptors have been reported in rabbit alveolar macrophages based on the presence of binding sites for [^{3}H]NECA with a K_d of 0.5 μmol/l, and the finding that activation of adenylate cyclase with NECA was ten times more potent than R-PIA. Alveolar macrophages produce mediators with procoagulant activity. Adenosine agonists suppressed procoagulant activity in vitro in rabbit alveolar macrophages, with a pharmacologic profile consistent with an A_2 adenosine receptor (HASDAY and SITRIN 1987); increasing endogenous adenosine levels with dipyridamole had a similar effect (HASDAY and SITRIN 1987).

III. Relevance to Asthma

1. Animal Studies

Airway hyperresponsiveness to adenosine is one of the features of asthma. When inhaled, adenosine produce little or no response in normal human subjects, but induces bronchoconstriction in asthmatic patients (CUSHLEY et al. 1983). Several animal models have been introduced in an attempt to understand the mechanism underlying adenosine-induced effects in the airways. The evidence that airway responsiveness to adenosine can be genetically controlled was demonstrated in the comparative study of two highly inbred rat strains, F344 and BDE. It was found that adenosine analogs administered intravenously produced bronchoconstriction in BDE but not in F344 rats (PAUWELS et al. 1995). The bronchoconstrictor effect of adenosine was attenuated by an anticholinergic agent atropine, suggesting involvement of cholinergic nerves. It should be noted that, in contrast to human airways, the major bronchoconstrictor mast cell mediator in the rat is serotonin. The nonselective serotonin receptor antagonist methysergide inhibited the bronchoconstrictor effect of adenosine, indicating a possible involvement of mast cells. Pharmacological analysis of adenosine-induced bronchoconstriction in this model suggested that multiple receptors, including A_1, A_{2B}, and A_3, may participate in mediation of this effect (PAUWELS and JOOS 1995).

A different approach to create an animal model of asthma was used by several groups and relies on actively sensitizing animals at birth. The study of isolated lungs from rats actively sensitized with ovalbumin demonstrated that adenosine analogs administered by intra-arterial infusion enhanced antigen-induced bronchoconstriction. However, adenosine analogs on their own did not produce bronchoconstriction (POST et al. 1990). A similar approach in guinea pigs yielded better results; bronchoconstrictor responses to adenosine analogs were greater in tissues from sensitized compared with unsensitized guinea pigs (THORNE et al. 1996). Immunization of neonatal rabbits with ragweed allergen is perhaps the animal model of asthma that has been studied most extensively. As in humans, normal rabbits do not experience bronchoconstriction after inhalation of adenosine. Actively sensitized rabbits show bronchoconstrictor response to the allergen as well as to inhaled adenosine. Pharmacological analysis revealed the involvement of A_1 receptors (ALI et al. 1994a,b; EL-HASHIM et al. 1996). Binding studies with [^{3}H] DPCPX demonstrated the presence of A_1 receptors in lung plasma membranes from allergic rabbits, while no specific binding was observed in membranes from normal rabbit lung (ALI et al. 1994b). Furthermore, the selective inhibition of the synthesis of A_1 receptors with DNA antisense attenuated airway constriction to adenosine and allergen in sensitized rabbits (NYCE and METZGER 1997). Study of the signal transduction pathways revealed that activation of phospholipase C may play an essential role in the mechanism of adenosine-induced bronchoconstriction in this allergic rabbit model (ABEBE and MUSTAFA 1998). The location of these allergen-inducible A_1 receptors in rabbit lung remains unclear. Because adenosine analogs contract epithelium-free isolated airway smooth muscle preparations from allergic rabbits, it has been suggested that specific adenosine receptors are located on the airway smooth muscle. On the other hand, the authors excluded the presence of mast cells in their preparations based on histological examination (ALI et al. 1994b). This is contrary to the observations made in human airway tissues. Mast cells can be seen under an electron microscope in the smooth muscle layer between the cells, both in the normal state and in diseases such as asthma (LAITINEN and LAITINEN 1988). More recently A_{2B} receptors have been shown to mediate bronchoconstriction in a sensitised Brown Norway rat model (FOZARD and HANNON 2000), underscoring the importance of species differences in adenosine-mediated bronchoconstriction.

2. Human Studies

Adenosine is increased in inflammatory conditions of the airways. High concentrations of adenosine have been measured in the BAL fluid of subjects with asthma and chronic bronchitis compared to normal controls (DRIVER et al.

1993). Several lines of evidence suggest a contribution of adenosine in asthma. Adenosine provokes dose-dependent bronchoconstriction when administered by inhalation to patients with asthma but not to normal subjects (CUSHLEY et al. 1983). In addition, inhalation of its parent nucleotides, AMP and ADP, produces almost identical effects on the airways (MANN et al. 1986) as both nucleotides are dephosphorylated to yield adenosine. AMP is commonly used instead of adenosine in inhalers because of its higher solubility, making it a more effective delivery system.

The ability of the adenosine uptake inhibitor, dipyridamole, to potentiate adenosine-induced bronchoconstriction indicated interaction of the nucleoside with cell surface purine receptors (CRIMI et al. 1988). Moreover, theophylline, which acts as an adenosine receptor antagonist, has been shown to produce a greater protection against adenosine than against histamine-induced bronchoconstriction (MANN and HOLGATE 1985). One difficulty with the concept that the antiasthmatic effect of reanthines is related to adenosine receptor antagonism was the observation that enprofylline was an effective antiasthmatic reanthine but was thought to lack adenosine receptor antagonist activity (CLARKE et al. 1989). However, it has recently been shown that this compound does, in fact, possess A_{2B} receptor antagonist properties (FEOKTISTOV and BIAGGIONI 1995).

It is proposed that mast cell activation contributes to adenosine-induced bronchoconstriction. In vitro studies have shown that adenosine, and the non-selective A_2 agonist NECA, markedly enhances histamine release (HOLGATE et al. 1991) and prostanoid generation (PEACHELL et al. 1988) from immunologically stimulated human lung mast cells. More direct evidence that adenosine-induced bronchoconstriction is the consequence of the release of preformed and newly generated mediators from airway mast cells comes from a series of studies in which a significant rise in plasma levels of histamine is found after AMP challenge (PHILLIPS et al. 1990). It has been also shown that endobronchial challenge with AMP produces high concentrations of mast cell-derived histamine and tryptase in the BAL fluid of asthmatic patients (POLOSA et al. 1995). Tryptase is a highly specific marker for mast cells (SCHWARTZ 1990) and provides strong evidence that these cells are activated by adenosine in vivo. Nasal provocation with AMP mimics many of the symptoms of rhinitis and releases histamine into the nasal cavity (CRIMI et al. 1993; POLOSA et al. 1994). Furthermore, the potent and specific H_1 receptor antagonist cetirizine abolished nasal symptoms in all subjects studied (CRIMI et al. 1993). Similarly, adenosine-induced bronchoconstriction in asthmatics can be blocked by H_1 receptor antagonists (RAFFERTY et al. 1987; PHILLIPS et al. 1987) and prevented by cromoglycate and nedocromil sodium, drugs that inhibit mast cell degranulation (PHILLIPS et al. 1989).

Given that inhaled adenosine affects only asthmatics but has no effect in normal subjects, there appears to be an intrinsic difference in the way adenosine interacts with mast cells from asthmatics. The in vitro response produced by A_{2B} receptors in HMC-1 cells and in canine BR mastocytoma cells appears to mimic the in vivo responses to inhaled adenosine in asthmatics, inasmuch as adenosine provokes mast cells activation in these cell lines as it does in asthmatics. On the other hand, the in vitro response of mouse bone marrow-derived mast cells to adenosine challenge resembles that observed in mast cells from normal human lung; in both cases adenosine potentiates mast cells activation but does not evoke direct activation. The molecular mechanisms behind these differential A_{2B}-mediated responses in asthmatic vs normal mast cells, and in HMC-1 cells vs mouse bone marrow-derived mast cells, remain to be elucidated. Defining the adenosine receptor subtype(s) involved in asthma will provide the rationale for the development of novel antiasthmatic agents (Feoktistov et al. 1998a,b).

IV. Relevance to Acute Lung Injury and other Inflammatory Processes

Because of the potent anti-inflammatory actions of adenosine, there is interest in determining its potential role in adult respiratory distress syndrome (ARDS), an often fatal complication of sepsis. In a study in pigs exposed to endotoxemia, treatment with intravenous adenosine 30 min before endotoxemia increased cardiac output, prevented pulmonary vasoconstriction, and improved left ventricular function. Despite significant pulmonary vasodilation, there was a slight improvement of gas exchange, and systemic oxygen delivery almost doubled. On the other hand, adenosine did not prevent the increase in TNF-α, used as an index of endotoxemia, implying that the improvement in hemodynamics and oxygenation were not due to prevention of the inflammatory process (Thiel et al. 1998).

GP-1-515, a novel inhibitor of adenosine kinase, was studied in two models of septic shock. Adenosine kinase inhibition should result in increased adenosine levels. GP-1-515 significantly decreased mortality in mice that received a lethal injection of endotoxin. The beneficial effect was accompanied by decreased neutrophil accumulation in the lungs and was reversed by an adenosine receptor antagonist, implying that the effects were mediated by endogenous adenosine. Plasma levels of TNF-α, but not IL-1a or IL-6, were lower in the GP-1-515-treated animals. In a second model of sepsis, GP-1-515 increased survival from bacterial peritonitis in rats (Firestein et al. 1994). If these observations are confirmed, adenosine kinase inhibitors may represent a novel therapeutic approach to the treatment of inflammatory diseases if their effects on the microvasculature can be overcome (Erion et al. 2000).

Similarly, adenosine provides protection in isolated rat lungs subjected to fat emulsion damage, a model simulating adult respiratory distress syndrome. In this model, adenosine reduced pulmonary vascular resistance and fluid filtration rate, and prevented cell damage (Jolin et al. 1994). Adenosine also prevented the increase in pulmonary vascular resistance and capillary permeability (an index of pulmonary edema) induced by phorbol 12-myristate 13-acetate (PMA) in isolated rabbit lungs (Bradley et al. 1991). This beneficial effect was not associated with prevention of the increase in circulating TNF-α or leukotrienes induced by PMA (Bradley et al. 1991).

Despite the potential beneficial effects of adenosine in these animal models of inflammation, no beneficial effect of dipyridamole was observed in a double-blind study involving 40 patients with acute respiratory distress syndrome associated with sepsis (Vincent et al. 1985).

F. Summary

Adenosine receptors are widely distributed and it is not surprising, therefore, that they would be found in most, if not all, cells present in the lung. There is a substantial body of evidence suggesting that adenosine modulates many physiologic and pathologic processes in the lung. Adenosine receptors are present in epithelial cells and modulate the secretory function of these cells. The importance of endogenous adenosine on epithelial function is not completely defined, but the observation that adenosine antagonists may activate Cl^- efflux in epithelial cells expressing the cystic fibrosis defect, if confirmed, may lead to a novel therapeutic approach for this disease.

Adenosine modulates ventilatory drive at several levels, often resulting in opposite actions. Intravenous infusion of adenosine in human subjects elicits an uncontrollable urge to breath that often overshadows its cardiovascular actions. This phenomenon is explained by arterial chemoreceptor activation and, in animals, is mediated by A_2 receptors. These arterial chemoreceptors located in the carotid body are important as a defense mechanism against hypoxia, responding with an increase in ventilation. Given that adenosine is considered a retaliatory autacoid against cellular hypoxia, it is appealing to believe that it serves a similar role if the whole animal is made hypoxic. However, we do not know with certainty that endogenous adenosine is a metabolic signal that activates arterial chemoreceptor in response to hypoxia. This possibility could be explored in animals using specific adenosine receptor antagonists that cannot cross the blood brain barrier (e.g., DPSPX). There is also evidence that adenosine activates pulmonary afferent C fibers that induce respiratory depression in animals. This effect appears to be mediated by A_1 receptors but its relevance to humans has not been explored. When initially proposed, activation of arterial chemoreceptors by adenosine seemed like a paradoxical effect, because of the known depressant effect of adenosine in most neurons. There is now evidence that adenosine activates a variety of neural fibers that form the afferent limb of autonomic/respiratory reflexes.

In contrast, adenosine acts on respiratory neurons in the brainstem to depress respiration. It is unclear if this effect is mediated in more than one respiratory center and if the receptor subtype involved (A_1?) is the same in every case. Nonetheless, there is substantial evidence from animal studies that adenosine is released in the brainstem during hypoxia and contributes to the respiratory depression associated with severe and prolonged hypoxia. This process is perhaps more important early in life and provides a rationale for the use of caffeine and theophylline in neonatal apnea.

Vasodilation was among the first actions of adenosine to be recognized, but it has been only recently that the pulmonary vasodilatory effects of adenosine were explored. There is no agreement on the degree by which the endothelium contributes to adenosine-induced pulmonary vasodilation, but it is clear that adenosine can induce vasodilation even in the absence of endothelium. The receptor subtype involved, in the few studies where the issue has been explored, appears to be A_{2B}. It is appealing to propose intravenous adenosine as a selective pulmonary vasodilator because it would be cleared by the lung before it reaches the systemic circulation. Adenosine is indeed an effective pulmonary vasodilator in patients with primary and secondary forms of pulmonary hypertension and is commonly used to predict which patients will respond to oral vasodilators. However, in humans intravenous adenosine is not a truly selective pulmonary vasodilator because doses required to produce pulmonary vasodilation also increase cardiac output and decrease systemic vascular resistance. This occurs despite much higher concentrations of adenosine achieved in the pulmonary circulation compared to the systemic circulation and could be explained if the pulmonary circulation is less sensitive to the vasodilator effects of adenosine, for example, because it is mediated by low affinity of A_{2B} receptors.

There is substantial evidence that adenosine plays a role in asthma. Inhaled adenosine provokes bronchoconstriction in asthmatic but not in normal subjects; acute bronchoconstriction results in accumulation of adenosine in bronchoalveolar lavage fluid, and blockade of adenosine receptors with theophylline and enprofylline is effective in the treatment of asthma. Adenosine-induced bronchoconstriction is associated with mast cell activation in humans. The appreciation of the potential role of A_{2B} receptors in the pathogenesis of asthma raises the possibility that A_{2B} receptors could the target for drug development. More studies are needed to determine the relative contribution of A_1 and A_{2B} receptors in asthma.

References

Abebe W, Mustafa SJ (1998) A_1 adenosine receptor-mediated Ins(1,4,5)P3 generation in allergic rabbit airway smooth muscle. Am J Physiol 275:L990–L997

Ali S, Metzger WJ, Olanrewaju HA, Mustafa SJ (1997) Adenosine receptor-mediated relaxation of rabbit airway smooth muscle: a role for nitric oxide. Am J Physiol 273:L581–L587

Ali S, Mustafa SJ, Metzger WJ (1994a) Adenosine receptor-mediated bronchoconstriction and bronchial hyperresponsiveness in allergic rabbit model. Am J Physiol 94:L271–L277

Ali S, Mustafa SJ, Metzger WJ (1994b) Adenosine-induced bronchoconstriction and contraction of airway smooth muscle from allergic rabbits with late-phase airway

obstruction: evidence for an inducible adenosine A_1 receptor. J Pharmacol Exp Ther 94:1328–1334
Aranda JV, Turmen T (1979) Methylxanthines in apnea of prematurity. Clin Perinatol 6:87–108
Auchampach JA, Jin X, Wan TC, Caughey GH, Linden J (1997) Canine mast cell adenosine receptors: Cloning and expression of the A_3 receptors and evidence that degranulation is mediated by the A_{2B} receptor. Mol Pharmacol 52:846–860
Avital A, Jansen AH, Sitar DS, Chernick (1993) Influence of prolonged adenosine receptor blockade on fetal sleep and breathing patterns. Respir Physiol 91:227–236
Bakhle YS, Chelliah R (1983a) Metabolism and uptake of adenosine in rat isolated lung and its inhibition. Br J Pharmacol 79:509–515
Bakhle YS, Chelliah R (1983b) Effect of streptozotocin-induced diabetes on the metabolism of ADP, AMP and adenosine in the pulmonary circulation of rat isolated lung. Diabetologia 24:455–459
Barker PM, Gatzy JT (1998) Effects of adenosine, ATP, and UTP on chloride secretion by epithelia explanted from fetal rat lung. Pediatric Research 43:652–659
Barnes PJ, Chung KF, Page CP (1998) Inflammatory mediators of asthma: An update. Pharmacol Rev 50:515–596
Barraco RA, El-Ridi MR, Ergene E, Phillis JW (1991) Adenosine receptor subtypes in the brainstem mediate distinct cardiovascular response patterns. Brain Res Bull 26:59–84
Barraco RA, El-Ridi MR, Parizon M (1990) The adenosine analog, 5′-*N*-ethylcarboxamidoadenosine, exerts mixed agonist action on cardiorespiratory parameters in the intact but not decerebrate rat following microinjections into the nucleus tractus solitarius. Brain Res 530:54–72
Barraco RA, Janusz CA, Schoener EP, Simpson LL (1990) Cardiorespiratory function is altered by picomole injections of 5′-*N*-ethylcarboxamidoadenosine into the nucleus tractus solitarius of rats. Brain Res 507:234–246
Barraco RA, Schoener EP, Polasek PM, Simpson LL, Janusz, CJ, Parizon M (1988) Respiratory effects of cyclic AMP following injection into the nucleus tractus solitarius of rats. Neuropharmacol 27:1285–1294
Biaggioni I (1992) Contrasting excitatory and inhibitory effects of adenosine in blood pressure regulation. Hypertension 20:457–465
Biaggioni I, Killian TJ, Mosqueda-Garcia R, Robertson RM, Robertson D (1991) Adenosine increases sympathetic nerve traffic in humans. Circulation 83: 1668–1675
Biaggioni I, King LS, Enayat N, Robertson D, Newman JH (1989) Adenosine produces pulmonary vasoconstriction in sheep. Evidence for thromboxane A_2 prostaglandin endoperoxide receptor activation. Circ Res 65:1516–1525
Biaggioni I, Mosqueda-Garcia R (1995) Adenosine in cardiovascular homeostasis and the pharmacological control of its activity. In Hypertension: Pathophysiology, Management and Diagnosis (Edited by Laragh J. and Brenner B.M.), p 1125. Raven Press, New York
Biaggioni I, Olafsson B, Robertson RM, Hollister AS, Robertson D (1987) Cardiovascular and respiratory effects of adenosine in conscious man: Evidence for chemoreceptor activation. Circ Res 61:779–786
Bissonnette JM, Hohimer AR, Chao CR, Knopp SJ, Notoroberto NF (1990) Theophylline stimulates fetal breathing movements during hypoxia. Pediatric Research 28:83–86
Bissonnette JM, Hohimer AR, Knopp SJ (1991) The effect of centrally administered adenosine on fetal breathing movements. Respiration Physiology 84:273–285
Björck T, Gustafsson LE, Dahlën SE (1992) Isolated bronchi from asthmatics are hyperresponsive to adenosine, which apparently acts indirectly by liberation of leukotrienes and histamine. Am Rev Respir Dis 145:1087–1091
Blanchard PW, Cote A, Hobbs S, Foulon P, Aranda JV, Bureau MA (1986) Abolition of ventilatory response to caffeine in chemodenervated lambs. J Appl Physiol 61:133–137

Bottiglieri DF, Kost CK, Jr, Jackson EK (1990) Angiotensin II-induced [^{3}H]adenosine release from in situ rat lung. J Cardiovasc Pharmacol 16:101–106
Bottiglieri DF, Robertson D, Jackson EK (1988) A comparison of nucleoside transport and metabolism in hypertensive and normotensive rats. Hypertension 12:102–107
Bowlin TL, Borcherding DR, Edwards CK, McWhinney CD (1997) Adenosine A_3 receptor agonists inhibit murine macrophage tumor necrosis factor-alpha production in vitro and in vivo. Cell Mol Biol 43:345–349
Bradley JD, Zanaboni PB, Dahms TE, Webster RO (1991) Vascular resistance and Kf in normal and PMA-injured rabbit lungs: effects of adenosine. J Appl Physiol 71:417–424
Bradley JD, Zanaboni PB, Webster RO, Baudendistel LJ, Dahms TE (1991) Adenosine prevents phorbol ester injury in rabbit lungs: role of leukotrienes and TNF. J Appl Physiol 71:1949–1955
Burr D, Sinclair JD (1988) The effect of adenosine on respiratory chemosensitivity in the awake rat. Respiration Physiology 72:47–57
Cheng DY, DeWitt BJ, Suzuki F, Neely CF, Kadowitz PJ (1996) Adenosine A_1 and A_2 receptors mediate tone-dependent responses in feline pulmonary vascular bed. Am J Physiol 270:H200–H207
Church MK, Hughes PJ, Vardey CJ (1986) Studies on the receptor mediating cyclic AMP-independent enhancement by adenosine by IgE-dependent mediator release from rat mast cells. Br J Pharmacol 87:233–242
Clarke H, Cushley MJ, Persson CG, Holgate ST (1989) The protective effects of intravenous theophylline and enprofylline against histamine- and adenosine 5′-monophosphate-provoked bronchoconstriction: implications for the mechanisms of action of xanthine derivatives in asthma. Pulm Pharmacol Ther 2:147–154
Costa F, Biaggioni I (1993) Adenosine activates forearm afferents producing sympathetic stimulation in humans. J Pharmacol Exp Ther 267:1369–1374
Costa F, Biaggioni I (1994) Role of adenosine in the sympathetic activation produced by isometric exercise in humans. J Clin Invest 93:1354–1660
Costa F, Biaggioni I (1998) Role of nitric oxide in adenosine-induced vasodilation in humans. Hypertension 31:1061–1064
Crimi N, Palermo F, Oliveri R, Maccarrone C, Palermo B, Vancheri C, Polosa R, Mistretta A (1988) Enhancing effect of dipyridamole inhalation on adenosine-induced bronchospasm in asthmatic patients. Allergy 88:179–183
Crimi N, Palermo F, Oliveri R, Polosa R, Settinieri I, Mistretta A (1992) Protective effects of inhaled ipratropium bromide on bronchoconstriction induced by adenosine and methacholine in asthma. Eur Respir J 5:560–565
Crimi N, Polosa R, Magri S, Prosperini G, Milazzo V L, Santonocito G, Mistretta A (1995) Inhaled lysine acetylsalicylate (L-ASA) attenuates the bronchoconstrictor response to adenosine 5′-monophosphate (AMP) in asthmatic subjects. Eur Respir J 95:905–912
Crimi N, Polosa R, Mistretta A (1993) Mechanisms of the effector function of adenosine in asthma and rhinitis. Drug Develop Res 28:322–327
Cushley MJ, Tattersfield AE, Holgate ST (1983) Inhaled adenosine and guanosine on airway resistance in normal and asthmatic subjects. Br J Clin Pharmacol 83:161–165
Darnall RA, Bruce RD (1987) Effects of adenosine and xanthine derivatives on breathing during acute hypoxia in the anesthetized newborn piglet. Pediatr Pulmonol 3:110–116
Dong XW, Feldman JL (1995) Modulation of inspiratory drive to phrenic motoneurons by presynaptic adenosine A1 receptors. J Neurosci 15:3458–3467
Driver AG, Kukoly CA, Ali S, Mustafa SJ (1993) Adenosine in bronchoalveolar lavage fluid in asthma. Am Rev Respir Dis 148:91–97
Drury AW, Szent-Györgyi A (1929) The physiological activity of adenosine compounds with a special reference to their action upon the mammalian heart. J Physiol 68:213–237

Eidelman O, Guay-Broder C, van Galen PJ, Jacobson KA, Fox C, Turner RJ, Cabantchik ZI, Pollard HB (1992) A_1 adenosine-receptor antagonists activate chloride efflux from cystic fibrosis cells. Proc Natl Acad Sci 89:5562–5566
el-Hashim A, D'Agostino B, Matera MG, Page C (1996) Characterization of adenosine receptors involved in adenosine-induced bronchoconstriction in allergic rabbits. Br J Pharmacol 119:1262–1268
Eldridge FL, Millhorn DE, Kiley JP (1985) Antagonism by theophylline of respiratory inhibition induced by adenosine. J Appl Physiol 59:1428–1433
Elnazir B, Marshall JM, Kumar P (1996) Postnatal development of the pattern of respiratory and cardiovascular response to systemic hypoxia in the piglet: the roles of adenosine. J Physiol 492:573–585
Erion MD, Weisner JD, Rosengen S, Ugarkar B, Boyer SH, Tsuchiya M, Nakane M, Pettersen BA, Nagahisa A. Therapeutic potential of adenosine kinase inhibitors as analgesic agents. Drug Dev Res 50:22
Farmer SG, Canning BJ, Wilkins DE (1988) Adenosine receptor-mediated contraction and relaxation of guinea-pig isolated tracheal smooth muscle: effects of adenosine antagonists. Br J Pharmacol 95:371–378
Feoktistov I, Biaggioni I (1993) Characterization of adenosine receptors in human erythroleukemia cells. Further evidence for heterogeneity of adenosine A_2 receptors. Mol Pharmacol 43:909–914
Feoktistov I, Biaggioni I (1995) Adenosine A_{2b} receptors evoke interleukin-8 secretion in human mast cells. An enprofylline-sensitive mechanism with implications for asthma. J Clin Invest 96:1979–1986
Feoktistov I, Biaggioni I (1996) Role of adenosine in asthma. Drug Dev Res 39:333–336
Feoktistov I, Biaggioni I (1997) Adenosine A_{2B} receptors. Pharmacol Rev 49:381–402
Feoktistov I, Goldstein AE, Biaggioni I (1999) Role of p38 mitogen-activated protein kinase and extracellular signal-regulated protein kinase kinase in adenosine A_{2B} receptor-mediated interleukin-8 production in human mast cells. Mol Pharmacol 55:726–735
Feoktistov I, Polosa R, Holgate ST, Biaggioni I (1998a) Adenosine A_{2B} receptors – a novel therapeutic target in asthma? Trends Pharmacol Sci 19:148–153
Feoktistov I, Wells JN, Biaggioni I (1998b) Adenosine A_{2B} receptors as therapeutic targets. Drug Dev Res 45:198–206
Firestein GS, Boyle D, Bullough DA, Gruber HE, Sajjadi FG, Montag A, Sambol B, Mullane KM (1994) Protective effect of an adenosine kinase inhibitor in septic shock. J Immunol 94:5853–5859
Fozard JR, Pfannkuche HJ, Schuurman HJ (1996) Mast cell degranulation following adenosine A_3 receptor activation in rats. Eur J Pharmacol 298:293–297
Fozard JR, Hannon JP (2000) Species differences in adenosine receptor-mediated bronchoconstricter respinses. Clin Exp Allergy 30:1213–1220
Fredholm BB, Brodin K, Strandberg K (1979) On the mechanism of relaxation of tracheal muscle by theophylline and other cyclic nucleotide phosphodiesterase inhibitors. Acta Pharmacol Toxicol 45:336–344
Fullerton DA, Kirson LE, Jones SD, McIntyre RC, Jr. (1996) Adenosine is a selective pulmonary vasodilator in cardiac surgical patients. Chest 109:41–46
Ghai G, Zimmerman MB, Hopkins MF (1987) Evidence for A_1 and A_2 adenosine receptors in guinea pig trachea. Life Sci 41:1215–1224
Gilfillan AM, Rooney SA (1987) Functional evidence for adenosine A_2 receptor regulation of phosphatidylcholine secretion in cultured type II pneumocytes. J Pharmacol Exp Ther 241:907–914
Gobran LI, Rooney SA (1990) Adenosine A_1 receptor-mediated inhibition of surfactant secretion in rat type II pneumocytes. Am J Physiol 258:L45–L51
Gottlieb JE, Peake MD, Sylvester JT (1984) Adenosine and hypoxic pulmonary vasodilation. Am J Physiol 247:H541–H547

Griese M, Gobran LI, Douglas JS, Rooney SA (1991) Adenosine A_2-receptor-mediated phosphatidylcholine secretion in type II pneumocytes. Am J Physiol 260:L52–L60
Griffiths TL, Warren SJ, Chant AD, Holgate ST (1990) Ventilatory effects of hypoxia and adenosine infusion in patients after bilateral carotid endarterectomy. Clin Sci 78:25–31
Grundstrom N, Andersson RG, Wikberg JE (1981) Investigations on possible presynaptic effect of adenosine and noradrenaline on cholinergic neurotransmission in guinea pig trachea. Acta Pharmacol Toxicol 49:158–160
Hales CA, Kazemi H (1974) Hypoxic vascular response of the lung: effect of aminophylline and epinephrine. Am Rev Respir Dis 110:126–132
Hannon JP, Pfannkuche HJ, Fozard JR (1995) A role for mast cells in adenosine A_3 receptor-mediated hypotension in the rat. Br J Pharmacol 115:945–952
Hasday JD, Sitrin RG (1987) Adenosine receptors on rabbit alveolar macrophages: binding characteristics and effects on cellular function. J Lab Clin Med 110: 264–273
Hasday JD, Sitrin RG (1987) Dipyridamole stimulates urokinase production and suppresses procoagulant activity of rabbit alveolar macrophages: a possible mechanism of antithrombotic action. Blood 69:660–667
Haynes J Jr, Obiako B, Thompson WJ, Downey J (1995) Adenosine-induced vasodilation: receptor characterization in pulmonary circulation. Am J Physiol 268:H-1862–H-1868
Haywood GA, Sneddon JF, Bashir Y, Jennison SH, Gray HH, McKenna WJ (1992) Adenosine infusion for the reversal of pulmonary vasoconstriction in biventricular failure. A good test but a poor therapy. Circulation 86:896–902
Headrick JP, Berne RM (1990) Endothelium-dependent and -independent relaxations to adenosine in guinea pig aorta. Am J Physiol 259:H62–H67
Hedner T, Hedner J, Wessberg P, Jonasson T (1982) Regulation of breathing in the rat: Indications for a role of central adenosine mechanisms. Neurosci Lett 33: 147–151
Hellewell PG, Pearson JD (1983) Metabolism of circulating adenosine by the porcine isolated perfused lung. Circ Res 53:1–7
Herlenius E, Lagercrantz H, Yamamoto Y (1997) Adenosine modulates inspiratory neurons and the respiratory pattern in the brainstem of neonatal rats. Pediatric Research 42:46–53
Hirasawa N, Santini F, Beaven MA (1995) Activation of the mitogen-activated protein kinase/cytosolic phospholipase A_2 pathway in a rat mast cell line. Indications of different pathways for release of arachidonic acid and secretory granules. J Immunol 154:5391–5402
Holgate ST, Church MK, Polosa R (1991) Adenosine: a positive modulator of airway inflammation in asthma. Ann N Y Acad Sci 629:227–236
Hong JL, Ho CY, Kwong K, Lee LY (1998) Activation of pulmonary C fibres by adenosine in anaesthetized rats: role of adenosine A_1 receptors. J Physiol 508: 109–118
Horowitz DA, Jablonski WJ, Mehta KA (1982) Apnea associated with theophylline withdrawal in a term neonate. American Journal of Diseases of Children 136:73–74
Hughes PJ, Holgate ST, Roath S, Church MK (1983) The relationship between cyclic AMP changes and histamine release from basophil-rich human leucocytes. Biochem Pharmacol 32:2557–2563
Hunt CE, Brouillette RT, Hanson D (1983) Theophylline improves pneumogram abnormalities in infants at risk for sudden infant death syndrome. Journal of Pediatrics 103:969–974
Ishizuka T, Kawasome H, Terada N, Takeda K, Gerwins P, Keller GM, Johnson GL, Gelfand EW (1998) Stem cell factor augments Fc epsilon RI-mediated TNF-alpha production and stimulates MAP kinases via a different pathway in MC/9 mast cells. J Immunol 161:3624–3630

Jacobson KA, Park KS, Jiang JL, Kim YC, Olah ME, Stiles GL, Ji XD (1997) Pharmacological characterization of novel A_3 adenosine receptor-selective antagonists. Neuropharmacol 36:1157–1165
Jacobson KA, Suzuki F (1996) Recent developments in selective agonists and antagonists acting at purine and pyrimidine receptors. Drug Dev Res 39:289–300
Javaheri S, Parker TJ, Wexler L, Liming JD, Lindower P, Roselle GA (1996) Effect of theophylline on sleep-disordered breathing in heart failure. N Engl J Med 335: 562–567
Jolin A, Myklebust R, Olsen R, Bjertnaes LJ (1994) Adenosine protects ultrastructure of isolated rat lungs against fat emulsion injury. Acta Anaesthesiol Scand 38: 75–81
Kawakami Y, Hartman SE, Holland PM, Cooper JA, Kawakami T (1998) Multiple signaling pathways for the activation of JNK in mast cells: involvement of Bruton's tyrosine kinase, protein kinase C, and JNK kinases, SEK1 and MKK7. J Immunol 161:1795–1802
Kinet JP (1999) The high-affinity IgE receptor (Fc_gRI): from physiology to pathology. Annual Review of Immunology 17:931–972
Knight D, Zheng X, Rocchini C, Jacobson M, Bai T, Walker B (1997) Adenosine A_3 receptor stimulation inhibits migration of human eosinophils. J Leukocyte Biol 62:465–468
Kohno Y, Ji X, Mawhorter SD, Koshiba M, Jacobson KA (1996) Activation of A_3 adenosine receptors on human eosinophils elevates intracellular calcium. Blood 88:3569–3574
Konduri GG, Garcia DC, Kazzi NJ, Shankaran S (1996) Adenosine infusion improves oxygenation in term infants with respiratory failure. Pediatrics 97:295–300
Konduri GG, Gervasio CT, Theodorou AA (1993) Role of adenosine triphosphate and adenosine in oxygen-induced pulmonary vasodilation in fetal lambs. Pediatric Research 33:533–539
Koos BJ, Chao A, Doany W (1992) Adenosine stimulates breathing in fetal sheep with brain stem section. J Appl Physiol 72:94–99
Koos BJ, Chau A (1998) Fetal cardiovascular and breathing responses to an adenosine A_{2a} receptor agonist in sheep. Am J Physiol 274:R152–R159
Koos BJ, Doany W (1991) Role of plasma adenosine in breathing responses to hypoxia in fetal sheep. J Devel Physiol 16:81–85
Koos BJ, Mason BA, Punla O, Adinolfi AM (1994) Hypoxic inhibition of breathing in fetal sheep: relationship to brain adenosine concentrations. J Appl Physiol 77:2734–2739
Koos BJ, Matsuda K (1990) Fetal breathing, sleep state, and cardiovascular responses to adenosine in sheep. J Appl Physiol 68:489–495
Kwong K, Hong JL, Morton RF, Lee LY (1998) Role of pulmonary C fibers in adenosine-induced respiratory inhibition in anesthetized rats. J Appl Physiol 84:417–424
Lagercrantz H, Yamamoto Y, Fredholm BB, Prabhakar NR, von Euler C (1984) Adenosine analogues depress ventilation in rabbit neonates. Theophylline stimulation of respiration via adenosine receptors? Pediatric Research 18:387–390
Laitinen A, Laitinen LA (1994) Airway morphology: epithelium/basement membrane. Am J Resp Crit Care Med 150:S14–S17
Laitinen LA, Laitinen A (1988) Mucosal inflammation and bronchial hyperreactivity. Eur Respir J 1:488–489
Lazarowski ER, Mason SJ, Clarke L, Harden TK, Boucher RC (1992) Adenosine receptors on human airway epithelia and their relationship to chloride secretion. Br J Pharmacol 106:774–782
LeMessurier DH (1936) The site of action of caffeine as a respiratory stimulant. J Pharmacol Exp Ther 57:458–463
Lichtenstein LM, Bochner BS (1991) The role of basophils in asthma. Ann N Y Acad Sci 629:48–61

Lopes JM, Davis GM, Mullahoo K, Aranda JV (1994) Role of adenosine in the hypoxic ventilatory response of the newborn piglet. Pediatric Pulmonol 17:50–55
Losinski A, Alexander SP (1995) Adenosine receptor-mediated relaxation of guinea-pig precontracted, isolated trachea. Br J Pharmacol 116:2425–2428
Luttrell LM, Daaka Y, Lefkowitz RJ (1999) Regulation of tyrosine kinase cascades by G-protein-coupled receptors. Curr Opin Cell Biol 11:177–183
Mann JS, Cushley MJ, Holgate ST (1985) Adenosine-induced bronchoconstriction in asthma. Role of parasympathetic stimulation and adrenergic inhibition. Am Rev Respir Dis 85:1–6
Mann JS, Holgate ST (1985) Specific antagonism of adenosine-induced bronchoconstriction is asthma by oral theophylline. Br J Clin Pharmacol 19:685–692
Mann JS, Holgate ST, Renwick AG, Cushley MJ (1986) Airway effects of purine nucleosides and nucleotides and release with bronchial provocation in asthma. J Appl Physiol 86:1667–1676
Manzini S, Ballati L (1990) 2-chloroadenosine induction of vagally-mediated and atropine-resistant bronchoconstrictor responses in anaesthetised guinea pigs. Br J Pharmacol 100:251–256
Marone G, Cirillo R, Genovese A, Marino O, Quattrin S (1989) Human basophil/mast cell releasability. VII. Heterogeneity of the effect of adenosine on mediator secretion. Life Sci 45:1745–1754
Marone G, Vigorita S, Antonelli C, Torella G, Genovese, Condorelli M (1985) Evidence for an adenosine A_2/Ra receptor on human basophils. Life Sci 36:339–345
Marquardt DL, Parker CW, Sullivan TJ (1978) Potentiation of mast cell mediator release by adenosine. J Immunol 120:871–878
Marquardt DL, Walker LL (1990) Modulation of mast cell responses to adenosine by agents that alter protein kinase C activity. Biochem Pharmacol 39:1929–1934
Marquardt DL, Walker LL (1994) Inhibition of protein kinase A fails to alter mast cell adenosine responsiveness. Agents Actions 43:7–12
Marquardt DL, Walker LL, Heinemann S (1994) Cloning of two adenosine receptor subtypes from mouse bone marrow-derived mast cells. J Immunol 152:4508–4515
Maxwell DL, Fuller RW, Conradson T-B, Dixon CMS, Aber V, Hughes JMB, Barnes PJ (1987) Contrasting effects of two xanthines, theophylline and enprofylline, on the cardiorespiratory stimulation of infused adenosine in man. Acta Physiol Scand 131:459–465
McCormack DG, Clarke B, Barnes PJ (1989) Characterization of adenosine receptors in human pulmonary arteries. Am J Physiol 256:H41–H46
McCoy DE, Schwiebert EM, Karlson KH, Spielman WS, Stanton BA (1995) Identification and function of A_1 adenosine receptors in normal and cystic fibrosis human airway epithelial cells. Am J Physiol 268:C1520–C1527
McQueen DS, Ribeiro JA (1986) Pharmacological characterization of the receptor involved in chemoexcitation induced by adenosine. Br J Pharmacol 88:615–620
Meade CJ, Mierau J, Leon I, Ensinger HA (1996) In vivo role of the adenosine A_3 receptor: N6-2-(4-aminophenyl)ethyladenosine induces bronchospasm in BDE rats by a neurally mediated mechanism involving cells resembling mast cells. J Pharmacol Exp Ther 279:1148–1156
Millhorn DE, Eldridge FL, Kiley JP, Waldrop TG (1984) Prolonged inhibition of respiration following acute hypoxia in glomectomized cats. Respiration Physiology 57:331–340
Monteiro EC, Ribeiro JA (1987) Ventilatory effects of adenosine modulated by carotid body chemoreceptors in the rat. Naunyn-Schmiedeberg's Arch Pharmacol 335: 143–148
Morgan JM, McCormack DG, Griffiths MJ, Morgan CJ, Barnes PJ, Evans TW (1991) Adenosine as a vasodilator in primary pulmonary hypertension. Circulation 84:1145–1149
Moser GH, Schrader J, Deussen A (1989) Turnover of adenosine in plasma of human and dog blood. Am J Physiol 256:C799–C806

Mosqueda-Garcia R, Tseng CJ, Appalsamy M, Beck C, Robertson D (1991) Cardiovascular excitatory effects of adenosine in the nucleus of the solitary tract. Hypertension 18:494–502
Neely CF, Haile DM, Cahill BE, Kadowitz PJ (1991) Adenosine and ATP produce vasoconstriction in the feline pulmonary vascular bed by different mechanisms. J Pharmacol Exp Ther 258:753–761
Neely CF, Kadowitz P, Lippton H, Neiman M, Hyman A (1989) Adenosine does not mediate the pulmonary vasodilator response to ATP in the feline pulmonary vascular bed. J Pharmacol Exp Ther 250:170–170
Neely CF, Pellack D, Kadowitz P (1989) A mechanism to explain the tone-dependent responses of the feline pulmonary vascular bed to adenosine. FASEB J. 3, A280(Abstract)
Nussbacher A, Arie S, Kalil R, Horta P, Feldman MD, Bellotti G, Pileggi F, Ellis M, Johnson WH, Camarano GB (1995) Mechanism of adenosine-induced elevation of pulmonary capillary wedge pressure in humans. Circulation 92:371–379
Nyce JW, Metzger WJ (1997) DNA antisense therapy for asthma in an animal model. Nature 97:721–725
Patole S, Lee J, Buettner P, Whitehall J (1998) Improved oxygenation following adenosine infusion in persistent pulmonary hypertension of the newborn. Biology of the Neonate 74:345–350
Paty PS, Sherman PF, Shepard JM, Malik AB, Kaplan JE (1992) Role of adenosine in platelet-mediated reduction in pulmonary vascular permeability. Am J Physiol 262:H771–H777
Pauwels R, Van Der Straeten M (1987) An animal model for the adenosine induced bronchoconstriction. Am Rev Respir Dis 136:374–378
Pauwels RA, Germonpre PR, Kips JC, Joos GF (1995) Genetic control of indirect airway responsiveness in the rat. Clin Exp Allergy 25:55–60
Pauwels RA, Joos GF (1995) Characterization of the adenosine receptors in the airways. Arch Int Pharmacodyn Ther 329:151–160
Peachell PT, Columbo M, Kagey-Sobotka A, Lichtenstein LM, Marone G (1988) Adenosine potentiates mediator release from human lung mast cells. Am Rev Respir Dis 138:1143–1151
Peachell PT, Lichtenstein LM, Schleimer RP (1991) Differential regulation of human basophil and lung mast cell function by adenosine. J Pharmacol Exp Ther 256:717–726
Peachell PT, MacGlashan DW J, Lichtenstein LM, Schleimer RP (1988) Regulation of human basophil and lung mast cell function by cyclic adenosine monophosphate. J Immunol 140:571–579
Phillips GD, Ng WH, Church MK, Holgate ST (1990) The response of plasma histamine to bronchoprovocation with methacholine, adenosine 5′-monophosphate, and allergen in atopic nonasthmatic subjects. Am Rev Respir Dis 90:9–13
Phillips GD, Rafferty P, Beasley R, Holgate ST (1987) Effect of oral terfenadine on the bronchoconstrictor response to inhaled histamine and adenosine 5′-monophosphate in non-atopic asthma. Thorax 87:939–945
Phillips GD, Scott VL, Richards R, Holgate ST (1989) Effect of nedrocromil sodium and sodium cromoglycate against bronchoconstriction induced by inhaled adenosine 5′-monophosphate. Eur Respir J 2:210–217
Phillis JW, Scislo TJ, O'Leary DS (1997) Purines and the nucleus tractus solitarius: effects on cardiovascular and respiratory function. Clin Exp Pharmacol Physiol 24:738–742
Polosa R, Crimi N, Raccuglia D, Pagano C, Walters C, Church MK, Mistretta A (1994) Histamine changes to nasal provocation with AMP and GMP in rhinitic and normal subjects. J Allergy Clin Immunol 93:329
Polosa R, Ng WH, Crimi N, Vanceri C, Holgate ST, Church MK, Mistretta A (1995) Release of mast cell-derived mediators after endobronchial adenosine challenge in asthma. Am J Resp Crit Care Med 151:624–629

Polosa R, Rajakulasingam K, Church MK, Holgate ST (1992) Repeated inhalation of bradykinin attenuates adenosine 5′-monophosphate (AMP) induced bronchoconstriction in asthmatic airways. Eur Respir J 92:700–706

Polosa R, Santonocito G, Magri S, Paolino G, Armato F, Pagano C, Crimi N (1997) Neutral endopeptidase inhibition with inhaled phosphoramidon: no effect on bronchial responsiveness to adenosine 5′-monophosphate (AMP) in asthma. Eur Respir J 10:2460–2464

Post MJ, Te Biesebeek JD, Wemer J, van Rooij HH, Porsius AJ (1990) Adenosine enhances antigen-induced bronchoconstriction and histamine release in rat isolated lungs. Agents Actions 30:30–33

Pradalier A (1993) Late-phase reaction in asthma: basic mechanisms. Int Arch Allergy Immunol 101:322–325

Rafferty P, Beasley CR. Holgate ST (1987) The contribution of histamine to immediate bronchoconstriction provoked by inhaled allergen and adenosine-5′-monophosphate in atopic asthma. Am Rev Respir Dis 136:369–373

Ramkumar V, Stiles GL, Beaven MA, Ali H (1993) The A_3 adenosine receptor is the unique adenosine receptor which facilitates release of allergic mediators in mast cells. J Biol Chem 268:16887–16890

Redington AE, Polosa R, Walls AF, Howarth PH, Holgate ST (1995) Role of mast cells and basophils in asthma. Chemical Immunology 62:22–59

Reis DJ, Granata AR, Perrone MH, Talman W (1981) Evidence that glutamic acid is the neurotransmitter of baroreceptor afferents terminating in the nucleus tractus solitarius (NTS). J Auton Nerv Syst 3:321–334

Riordan JR, Rommens JM, Kerem B, Alon N, Rozmahel R, Grzelczak Z, Zielenski J, Lok S, Plavsic N, Chou JL (1989) Identification of the cystic fibrosis gene: cloning and characterization of complementary DNA. Science 245:1066–1073

Robeva AS, Woodard RL, Jin X, Gao Z, Bhattacharya S, Taylor HE, Rosin DL, Linden J (1996) Molecular characterization of recombinant human adenosine receptors. Drug Dev Res 39:243–252

Rooney SA, Gobran LI (1988) Adenosine and leukotrienes have a regulatory role in lung surfactant secretion in the newborn rabbit. Bioch Biophys Acta 960:98–106

Rugolo M, Mastrocola T, Whorle C, Rasola A, Gruenert DC, Romeo G, Galietta LJ (1993) ATP and A_1 adenosine receptor agonists mobilize intracellular calcium and activate K^+ and Cl^- currents in normal and cystic fibrosis airway epithelial cells. J Biol Chem 268:24779–24784

Runold M, Cherniack NS, Prabhakar NR (1990a) Effect of adenosine on isolated and superfused cat carotid body activity. Neurosci Lett 113:111–114

Runold M, Cherniack NS, Prabhakar NR (1990b) Effect of adenosine on chemosensory activity of the cat aortic body. Resp Physiol 80:299–306

Sakai N, Tamaoki J, Kobayashi K, Katayama M, Takizawa T (1989) Adenosine potentiates neurally- and histamine-induced contraction of canine airway smooth muscle. Int Arch Allergy Appl Immunol 90:280–284

Schmidt C, Bellingham MC, Richter DW (1995) Adenosinergic modulation of respiratory neurones and hypoxic responses in the anaesthetized cat. J Physiol 483:769–781

Schrader BJ, Inbar S, Kaufmann L, Vestal RE, Rich S (1992) Comparison of the effects of adenosine and nifedipine in pulmonary hypertension. J Am Coll Cardiol 19:1060–1064

Schwartz LB (1990) Tryptase, a mediator of human mast cells. J Allergy Clin Immunol 86:594–598

Steinhorn RH, Morin FC, Van Wylen DG, Gugino SF, Giese EC, Russell JA (1994) Endothelium-dependent relaxations to adenosine in juvenile rabbit pulmonary arteries and veins. Am J Physiol 266:H2001–H2006

Strek ME, Leff AR (1997) Eosinophils. In Asthma (Edited by Barnes PJ, Grunstein MM, Leff AR, Woolcock AJ), p. 399. Lippincott-Raven, Philadelphia

Strohl KP, Haxhiu MA, Dick TE (1998) Respiratory Control. In Pulmonary and Critical Care Medicine (Edited by Bone R.C.Mosby-Year Book, Inc., St. Louis
Suzuki H, Takei M, Nakahata T, Fukamachi H (1998) Inhibitory effect of adenosine on degranulation of human cultured mast cells upon cross-linking of Fc epsilon RI. Biochem Biophys Res Commun 242:697–702
Szentmiklosi AJ, Ujfalusi A, Cseppento A, Nosztray K, Kovacs P, Szabo JZ (1995) Adenosine receptors mediate both contractile and relaxant effects of adenosine in main pulmonary artery of guinea pigs. Naunyn-Schmiedebergs Arch Pharmacol 351:417–425
Thiel M, Kreimeier U, Holzer K, Moritz S, Peter K, Messmer K (1998) Effects of adenosine on cardiopulmonary functions and oxygen-derived variables during endotoxemia. Critical Care Medicine 26:322–337
Thomas T, Marshall JM (1993) The role of adenosine in hypoxic pulmonary vasoconstriction in the anaesthetized rat. Exp Physiol 78:541–543
Thorne JR, Danahay H, Broadley KJ (1996) Analysis of the bronchoconstrictor responses to adenosine receptor agonists in sensitized guinea-pig lungs and trachea. Eur J Pharmacol 316:263–271
Tseng CJ, Biaggioni I, Appalsamy M, Robertson D (1988) Purinergic receptors in the brainstem mediate hypotension and bradycardia. Hypertension 11:191–197
Utterback DB, Staples ED, White SE, Hill JA, Belardinelli L (1994) Basis for the selective reduction of pulmonary vascular resistance in humans during infusion of adenosine. J Appl Physiol 76:724–730
Vincent JL, Brimioulle S, Berre J, Kahn RJ (1985) Prevention of the adult respiratory distress syndrome with dipyridamole. Critical Care Medicine 13:783–785
Vogi A (1927) Euphylline. Wien Klin Worchenschr 40:105
Walday P, Aas P (1991) Prejunctional stimulation of cholinergic nerves in rat airway smooth muscle by an adenosine analogue. Pulm Pharmacol 4:114–119
Walker BA, Jacobson MA, Knight DA, Salvatore CA, Weir T, Zhou D, Bai TR (1997) Adenosine A_3 receptor expression and function in eosinophils. Am J Resp Cell Mol Biol 16:531–537
Watt AH, Buss DC, Routledge PA (1986) Comparison of respiratory effects of intravenous adenosine in neonatal and adult rabbits. Life Sci 39:1617–1622
Watt AH, Reid PG, Stephens MR, Routledge PA (1987) Adenosine-induced respiratory stimulation in man depends on site of infusion. Evidence for an action on the carotid body? Br J Clin Pharmacol 23:486–490
Watt AH, Routledge PA (1985) Adenosine stimulates respiration in man. Br J Clin Pharmacol 20:503–506
Wiklund NP, Cederqvist B, Gustafsson LE (1989) Adenosine enhancement of adrenergic neuroeffector transmission in guinea-pig pulmonary artery. Br J Pharmacol 96:425–433
Wiklund NP, Cederqvist B, Matsuda H, Gustafsson LE (1987) Adenosine can excite pulmonary artery. Acta Physiol Scand 131:477–478
Xaus J, Mirabet M, Lloberas J, Soler C, Lluis C, Franco R, Celada A (1999) IFN-gamma up-regulates the A_{2B} adenosine receptor expression in macrophages: a mechanism of macrophage deactivation. J Immunol 162:3607–3614
Yan S, Laferriere A, Zhang C, Moss IR (1995) Microdialyzed adenosine in nucleus tractus solitarii and ventilatory response to hypoxia in piglets. J Appl Physiol 95:405–410
Yukawa T, Kroegel C, Chanez P, Dent G, Ukena D, Chung KF, Barnes PJ (1989) Effect of theophylline and adenosine on eosinophil function. Am Rev Respir Dis 140:327–333
Zhou Q-Y, Li C, Olah ME, Johnson RA, Stiles GL, Civelli O (1992) Molecular cloning and characterization of an adenosine receptor: The A_3 adenosine receptor. Proc Natl Acad Sci 89:7432–7436

CHAPTER 23

P2 Receptors in the Respiratory System

C. D. WEGNER

A. Introduction

As in other tissues, it is now well established that nucleotides are released into the extracellular fluid of the respiratory tract during normal homeostatic as well as pathologic conditions to act on cell surface P2 purinergic receptors. Both the ligand gated ion channel P2X subtype and the G-protein coupled P2Y subtype have been identified and shown to provide important regulatory functions in essentially all cell components of the respiratory system. Taken together these functions appear to enhance respiratory host defense by accelerating mucociliary clearance, facilitating inflammation and possibly aiding tissue repair (Fig. 1). The evidence that supports each of these functions is reviewed below.

B. Respiratory Epithelium

In addition to providing a physical barrier between the external environment and internal organs, the respiratory epithelium confers three additional crucial functions:

1. The execution and regulation of airway "cleansing" via mucociliary clearance
2. The regulation of inflammation during infection, allergen inhalation and injury
3. The maintenance of open and "lubricated" airspaces via the release of surfactant

P2 receptor activation impacts each of these functions by effects on epithelial ion transport, mucus secretion, ciliary beat frequency, and phosphatidylcholine (surfactant) release. In addition, recruitment of cells from the immune system (macrophages, etc.) have the ability to introduce additional P2 receptor subtypes representing a dynamic situation (DUBYAK, Chap. 25, this volume).

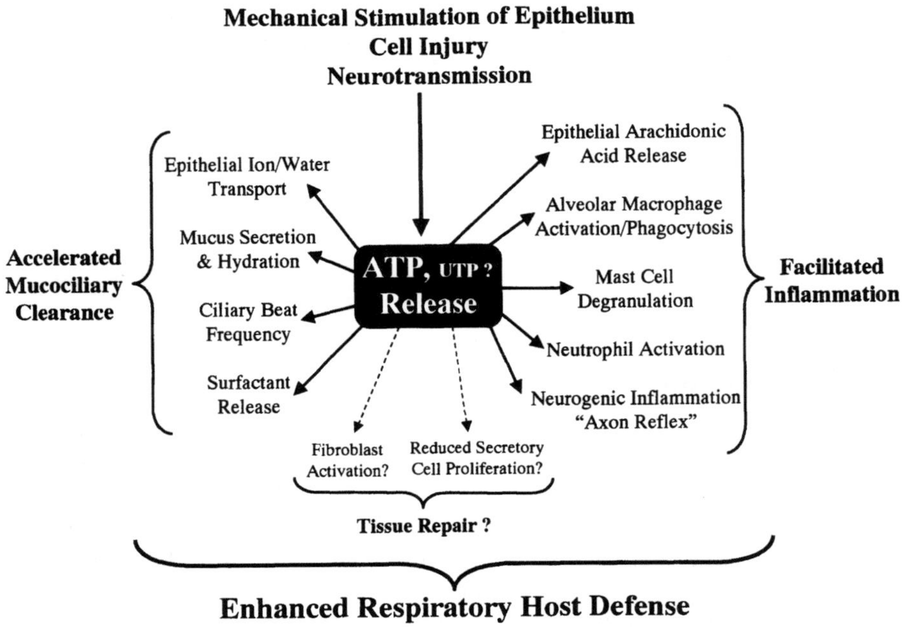

Fig. 1. Stimuli for nucleotide release and summary of functions in the respiratory system

I. Epithelial Ion Transport

1. Cystic Fibrosis

The importance of epithelial ion transport to mucociliary clearance, as well as possibly lung inflammation and defense, have been elucidated in studies comparing respiratory tissue and cells from patients with cystic fibrosis (CF) to those of patients without. The pulmonary symptoms of CF, impaired mucus clearance, neutrophilic inflammation, and chronic residual bacterial infection, result from congenital mutations in a single gene product, the cystic fibrosis transmembrane regulator (CFTR). These mutations result in the inability of the CFTR to incorporate (and, hence, function) into the apical membrane of polarized epithelia. In addition to acting as an adenylate cyclase (cAMP) – mediated chloride ion channel, CFTR seems to be connected to intracellular signal transduction pathway(s) (STUTTS and BOUCHER 1999). The results, whether via impaired chloride conductance, resultant changes in transmembrane potential or intracellular ion composition, and/or altered signal transduction, are critical defects in epithelial function. These defects include:

1. Enhanced water absorption due to decreased Cl^- secretion and increased Na^+ absorption (MALL et al. 1998; MATSUI et al. 1998)
2. Increased production of the potent neutrophil chemoattractant, IL-8 (ZAHM et al. 1997; TABARY et al. 1998)

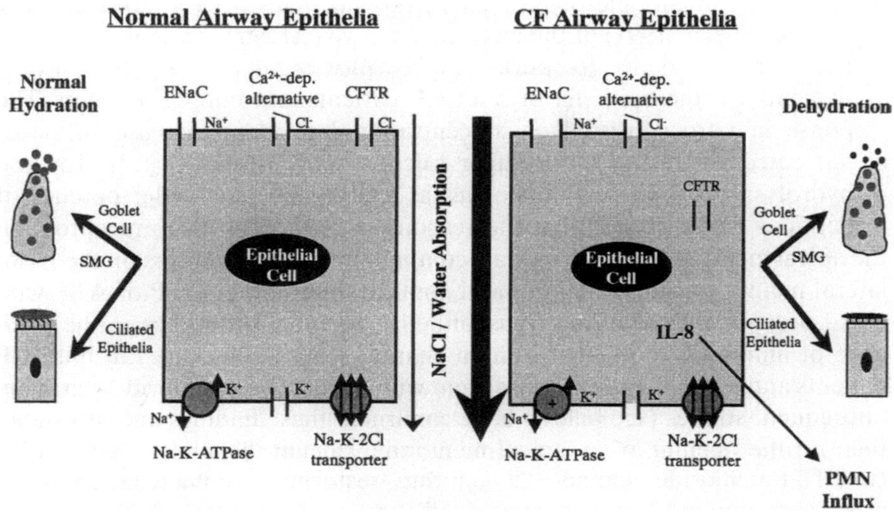

Fig. 2. Key ion channels and transporters as well as their relative function in normal vs cystic fibrosis (CF) airway epithelia. Mutant CFTR is produced but not expressed at the membrane in CF cells. This results in no CFTR function (reduced apical membrane chloride conductance), opening of ENaC, enhanced Na-K-ATPase activity (amplified Na^+ absorption), and increased IL-8 production. The consequences are heightened water absorption and neutrophil infiltration, the former leading to reduced mucus hydration and clearance. See text for details and definitions of abbreviations. CFTR, cystic fibrosis transmembrane regulator; ENaC, epithelial sodium channel; PMN, polymorphonuclear neutrophil; SMG, submucosal glands

3. Decreased release and/or function of endogenous epithelial-derived antibiotic/antiviral products (defensins and nitric oxide) (Smith et al. 1996; Goldman et al. 1997; Kelley and Drumm 1998; Meng et al. 1998)

Enhanced water absorption impairs mucociliary clearance in two ways. First, secreted mucus is less hydrated, making it more difficult to be propelled by cilia, as well as more viscous and glutinous such that it clogs glands, sinuses, airways, and Eustachian tubes (Inglis et al. 1997). Second, the height (depth) of the liquid layer in which the cilia beat is lowered. The optimal height of this so-called "periciliary liquid layer" between the apical membrane of the epithelium and the mucus gel is critical to the efficient propulsion of the mucus gel by cilia (Boucher 1999; Matsui et al. 1998; Stutts and Boucher 1999). Hence, as dramatically emphasized in CF, epithelial ion transport is critical to respiratory defense and function (Fig. 2).

2. P2 Purinergic Receptors Stimulation Increases Water Secretion

Application of nucleotides (e.g., ATP, UTP, ADP and their analogues) to either the apical (also referred to as luminal or mucosal) or basolateral (also referred to as serosal) membrane of primary cultures of human nasal epithelium in vitro (MASON et al. 1991) or the nasal cavity in vivo (KNOWLES et al. 1991) provided the first evidence for purinergic receptor regulation of ion transport equivalently in the epithelia of both CF patients and non-CF patients. The response in vitro consisted of concentration-dependent increases in short circuit current (I_{sc}) and intracellular calcium concentration $[Ca^{2+}]_i$. The use of hydrolysis-resistant ATP analogues as well as the rank order potency of nucleotides demonstrated that the response was through $P2Y_2$ receptors on the apical membrane as well as via a combination of P2 receptors on the basolateral membrane. In vivo, intranasal application of either UTP or ATP, with equal potency, increased the transepithelial potential difference in the presence of amiloride (epithelial sodium channel, ENaC, blockage) in both CF patients and non-CF patients, consistent with an increase in chloride secretion. Subsequent studies (see below) have confirmed these findings and attributed them to the opening of an apical membrane calcium dependent, alternative (to CFTR) chloride channel ($Cl^-_{Ca,a}$); thus, restoring or enhancing epithelial water secretion in CF and normal respiratory epithelia, respectively.

However, as is true for the CFTR defect in CF, apical chloride conductance seems to be only part of the functional effects of P2 receptor stimulation on respiratory epithelia. In primary cultures of human bronchial epithelia from CF patients and non-CF patients, increases in $[Ca^{2+}]_i$ produced by UTP (or other stimuli) applied to either membrane also induced an initial increase followed by a prolonged decrease in I_{Na} (Na^+ absorption), the decrease at least partially due to a long-term downregulation of basolateral K^+ conductance (DEVOR and PILEWSKI 1999). This secondary and prolonged decrease in I_{Na} (Na^+ absorption) could maintain the diminished transepithelial water absorption initiated by the transient increase in I_{Cl} (Cl^- secretion). In a polarized human bronchial epithelial cell line, 16HBE14o-, the rank order potency of nucleotides applied to either membrane indicated the presence of a $P2Y_2$ receptor, activation of which resulted in apical Cl^- secretion mainly due to activation of basolateral K^+ channels (KOSLOWSKY et al. 1994). Hypotonic swelling also opened transiently both of these Cl^- and K^+ channels. In cultures of CF and normal nasal epithelium, basolateral ATP increased Cl^- secretion indirectly by activating basolateral K^+ conductance in normal, but not CF, while luminal ATP activated apical membrane Cl^- conductance (opened $Cl^-_{Ca,a}$) both in normal and CF equivalently (CLARKE and BOUCHER 1992). In the polarized human intestinal epithelial cell line, Caco-2, opening of a calcium activated K^+ channel by increase in $[Ca^{2+}]_i$ was also shown to contribute a driving force for apical Cl^- secretion (INQUE et al. 1997). In primary cultures of rat tracheal epithelia, apical ATP stimulated Cl^- secretion through $P2Y_2$ receptors via both $[Ca^{2+}]_i$-dependent and independent signaling pathways opening outward rectifying Cl^- channels (ORCCs) and CFTR in addition to $Cl^-_{Ca,a}$ (HWANG et al. 1996). Basolateral ATP stimulated Cl^- secretion via a combination of receptor subtypes ($P2Y_2$, likely $P2Y_1$, and possibly P2X receptors) and independent of $[Ca^{2+}]_i$ or cAMP signaling by opening only CFTR. In primary cultures in dog tracheal epithelia, apical UTP caused phosphorylation activation of the basolateral Na-K-Cl cotransporter secondary to both a decrease in intracellular chloride concentration and cell shrinkage (HAAS et al. 1995). Finally, in normal and CF nasal epithelium, as well as in several epithelial cell lines, ATP and UTP induced increases in $[Ca^{2+}]_i$ resulted in the release of arachidonic acid consistent with activation of phospholipase A_2, PLA_2 (LAZAROWSKI et al. 1994). Whether these effects of P2 receptor stimulation also normalize CF epithelia IL-8 production and/or endogenous antibiotic/antiviral function still needs to be determined.

Regarding the purinergic receptor subtypes and signal transduction pathways that mediate these effects, the following additional data have been reported. In polarized cultures of human nasal epithelia, mucosal UDP increased inositol phosphate, $[Ca^{2+}]_i$, and I_{Cl^-} with a maximal effect half that of mucosal UTP (LAZAROWSKI et al. 1997). Serosal UDP had no effect, while serosal UTP produced responses 50% greater than those of mucosal UTP. Mucosal UDP desensitized on repeated application, but did not cross-desensitize mucosal UTP. Hence, the presence of a mucosal UDP receptor distinct from $P2Y_2$. In the human lung epithelial carcinoma cell line, A549, rank order nucleotide potency support a $P2Y_2$ pertussis toxin-insensitive receptor pathway with calcium mobilization via phospholipase C (PLC)-dependent release from an inositol triphosphate (IP_3)-sensitive pool (CLUNES et al. 1997; CLUNES and KEMP 1996). In the human goblet cell-like clone, HT-29.cll6E, the ATP-induced increase in Cl^- conductance was not diminished when $[Ca^{2+}]_i$ was clamped, but did require some intracellular calcium; that is, it did not occur in the presence of a calcium chelator (GUO et al. 1995).

Results in non-human and/or non-respiratory epithelia support and extend these findings. In primary cultures of rabbit tracheal epithelia, ATP produced a rapid transient (pulse) increase followed by sustained oscillations in $[Ca^{2+}]_i$ (WOODRUFF et al. 1999). Protein kinase C (PKC) inhibition (bisindolylmaleimide) attenuated both the pulse increase and oscillations. In rabbit non-ciliated terminal bronchiolar epithelia (Clara cells), apical, but not basal, UTP and ATP increased $[Ca^{2+}]_i$ and I_{sc}, while decreasing transepithelial resistance; UTP and ATP cross-desensitized (VAN SCOTT et al. 1995). In isolated and cultured bovine ciliated epithelial cells, nucleotide rank order potency and pertussis toxin sensitivity supported $P2Y_2$ activation for the ATP-induced increase in $[Ca^{2+}]_i$ (SHAHIDULLAH and WILSON 1997). Repeated application of ATP resulted in desensitization that was augmented by phorbol-myristate-acetate (PKC activator) and abolished by staurosporine (PKC inhibitor). In cultured murine gallbladder epithelia, transepithelial anion secretion as well as increase in $[Ca^{2+}]_i$ and inositol phosphate (IP) were all desensitized by repeated UTP stimulation in a concentration and time dependent manner (CLARKE et al. 1999). In equine sweat gland epithelia, a calcium chelator (BAPTA) attenuated the anion (Cl^- and HCO^-_3) secretion to apical ATP more than to apical UTP (KO et al. 1997). Cross-desensitization experiments confirmed two populations of UTP receptors, one sensitive to ATP and UTP

($P2Y_2$) and the other only to UTP; UDP appeared to activate the ATP-insensitive receptors. In Caco-2 cells, nucleotide stimulation, I_{sc} measurement and cross-desensitization experiments demonstrated:

1. Two different apical purinoceptor subtypes: $P2Y_2$ and a uridine nucleotide (UTP > UDP) preferring receptor
2. Two different basolateral membrane P2 receptors: $P2Y_2$ and one with an unusual profile in that it reacted with both 2-MeSATP and ADP (2-MeSATP > ADP >> ATP) (INQUE et al. 1997)

Finally, in nasal epithelial cell lines from wild type and $P2Y_{2-/-}$ mice, mechanical stimulation of a single cell resulted in a cell to cell propagation of cytosolic calcium transients (intercellular calcium wave) (HOMOLYA et al. 1998). Inhibition of nucleotide hydrolysis (apyrase, a tri- and diphosphatase) greatly enhanced, and desensitization of P2 receptors by ATP abolished, this wave. In $P2Y_{2-/-}$ cells, PAPS (adenosine 3′-phosphate 5′-phosphosulfate, a specific $P2Y_1$ receptor antagonist) blocked wave propagation. Hence, wave propagation was mediated by autocrine/paracrine activity of secreted ATP on $P2Y_2$ and $P2Y_1$ receptors.

In summary, nucleotide stimulation of either the apical or basolateral side of respiratory epithelium leads to increased ion/water secretion (decreased absorption) via changes in several ion channels and/or pumps independent of CFTR. These effects are mediated via P2 receptors with $P2Y_2$ playing a prominent role on both membranes. The most prominent pathways are depicted in Fig. 3.

II. Mucus Secretion

The involvement of P2 receptors in mucus secretion is similar to that in ion/water transport and regulation of the depth of the periciliary liquid layer. Secreted mucus must be optimally hydrated to function effectively in the respiratory tract. Secretion and hydration are executed by epithelia and goblet cells in the airspace lining epithelium as well as serous and mucus cells in the submucosal glands. In CF, secreted mucus is markedly less hydrated due to decreased Cl^- secretion and increased Na^+ absorption resulting from the lack of CFTR expression; submucosal glands are the most abundant sites of CFTR expression in the normal human bronchus (ENGELHARDT et al. 1992). Additionally, as described above, nucleotides acting via P2 receptors, especially the $P2Y_2$, enhance both the secretion and hydration of mucus.

Nucleotide enhancement of mucus secretion appears to follow the same pattern as that described above for receptor activation, increase in $[Ca^{2+}]_i$, and ion transport. In both CF and normal human airway (nasal and tracheobronchial) explants, mucosal ATP and UTP, but not 2-MeSATP, induced a biphasic secretory response as assessed by video microscopy of single goblet cells as well as a mucin ELISA of the perfusate (LETHEM et al. 1993). In iso-

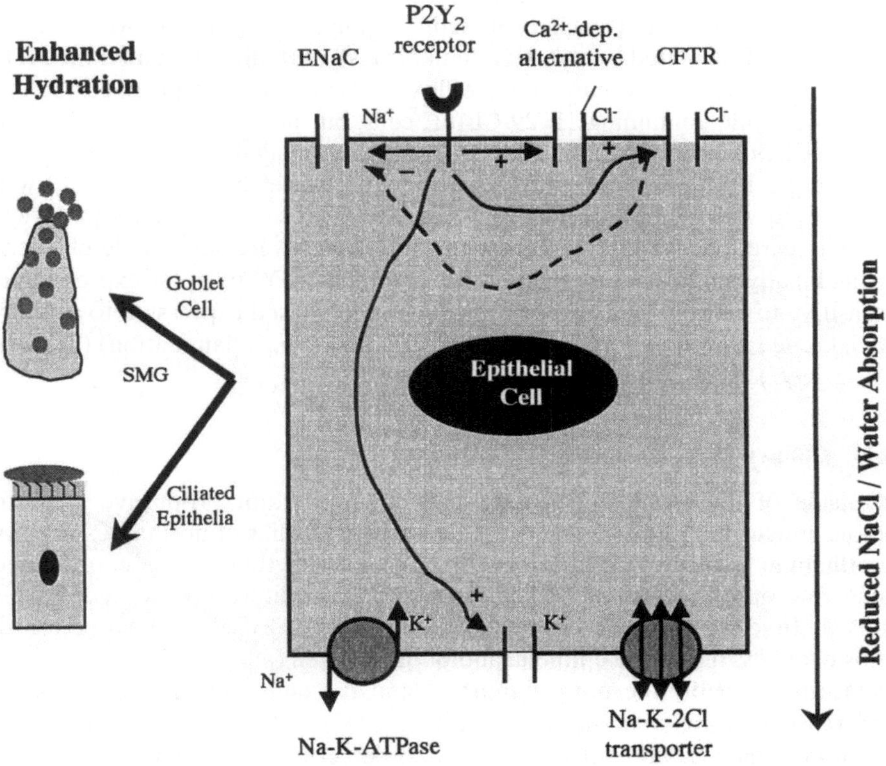

Fig. 3. Key ion channel effects of $P2Y_2$ receptor stimulation on the apical membrane of airway epithelia. See text for details. CFTR, cystic fibrosis transmembrane regulator; ENaC, epithelial sodium channel; SMG, submucosal glands

lated canine tracheal epithelial goblet cells, luminal ATP, but not ADP, also stimulated a biphasic (burst followed by a plateau) secretory response (DAVIS et al. 1992). Serosal ATP and ADP elicited a more complex set of highly variable responses. In primary cultures of human tracheal submucosal gland cells, nucleotide application in a rank order potency consistent with $P2Y_2$ receptor activation induced a Cl^- secretion (increase in I_{sc}) dependent on increase in $[Ca^{2+}]_i$ and activation of $[Ca^{2+}]_i$-dependent K^+ channels (YAMAYA et al. 1996). In similar preparations from CF patients and non-CF patients and using a capacitance probe technique to monitor fluid transport, gland fluid secretion was increased by cAMP mediated primary via CFTR, UTP entirely via $Cl^-_{Ca,a}$, and methacholine via activation of both channels (JIANG et al. 1997). CF glands showed a normal secretory response to UTP, reduced to methacholine and no response to cAMP. In an attempt to link ion transport (chloride conductance) and mucin secretion, two different subclones derived from the human adenocarcinoma cell line HT-29 were studied – one with mucin granules, HT29-

Cl.16E, and the other without, HT29-Cl.19A. Application of granule fusion inhibitors, phalloidin (blocker of actin depolymerization) or wortmannin (inhibitor of regulated exocytosis in leukocytes), partially attenuated the ATP, but not cAMP, induced luminal Cl^- efflux (50% and 70%, respectively) in the mucin granule containing HT29-Cl.16E cells, but not in the granule lacking HT29-Cl.19A cells (MERLIN et al. 1996). Hence, a significant portion of the purinergic, but not cAMP, activated Cl^- conductance (channel opening) was associated with granule fusion; however, the P2 receptor initiated pathways for Cl^- secretion and granule fusion (mucus release) were not completely identical. Finally, studies in rats suggest that glandular $P2Y_2$ receptor expression is sensitive to pathological changes, resulting in increased expression to support heightened secretion in response to perturbation (e.g., inflammation) (TURNER et al. 1997).

III. Ciliary Beat Frequency

Outside of the cough and sneeze, cilia are the major propulsive force to move mucus mouthward, cleansing the airways. Studies of normal respiratory epithelium indicate that cilia beat in a normal rhythm that is coordinated and rate optimized for clearance vs energy expenditure (WONG and YEATES 1997). In disease states, ciliary beat frequency (CBF) can be decreased; however, loss of ciliated epithelia and/or disoriented (and discoordinated) cilia beat are probably the more important and frequent causes of cilia-related decrease in mucus clearance (RAYNER et al. 1995). In CF, unlike ion and mucus secretion, there is no evidence for a basal defect in CBF (ADAM et al. 1997). However, like ion and mucus secretion, P2 receptor stimulation, again through similar receptor subtypes and signal transduction pathways, accelerates CBF.

In frog esophageal epithelium (the long-used, classical preparation for studying CBF), apical ATP-induced biphasic increases in $[Ca^{2+}]_i$, an initial spike followed by a plateau, and CBF (LEVIN et al. 1997). PKC inhibition did not effect the initial response but caused a rapid decay of the plateau increases in $[Ca^{2+}]_i$ and CBF. Studies with PKC activation, voltage-operated Ca^{2+} channel blockage, or PLC inhibition indicated that the initial spike in $[Ca^{2+}]_i$ resulted from calcium release from intracellular stores (G-protein, PLC, and IP_3 pathway), while the plateau increases in $[Ca^{2+}]_i$ and CBF came from a sustained calcium influx through non-voltage-operated channels. The opening of calcium-dependent K^+ channels also contributed to ATP-induced membrane hyperpolarization (ALFAHEL et al. 1996). Similar results have been generated in rabbit airway ciliated cells with a secondary autocrine action of nitric oxide also proposed as a potentiator of the plateau elevation in CBF (KORNGREEN and PRIEL 1996; UZLANER and PRIEL 1999). Finally, topical application of ATP to the trachea of ventilated, barbiturate-anesthetized dogs stimulated CBF and adenosine blunted this increase (WONG and YEATES 1992). In cultured rabbit tracheal epithelia, adenosine has been reported to depress

CBF via A_1 receptors and a decrease in intracellular cAMP (Tamaoki et al. 1989).

IV. Surfactant Release

Surfactant, released by Clara (non-ciliated epithelial) cells in the terminal airways and type II pneumocytes in the alveoli, lowers the air-liquid surface tension, a function which is essential for adequate gas exchange and breathing mechanics. Lowered surface tension maintains the airspaces open by preventing collapse as well as facilitating reopening, if necessary, at low transmural pressures. Surfactant also seems to provide a substantial anti-oxidant, anti-inflammatory protection to the airways (Baker et al. 1989; Lewis et al. 1991). Finally, surfactant likely "lubricates" the interface between the mucus gel and periciliary liquid layer in respiratory airspaces to facilitate clearance via cough, sneeze, and/or cilia. As in other respiratory epithelia, in type II pneumocytes, $P2Y_2$ receptor stimulation and signal transduction pathways similar to those described above exist and result in enhanced secretion of at least the primary lipid component of surfactant, phosphatidylcholine, PC (Rooney 1998).

Primary cultures of rat type II pneumocytes are the primary preparation to study surfactant release. $P2Y_2$ receptor activation with either UTP or ATP is coupled to a phosphoinositide-specific PLC that activates phospholipase D (PLD) to form PC (Gobran et al. 1994; Gobran and Rooney 1997). ATP also acts through two additional pathways:

1. Adenylate cyclase (Gobran and Rooney 1997; Gobran et al. 1998) probably resulting from the hydrolysis of ATP to adenosine and then stimulation of the A_{2A} receptor (Griese et al. 1991, 1993)
2. Possibly a $[Ca^{2+}]_i$, calmodulin-dependent pathway (Griese et al. 1993)

Activation of adenosine A_1 receptors inhibits PC secretion (Gobran and Rooney 1990). Finally, while a potent secretor of PC, ATP does not appear to increase surfactant protein-A, SP-A, release (Rooney et al. 1993). SP-A is an important anti-infective, anti-inflammatory component of normal pulmonary surfactant (LeVine et al. 1999).

V. Summary: Therapeutic Implications

The nucleotides ATP and UTP activate P2 receptors on the surface of respiratory epithelia to enhance mucociliary clearance (airway cleansing) and host defense via four key events:

1. Enhancing the depth of the periciliary liquid layer through enhanced anion secretion and/or diminished Na^+ absorption
2. Increasing the secretion of well hydrated mucus
3. Accelerating ciliary beat frequency
4. Potentiating surfactant (PC) release

Given that release of nucleotides into the extracellular space adjacent to respiratory epithelium is produced by mechanical (e.g., by an inhaled particle) stimulation of the epithelia (HOMOLYA et al. 1998; LAZAROWSKI et al. 1997) as well as cellular injury (e.g., secondary to infection or inflammation), these effects are phylogenically sound (i.e., appropriate physiologic responses) and emphasize an important role for nucleotides in regulating homeostasis in the respiratory system. Since $P2Y_2$ receptors play a prominent role in transducing these effects, a specific $P2Y_2$ agonist offers an attractive target for the therapy of lung diseases where mucociliary clearance is impaired. CF is of particular interest, since $P2Y_2$ stimulation reverses many, if not all, of the respiratory defects produced by CFTR mutation. Indeed in CF patients, non-CF patients, and patients with chronic bronchitis, inhaled UTP has been shown to accelerated mucociliary clearance (BENNETT et al. 1996, 1999; OLIVIER et al. 1996). However, as noted below (see Sect. D), $P2Y_2$ receptor stimulation also appears to potentiate the activation of macrophages, mast cells, and neutrophils which may exacerbate an underlying inflammation and, consequently, enhance disease progression including tissue injury and remodeling.

C. Airway Smooth Muscle

Data on the effects of nucleotides on airway muscle tone have been contradictory (FEDAN 1997). In general, contractile responses are produced in unconstricted airway preparations, while relaxant responses are elicited in precontracted airways. This pattern is particularly true in isolated guinea pig trachea preparations (ADVENIER et al. 1982), where biphasic responses to ATP, an initial contraction followed by relaxation, have also been reported (KAMIKAWA and SHIMO 1976). Variability in nucleotide hydrolysis between preparations as well as species differences in the balance of constrictor vs relaxant prostaglandins and thromboxanes produced by the aforementioned nucleotide-stimulated release of arachidonic acid from airway epithelia (LAZAROWSKI et al. 1994) probably explain many of the inconsistencies. However, focussing solely on studies that used hydrolysis-resistant nucleotides as stimulants, the presence of P2 receptors on airway smooth muscle is indicated.

In guinea pig isolated trachea preparations precontracted with methacholine, nonhydrolyzable ATP analogues were potent relaxants (WELFORD and ANDERSON 1988; FEDAN et al. 1993a); in contrast, hydrolyzable ATP and UTP caused contraction (FEDAN et al. 1993b, 1994). In isolated rabbit trachealis muscle contracted with acetylcholine, ATP and ADP, as well as ATP analogues stable to hydrolysis, produced an epithelium- and partially indomethacin-dependent relaxation that was blocked by the putative P2Y receptor antagonist, Reactive Blue 2 (AKSOY and KELSEN 1994). Finally, in primary cultures of isolated swine or rat tracheal smooth muscle cells, nucleotides induced increases in $[Ca^{2+}]_i$ with a rank order potency consistent with $P2Y_2$ receptor activation (SAWAI et al. 1996; MICHOUD et al. 1997). These increases in $[Ca^{2+}]_i$ were generated independent of extracellular calcium and via a PLC-dependent pathway.

D. Resident and Infiltrating Leukocytes

Nucleotides released into extracellular fluid by mechanical stimulation of epithelia, cell injury, and/or neurotransmission not only act via P2 receptor stimulation to accelerate mucociliary clearance (see above), but also likely recruit and potentiate the activation of macrophages, mast cells, and neutrophils to engulf and remove the insult. While such effects on leukocytes are covered in detail in other chapters (BIAGGIONI and FEOKTISTOV, Chap. 22, and MONTESINOS and CRONSTEIN, Chap. 24, both this volume), they are also briefly included here because of the key role they likely contribute to respiratory defense.

I. Alveolar Macrophages

Alveolar macrophages are the most plentiful resident leukocyte within the respiratory tract and play a critical role in host defense by vacuuming especially the terminal airways and alveoli, engulfing foreign particles and cell debris they find in their path. Extracellular nucleotides have been recently shown to assist in this function via stimulating macrophage P2 receptors.

ATP analogues, in a rank order potency consistent with $P2Y_2$ receptor activation, enhance the phagocytosis of mouse peritoneal macrophages (ICHINOSE 1999). In human macrophages, ATP and UTP potentiate the superoxide generation triggered by opsonized zymosan via opening of a charybdotoxin-sensitive, calcium-dependent outward K^+ channel (SCHMID-ANTOMARCHI et al. 1997). ATP also enhances microbial killing by human macrophages via two P2 receptor pathways:

1. Via $P2X_7$ stimulation without an increase in reactive-oxygen-species (ROS) or nitric oxide (NO) generation (SIKORA et al. 1999; DI VIRGILIO et al., Chap. 26, this volume)
2. Via a P2Y receptor stimulation with increases in ROS and NO (LAMMAS et al. 1997).

ATP also triggers release of the pro-inflammation cytokine, IL-1, via $P2X_7$ receptor activation in human macrophages (FERRARI et al. 1997), while UTP via P2Y receptor activation causes arachidonic acid release (LIN and CHEN 1998) and potentiates bacterial endotoxin-induced NO production (CHEN et al. 1998) in several macrophage cell lines. Hence, extracellular nucleotides acting via $P2X_7$ and P2Y receptors induced notable effects that likely augment the effectiveness of alveolar macrophages in their crucial contribution to respiratory defense.

II. Mast Cells

Mast cells play an important role in the respiratory response to allergens by releasing mediators that cause bronchoconstriction, leukocyte recruitment, mucus secretion, and accelerated mucociliary clearance. All of these responses act to limit and remove the inhaled allergen. Here again, recent evidence indicates that P2 receptors meaningfully regulate mast cell responses.

In human lung mast cells, UTP, ATP, and ATP analogues enhanced anti-IgE, but not the calcium ionophore, A23187, and induced histamine release with rank order potency consistent with a P2Y receptor subtype (SCHULMAN et al. 1999). RT-PCR using mRNA encoding for purinergic receptor subtypes demonstrated the presence of $P2Y_1$ and $P2Y_2$, but not $P2X_7$ receptors on these cells. $P2X_7$ had been previously implicated in ATP-induced histamine release from rodent peritoneal mast cells. ADP, ATP, and UTP (as well as non-hydrolyzable derivatives) were chemotactic for rat bone marrow cultured mast cells with nucleotide selectivity and pertussis toxin sensitivity indicative of $P2Y_2$ activation (MCCLOSKEY et al. 1999). Altering calcium entry, depletion of intracellular calcium stores, or blocking K^+ channels all inhibited migration. In mature rat mucosal mast cells, P2 purinoceptor stimulation mobilized cytosolic calcium and activated K^+ conductance. Finally, rat basophilic leukemia cells, peritoneal mast cells, and mucosal mast cells all release ATP on stimulation and have a P2-type purinergic receptor that amplifies the $[Ca^{2+}]_i$ signals that initiate degranulation (OSIPCHUK and CAHALAN 1992). Hence, ATP release from a centennial mast cell can act in a positive feedforward manner on neighboring mast cells, spreading and amplifying the histamine release.

III. Neutrophil

Like macrophage and mast cell, neutrophil contributions to respiratory defense seems to be augmented by nucleotide stimulation via P2 receptors. On human neutrophils, ATP and UTP apparently acting via $P2Y_2$ activation induced chemotaxis as well as actin polymerization and shape change, both requirements for motility (VERGHESE et al. 1996). ATP increases human neutrophil $[Ca^{2+}]_i$ to cause degranulation, enzyme release, potentiate the oxidative burst, and enhance adhesion to endothelium (FREDHOLM 1997), also likely via $P2Y_2$ activation (DAWICKI et al. 1995). Enhanced expression of the adhesion molecule, Mac-1, by ATP occurred through a PKC-dependent pathway (AKBAR et al. 1997). Finally, akin to recent reports for epithelia, human neutrophils responded to diadenosine polyphosphates with increases in $[Ca^{2+}]_i$ that were 50–70% of the response to ATP (GASMI et al. 1997). Desensitization experiments demonstrated two P2 receptors, a high affinity, probably $P2Y_2$, and lower affinity, possibly Ap4A, receptor. ATP bound and stimulated both, whereas ApnA only the latter.

E. Fibroblasts

While data is presently scant, nucleotides also appear to bind to and activate fibroblast function. In human skin fibroblasts, ATP caused striking morphological changes, increased formation of cytoplasmic microvessels, and, when combined with PMA and bacterial endotoxin, the release of the pro-inflammatory cytokine, IL-6 (SOLINI et al. 1999). These ATP effects were mediated through $P2X_7$ and a P2Y subtype exhibiting a peculiar pharmacologic profile; CTP was the preferred agonist. In chick fibroblasts, ATP-induced retraction via an actin/myosis stress fiber sliding mechanism that was independent of microtubules and facilitated by intermediate filaments. The exact role of these ATP-induced, P2 receptor mediated effects are not clear, but they likely benefit fibroblast function in tissue remodeling.

F. Pulmonary Nerves

Given the prominent contribution of nucleotides and P2 receptors to nerve (especially sensory) traffic in other organs, it seems likely that such contributions also exist in the respiratory tract. However, precious little has been done (or at least reported) to study and characterized such pathways. However, two impressive studies do present convincing evidence that sensory C-fiber terminals in pulmonary airways are activated via P2X receptors.

In pentobarbital anesthetized dogs, injection of ATP or capsaicin (but not adenosine, AMP, or ADP) into the right atrium (or pulmonary artery) elicited a transient burst of action potentials in cervical vagal fibers (PELLEG and HURT 1996). The same response and fibers were triggered by a deep inflation of the lungs to 2–3 times tidal volume. α,β-Methylene ATP (α,β-MeATP, a non-degradable analogue of ATP) was ten times more potent than ATP, while β,γ-MeATP was inactive. The response was blocked by PPADS (P2X antagonist) but not Reactive Blue 2 (a P2Y antagonist) or pertussis toxin. These results indicate a P2X purinoceptor mediated activation of pulmonary sensory C-fiber nerve terminals and have now also been confirmed in rats using nearly an identical protocol (MCQUEEN et al. 1998).

In animals, activation of these sensory C-fibers by deep lung inflation causes apnea followed by rapid shallow breathing and reflex bronchoconstriction. As these same nerve terminals are sensitized, made more excitable, by ozone exposure in rats (HO and LEE 1998), C-fiber activation in humans is thought to induced the same sensory aberrations as those associated with ozone exposure. These include changes in breathing pattern (rapid shallow breathing), chest discomfort on deep inspiration, and cough. In addition, stimulation of airway C-fiber nerve terminals has also been shown to cause the release of tachykinins directly as well as through an antidromic "axon reflex" (BARNES 1986). Tachykinins cause plasma extravasation that activates airway rapidly adapting receptors (RAR nerve terminals), which in-turn elicits cough

(WIDDICOMBE 1998). Thus, activation of airway C-fiber nerve terminals results in impressionable sensations of lung congestion.

G. Lung Cancer

While certainly not a major focus for this chapter, P2 receptors have also been shown to play a role in cell growth and differentiation. In addition, agonists that stimulate calcium mobilization seem to play a critical growth/differentiation regulatory role in secretory cell types. Hence, a brief review of data possibly relevant to lung cancer are presented below.

In several human androgen-independent prostate (secretory) carcinoma cell lines, ATP and hydrolysis-resistant ATP analogues stimulated PLC-dependent phosphatidylinositol turnover leading to Ca^{2+} mobilization and, more importantly, >90% inhibition of growth in all the lines tested (FANG et al. 1992). In a human metastatic lung carcinoma cell line, PG, ATP, and bombesin induced rapid transient increases in $[Ca^{2+}]_i$ via IP_3 activation and G-protein (cholera toxin sensitive), PLC-linked receptors (FANG and WU 1993). UTP also mobilized intracellular calcium in these cells. However, these agents had differential effects on cell growth. Bombesin stimulated PG growth, ATP (and hydrolysis-resistant analogues) significantly inhibited growth, and UTP had no effect. These interesting findings require further confirmation and characterization in additional lung carcinoma cell lines before their meaning and value can be interpreted. In advanced stage cancer patients with non-small-cell lung cancer, ATP infusion reduced the weight loss normally associated with disease progression, reducing cachexia and improving quality of life (AGTERESCH et al. 2000).

H. Summary

Extracellular nucleotides acting via various cell membrane purinergic P2 receptor subtypes have notable effects on essentially all the major cell components of the respiratory system. The effects are predominantly stimulatory, as opposed to inhibitory, to cell function, with the notable exception being the unconfirmed effect on cell growth and differentiation. On the polarized respiratory epithelium, nucleotides applied to either, but especially the apical, membrane alter ion transport to promote water secretion, enhance mucus secretion, accelerate ciliary beat frequency, and potentiate surfactant, specifically phophatidylcholine, release. $P2Y_2$ receptors play a primary role in these effects, but other P2Y receptors are certainly also involved. Nucleotides act on airway smooth muscle to induced either contraction or dilation, directly or indirectly (through arachidonic acid metabolites released by epithelia), depending on the preparation studied. Since clinical studies with inhaled UTP have not found significant effects on airway tone (OLIVIER et al. 1996), the in vivo importance of at least P2Y receptor mediated actions on airway smooth muscle seems to be minimal. The same may not be true for P2 receptor transduced effects on resident and infiltrating leukocytes. At least in vitro, nucleotides acting principally through $P2Y_2$, other P2Y, and $P2X_7$ receptors notably induce potentiation of macrophage, mast cell, and neutrophil activation. Such effects may augment and facilitate the crucial contribution of these cells to respiratory defense. Finally, while data is still scanty, ATP acting via stimulation of a P2X subtype receptor seems to stimulate airway sensory C-fiber nerve terminals. These fibers have been associated with the sensation of lung congestion and may also potentiate inflammation as well as cough. Thus, in conclusion, extracellular nucleotides acting through cell surface purinergic P2 receptors appear to contribute important regulatory functions via actions on essentially all cell components of the respiratory system. The ability of purine nucleotides to stimulate $P2Y_2$ receptors and increase ciliary function may offer a new approach to the treatment of respiratory tract infections either on their own or in combination with low doses of traditional antibiotics.

References

Adam EC, Schumacher DU, Schumacher U (1997) Cilia from a cystic fibrosis patient react to the ciliotoxic *Pseudomonas aeruginosa* II lectin in a similar manner to normal control cilia – a case report. J Laryngol Otol 111:760–762

Advenier C, Bidet D, Floch-Saint-Aubin A, Renier A (1982) Contribution of prostaglandins and thromboxanes to the adenosine and ATP-induced contraction of guinea-pig isolated trachea. Br J Pharmacol 7:39–44

Agteresch HJ, Dagnelie PC, van der Gaast A, Stijnen T, Wilson JH (2000) Randomized clinical trial of adenosine 5′-triphosphate in patients with advanced non-small-cell lung cancer. J Natl Cancer Inst 92(4):321–328

Akbar GK, Mills DC, Kunapuli SP (1997) Characterization of extracellular nucleotide-induced Mac-1 (αM $\beta 2$ integrin) surface expression on peripheral blood leukocytes. Biochem Biophys Res Commun 233:71–75

Aksoy MO, Kelsen SG (1994) Relaxation of rabbit tracheal smooth muscle by adenine nucleotides: mediation by P2-purinoceptors. Am J Respir Cell Molecular Biol 10:230–236

Alfahel E, Korngreen A, Parola AH, Priel Z (1996) Purinergically induced membrane fluidization in ciliary cells characterization and control by calcium and membrane potential. Biophys J 70:1045–1053

Baker RR, Holm BA, Panus PC, Matalon S (1989) Development of O2 tolerance in rabbits with no increase in antioxidant enzymes. J Appl Physiol 66:1679–1684

Barnes PJ (1986) Asthma as an axon reflex. Lancet 1:242–244

Bennett WD, Olivier KN, Zeman KL, Hohneker KW, Boucher RC, Knowles MR (1996) Effect of uridine 5′ triphosphate plus amiloride on mucociliary clearance in adult cystic fibrosis. Am J Respir Crit Care Med 153:1796–1801

Bennett WD, Zeman KL, Foy C, Shaffer C, Johnson F, Sannuti A, Johnson J (1999) Aerosolized Prolucin (uridine 5′-triphosphate, UTP) acutely enhances mucociliary clearance in mild chronic bronchitis (CB). Am J Respir Crit Care Med 159:A805 (Abstract)

Boucher RC (1999) Molecular insights into the physiology of the "thin film" of airway surface liquid. J Physiol 516:631–638

Chen BC, Chou CF, Lin WW (1998) Pyrimidinoceptor-mediated potentiation of inducible nitric-oxide synthase induction in J774 macrophages. Role of intracellular. J Biol Chem 273:29754–29763

Clarke LL, Boucher RC (1992) Chloride secretory response to extracellular ATP in human normal and cystic fibrosis nasal epithelia. Am J Physiol 263:C348–C356

Clarke LL, Harline MC, Otero MA, Glover GG, Garrad RC, Krugh B, Walker NM, Gonzalez FA, Turner JT, Weisman GA (1999) Desensitization of $P2Y_2$ receptor-activated transepithelial anion secretion. Am J Physiol 276:C777–C787

Clunes MT, Kemp PJ (1996) P_{2u} purinoceptor modulation of intracellular Ca^{2+} in a human lung adenocarcinoma cell line: downregulation of Ca^{2+} influx by protein kinase C. Cell Calcium 20:339–346

Clunes MT, Olver RE, Kemp PJ (1997) $P2Y_2$ receptor occupancy releases stored Ca^{2+} via a single, pertussis toxin-insensitive signal transduction pathway in a human alveolar epithelial cell line (A549). J Physiol 505.P:66P

Davis CW, Dowell ML, Lethem M, Van Scott M (1992) Goblet cell degranulation in isolated canine tracheal epithelium: response to exogenous ATP, ADP, and adenosine. Am J Physiol 262:C1313–C1323

Dawicki DD, McGowan-Jordan J, Bullard S, Pond S, Rounds S (1995) Extracellular nucleotides stimulate leukocyte adherence to cultured pulmonary artery endothelial cells. Am J Physiol (Lung Cell Mol Physiol) 268:L666–L673

Devor DC, Pilewski JM (1999) UTP inhibits Na^+ absorption in wild-type and deltaF508 CFTR-expressing human bronchial epithelia. Am J Physiol 276:C827–C837

Engelhardt JF, Yankaskas JR, Ernst SA, Yang Y, Marino CR, Boucher RC, Cohn JA, Wilson JM (1992) Submucosal glands are the predominant site of CFTR expression in the human bronchus. Nat Genet 2:240–248

Fang W-G, Pirnia F, Bang Y-J, Myers CE, Trepel JB (1992) P_2-purinergic receptor agonists inhibit the growth of androgen-independent prostate carcinoma cells. J Clin Invest 89:191–196

Fang W-G, Wu B-Q (1993) Differential growth regulation of a metastatic human lung carcinoma cell line through activation of phosphatidyl inositol turnover signal transduction pathway. Clin Exp Metastasis 11:330–336

Fedan JS, Belt JJ, Yuan LX, Frazer DG (1993b) Contractile effects of nucleotides in guinea pig isolated, perfused trachea: involvement of respiratory epithelium, prostanoids and Na^+ and Cl^- channels. J Pharmacol Exp Ther 264:210–216

Fedan JS, Belt JJ, Yuan LX, Frazer DG (1993a) Relaxant effects of nucleotides in guinea pig isolated, perfused trachea: lack of involvement of prostanoids, Cl^- channels and adenosine. J Pharmacol Exp Ther 264:217–220

Fedan JS, Stem JL, Day B (1994) Contraction of the guinea pig isolated, perfused trachea to purine and pyrimidine agonists. J Pharmacol Exp Ther 268:1321–1327

Fedan JS (1997) Nucleotides and nucleosides in pulmonary function. In: Jacobson KA, Jarvis MF (eds)Purinergic approaches in experimental therapeutics. Wiley-Liss, New York, pp 333–356

Ferrari D, Chiozzi P, Falzoni S, Dal Susino M, Melchiorri L, Baricordi OR, Di Virgilio F (1997) Extracellular ATP triggers IL-1 β release by activating the purinergic P2Z receptor of human macrophages. J Immunol 159:1451–1458

Fredholm BB (1997) Purines and neutrophil leukocytes. General Pharmacol 28:345–350

Gasmi L, Mclennan Ag, Edwards SW (1997) Diadenosine polyphosphates induce intracellular Ca^{2+} mobilization in human neutrophils via a pertussis toxin sensitive G-protein. Immunol 90:154–159

Gobran LI, Rooney SA (1990) Adenosine A1 receptor-mediated inhibition of surfactant secretion in rat type II pneumocytes. Am J Physiol 258:L45–L51

Gobran LI, Rooney SA (1997) Adenylate cyclase-coupled ATP receptor and surfactant secretion in type II pneumocytes from newborn rats. Am J Physiol (Lung Cell Mol Physiol) 272:L187–L196

Gobran LI, Xu Z, Zhun L, Rooney SA (1994) P_{2u} purinoceptor stimulation of surfactant secretion coupled to phosphatidylcholine hydrolysis in type II cells. Am J Physiol 267:L625–L633

Gobran LI, Xu ZX, Rooney SA (1998) PKC isoforms and other signaling proteins involved in surfactant secretion in developing rat type II cells. Am J Physiol 274:L901–L907

Goldman MJ, Anderson GM, Stolzenberg ED, Kari UP, Zasloff M, Wilson JM (1997) Human β-defensin-1 is a salt-sensitive antibiotic in lung that is inactivated in cystic fibrosis. Cell 88:553–560

Griese M, Gobran LI, Rooney SA (1991) A2 and P2 purine receptor interactions and surfactant secretion in primary cultures of type II cells. Am J Physiol 261:L140–L147

Griese M, Gobran LI, Rooney SA (1993) Signal-transduction mechanisms of ATP-stimulated phosphatidylcholine secretion in rat type II pneumocytes: interactions between ATP and other surfactant secretagogues. Biochem Biophys Acta 1167:85–93

Guo XW, Merlin D, Harvey RD, Laboisse C, Hopfer U (1995) Stimulation of Cl^- secretion by extracellular ATP does not depend on increased cytosolic Ca^{2+} in HT-29.cl16E. Am J Physiol 269:C1457–C1463

Haas M, McBrayer D, Lytle C (1995) $[Cl^-]_i$-dependent phosphorylation of the Na-K-Cl cotransport protein of dog tracheal epithelial cells. J Biol Chem 270:28955–28961

Ho CY, Lee LY (1998) Ozone enhances excitabilities of pulmonary C fibers to chemical and mechanical stimuli in anesthetized rats. J Appl Physiol 85:1509–1515

Homolya L, Steinberg TH, Boucher RC (1998) Intercellular propagation of calcium waves is mediated by extracellular nucleotides in airway epithelial cells. Drug Devel Res 43:43–63(Abstract)

Hwang T-H, Schwiebert EM, Guggino WB (1996) Apical and basolateral ATP stimulates tracheal epithelial chloride secretion via multiple purinergic receptors. Am J Physiol 270:C1611–C1623

Ichinose M (1999) Modulation of phagocytosis by P_2-purinergic receptors in mouse peritoneal macrophages. Jap J Physiol 45:707–721

Inglis SK, Corboz MR, Taylor AE, Ballard ST (1997) Effect of anion transport inhibition on mucus secretion by airway submucosal. Am J Physiol 272 (2pt1):L372–L377

Inque CN, Woo JS, Schwiebert EM, Morita T, Hanaoka K, Guggino SE, Guggino WB (1997) Role of purinergic receptors in chloride secretion in Caco-2 cells. Am J Physiol 272:C1862–C1870

Jiang C, Finkbeiner WE, Widdicombe JH, Miller SS (1997) Fluid transport across cultures of human tracheal glands is altered in cystic fibrosis. J Physiol 501:637–647

Kamikawa Y, Shimo Y (1976) Mediation of prostaglandin E2 in the biphasic response to ATP of the isolated tracheal muscle of guinea-pigs. Br J Pharmacol 28:294–297

Kelley T, Drumm M (1998) Inducible nitric oxide synthase expression is reduced in cystic fibrosis murine and human airway epithelial cells. J Clin Invest 102:1200–1207

Knowles MR, Clarke LL, Boucher RC (1991) Activation by extracellular nucleotides of chloride secretion in the airway epithelia of patients with cystic fibrosis. N Engl J Med 8:533–538

Ko WH, Wilson SM, Wong PYD (1997) Purine and pyrimidine nucleotide receptors in the apical membranes of equine cultured epithelia. Br J Pharmacol 121:150–156

Korngreen A, Priel Z (1996) Purinergic stimulation of rabbit ciliated airway epithelia: control by multiple calcium sources. J Physiol 497:53–66

Koslowsky T, Hug T, Ecke D, Klein P, Greger R, Gruenert DC, Kunzelmann K (1994) Ca^{2+}- and swelling induced activation of ion conductances in bronchial epithelial cells. Pflügers Arch 428:597–603

Lammas DA, Stober C, Harvey CJ, Kendrick N, Panchalingam S, Kumararatne DS (1997) ATP-induced killing of mycobacteria by human macrophages is mediated by purinergic P2X($P2X_7$) receptors. Immunity 7:433–444

Lazaorowski, ER, Boucher RC, Harden TK (1994) Calcium-dependent release of arachidonic acid in response to purinergic receptor activation in airway epithelium. Am J Physiol 266:C406–C415

Lazarowski ER, Homolya L, Boucher RC, Harden TK (1997) Direct demonstration of mechanically induced release of cellular UTP and its implication for uridine nucleotide receptor activation. J Biol Chem 272:24348–24354

Lazarowski ER, Paradiso AM, Watt WC, Harden TK, Boucher RC (1997) UDP activates a mucosal-restricted receptor on human nasal epithelial cells that is distinct from the P2Y2 receptor. Proc Natl Acad Sci 94:2599–2603

Lethem MI, Dowell ML, Van Scott M, Yankaskas JR, Egan T, Boucher RC, Davis CW (1993) Nucleotide regulation of goblet cells in human airway epithelial explants: normal exocytosis in cystic fibrosis. Am J Respir Cell Molecular Biol 9:315–322

Levin R, Braiman A, Priel Z (1997) Protein kinase C induced calcium influx and sustained enhancement of ciliary beating by extracellular ATP. Cell Calcium 21:103–113

LeVine AM, Kurak KE, Wright JR, Watford WT, Bruno MD, Ross GF, Whitsett JA, Korfhagen TR (1999) Surfactant protein-A binds group B streptococcus enhancing phagocytosis and clearance from lungs of surfactant protein-A-deficient mice. Am J Respir Cell Molecular Biol 20:279–286

Lewis J, Ikegami M, Higuchi R, Jobe A, Absolom D (1991) Nebulized vs. instilled exogenous surfactant in an adult lung injury model. J Appl Physiol 71:1270–1276

Lin WW, Chen BC (1998) Pharmacological comparison of UTP- and thapsigargin-induced arachidonic acid release in mouse RAW 264.7 macrophages. Br J Pharmacol 123:1173–1181

Mall M, Bleich M, Greger R, Schreiber R, Kunzelmann K (1998) The amiloride-inhibitable Na^+ conductance is reduced by the cystic fibrosis transmembrane conductance regulator in normal but not in cystic fibrosis airways. J Clin Invest 102:15–21

Mason J, Paradiso AM, Boucher RC (1991) Regulation of transepithelial ion transport and intracellular calcium by extracellular ATP in human normal and cystic fibrosis airway epithelium. Am J Pharmacol 03:1649–1656

Matsui H, Grubb BR, Tarran R, Randell SH, Gatzy JT, Davis CW, Boucher RC (1998) Evidence for periciliary liquid layer depletion, not abnormal ion composition, in the pathogenesis of cystic fibrosis airways disease. Cell 95:1005–1015

McCloskey MA, Fan Y, Luther S (1999) Chemotaxis of rat mast cells toward adenine nucleotides. J Immunol 163:970–977

McQueen DS, Bond SM, Moores C, Chessell I, Humphrey PP, Dowd E (1998) Activation of P2X receptors for adenosine triphosphate evokes cardiorespiratory reflexes in anaesthetized rats. J Physiol 507:843–855

Meng QH, Springall DR, Bishop AE, Morgan K, Evans TJ, Habib S, Gruenert DC, Gyi KM, Hodson ME, Yacoub MH, Polak JM (1998) Lack of inducible nitric oxide synthase in bronchial epithelium: a possible mechanism of susceptibility to infection in cystic fibrosis. J Pathol 184:323–331

Merlin D, Guo X, Martin K, Laboisse C, Landis D, Dubyak G, Hopfer U (1996) Recruitment of purinergically stimulated Cl^- channels from granule membrane to plasma membrane. Am J Physiol 271:C612–C619

Michoud MC, Tolloczko B, Martin JG (1997) Effects of Purine nucleotides and nucleoside on cytosolic calcium levels in rat tracheal smooth muscle. Am J Respir Cell Molecular Biol 16:199–205

Olivier KN, Bennett WD, Hohneker KW, Zeman KL, Edwards LJ, Boucher RC, Knowles MR (1996) Acute safety and effects on mucociliary clearance of aerosolized uridine 5′-triphosphate ± amiloride in normal human adults. Am J Respir Crit Care Med 154:217–223

Osipchuk Y, Cahalan M (1992) Cell-to-cell spread of calcium signals mediated by ATP receptors in mast cells. Nature 359:241–244

Pelleg A, Hurt CM (1996) Mechanism of action of ATP on canine pulmonary vagal C fibre nerve terminals. J Physiol 490:265–275
Rayner CF, Rytman A, Dewar A, Cole PJ, Wilson R (1995) Ciliary disorientation in patients with chronic upper respiratory tract inflammation. Am J Respir Crit Care Med 151:800–804
Rooney SA (1998) Role of purinoceptors in the regulation of lung surfactant secretion. In: Turner JT, Weisman GA, Fedan JS (eds)The Receptors Series; The P2 nucleotide receptors. Humana Press, New Haven, pp291–318
Rooney SA, Gobran LI, Umstead TM, Phelps DS (1993) Secretion of surfactant protein a from rat type II pneumocytes. Am J Physiol 265:L586–L590
Sawai H, Wang R, Yamashita T, Kokubun S (1996) Effects of purinoceptor agonists on cytosolic Ca^{2+} concentration in swine tracheal smooth muscle cells in culture. Br J Pharmacol 119:539–544
Schmid-Antomarchi H, Schmid-Alliana A, Romey G, Ventura MA, Breittmayer V, Millet MA, Husson H, Moghrabi B, Lazdunski M, Rossi B (1997) Extracellular ATP and UTP control the generation of reactive oxygen intermediates in human macrophages through the opening of a charybdotoxin-sensitive $Ca2^{+}$ dependent K^{+} channel. J Immunol 159:6209–6215
Schulman ES, Glaum MC, Post T, Wang Y, Raible DG, Mohanty J, Butterfield JH, Pelleg A (1999) ATP modulates anti-IgE-induced release of histamine from human lung mast cells. Am J Respir Cell Molecular Biol 20:530–537
Shahidullah M, Wilson WS (1997) Mobilisation of intracellular calcium by $P2Y_2$ receptors in cultured, non transformed bovine ciliary epithelial cells. Cur Eye Res 16:1006–1016
Sikora A, Liu J, Brosnan C, Buell G, Chessel I, Bloom BR (1999) Cutting edge: purinergic signaling regulates radical-mediated bacterial killing mechanisms in macrophages through a $P2X_7$-independent mechanism. J Immunol 163:558–561
Smith JJ, Travis SM, Greenberg EP, Welsh MJ (1996) Cystic fibrosis airway epithelia fail to kill bacteria because of abnormal airway surface fluid. Cell 85:229–236
Solini A, Chiozzi P, Morelli A, Fellin R, DI Virgilio F (1999) Human primary fibroblasts in vitro express a purinergic $P2X_7$ receptor coupled to ion fluxes, microvesicle formation and IL-6 release J Cell Sci 122:297–305
Stutts MJ, Boucher RC (1999) Cystic fibrosis gene and functions of CFTR implications of dysfunctional ion transport for pulmonary pathogenesis. In: Yankaskas JR, Knowles MR (eds) Cystic fibrosis in adults. Lippincott-Raven, Philadelphia, pp3–25
Tabary O, Zahm J-M, Hinnrasky J, Couetil J-P, Cornillet P, Guenounou M, Gaillard D, Puchelle E, Jacquot J (1998) Selective up-regulation of chemokine IL-8 expression in cystic fibrosis bronchial gland cells in vivo and in vitro. Am J Pathol 153: 921–930
Tamaoki J, Kondo M, Takizawa T (1989) Adenosine-mediated cyclic AMP-dependent inhibition of ciliary activity in rabbit tracheal epithelium. Am Rev Respir Dis 139:441–445
Turner JT, Weisman GA, Camden JD (1997) Upregulation of $P2Y_2$ nucleotide receptors in rat salivary gland cells during short-term culture. Am J Physiol 273:C1100–C1107
Uzlaner N, Priel Z (1999) Interplay between the NO pathway and elevated $[Ca^{2+}]_i$ enhances ciliary activity in rabbit trachea. J Physiol 516:179–190
Van Scott MR, Chinet TC, Burnette AD, Paradiso AM (1995) Purinergic regulation of ion transport across nonciliated bronchiolar epithelial (clara) cells. Am J Physiol 269:L30–L37
Verghese NW, Kneisler TB, Boucheron JA (1996) P2U agonists induce chemotaxis and actin polymerization in human neutrophils and differentiated HL60 cells. J Biol Chem 271:15597–15601
Welford LA, Anderson WH (1988) Purine receptors and guinea-pig trachea: evidence for a direct action of ATP. Br J Pharmacol 95:689–694

Widdicombe JG (1998) Afferent receptors in the airways and cough. Respir Physiol 114:5–15
Wong LB, Yeates DB (1992) Luminal purinergic regulatory mechanisms of tracheal ciliary beat frequency. Am J Respir Cell Molecular Biol 7:447–454
Wong LB, Yeates DB (1997) Dynamics of the regulation of ciliated epithelial function. Comm Theor Biol 4:183–208
Woodruff ML, Chaban VV, Worley CM, Dirksen ER (1999) PKC role in mechanically induced Ca^{2+} waves and ATP-induced Ca^{2+}+ oscillations in airway epithelial cells. Am J Physiol 276:L669–L678
Yamaya M, Sekizawa L, Kakuta Y, Ohrui T, Sawai T, Sasaki H (1996) P2u-purinoceptor regulation of chloride secretion in cultured human tracheal submucosal glands. Am J Physiol 270:L979–L984
Zahm J-M, Gaillard D, Dupuit F, Hinnrasky J, Porteous D, Dorin JR, Puchelle E (1997) Early alterations in airway mucociliary clearance and inflammation of the lamina propria in CF mice. Am J Physiol 72:C853–C859

Section IV
Immune System

CHAPTER 24

Role of P1 Receptors in Inflammation

M.C. MONTESINOS and B.N. CRONSTEIN

A. Introduction

The first demonstration that adenosine had anti-inflammatory properties was made in 1983 (CRONSTEIN et al. 1983) and the purine was subsequently shown to mediate its anti-inflammatory effects via specific cell surface receptors (CRONSTEIN et al. 1985; ROBERTS et al. 1985). Since 1983 it has also been known that locally released adenosine may act as an endogenous anti-inflammatory mediator (CRONSTEIN et al. 1983) and it was subsequently demonstrated that promotion of endogenous adenosine release may mediate the anti-inflammatory effects of several potent agents currently in use for the treatment of inflammatory diseases like rheumatoid arthritis (RA).

In the present chapter, the mechanism by which adenosine is generated at sites of inflammation and the many receptor-mediated anti-inflammatory effects of adenosine are reviewed. The role of ATP and other purines in immune system function and the inflammatory response is covered elsewhere (DI VIRGILIO et al. Chap. 26, and DUBYAK et al. Chap. 25, both this volume).

B. Generation of Adenosine at Inflammation and Injured Sites

Adenosine may be generated intracellularly or extracellularly as a result of the ATP catabolism that occurs in ischemic or inflamed tissues. Adenine nucleotides are released into the extracellular fluid where they are converted first to AMP by various phosphatases and then to adenosine by the action of ecto-5′-nucleotidase (KITAKAZE et al. 1993; LENNON et al. 1998; MORABITO et al. 1998; NODE et al. 1997; PEARSON and GORDON 1979).

The extracellular concentration of adenosine is regulated by both cellular uptake of adenosine and adenosine catabolism by adenosine deaminase (ADA) (RESTA et al. 1997; VAN BELLE 1993). Oxidants and proteases released by inflammatory cells may contribute to increased adenosine concentration by inactivating the catabolic enzyme ADA (VAN WAEG and VAN DEN BERGHE 1991). The complex regulation of adenosine concentration may vary depend-

Table 1. Modulation of inflammatory cell function by adenosine receptors

	A1 Receptor	A2A Receptor	A2B Receptor	A3 Receptor
Neutrophil	↑Adhesion ↑Migration ↑Phagocytosis	↓Adhesion ↓O_2-Generation ↓Degranulation ↓Phagocytosis ↓Apoptosis ↓Phlogiston release		↓Degranulation
Monocyte/ macrophage	↑Phagocytosis ↑Giant cell formation	↓O_2– Generation ↓Phagocytosis ↓Cytokine release? ↓Giant cell formation ↓C2 Synthesis ↓Procoagulant secretion ↑Plasminogen activator		↓Cytokine release ↑IL-10 ↓NOS and NO
Eosinophil		↓O_2– Generation?		↓Migration
Mast cell		↓Degranulation?	↑IL-8 Secretion	↑Degranulation ↑Histamine release
Endothelial cell		↓Cytokine release ↑Migration ↑Proliferation ↑Angiogenesis		

ing on type of cell present, the number of necrotic cells, and the generation of other inflammatory mediators. Moreover, several pharmacological agents commonly used for the treatment of inflammatory conditions also increase the adenosine concentration at inflamed sites by mechanisms which have not been fully established but which clearly involve the extracellular dephosphorylation of AMP by ecto-5′-nucleotidase (Morabito et al. 1998). The increase in extracellular adenosine, acting at specific receptors on cell surface, then suppresses the function of the different cell types involved in the inflammatory process (Tables 1 and 2).

C. The Effects of Adenosine on Inflammatory Cells

I. Neutrophils

The first step in the development of inflammation is the recruitment of leukocytes into the inflamed or injured site. Neutrophils, polymorphonuclear leukocytes (PMN), are the most abundant leukocytes in the blood stream. Neutrophils are the first leukocytes recruited to the inflamed site and com-

Table 2. Adenosine receptors modulate phlogiston release

	A_{2A} Receptor	A_{2B} Receptor	A_3 Receptor
Neutrophil	↓TNF-α ↓LTB4		
Monocyte/macrophage	↓TNF-α ↓C2 ↓Procoagulant ↑Plasminogen activator		↑IL-10 ↓TNF-α ↓IL-6 ↓IL-12
Mast cell		↑IL-8	
Endothelial cell	↓IL-6 ↓IL-8		

prise the histological determinant of acute inflammation. The initial step in recruitment of PMN is adhesion of the cells to the endothelium of post-capillaries venules, a process mediated by adhesion molecules present on both neutrophils and endothelial cells (Cronstein and Weissmann 1993). The extravasation of neutrophils then occurs by directed migration or chemotaxis in response to chemotactic gradients. Once in the extravascular space, neutrophils avidly ingest foreign organisms and cellular debris. Although neutrophils are clearly essential elements in host defense, they are also capable of destroying normal undamaged tissue surrounding the injured or infected site by releasing oxygen radicals and proteolytic enzymes. To limit neutrophil accumulation at inflamed sites and to decrease their secretion of potentially injurious oxidants, a variety of endogenous mechanisms have evolved. Adenosine, acting through its specific receptors, is quite prominent among those molecules present at acutely inflamed sites that might be involved in limiting neutrophil-mediated injury.

The first attempt to study the anti-inflammatory effect of adenosine on neutrophils was performed by Marone et al. (1980) who reasoned that, since β-adrenergic agents blocked stimulated neutrophil degranulation, presumably via increasing intracellular cAMP, adenosine, which increases cAMP in other cells, should also inhibit neutrophil degranulation. Surprisingly, adenosine did not affect the chemoattractant formyl-leucyl-phenylalanine (fMLP)-stimulated neutrophil degranulation and it was not until 1983 that Cronstein et al. (1983) first demonstrated that adenosine inhibited superoxide anion generation by neutrophils stimulated by the chemoattractant fMLP, the complement factor C5a, and the calcium ionophore, A23187. Only extracellular adenosine could evoke this inhibitory effect, since dipyridamole, an adenosine uptake blocker, enhanced the effect of adenosine on superoxide anion generation. Subsequent studies have confirmed these initial results and have demonstrated that adenosine inhibits stimulated neutrophil superoxide anion generation by occupying adenosine A_{2A} receptors on the neutrophil surface (Cronstein et al. 1985, 1990; de la Harpe and Nathan 1989; Fredholm et al. 1996; Roberts et al. 1985; van Calker et al. 1991; Zalavary and Bengtsson 1998b).

The effects of adenosine on stimulated degranulation and aggregation are more controversial than its inhibitory effect on superoxide anion generation. The original observation of MARONE et al. (1980), that adenosine does not inhibit azurophil granule release, has been confirmed by several laboratories (CRONSTEIN et al. 1983, 1988; GRINSTEIN and FURUYA 1986; MCGARRITY et al. 1989; WALKER et al. 1989). However, other studies showed that adenosine inhibits stimulated degranulation (RICHTER 1992; SCHMEICHEL and THOMAS 1987), probably via interaction with both A_{2A} and A_3 receptors (BOUMA et al. 1997). In a similar manner, the effects of adenosine on neutrophil aggregation remain controversial (BOUMA et al. 1997; CRONSTEIN et al. 1983, 1988; GRINSTEIN and FURUYA 1986; MCGARRITY et al. 1989; RICHTER 1992; SCHMEICHEL and THOMAS 1987; WALKER et al. 1989). These discrepancies may be explained, in part, by the fact that adenosine has different effects on neutrophil function when different stimuli are used (CRONSTEIN et al. 1983), and by the use of different experimental techniques (CRONSTEIN et al. 1986; HASLETT et al. 1985).

The effects of adenosine on chemotaxis, phagocytosis, and adhesion to vascular endothelial cells are more complex. At low concentration, adenosine promotes neutrophil migration in response to the chemoattractant fMLP and the complement factor C5a (ROSE et al. 1988), induces phagocytosis of immunoglobulin-coated particles (SALMON and CRONSTEIN 1990), and promotes adhesion of stimulated neutrophils to endothelial cells and some surfaces (CRONSTEIN et al. 1986, 1992b). The order of agonist potency and the concentration necessary to obtain these proinflammatory effects on neutrophil function suggest that the adenosine A_1 receptor is involved. In contrast, at higher concentrations, adenosine interacts with adenosine A_{2A} receptors inhibiting these same functions (CRONSTEIN et al. 1992b; SALMON and CRONSTEIN 1990; ZALAVARY et al. 1994). A recent study showed that adenosine inhibits actin polymerization in stimulated neutrophils, reducing their ability to extend pseudopods during phagocytosis (ZALAVARY and BENGTSSON 1998a). Adenosine has been reported to inhibit both β_2-integrin-mediated (CRONSTEIN et al. 1992b) and L-selectin-mediated neutrophil adhesion to vascular endothelium (FIRESTEIN et al. 1995). Subsequent studies demonstrated that adenosine inhibits the upregulation of β_2-integrin and shedding of L-selectin by stimulated neutrophils (THIEL et al. 1996; WOLLNER et al. 1993; ZALAVARY and BENGTSSON 1998a).

A more recently discovered function of neutrophils is the regulation of the synthesis of a variety of inflammatory mediators or cytokines such as leukotriene B_4 (LTB_4) and tumor necrosis factor (TNF-α). High concentrations of TNF-α stimulate endothelial cell expression of adhesion molecules promoting further neutrophil accumulation, and increase the neutrophil's destructive and bactericidal potential by "priming" them. Recent studies suggest that adenosine, via activation of its receptors, inhibits the generation and release of these phlogistons. Judging from the effects of various receptor-specific agonists and antagonists, the receptors involved are A_2-like, most likely A_{2A} (KRUMP et al. 1996; THIEL and CHOUKER 1995).

Priming of neutrophils after exposure to bacterial products (lipopolysaccharide), lipid products (platelet activating factor [PAF]) and cytokines (TNF-α) results in a greater generation of superoxide anion and other oxygen metabolites, and a greater adherence of neutrophils to the endothelium. Adenosine, acting at A_2-like receptors, also inhibits neutrophil priming, although this inhibition is both stimulus- and function-specific (DE LA HARPE and NATHAN 1989; STEWART and HARRIS 1993; SULLIVAN et al. 1990).

The inhibition of neutrophil functions by adenosine may result in diminished bactericidal capacity (HARDART et al. 1991). Indeed, some microorganisms have adopted the release of adenosine into the extracellular milieu as a mechanism to enhance their own survival (SMAIL et al. 1992).

Although inhibition of neutrophil microbiocidal activity may be deleterious, inhibition of the capacity of neutrophils to injure healthy tissue surrounding injured or inflamed sites is clearly beneficial. CRONSTEIN et al. (1986) first reported that adenosine, whether added exogenously or generated endogenously, prevents stimulated neutrophil-mediated injury to endothelial cells. In later studies a number of laboratories confirmed and expanded this original finding (BOISSEAU et al. 1996; JORDAN et al. 1997; LENNON et al. 1998; MINAMINO et al. 1996; VINTEN-JOHANSEN et al. 1995; ZHAO et al. 1996). These studies show that the receptors involved in inhibition of neutrophil-mediated endothelial injury are A_2-like receptors. In addition, adenosine also inhibits stimulated neutrophil-mediated injury of other cell types such as cardiac myocytes (BULLOUGH et al. 1995).

Adenosine modulates neutrophil function via interaction with specific adenosine receptors on the neutrophil surface. Of the four subtypes of adenosine receptors known, only the A_1, A_{2A}, and A_{2B} have been implicated as playing a role in affecting neutrophil function. Signal transduction at neutrophil adenosine A_1 receptors appears to be similar to other cell types, via pertussis toxin-sensitive G proteins. Adenosine A_2-like receptors were first described on the basis of their capacity to stimulate adenylate cyclase and, thus, increase cellular cAMP (LONDOS et al. 1980; VAN CALKER et al. 1979). Adenosine A_2-like receptors on neutrophils also appear to be linked to adenylate cyclase (CRONSTEIN et al. 1988; IANNONE et al. 1985, 1989), however, the role of cAMP as the second intracellular messenger for inhibition of neutrophil function is not well established. Although the effects of cAMP analogs on stimulated neutrophils parallel those of adenosine A_2-like receptor activation, the kinetics of the adenosine mediated increase in cAMP does not correlate with the inhibition of superoxide anion generation (CRONSTEIN et al. 1988; FREDHOLM et al. 1996; IANNONE et al. 1985, 1989). Moreover, inhibitors of protein kinase A, the downstream signaling element for cAMP, reverse the effects of cAMP analogs, but not of adenosine agonists, on superoxide anion generation stimulated by chemoattractants (CRONSTEIN et al. 1992a; REVAN et al. 1996).

Chemoattractant stimulation of neutrophils increases phospholipid turnover, resulting in generation of the active lipid intermediates diacylglycerol (DAG) and inositol 1,4,5-triphosphate (IP_3). Adenosine receptor activation affects neither the early wave of DAG production (CRONSTEIN and HAINES 1992), nor the stimulated generation of IP_3 (WALKER et al. 1990). Similarly, adenosine does not affect the early and rapid increment in intracellular free calcium $[Ca^{2+}]_i$ mediated by IP_3, but markedly inhibits the late sustained fluxes of $[Ca^{2+}]_i$ (THIEL and BARDENHEUER 1992; WALKER et al. 1990), as well as the second wave in DAG production (CRONSTEIN and HAINES 1992).

Another possible signal transduction mechanism for neutrophil adenosine A_{2A} receptors is the phosphorylation or dephosphorylation of various proteins (BENGIS-GARBER and GRUENER 1992). More recently, REVAN et al. (1996) have suggested a novel mechanism. In this study the authors show that inhibitors of protein phosphatases completely reverse the inhibitory effect of adenosine on stimulated superoxide anion, and adenosine receptor activation stimulates a marked increase in plasma membrane-associated serine/threonine protein phosphatase activity. This activated phosphatase may dephosphorylate the receptor itself inhibiting the association of chemoattractant receptors with the cytoskeleton, resulting in receptor desensitization (BURKEY and WEBSTER 1993; CRONSTEIN and HAINES 1992; CRONSTEIN et al. 1991b, 1992a).

The presence of at least two different adenosine receptors on neutrophils has been demonstrated pharmacologically. When activated, these two adenosine receptors mediate opposing effects on stimulated neutrophil function: A_1 receptors enhance neutrophil function at very low concentrations of adenosine, whereas A_2 receptors inhibit neutrophil function at higher adenosine concentration, likely to be obtained in cellular areas in which an excess of neutrophil activity would be detrimental.

II. Monocytes/Macrophages

Monocytes, like neutrophils, migrate from the blood into injured, infected, or inflamed sites by a mechanism that involves endothelial cells and adhesion molecules. Unlike neutrophils, monocytes, once in the extravascular space, differentiate into macrophages and after 12–24h become the predominant histological feature of the inflammatory infiltrate. Macrophages take on different functions characteristic of the tissue where they mature. These large long-lived cells synthesize and release into the environment cytokines and other proteins that play a central role in the development of acute and chronic inflammation and wound healing. Therefore, modulation of monocyte/macrophage functions by pharmacological agents may have a greater impact in both the initiation and the resolution of the inflammatory response, than expected from the local effect on monocyte/macrophage alone.

Adenosine modulates monocyte/macrophage function by activating adenosine receptors on the cell surface. Generally, similar results have been obtained using human monocytes, murine macrophages, and human and murine macrophage cell lines. Like neutrophils, monocytes phagocytose debris and immunoglobulin-coated particles, and A_1 receptors, when activated, enhance phagocytosis, whereas A_2-like receptor activation inhibits phagocytosis (Salmon et al. 1993), and inhibits stimulated generation of superoxide anion generation (Elliott et al. 1986; Leonard et al. 1987). Adenosine also inhibits the synthesis of NO synthase and NO production by stimulated macrophages (Hasko et al. 1996; Hon et al. 1997), probably through activation of A_{2B} receptors (Xaus et al. 1999).

Under certain conditions, such as infection with *Mycobacterium tuberculosis* or in rheumatoid nodules, macrophages fuse into multinucleated giant cells. Merrill et al. (1997) showed that A_1 receptor activation promotes giant cell formation, whereas A_{2A} receptor activation inhibits it.

During resolution of the inflammatory response, macrophages undergo apoptosis. The adenosine receptor agonist 2-CADO induces apoptosis in human peripheral blood monocytes by a complex mechanism that involves A_{2A} and A_3 receptors and adenosine uptake (Barbieri et al. 1998).

Adenosine also modulates protein synthesis and secretion by stimulated monocyte/macrophages. In 1984, it was first demonstrated that adenosine, acting at A_2-like receptors, inhibits synthesis and release of the complement component C2 by stimulated monocytes (Lappin and Whaley 1984). Subsequent studies showed that A_2 receptors are involved in the inhibition of synthesis and release of procoagulant; by contrast, adenosine, acting at a poorly characterized receptor (likely A_3 receptor) promotes the synthesis and secretion of plasminogen activator (Colli and Tremoli 1991; Hasday and Sitrin 1987).

Endogenous adenosine, as well as exogenously added adenosine, inhibits the secretion of the proinflammatory cytokine TNF-α by stimulated monocyte/macrophages, although some controversy exists regarding the receptor involved. Some investigators reported that A_2 receptor activation inhibits the stimulated secretion of TNF-α (Eigler et al. 1997; Prabhakar et al. 1995; Ritchie et al. 1997; Xaus et al. 1999), whereas others reported that A_3 receptors are responsible for the inhibition of TNF-α production (Sajjadi et al. 1996; Hasko et al. 1996; McWhinney et al. 1996). The effect of adenosine A_3 receptor activation on TNF-α synthesis and secretion appears to occur at a transcriptional level, mediated by either NFκB (McWhinney et al. 1996; Hasko et al. 1996) or AP-1 (Sajjadi et al. 1996). Szabo et al. (1998) recently showed that activation of adenosine A_3 receptors, and to a lesser extent A_2-like receptors, suppresses the expression and production of the macrophage inflammatory protein-1 (MIP-1α), a chemokine which enhances neutrophil recruitment into inflammatory sites, in immunostimulated RAW macrophages. The A_3 receptor agonist IB-MECA also inhibited the production of IL-12 and IL-6. These authors also found that IB-MECA significantly reduced the severity of joint inflammation in an *in vivo* model of collagen-induced arthritis in mice.

Adenosine has also been reported to increase monocyte production of some cytokines. Acting at A_2-like receptors, adenosine increased the production of the inflammatory cytokines IL-6 and IL-8 (Le Vraux et al. 1993), an observation only partially confirmed (Ritchie et al. 1997). Interestingly, adenosine enhances the secretion of IL-10, a cytokine that possesses several anti-inflammatory properties, by stimulated monocytes, although the adenosine receptor involved has not been well characterized (HASKO et al. 1996; SZABO et al. 1998).

Recently, XAUS et al. (1999) have reported that interferon-γ (IFN-γ) up-regulates the expression of A_{2B} receptor in macrophages by induction of *de novo* synthesis. The up-regulated receptor is functional, its activation produces accumulation of cAMP, and inhibits the expression of MHC class II genes and NO synthase and proinflammatory cytokines in IFN-γ-stimulated macrophages. Therefore, Xaus and colleagues postulate that the up-regulation of A_{2B} receptor could be a feedback mechanism for macrophage deactivation.

Adenosine, acting at A_3 and/or A_{2A} receptors, promotes the production of the central anti-inflammatory cytokine, IL-10, and inhibits the synthesis of the central inflammatory cytokine TNF-α and the chemokine MIP-1α. Via its effects on cytokine and protein production, receptor mediated effects of adenosine are amplified and may lead to systemic inhibition of inflammation.

III. Eosinophils

Increased numbers of eosinophils in the circulation is characteristic of allergic states and parasitic infections. Eosinophils are commonly present in the lungs of asthmatic patients and at parasite invasion sites. However, the role that these cells play in other inflammatory/host response situations has not been well characterized. Little work has been done to evaluate the effects of adenosine on eosinophil function. WALKER et al. (1997) have demonstrated the presence of A_3 receptors on eosinophils by in situ hybridization and RT-PCR. Activation of these receptors by agonists increases stimulated fluxes in intracellular calcium (KOHNO et al. 1996) and inhibits stimulated eosinophil chemotaxis (KNIGHT et al. 1997). Pretreatment with high concentrations of adenosine primed the platelet activating factor-stimulated oxidant release by eosinophils (WALKER 1996).

IV. Mast Cells

Mast cells differentiate from blood basophils that migrate into the tissues, primarily in the gut and lung. These cells participate in acute hypersensitivity and allergic reactions by releasing mediators, such as histamine, that cause many of the manifestations of hypersensitivity. Mast cells play a critical role in the development of allergic asthma; histamine, released from the granules of stimulated mast cells, is a potent bronchoconstrictor and vasoactive agent (BIAGGIONI and FEOKTISTOV, Chap. 22, this volume; DUBYAK, Chap. 25, this volume).

Adenosine produces a rapid bronchoconstriction when inhaled by asthmatic patients. These effects of adenosine on bronchial smooth muscle are indirectly mediated via activation of mast cell degranulation, since histamine and leukotriene B_4 antagonists block adenosine-induced bronchoconstriction (BJORCK et al. 1992). Adenosine promotes degranulation of mast cells through the interaction with cell surface receptors (reviewed in HOLGATE et al. 1991).

Mast cells express all four known adenosine receptor subtypes (AUCHAMPACH et al. 1997; FEOKTISTOV and BIAGGIONI 1998; MARQUARDT 1998; RAMKUMAR et al. 1993), and therefore the response of mast cells must be complex and may involve more than one adenosine receptor subtype. Initial studies suggested that A_3 receptors are responsible for the promotion of mast cell degranulation (ALI et al. 1996; FOZARD et al. 1996; HANNON et al. 1995; JIN et al. 1997; MEADE et al. 1996; SAWYNOK et al. 1997; SHEPHERD et al. 1996). By contrast, FEOKTISTOV and BIAGGIONI (1995) demonstrated that adenosine provokes IL-8 secretion by mast cells via interaction with A_{2B} receptors, an effect that can be blocked by enprofylline acting as a specific antagonist of A_{2B} receptors. Subsequent studies support the hypothesis that activation of A_{2B} receptors results in mast cell degranulation accompanied by cAMP accumulation and calcium mobilization (AUCHAMPACH et al. 1997; FEOKTISTOV and BIAGGIONI 1998). Under certain experimental conditions, adenosine also can inhibit stimulated mast cell degranulation, an effect mimicked by A_{2A} selective agonists and blocked by specific A_{2A} antagonists (SUZUKI et al. 1998).

V. Endothelial Cells

For many years it was thought that the only role of endothelial cells in inflammation was as a cellular barrier that leukocytes had to traverse in order to reach the injured, infected, or inflamed site. It is now generally accepted that endothelial cells play an active role in the inflammatory response. Endothelial cells, after stimulation by cytokines and endotoxin, regulate the development of the inflammatory lesion by expressing adhesion molecules and releasing chemoattractants, such as platelet activating factor (PAF) and IL-8, that attract leukocytes out of the circulation and stimulate them to migrate into the extravascular space (reviewed in CRONSTEIN and WEISSMANN 1993). Moreover, the new formation of blood vessels, a process known as angiogenesis, participates in the resolution of inflammation in wound healing, and in the tissue destruction of chronic inflammatory pathologies such as the synovial tissue in rheumatoid arthritis.

Endothelial cells express all four known adenosine receptors (MONTESINOS et al. 1997) and the effects of adenosine on endothelial cell function during inflammation are a new area of interest. Adenosine prevents the increase in permeability in endothelial monolayers exposed to activated neutrophils or oxidants (HASELTON et al. 1993; RICHARD et al. 1998), promoting the endothelial barrier function through activation of A_{2B} receptors (LENNON et al. 1998). This could be a basic mechanism of endothelial resealing during leukocyte transmigration (LENNON et al. 1998).

Inflammatory mediators such as TNF-α stimulate tissue factor expression, a principal initiator of coagulation, in endothelial cells, an effect that can be inhibited by adenosine via interaction with A_{2A} receptors (DEGUCHI et al. 1998). Adenosine also diminishes stimulated endothelial expression of E-selectin and intercellular adhesion molecule-1 (ICAM-1) and secretion of the inflammatory cytokines IL-6 and IL-8 (BOUMA et al. 1996). Other studies have shown that adenosine induces endothelial cell migration and proliferation via interaction with specific adenosine receptors (most likely A_{2A} receptors) inducing angiogenesis (ETHIER et al. 1993; SCHIELE and SCHWABE 1994; SEXL et al. 1995; TAKAGI et al. 1996a,b). The proliferative effect of adenosine on endothelial cells is not mediated by cAMP. Interestingly, adenosine A_2-like receptors directly stimulate mitogen-activated protein kinase (MAPK) activity (SEXL et al. 1997), in contrast to the effect of A_{2A} receptor occupancy on MAPK activation in neutrophils (PILLINGER et al. 1996). In summary, adenosine acting at A_2-like receptors on endothelial cells inhibits the recruitment of leukocytes to inflamed sites and promotes angiogenesis – functions associated with resolution of inflammation and wound healing.

D. *In Vivo* Effects of Adenosine on Inflammation

Adenosine, acting at its specific receptors, is able to inhibit the function of inflammatory cells, except eosinophils and mast cells. However, the clinical utility of adenosine or its agonists in the treatment of inflammatory conditions is not feasible, due to the numerous physiological effects of adenosine in other organs. Several pharmacological agents have been developed that raise local adenosine concentration by increasing cellular release of its precursors or by inhibiting adenosine uptake. Thus endogenously released adenosine may be an important physiologic modulator of inflammation.

Adenosine antagonists potentiate inflammation in the hamster cheek pouch model of inflammation, suggesting an active anti-inflammatory effect of endogenously released adenosine *in vivo* (ROSENGREN et al. 1991). Similarly, endogenous adenosine downregulates ischemia-reperfusion injury in some systems (JORDAN et al. 1997; VINTEN-JOHANSEN et al. 1995; ZHAO et al. 1996). By contrast, in other experimental models, exogenous and endogenous adenosine, acting at A_1 receptors, stimulates endothelial expression of adhesion molecules enhancing neutrophil adhesion and neutrophil-mediated myocardial in jury (BECKER et al. 1992; RASCHKE and BECKER 1995; SCHWARTZ et al. 1993; ZAHLER et al. 1994).

Adenosine and adenosine receptor agonists have been shown to have anti-inflammatory properties in several animal experimental models. Surprisingly, A_1 receptor agonists are potent inhibitors of inflammation (LESCH et al. 1991; SCHRIER et al. 1990). One potential explanation for this unexpected finding is that adenosine presents a potent central nervous system (CNS) effect, acting at A_1 receptors on the spinal cord, on inflammation in the periphery. Adenosine administered into the spinal cord markedly reduced cutaneous inflammation, in part, by increasing local release of adenosine (BONG et al. 1996; POON and SAWYNOK 1998; REEVE and DICKENSON 1995). However, it is unlikely that this effect leads to the development of new anti-inflammatory strategies because of the many unwanted CNS effects of A_1 receptor activation. Adenosine, infused directly into the inflamed joints of animals with adjuvant-induced arthritis, markedly inhibited joint inflammation (GREEN et al. 1991), despite the short half-life (2–8 s) of adenosine in biological fluids (MOSER et al. 1989). In a recent study, SZABO et al. (1998) first reported that an adenosine A_3 receptor agonist inhibited joint inflammation in a murine model of collagen-induced arthritis. The authors attribute this particular action of the adenosine A_3 receptor agonist to its inhibitory effects on macrophage functions.

Pharmacological elevation of extracellular adenosine at inflamed sites can be achieved by several approaches: inhibition of adenosine uptake, inhibition of adenosine utilization, and modulation of adenine nucleotide metabolism. Two of these strategies are currently under study for the development of anti-inflammatory agents.

A number of studies have suggested that adenosine uptake blockers may prevent inflammatory injury during reperfusion or in other settings by increasing adenosine concentrations at the inflamed site (VAN BELLE 1993). However, none of the adenosine uptake inhibitors are currently in use to treat inflammatory conditions.

In most cell types, adenosine, generated within the cell as a result of purine nucleotide catabolism or taken up from the outside, is recycled to adenine nucleotides by the enzyme adenosine kinase. Several studies have demonstrated that a specific adenosine kinase inhibitor can diminish inflammation or protect against endotoxin shock, both *in vitro* and *in vivo*. This effect is clearly mediated by adenosine, since elimination of adenosine by ADA or administration of adenosine antagonists completely abrogates the effect of the adenosine kinase inhibitor on inflammation (CRONSTEIN et al. 1995; FIRESTEIN et al. 1993, 1995; ROSENGREN et al. 1995).

Two drugs commonly used for the treatment of rheumatoid arthritis (RA) and inflammatory bowel disease promote adenosine release in vitro and *in vivo* and evidence indicates that adenosine mediates the anti-inflammatory actions of these drugs. Low dose, weekly methotrexate is now one of the most widely used second line agents for the treatment of rheumatoid arthritis and sulfasalazine has recently been approved for the treatment of rheumatoid arthritis. Although it was originally thought that methotrexate diminishes the inflammation of rheumatoid arthritis via folate-dependent inhibition of lymphocyte proliferation, it is now clear that folate, either folic acid or reduced folate (leucovorin), can be administered to patients taking methotrexate without reversing the anti-inflammatory effects of methotrexate (but with diminished toxicity). The pharmacology of methotrexate has been well studied over the years; most of the administered methotrexate is taken up by cells and polyglutamated. Methotrexate polyglutamates are long-lived (weeks) metabolites that, in addition to modulation of folate metabolism, directly inhibit such enzymes as 5-aminoimidazole-4-carboxamide ribonucleotide (AICAR) transformylase. The intracellular accumulation of AICAR has previously been associated with increases in extracellular adenosine and both *in vitro* and *in vivo* studies indicate that, at doses similar to those used to treat rheumatoid arthritis, methotrexate treatment leads to marked increases in intracellular AICAR concentrations and to increases in extracellular adenosine concentrations in inflammatory exudates (CRONSTEIN et al. 1991a, 1993). Indeed, the anti-inflammatory effects of methotrexate are completely reversed by adenosine A_2-like receptor antagonists in *in vivo* models of acute inflammation (CRONSTEIN et al. 1993). Non-selective adenosine receptor antagonists (theophylline and caffeine), but not more selective adenosine receptor antagonists, completely reverse the anti-inflammatory effects of methotrexate in the rat adjuvant model of rheumatoid arthritis (YAP et al. 1997). Similarly, sulfasalazine directly inhibits AICAR transformylase *in vitro* and promotes intracellular AICAR accumulation and increased adenosine concentrations in inflammatory exudates (GADANGI et al. 1996). Evidence from our laboratory indicates that AICAR promotes the release of adenine nucleotide which is converted, ultimately, in the extracellular fluid to adenosine by the action of ecto-5′-nucleotidase (MORABITO et al. 1998). Thus, these potent antirheumatic agents diminish inflammation by promoting an increase in extracellular adenosine and adenosine, acting at its receptor, blocks the inflammatory response at the affected areas.

Aspirin, one of the oldest anti-inflammatory agents still in use, at low and moderate doses mediates its effects via inhibition of prostaglandin synthesis. However, much higher doses are usually used to treat inflammatory diseases such as rheumatic fever or inflammatory arthritis and it is clear that the doses of aspirin which are sufficient to maximally inhibit prostaglandin synthesis are inadequate in these clinical settings. Aspirin and sodium salicylate, which at high doses uncouple oxidative phosphorylation, diminish cellular ATP concentrations with an associated increase in extracellular adenosine concentration (CRONSTEIN et al. 1994). The adenosine released mediates the anti-inflammatory effects of aspirin (and sodium salicylate which does not inhibit prostaglandin synthesis) in a murine model of acute inflammation (CRONSTEIN et al. 1999). Both aspirin and sodium salicylate diminish inflammation even in mice lacking prostaglandin H synthase II (COX-2 knockout mice). Thus, the potent anti-inflammatory effects of high dose salicylates are mediated, at least in part, by adenosine. An increase in expression of COX-2 involving a P2Y-like receptor has been observed in reactive gliosis (BRAMBILLA et al. 2000).

E. Conclusion

Adenosine, acting via all four of its known receptors, modulates the function of nearly every cell involved in the inflammatory process. With some exceptions, adenosine generally suppresses inflammatory cell function. Although the direct use of adenosine receptor agonists may be limited by unwanted side effects mediated by adenosine receptors on uninvolved cells and tissues, it has become apparent that enhanced adenosine release at inflamed sites has been and remains a useful approach to the development of anti-inflammatory drugs.

References

Ali H, Choi OH, Fraundorfer PF, Yamada K, Gonzaga HM, Beaven MA (1996) Sustained activation of phospholipase D via adenosine A_3 receptors is associated with enhancement of antigen- and Ca(2+)-ionophore-induced secretion in a rat mast cell line. J Pharmacol Exp Ther 276:837–845

Auchampach JA, Jin X, Wan TC, Caughey GH, Linden J (1997) Canine mast cell adenosine receptors: cloning and expression of the A_3 receptor and evidence that degranulation is mediated by the A_{2B} receptor. Mol Pharmacol 52:846–860

Barbieri D, Abbracchio MP, Salvioli S, Monti D, Cossarizza A, Ceruti S, Brambilla R, Cattabeni F, Jacobson KA, Franceschi C (1998) Apoptosis by 2-chloro-2′-deoxy-adenosine and 2-chloro-adenosine in human peripheral blood mononuclear cells. Neurochem Int 32:493–504

Becker BF, Zahler S, Raschke P, Schwartz LM, Beblo S, Schrodl W, Kiesl D (1992) Adenosine enhances neutrophil sticking in the coronary system: a novel mechanism contributing to cardiac reperfusion damage. Pharm Pharmacol Letts 2:8–11

Bengis-Garber C, Gruener N (1992) Cross-talk between cAMP and formylmet-leu-phe in human neutrophils: phosphorylation of a 52,000 molecular weight protein. Cell Signal 4:247–260

Bjorck T, Gustafsson LE, Dahlen SE (1992) Isolated bronchi from asthmatics are hyperresponsive to adenosine, which apparently acts indirectly by liberation of leukotrienes and histamine. Am Rev Respir Dis 145:1087–1091

Boisseau MR, Pruvost A, Renard M, Closse C, Belloc F, Seigneur M, Maurel A (1996) Effect of buflomedil on the neutrophil-endothelial cell interaction under inflammatory and hypoxia conditions. Haemostasis 26 Suppl 4:182–188

Bong GW, Rosengren S, Firestein GS (1996) Spinal cord adenosine receptor stimulation in rats inhibits peripheral neutrophil accumulation. The role of *N*-methyl-D-aspartate receptors. J Clin Invest 98:2779–2785

Bouma MG, Jeunhomme TMMA, Boyle DL, Dentener MA, Voitenok NN, van den Wildenberg FAJM, Buurman WA (1997) Adenosine inhibits neutrophil degranulation in activated human whole blood; involvement of adenosine A_2 and A_3 receptors. J Immunol 158:5400–5408

Bouma MG, van den Wildenberg FAJM, Buurman WA (1996) Adenosine inhibits cytokine release and expression of adhesion molecules by activated human endothelial cells. Am J Physiol 39:C522–C529

Brambilla R, Ceruti S, Malorni W, Caltobeni F, Abbracchio MP (2000) A novel gliotic P2 receptor mediating cyclotexygenase-2 induction in rat and human astocytic. J Ant Nerv System 81:3–9

Bullough DA, Magill MJ, Firestein GS, Mullane KM (1995) Adenosine activates A_2 receptors to inhibit neutrophil adhesion and injury to isolated cardiac myocytes. J Immunol 155:2579–2586

Burkey TH, Webster RO (1993) Adenosine inhibits fMLP-stimulated adherence and superoxide anion generation by human neutrophils at an early step in signal transduction. Biochim Biophys Acta 1175:312–318

Colli S, Tremoli E (1991) Multiple effects of dipyridamole on neutrophils and mononuclear leukocytes: adenosine-dependent and adenosine-independent mechanisms. J Lab Clin Med 118:136–145

Cronstein BN, Duguma L, Nicholls D, Hutchison A, Williams M (1990) The adenosine/neutrophil paradox resolved. Human neutrophils possess both A_1 and A_2 receptors which promote chemotaxis and inhibit O_2- generation, respectively. J Clin Invest 85:1150–1157

Cronstein BN, Eberle MA, Gruber HE, Levin RI (1991a) Methotrexate inhibits neutrophil function by stimulating adenosine release from connective tissue cells. Proc Natl Acad Sci (USA) 88:2441–2445

Cronstein BN, Haines KA (1992) Adenosine A_2 receptor occupancy does not affect "triggering" but inhibits "activation" of human neutrophils by a mechanism independent of actin filament formation. Biochem J 281:631–635

Cronstein BN, Haines KA, Kolasinski SL, Reibman J (1991b) Gs linked receptors (Beta-adrenergic and adenosine A_2) uncouple chemoattractant receptors from G proteins. Clin Res 39:343 A
Cronstein BN, Haines KA, Kolasinski SL, Reibman J (1992a) Occupancy of G alpha s-linked receptors uncouples chemoattractant receptors from their stimulus-transduction mechanisms in the neutrophil. Blood 80:1052–1057
Cronstein BN, Kramer SB, Rosenstein ED, Korchak HM, Weissmann G, Hirschhorn R (1988) Occupancy of adenosine receptors raises cyclic AMP alone and in synergy with occupancy of chemoattractant receptors and inhibits membrane depolarization. Biochem J 252:709–715
Cronstein BN, Kramer SB, Weissmann G, Hirschhorn R (1983) Adenosine: a physiological modulator of superoxide anion generation by human neutrophils. J Exper Med 158:1160–1177
Cronstein BN, Levin RI, Belanoff J, Weissmann G, Hirschhorn R (1986) Adenosine: an endogenous inhibitor of neutrophil-mediated injury to endothelial cells. J Clin Invest 78:760–770
Cronstein BN, Levin RI, Philips MR, Hirschhorn R, Abramson SB, Weissmann G (1992b) Neutrophil adherence to endothelium is enhanced via adenosine A_1 receptors and inhibited via adenosine A_2 receptors. J Immunol 148:2201–2206
Cronstein BN, Montesinos MC, Weissmann G (1999) Salicylates and sulfasalazine, but not glucocorticoids, inhibit leukocyte accumulation by an adenosine-dependent mechanism that is independent of inhibition of prostaglandin synthesis and p105 of NFkB. Proc Natl Acad Sci USA 96:6377–6381
Cronstein BN, Naime D, Firestein G (1995) The antiinflammatory effects of an adenosine kinase inhibitor are mediated by adenosine. Arthritis Rheum 38:1040–1045
Cronstein BN, Naime D, Ostad E (1993) The antiinflammatory mechanism of methotrexate: increased adenosine release at inflamed sites diminishes leukocyte accumulation in an *in vivo* model of inflammation. J Clin Invest 92:2675–2682
Cronstein BN, Rosenstein ED, Kramer SB, Weissmann G, Hirschhorn R (1985) Adenosine; a physiologic modulator of superoxide anion generation by human neutrophils. Adenosine acts via an A_2 receptor on human neutrophils. J Immunol 135:1366–1371
Cronstein BN, Van de Stouwe M, Druska L, Levin RI, Weissmann G (1994) Nonsteroidal antiinflammatory agents inhibit stimulated neutrophil adhesion to endothelium: adenosine dependent and independent mechanisms. Inflamm 18:323–335
Cronstein BN, Weissmann G (1993) The adhesion molecules of inflammation. Arth Rheum 36:147–157
de la Harpe J, Nathan CF (1989) Adenosine regulates the respiratory burst of cytokine-triggered human neutrophils adherent to biologic surfaces. J Immunol 143:596–602
Deguchi H, Takeya H, Urano H, Gabazza EC, Zhou H, Suzuki K (1998) Adenosine regulates tissue factor expression on endothelial cells. Thromb Res 91:57–64
Eigler A, Greten TF, Sinha B, Haslberger C, Sullivan GW, Endres S (1997) Endogenous adenosine curtails lipopolysaccharide-stimulated tumour necrosis factor synthesis. Scand J Immunol 45:132–139
Elliott KRF, Stevenson HC, Miller PJ, Leonard EJ (1986) Synergistic action of adenosine and formyl-met-leu-phe in raising cyclic AMP content of purified human monocytes. Biochem Biophys Res Comm 138:1376–1382
Ethier MF, Chander V, Dobson JG, Jr. (1993) Adenosine stimulates proliferation of human endothelial cells in culture. Am J Physiol 265:H131–H138
Feoktistov I, Biaggioni I (1995) Adenosine A2b receptors evoke interleukin-8 secretion in human mast cells. An enprofylline-sensitive mechanism with implications for asthma. J Clin Invest 96:1979–1986
Feoktistov I, Biaggioni I (1998) Pharmacological characterization of adenosine A_{2B} receptors: studies in human mast cells co-expressing A_{2A} and A_{2B} adenosine receptor subtypes. Biochem Pharmacol 55:627–633

Firestein GS, Bullough D, Erion M, Ugarkar B, Browne G, Barankiewicz J, Gruber H, Mullane K (1993) Adenosine regulating agents: a novel approach to acute inflammation and inflammatory arthritis. Clin Res 41:170A

Firestein GS, Bullough DA, Erion MD, Jimenez R, Ramirez-Weinhouse M, Barankiewicz J, Smith CW, Gruber HE, Mullane KM (1995) Inhibition of neutrophil adhesion by adenosine and an adenosine kinase inhibitor: the role of selectins. J Immunol 154:326–334

Fozard JR, Pfannkuche HJ, Schuurman HJ (1996) Mast cell degranulation following adenosine A_3 receptor activation in rats. Eur J Pharmacol 298:293–297

Fredholm BB, Zhang Y, van der Ploeg I (1996) Adenosine A_{2A} receptors mediate the inhibitory effect of adenosine on formyl-Met-Leu-Phe-stimulated respiratory burst in neutrophil leucocytes. Naunyn-Schmiedebergs Arch Pharmacol 354:262–267

Gadangi P, Longaker M, Naime D, Levin RI, Recht PA, Montesinos MC, Buckley MT, Carlin G, Cronstein BN (1996) The antiinflammatory mechanism of sulfasalazine is related to adenosine release at inflamed sites. J Immunol 156:1937–1941

Green PG, Basbaum AI, Helms C, Levine JD (1991) Purinergic regulation of bradykinin-induced plasma extravasation and adjuvant-induced arthritis in the rat. Proc Natl Acad Sci USA 88:4162–4165

Grinstein S, Furuya W (1986) Cytoplasmic pH regulation in activated human neutrophils: effects of adenosine and pertussis toxin on Na+/H+ exchange and metabolic acidification. Biochim Biophys Acta 889:301–309

Hannon JP, Pfannkuche HJ, Fozard JR (1995) A role for mast cells in adenosine A_3 receptor-mediated hypotension in the rat. Br J Pharmacol 115:945–952

Hardart GE, Sullivan GW, Carper HT, Mandell GL (1991) Adenosine and 2-phenylaminoadenosine (CV-1808) inhibit human neutrophil bactericidal function. Infect Immun 59:885–889

Hasday JD, Sitrin RG (1987) Adenosine receptors on rabbit alveolar macrophages: binding characteristics and effects on cellular function. J Lab Clin Med 110:264–273

Haselton FR, Alexander JS, Mueller SN (1993) Adenosine decreases permeability of *in vitro* endothelial monolayers. J Appl Physiol 74:1581–1590

Hasko G, Szabo C, Nemeth ZH, Kvetan V, Pastores SM, Vizi ES (1996) Adenosine receptor agonists differentially regulate Il-10, TNF-α, and nitric oxide production in RAW 264.7 macrophages and in endotoxemic mice. J Immunol 157:4634–4640

Haslett C, Guthrie LA, Kopaniak MM, Johnston RB, Jr., Henson PM (1985) Modulation of multiple neutrophil functions by preparative methods or trace concentrations of bacterial lipopolysaccharide. Am J Path 119:101–110

Holgate ST, Church MK, Polosa R (1991) Adenosine: a positive modulator of airway inflammation in asthma. Ann New York Acad Sci 629:227–236

Hon WM, Moochhala S, Khoo HE (1997) Adenosine and its receptor agonists potentiate nitric oxide synthase expression induced by lipopolysaccharide in RAW 264.7 murine macrophages. Life Sci 60:1327–1335

Iannone MA, Reynolds-Vaughn R, Wolberg G, Zimmerman TP (1985) Human neutrophils possess adenosine A_2 receptors. Fed Proc 44:580

Iannone MA, Wolberg G, Zimmerman TP (1989) Chemotactic peptide induces cAMP elevation in human neutrophils by amplification of the adenylate cyclase response to endogenously produced adenosine. J Biol Chem 264:20177–20180

Jin X, Shepherd RK, Duling BR, Linden J (1997) Inosine binds to A_3 adenosine receptors and stimulates mast cell degranulation. J Clin Invest 100:2849–2857

Jordan JE, Zhao ZQ, Sato H, Taft S, Vinten-Johansen J (1997) Adenosine A_2 receptor activation attenuates reperfusion injury by inhibiting neutrophil accumulation, superoxide generation and coronary endothelial adherence. J Pharmacol Exp Ther 280:301–309

Kitakaze M, Hori M, Morioka T, Takashima S, Minamino T, Sato H, Inoue M, Kamada T (1993) Attenuation of ecto-5′-nucleotidase activity and adenosine release in activated human polymorphonuclear leukocytes. Circ Res 73:524–533

Knight D, Zheng X, Rocchini C, Jacobson M, Bai T, Walker B (1997) Adenosine A_3 receptor stimulation inhibits migration of human eosinophils. J Leukoc Biol 62:465–468

Kohno Y, Ji X, Mawhorter SD, Koshiba M, Jacobson KA (1996) Activation of A_3 adenosine receptors on human eosinophils elevates intracellular calcium. Blood 88:3569–3574

Krump E, Lemay G, Borgeat P (1996) Adenosine A_2 receptor-induced inhibition of leukotriene B4 synthesis in whole blood ex vivo. Brit J Pharm 117:1639–1644

Lappin D, Whaley K (1984) Adenosine A_2 receptors on human monocytes modulate C2 production. Clin Exp Immunol 57:454–460

Le Vraux V, Chen YL, Masson I, De Sousa M, Giroud JP, Florentin I, Chauvelot-Moachon L (1993) Inhibition of human monocyte TNF production by adenosine receptor agonists. Life Sci 52:1917–1924

Lennon PF, Taylor CT, Stahl GL, Colgan SP (1998) Neutrophil-derived 5′-adenosine monophosphate promotes endothelial barrier function via CD73-mediated conversion to adenosine and endothelial A_{2B} receptor activation. J Exp Med 188:1433–1443

Leonard EJ, Shenai A, Skeel A (1987) Dynamics of chemotactic peptide-induced superoxide generation by human monocytes. Inflammation 11:229–240

Lesch ME, Ferin MA, Wright CD, Schrier DJ (1991) The effects of (*R*)-*N*-(1-methyl-2-phenylethyl) adenosine (L-PIA), a standard A_1-selective adenosine agonist on rat acute models of inflammation and neutrophil function. Agents Actions 34:25–27

Londos C, Cooper DMF, Wolff J (1980) Subclasses of external adenosine receptors. Proc Natl Acad Sci USA 77:2551–2554

Marone G, Thomas L, Lichtenstein L (1980) The role of agonists that activate adenylate cyclase in the control of cAMP metabolism and enzyme release by human polymorphonuclear leukocytes. J Immunol 125:2277–2283

Marquardt DL (1998) Mast cell adenosine receptor characteristics and signaling. Adv Exp Med Biol 431:79–82

McGarrity ST, Stephenson AH, Webster RO (1989) Regulation of human neutrophil functions by adenine nucleotides. J Immunol 142:1986–1994

Meade CJ, Mierau J, Leon I, Ensinger HA (1996) *In vivo* role of the adenosine A_3 receptor: N6-2-(4-aminophenyl)ethyladenosine induces bronchospasm in BDE rats by a neurally mediated mechanism involving cells resembling mast cells. J Pharmacol Exp Ther 279:1148–1156

Merrill JT, Shen C, Schreibman D, Coffey D, Zakharenko O, Fisher R, Lahita RG, Salmon J, Cronstein BN (1997) Adenosine A_1 receptor promotion of multinucleated giant cell formation by human monocytes: a mechanism for methotrexate-induced nodulosis in rheumatoid arthritis. Arth Rheum 40:1308–1315

Minamino T, Kitakaze M, Node K, Funaya H, Inoue M, Hori M, Kamada T (1996) Adenosine inhibits leukocyte-induced vasoconstriction. Amer J Physiol 271: H2622–2628

Montesinos MC, Gadangi P, Longaker M, Sung J, Levine J, Nilsen D, Reibman J, Li M, Jiang CK, Hirschhorn R, Recht PA, Ostad E, Levin RI, Cronstein BN (1997) Wound healing is accelerated by agonists of adenosine A_2 (G alpha s-linked) receptors. J Exp Med 186:1615–1620

Morabito L, Montesinos MC, Schreibman DM, Balter L, Thompson LF, Resta R, Carlin G, Huie MA, Cronstein BN (1998) Methotrexate and sulfasalazine promote adenosine release by a mechanism that requires ecto-5′-nucleotidase-mediated conversion of adenine nucleotides. J Clin Invest 101:295–300

Moser GH, Schrader J, Deussen A (1989) Turnover of adenosine in plasma of human and dog blood. Am J Physiol 256:C799–C806

Node K, Kitakaze M, Minamino T, Tada M, Inoue M, Hori M, Kamada T (1997) Activation of ecto-5′-nucleotidase by protein kinase C and its role in ischaemic tolerance in the canine heart. Br J Pharmacol 120:273–281

Pearson JD, Gordon JL (1979) Vascular endothelial and smooth muscle cells in culture selectively release adenine nucleotides. Nature 281:384–386

Pillinger MH, Feoktistov AS, Capodici C, Solitar B, Levy J, Oei TT, Philips MR (1996) Mitogen-activated protein kinase in neutrophils and enucleate neutrophil cytoplasts: evidence for regulation of cell-cell adhesion. J Biol Chem 271:12049–12056

Poon A, Sawynok J (1998) Antinociception by adenosine analogs and inhibitors of adenosine metabolism in an inflammatory thermal hyperalgesia model in the rat. Pain 74:235–245

Prabhakar U, Brooks DP, Lipshlitz D, Esser KM (1995) Inhibition of LPS-induced TNF alpha production in human monocytes by adenosine (A_2) receptor selective agonists. Inte J Immunopharmacol 17:221–224

Ramkumar V, Stiles GL, Beaven MA, Ali H (1993) The A_3 adenosine receptor is the unique adenosine receptor which facilitates release of allergic mediators in mast cells. J Biol Chem 268:16887–16890

Raschke P, Becker BF (1995) Adenosine and PAF dependent mechanisms lead to myocardial reperfusion injury by neutrophils after brief ischemia. Cardiovas Res 29:569–576

Reeve AJ, Dickenson AH (1995) The roles of spinal adenosine receptors in the control of acute and more persistent nociceptive responses of dorsal horn neurones in the anaesthetized rat. Br J Pharmacol 116:2221–2228

Resta R, Hooker SW, Laurent AB, Jamshedur Rahman SM, Franklin M, Knudsen TB, Nadon NL, Thompson LF (1997) Insights into thymic purine metabolism and adenosine deaminase deficiency revealed by transgenic mice overexpressing ecto-5′-nucleotidase (CD73). J Clin Invest 99:676–683

Revan S, Montesinos MC, Naime D, Landau S, Cronstein BN (1996) Adenosine A_2 receptor occupancy regulates stimulated neutrophil function via activation of a serine/threonine protein phosphatase. J Biol Chem 271:17114–17118

Richard LF, Dahms TE, Webster RO (1998) Adenosine prevents permeability increase in oxidant-injured endothelial monolayers. Am J Physiol 274:H35–H42

Richter J (1992) Effect of adenosine analogues and cAMP-raising agents on TNF-, GM-CSF-, and chemotactic peptide-induced degranulation in single adherent neutrophils. J Leukoc Biol 51:270–275

Ritchie PK, Spangelo BL, Krzymowski DK, Rossiter TB, Kurth E, Judd AM (1997) Adenosine increases interleukin 6 release and decreases tumour necrosis factor release from rat adrenal zona glomerulosa cells, ovarian cells, anterior pituitary cells, and peritoneal macrophages. Cytokine 9:187–198

Roberts PA, Newby AC, Hallett MB, Campbell AK (1985) Inhibition by adenosine of reactive oxygen metabolite production by human polymorphonuclear leucocytes. Biochem J 227:669–674

Rose FR, Hirschhorn R, Weissmann G, Cronstein BN (1988) Adenosine promotes neutrophil chemotaxis. J Exp Med 167:1186–1194

Rosengren S, Arfors KE, Proctor KG (1991) Potentiation of leukotriene B4-mediated inflammatory response by the adenosine antagonist, 8-phenyl theophylline. Intl J Microcirc Clin Exp 10:345–357

Rosengren S, Bong GW, Firestein GS (1995) Anti-inflammatory effects of an adenosine kinase inhibitor: decreased neutrophil accumulation and vascular leakage. J Immunol 154:5444–5451

Sajjadi FG, Takabayashi K, Foster AC, Domingo RC, Firestein GS (1996) Inhibition of TNF-alpha expression by adenosine: role of A_3 adenosine receptors. J Immunol 156:3435–3442

Salmon JE, Brownlie C, Brogle N, Edberg JC, Chen B-X, Erlanger BF (1993) Human mononuclear phagocytes express adenosine A_1 receptors: a novel mechanism for differential regulation of Fc-gamma receptor function. J Immunol 151:2775–2765

Salmon JE, Cronstein BN (1990) Fc gamma receptor-mediated functions in neutrophils are modulated by adenosine receptor occupancy. A_1 receptors are stimulatory and A_2 receptors are inhibitory. J Immunol 145:2235–2240

Sawynok J, Zarrindast MR, Reid AR, Doak GJ (1997) Adenosine A_3 receptor activation produces nociceptive behaviour and oedema by release of histamine and 5-hydroxytryptamine. Eur J Pharmacol 333:1–7

Schiele JO, Schwabe U (1994) Characterization of the adenosine receptor in microvascular coronary endothelial cells. Eur J Pharmacol 269:51–58

Schmeichel CJ, Thomas LL (1987) Methylxanthine bronchodilators potentiate multiple human neutrophil functions. J Immunol 138:1896–1903

Schrier DJ, Lesch ME, Wright CD, Gilbertsen RB (1990) The antiinflammatory effects of adenosine receptor agonists on the carrageenan-induced pleural inflammatory response in rats. J Immunol 145:1874–1879

Schwartz LM, Raschke P, Becker BF, Gerlach E (1993) Adenosine contributes to neutrophil-mediated loss of myocardial function in post-ischemic guinea-pig hearts. J Mol Cell Cardiol 25:927–938

Sexl V, Mancusi G, Baumgartner-Parzer S, Schutz W, Freissmuth M (1995) Stimulation of human umbilical vein endothelial cell proliferation by A_2-adenosine and beta 2-adrenoceptors. Br J Pharmacol 114:1577–1586

Sexl V, Mancusi G, Holler C, Gloria-Maercker E, Schutz W, Freissmuth M (1997) Stimulation of the mitogen-activated protein kinase via the A_{2A}-adenosine receptor in primary human endothelial cells. J Biol Chem 272:5792–5799

Shepherd RK, Linden J, Duling BR (1996) Adenosine-induced vasoconstriction *in vivo*. Role of the mast cell and A_3 adenosine receptor. Circ Res 78:627–634

Smail EH, Cronstein BN, Meshulam T, Esposito AL, Ruggeri RW, Diamond RD (1992) *In vitro*, *Candida albicans* releases the immune modulator adenosine and a second, high-molecular weight agent that blocks neutrophil killing. J Immunol 148: 3588–3595

Stewart AG, Harris T (1993) Adenosine inhibits platelet-activating factor, but not tumour necrosis factor-alpha-induced priming of human neutrophils. Immunology 78:152–158

Sullivan GW, Linden J, Hewlett EL, Carper HT, Hylton JB, Mandell GL (1990) Adenosine and related compounds counteract tumor necrosis factor-alpha inhibition of neutrophil migration: implication of a novel cyclic AMP-independent action on the cell surface. J Immunol 145:1537–1544

Suzuki H, Takei M, Nakahata T, Fukamachi H (1998) Inhibitory effect of adenosine on degranulation of human cultured mast cells upon cross-linking of Fc epsilon RI. Biochem Biophys Res Commun 242:697–702

Szabo C, Scott GS, Virag L, Egnaczyk G, Salzman AL, Shanley TP, Hasko G (1998) Suppression of macrophage inflammatory protein (MIP)-1 alpha production and collagen-induced arthritis by adenosine receptor agonists. Br J Pharmacol 125: 379–387

Takagi H, King GL, Ferrara N, Aiello LP (1996a) Hypoxia regulates vascular endothelial growth factor receptor *KDR/Flk* gene expression through adenosine A_2 receptors in retinal capillary endothelial cells. Invest Ophth Vis sci 37:1311–1321

Takagi H, King GL, Robinson GS, Ferrara N, Aiello LP (1996b) Adenosine mediates hypoxic induction of vascular endothelial growth factor in retinal pericytes and endothelial cells. Invest Ophthalmol Vis Sci 37:2165–2176

Thiel M, Bardenheuer H (1992) Regulation of oxygen radical production of human polymorphonuclear leukocytes by adenosine: the role of calcium. Pflugers Arch 420:522–528

Thiel M, Chambers JD, Chouker A, Fischer S, Zourelidis C, Bardenheuer HJ, Arfors KE, Peter K (1996) Effect of adenosine on the expression of beta(2) integrins and L-selectin of human polymorphonuclear leukocytes in vitro. J Leukoc Biol 59:671–682

Thiel M, Chouker A (1995) Acting via A_2 receptors, adenosine inhibits production of tumor necrosis factor-alpha of endotoxin-stimulated polymorphonuclear leukocytes. J Lab Clin Med 126:275–282

Van Belle H (1993) Nucleoside transport inhibition: a therapeutic approach to cardioprotection via adenosine? Cardiovascular Res 27:68–76

van Calker D, Muller M, Hamprecht B (1979) Adenosine regulates, via two different types of receptors, the accumulation of cyclic AMP in cultured brain cells. J Neurochem 33:999–1005

van Calker D, Steber R, Klotz KN, Greil W (1991) Carbamazepine distinguishes between adenosine receptors that mediate different second messenger responses. E J Pharmacol. 206:285–290

van Waeg G, Van den Berghe G (1991) Purine catabolism in polymorphonuclear neutrophils; phorbol myristate acetate-induced accumulation of adenosine owing to inactivation of extracellularly released adenosine deaminase. J Clin Invest 87:305–312

Vinten-Johansen J, Zhao ZQ, Sato H (1995) Reduction in surgical ischemic-reperfusion injury with adenosine and nitric oxide therapy. Ann Thorac Surg 60:852–857

Walker BA (1996) Effects of adenosine on guinea pig pulmonary eosinophils. Inflammation 20:11–21

Walker BA, Jacobson MA, Knight DA, Salvatore CA, Weir T, Zhou D, Bai TR (1997) Adenosine A_3 receptor expression and function in eosinophils. Am J Respir Cell Mol Biol 16:531–537

Walker BAM, Cunningham TW, Freyer DR, Todd RF, III, Johnson KJ, Ward PA (1989) Regulation of superoxide responses of human neutrophils by adenine compounds. Independence of requirement for cytoplasmic granules. Lab Invest 61:515–521

Walker BAM, Hagenlocker BE, Douglas VK, Ward PA (1990) Effects of adenosine on inositol 1,4,5-trisphosphate formation and intracellular calcium changes in formyl-met-leu-phe-stimulated human neutrophils. J Leukoc Biol 48:281–283

Wollner A, Wollner S, Smith JB (1993) Acting via A_2 receptors, adenosine inhibits the upregulation of Mac-1 (CD11b/CD18) expression on FMLP-stimulated neutrophils. Am J Resp Cell Mol Biol 9:179–185

Xaus J, Mirabet M, Lloberas J, Soler C, Lluis C, Franco R, Celada A (1999) IFN-gamma up-regulates the A_{2B} adenosine receptor expression in macrophages: a mechanism of macrophage deactivation. J Immunol 162:3607–3614

Yap JS, Montesinos MC, McCrary CT, Cronstein BN (1997) Theophylline reverses the effect of methotrexate on adjuvant arthritis: evidence that adenosine mediates the antiinflammatory effects of methotrexate. Arth Rheum 40:S98

Zahler S, Becker BF, Raschke P, Gerlach E (1994) Stimulation of endothelial adenosine A_1 receptors enhances adhesion of neutrophils in the intact guinea pig coronary system. Cardiovasc Res 28:1366–1372

Zalavary S, Bengtsson T (1998a) Adenosine inhibits actin dynamics in human neutrophils: evidence for the involvement of cAMP. Eur J Cell Biol 75:128–139

Zalavary S, Bengtsson T (1998b) Modulation of the chemotactic peptide- and immunoglobulin G-triggered respiratory burst in human neutrophils by exogenous and endogenous adenosine. Eur J Pharmacol 354:215–225

Zalavary S, Stendahl O, Bengtsson T (1994) The role of cyclic AMP, calcium and filamentous actin in adenosine modulation of Fc receptor-mediated phagocytosis in human neutrophils. Biochim Biophys Acta 1222:249–256

Zhao ZQ, Sato H, Williams MW, Fernandez AZ, Vinten-Johansen J (1996) Adenosine A_2-receptor activation inhibits neutrophil-mediated injury to coronary endothelium. Am J Physiol. 271:H1456–H1464

CHAPTER 25

Role of P2 Receptors in the Immune System

G. R. DUBYAK

A. Introduction

In many tissues, stressed or damaged cells can locally release nucleotides which, in turn, activate autocrine or paracrine signaling cascades that permit those tissues to adapt to the particular stress or injury that precipitated the release of cellular nucleotides (SCHWIEBERT 1999). In complex organisms, the introduction of foreign material – living or nonliving – into any organ or tissue constitutes a major stress that triggers the powerful adaptive responses of the immune system. Pathogenic microorganisms that injure or kill cells of the host organism often elicit these responses. In turn, the response of the host organism often involves the killing and removal of the invading microorganisms. The immune system also provides a surveillance mechanism for identifying and eliminating any host cells that might be irreversibly damaged or diseased.

The overall immune response involves three distinct groups of integrated host responses: (i) native immunity (also known as natural or intrinsic immunity), (ii) acquired immunity, and (iii) inflammation. Native immunity comprises those responses that permit the recognition and elimination of foreign organisms and material in the absence of any prior exposure of the host to such foreign matter. When trauma due to wounds or disease causes a breakdown of barrier tissues (skin and mucosal layers of the gut and airways), invading pathogens are targeted by immune effector cells that either reside in normal tissues or are rapidly recruited to damaged tissues (Fig. 1). These include monocytes, macrophages, neutrophils, and natural killer cells that can recognize microorganisms by antigen-independent mechanisms and eliminate them via phagocytosis or secretion of cytotoxic molecules. In addition, microbial-derived macromolecules, such as bacterial lipopolysaccharide or lipopeptides, can bind to specific receptors on these effector cells, resulting in the synthesis and release of cytokines that amplify the natural immune response or potentiate activation of the lymphocytes responsible for acquired immunity. Acquired or specific responses of the host organism to foreign material are the mechanisms generally termed "immunity" or "immune response." These represent defensive responses by the host organism to foreign substances (antigens) to which the organism has been previously exposed. As with

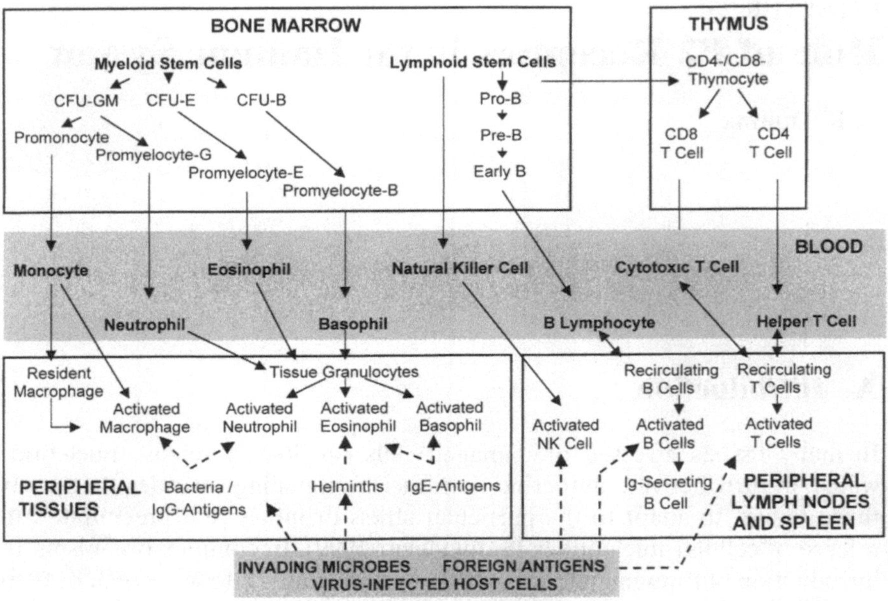

Fig. 1. Pathways of leukocyte development, tissue trafficking, and activation. The pathways for intra-tissue development and inter-tissue trafficking are shown as *solid arrows*. Pathways for the activation of leukocytes by various microbial, immune and, inflammatory stimuli are shown as *dashed arrows*. *CFU*, colony-forming unit; *GM*, granulocyte/macrophage; *E*, eosinophil; *B*, basophil; *Ig*, immunoglobulin

natural immunity, acquired immunity involves both immune effector cells, of which T and B lymphocytes are central players, and circulating macromolecules in the form of antibodies secreted by B lymphocytes. Particular foreign antigens act as selective ligands for cell surface receptors [the membrane immunoglobulin receptor (mIg) and the T cell receptor (TCR) in B or T cells, respectively] on clonal lymphocytes resulting in activation and proliferation of those particular lymphocytes. Finally, conditions that trigger immune responses often elicit inflammatory responses that facilitate recognition and repair of injured host tissues. Indeed, inflammation and immunity involve many common effector cells such as monocytes, macrophages, lymphocytes, and mast cells. Since stress, injury, and death at the cellular level represent overlapping aspects of immune/inflammatory responsiveness and nucleotide-based signaling, considerable attention has been directed towards characterizing the role of extracellular nucleotides and P2 receptors in modulating various immune and inflammatory responses.

This chapter summarizes recent progress in defining key aspects of nucleotide signaling in immune and inflammatory effector cells. Certain cells, such as monocyte/macrophages play multiple, often overlapping, roles in natural immunity, acquired immunity, and inflammation. Others, such as natural killer cells (NK cells) and eosinophils, are involved in more specialized aspects of natural or acquired immune responses. As a framework (Fig. 1) for systematic presentation of the many distinct pathways and cell types, immune effector cells have been "histologically" grouped as *mononuclear phagocytes* (monocytes and the various types of monocyte-derived macrophages), *granulocytes* (neutrophils, eosinophils, basophils, and mast cells), and *lymphocytes* (T cells, B cells, and NK cells). Immune and inflammatory responses are also strongly modulated by signals from specialized cell types which have other important functions unrelated to immunity and inflammation. Such immune "accessory" cells include endothelial cells, most types of epithelial cells, platelets, connective tissue cells, and smooth muscle cells (vascular and visceral). However, other chapters in this volume describe in detail the expression and functions of P1 and P2 receptors in these cells.

B. Mononuclear Phagocytes: Monocytes, Macrophages, Microglial Cells, and Dendritic Cells

Myeloid leukocytes which include neutrophils, monocytes, and the many subtypes of macrophages play central roles in natural immunity. Circulating blood monocytes are the progenitors of all macrophages (Fig. 1). Even in the absence of microbial infection, some monocytes are recruited to specific tissues wherein they differentiate into long-lived, resident macrophages, such as skin macrophages, alveolar macrophages. liver macrophages (Kupfer cells), and brain macrophages (microglial cells). However, large numbers of monocytes are locally recruited as needed to sites of microbial invasion; these acutely recruited monocytes rapidly acquire the morphological and functional characteristics of macrophages. Most analyses of P2 receptor expression have utilized human monocyte/macrophages, murine macrophages, and murine microglial cells; a limited number of studies have used rat or guinea pig macrophages and murine skin-derived dendritic cell lines (which have a macrophage lineage). In general, similar patterns of P2 subtype expression have been observed in macrophages from all of these species.

I. Expression of P2 Receptor Subtypes in Mononuclear Phagocytes

1. P2Y Receptors

Functional studies have indicated that ATP and UTP are equipotent agonists in activating mobilization of intracellular Ca^{2+} stores in human blood monocytes (Cowen et al. 1991; Falzoni et al. 1995), human monocyte-derived macrophages (Falzoni et al. 1995), human bone marrow fractions that contain monocyte progenitor cells (Cowen et al. 1991), murine peritoneal macrophages (Alonso-Torre and Trautmann 1993), murine macrophage cell lines (Greenberg et al. 1988; Chen et al. 1998), and murine antigen-presenting dendritic cells (Coutinho-Silva et al. 1999; Mutini et al. 1999). This suggests that all of these cells express a $P2Y_2$-like receptor. Subsequent RT-PCR analyses has verified expression of $P2Y_2$ mRNA in human monocytes (Jin et al. 1998), human bone marrow cells (Clifford et al. 1997), and THP-1 human leukemic monocytes (Martin et al. 1997). $P2Y_2$ receptor mRNA and function is strongly down-regulated when human THP-1 monocytic leukemia cells are treated with phorbol esters (Martin et al. 1997) or with physiological proinflammatory agents such as interferon-γ (IFN-γ) and LPS/endotoxin (Humphreys and Dubyak 1996).

Transcripts for the ADP-selective $P2Y_1$, the UTP-selective $P2Y_4$, and the UDP-selective $P2Y_6$ receptors are also present in human blood monocytes (Jin et al. 1998). Although this latter finding indicates that the $P2Y_1$, $P2Y_4$, and $P2Y_6$ genes are transcriptionally active in monocytes, there are no studies to indicate that these latter receptor subtypes are expressed in functionally significant numbers in the majority of circulating monocytes. However, functional studies have demonstrated that micromolar levels of $P2Y_1$ agonists, such as ADP and 2-methylthio-ATP, can modulate GTPase activity and Ca^{2+} mobilization in murine macrophages (Denlinger et al. 1996), K^+ channel activity in murine microglial cells (Langosch et al. 1994), Ca^{2+} influx in vitamin D_3-differentiated human U937 monocytes (Ventura and Thomopoulos 1995), and a guanine nucleotide-sensitive inward current in rat macrophages (Naumov et al. 1995). Significantly, $P2Y_1$ receptor mRNA and functional $P2Y_1$ receptors are expressed in human KG-1 myeloblasts which are similar to the common blood marrow progenitors of both monocytes and neutrophils (Clifford et al. 1997). These findings suggest that $P2Y_1$ receptor expression is suppressed as bone marrow progenitors differentiate into circulating monocytes, but is subsequently upregulated by posttranscriptional mechanisms as monocytes begin to differentiate into macrophages. Posttranscriptional mechanisms may be also employed to increase the expression of the uridine nucleotide-selective $P2Y_4$ and $P2Y_6$ subtypes following monocyte recruitment to different tissues. Consistent with this possibility are functional studies showing that UTP is significantly more efficacious than ATP as a positive modulator of LPS/endotoxin-dependent signaling in several types of murine macrophages (Chen et al. 1998; Lin 1997). In summary, at least four P2Y subtypes can be expressed in different populations of mononuclear phagocytic leukocytes.

2. P2X Receptors

The expression and function of the $P2X_7/P_{2Z}$ receptor has been extensively characterized in many types of mononuclear phagocytes (Di Virgilio et al., Chap. 26, this volume). This particular ATP-gated ion channel is very strongly expressed (mRNA, protein, and function) in human monocyte-derived macrophages (Blanchard et al. 1991, 1995; Falzoni et al. 1995; Hickman et al. 1994; Rassendren et al. 1997; Spranzi et al. 1993), mouse macrophages

(ALONSO-TORRE and TRAUTMANN 1993; CHIOZZI et al. 1997; COUTINHO-SILVA and PERESCHINI 1997; GREENBERG et al. 1988; NUTTLE and DUBYAK 1994; SUNG et al. 1985; SURPRENANT et al. 1996), mouse microglial cells (CHESSEL et al. 1997; FERRARI et al. 1996, 1999a), rat macrophages (COLLO et al. 1997), rat microglia (SURPRENANT et al. 1996), and mouse skin-derived dendritic cells (COUTINHO-SILVA et al. 1999; MUTINI et al. 1999). Prior to the cloning of the $P2X_7$ receptor, functional studies indicated that no significant $P2X_7/P_{2Z}$ receptor activity is observed in freshly isolated human monocytes (BLANCHARD et al. 1991; FALZONI et al. 1995; HICKMAN et al. 1994) but that receptor activity is rapidly induced when the monocytes are placed in primary tissue culture. However, a subsequent analysis using monoclonal antibodies specific for the human $P2X_7$ receptors has verified that this receptor is expressed as a cell surface protein in human monocytes (BUELL et al. 1998). Adherence of freshly isolated human monocytes to tissue culture surfaces appears to elevate $P2X_7$ function (BLANCHARD et al. 1991; FALZONI et al. 1995; HICKMAN et al. 1994; RASSENDREN et al. 1997) and the rate of this increase in receptor function is strongly potentiated by IFN-γ (BLANCHARD et al. 1991; FALZONI et al. 1995). When THP-1 human monocytic leukemia cells are stimulated by IFN-γ, LPS/endotoxin, or TNFα, there is a strong correlation between increased expression of $P2X_7$ receptors (mRNA and function) and the induction of other proinflammatory gene products, such as interleukin-1β and TNFα (HUMPHREYS and DUBYAK 1996, 1998). Elevation of cyclic AMP markedly attenuates the upregulation of $P2X_7$ receptor function in THP1 monocytes by a mechanism that does not involve major decreases in steady-state receptor mRNA levels (HUMPHREYS and DUBYAK 1998).

A smaller number of studies have addressed whether P2X subtypes other than $P2X_7$ are expressed in monocytes and monocyte-derived cells. The expression of $P2X_1$ receptor mRNA, protein, and function is greatly upregulated when HL-60 human myeloid leukemia cells are differentiated by phorbol ester treatment into macrophage-like cells (BUELL et al. 1996). Immunoreactive $P2X_1$ receptors are similarly expressed in THP-1 human monocytic leukemia cells (CLIFFORD et al. 1998). While these observations indicate that cells with monocytic lineage have the capacity to express the $P2X_1$ receptor gene, other analyses have revealed no significant expression of immunoreactive $P2X_1$ receptor protein or $P2X_1$-like channel activity in either normal human monocytes or murine macrophage cell lines (CLIFFORD et al. 1998). It is possible that the $P2X_1$ gene is expressed by monocyte-lineage leukocytes only at earlier developmental stages as in the normal bone marrow progenitors of blood monocytes or in leukemic monocytes that retain many phenotypic characteristics of the less differentiated progenitor cells. Finally, a human tissue screen revealed high levels of $P2X_4$ receptor mRNA in the human peripheral blood leukocyte (PBL) fraction which includes monocytes in addition to lymphocytes and various granulocytes (SOTO et al. 1996). However, patch-clamp analysis of freshly isolated human monocytes has indicated that these cells show no significant inward current when stimulated with 100 μmol/l ATP which should be sufficient for maximal activation of all P2X receptor subtypes other than $P2X_7$ (CLIFFORD et al. 1998).

II. Ecto-Nucleotidase Expression and Function in Monocyte/Macrophages

As reviewed (ZIMMERMANN, Chap. 8, first volume), the local metabolism of extracellular ATP, ADP, and other nucleotide tri-/diphosphates by CD39-family ecto-ATPase/ecto-apyrases and by CD73 ecto-5′-nucleotidase (which hydrolyzes AMP and other nucleotide monophosphates) will significantly modulate the magnitude and duration of P2 receptor signaling in any given tissue context. The activation of P2 receptors in monocytes and macrophages activates or potentiates a variety of antimicrobial and proinflammatory functions of these cells. Because some of these functional responses (e.g., release of reactive oxygen or nitrogen radicals) can also damage the host cells, extracellular catabolism of released nucleotides is required to minimize the activation of monocyte/macrophage P2 receptors in inappropriate biological settings. The unintentional activation of the UTP/ATP-activated $P2Y_2$ receptors in circulating blood monocytes is unlikely given the strong expression of CD39 ecto-apyrase in most endothelial cells (KANSAS et al. 1991; ROBSON et al. 1997). Significantly, a substantial downregulation of CD39 ecto-apryase activity is observed when human endothelial cells are activated by inflammatory stimuli such as TNFα (ROBSON et al. 1997). This decrease in endothelial ecto-apyrase activity has been associated with increased adherence of blood leukocytes (SUD'INA et al. 1998). This is consistent with data supporting a role for leukocyte P2Y receptors in promoting adherence to activated endothelial surfaces (ALTIERI et al. 1990; DAWICKI et al. 1995).

After monocytes have been recruited to normal or inflammatory tissue spaces, efficient catabolism of local nucleotides will depend on the ecto-nucleotidases expressed by monocyte/macrophages per se and on the endogenous ecto-nucleotidase activity within the particular tissue locus. Initial FACS analysis of CD39 (prior to its identification as an ecto-apyrase) in various human blood cells revealed no significant expression in circulating monocytes (KANSAS et al. 1991). In contrast, CD39 levels were significant in resident macrophages within lymph nodes (KANSAS et al. 1991). Significant ecto-apyrase activity and CD39 mRNA levels have been reported in HL-60 myeloid leukemia leukocytes, THP1 leukemic monocytes, and murine macrophage cell lines (CLIFFORD et al. 1997). In contrast, no significant expression of CD73 mRNA or ecto-AMPase activity was observed in these leukocytes. The differentiation of phorbol ester-treated HL-60 cells into macrophage-like cells was accompanied by a modest upregulation of CD39 mRNA and ecto-apyrase activity and a major induction of CD73 mRNA and ecto-AMPase activity (CLIFFORD et al. 1997). Thus, several lines of evidence indicate that there is a progressive upregulation of two major ecto-nucleotidases when monocytes exit the circulation and assume a macrophage phenotype. The recent identification of another human CD39 family gene (CD39-L4) that encodes a secreted form of apyrase suggests another significant and, perhaps, unique role for macrophages in the regulation of extracellular nucleotide levels (MULERO et al. 1999). The CD39-L4 cDNA was isolated from a macrophage library and macrophages were the only cell/tissue type that significantly expressed this gene product. The secretion of scavenger nucleotidases may play an important protective role in counteracting either the desensitization of P2 receptors in various immune effector cells or the hyperactivation of the potentially self-lethal $P2X_7$ receptors expressed by macrophages.

III. Sources of Extracellular Nucleotides for Activation of Monocyte/Macrophages

As for most cells within blood, the P2 receptors on circulating monocytes are likely stimulated by the ATP and other nucleotides released from degranulating platelets or activated endothelial cells (HOURANI, Chap. 20, and RALEVIC, Chap. 19, both this volume). Although the sources of nucleotides used for the activation of tissue monocytes and macrophages are less certain, four possibilities may be suggested based on different observations and/or speculations: (i) injured or stressed host cells; (ii) invading microbial pathogens; (iii) activated macrophages per se; and (iv) other immune effector cells, such as lymphocytes, that form transient, but high-affinity, cell–cell contacts with macrophages. At sites of tissue injury, host cells may release nucleotides due to direct mechanical lysis or the activity of as yet uncharacterized nucleotide transport systems that are stimulated by metabolic, osmotic, or xenobiotic stresses (SCHWIEBERT 1999). Likewise, microbial pathogens can release cytoplasmic nucleotides at invasion sites via either nonselective lytic mechanisms or facilitative nucleotide transporters. Recent studies have suggested that *Candida albicans*, a fungal microbial pathogen, will release nucleotides via nonlytic mechanisms in response to host-derived protective factors such as salivary histatins (KOSHLUKOVA et al. 1999). Host macrophages themselves may also release nucleotides as a consequence of certain types of microbial invasion. Infection of tissue-cultured macrophages with *Mycobacteria tuberculosis* (MTB) induces increased ATP levels within the extracellular medium via a mechanism which may involve nonselective release from macrophages killed during the lytic cycles of progressive MTB infection (SIKORA et al. 1999). Two studies have reported that direct stimulation of macrophages (Sperlagh et al. 1998) or microglial cells (Ferrari et al. 1997b) with purified LPS/endotoxin (derived from Gram-negative bacteria) can trigger ATP release which might be used for autocrine activation of P2 receptors and consequent potentiation of endotoxin signaling pathways. While this is an intriguing possibility, another group was unable to observe any significant ATP release from human monocytes and macrophages stimulated with LPS/endotoxin (Grahames et al. 1999). Finally, ATP may be secreted into the very restricted intercellular space that form when macrophages specifically bind to other immune effector cells. For example, when cytotoxic T-lymphocytes (CTL) form tight cell-to-cell contacts with superantigen-presenting macrophages, the CTL secrete a variety of factors that can induce lysis of the bound macrophage. One study has demonstrated that this lysis of bound macrophages by CTL can be significantly attenuated by inclusion of hexokinase in the tissue culture medium (Blanchard et al. 1995). Because extracellular hexokinase can rapidly scavenge locally released ATP, this finding suggests that CTL may release ATP and thereby activate the cytolytic $P2X_7$ receptors on the bound macrophage.

IV. Specific Roles for P2 Receptors in Monocyte/ Macrophage Function

1. Cell Adhesion and Cell Fusion

Monocytes and macrophages engage in multiple types of cell–cell interactions during natural and acquired immune responses. Circulating blood monocytes interact with endothelial cells during recruitment to sites of injury or microbial invasion. This involves upregulation or activation of adhesion molecules on both the monocytes and endothelial cell. Activation of Ca^{2+}-mobilizing $P2Y_2$ and/or $P2Y_1$ receptors in monocytes and neutrophils by ATP, ADP, or UTP increases the adhesion of such blood leukocytes to endothelial cells (Dawicki et al. 1995). This increased adhesion is mediated in part by acute activation of the CD11b/CD18 adhesion receptor, an mβ2 integrin also termed Mac-1 (Altieri et al. 1990; Dawicki et al. 1995; Freyer et al. 1988). At sites of pathological chronic inflammation, monocytic leukocytes can bind to each other in a homotypic cell–cell interaction that eventually results in cell fusion and the generation of multinucleate giant cells (MGC) that die shortly thereafter. While the function of MGC is not certain, their formation may reflect a mechanism for reducing the number of viable monocytes that would otherwise sustain and amplify the chronic inflammation. Significantly, $P2X_7$ receptor antagonists or prior downregulation of $P2X_7$ receptors in monocyte/macrophages significantly decreases their ability to fuse into MGC (Chiozzi et al. 1997). This suggests that $P2X_7$ receptors may act as homotypic adhesion molecules or that local stimulation of $P2X_7$ receptors induces signals that promote cell–cell fusion. With regard to this latter possibility, the activation of $P2X_7$ receptors in murine macrophages (El-Moatassim and Dubyak 1992) or human THP1 monocytes (Humphreys and Dubyak 1996) triggers a massive stimulation of phospholipase D (PLD) activity that results in accumulation of phosphatidic acid, a profusogenic lipid.

2. Modulation of Transcription Factors Involved in Immune and Inflammatory Gene Regulation

Recent studies have demonstrated that the activation of both P2Y and P2X-family receptors in monocyte/macrophages can either activate or modulate several transcription factors that play key roles in immune and inflammatory signaling. Activation of the $P2X_7$ receptor in murine microglial cells triggers a very rapid (within minutes) activation of NFAT (*n*uclear *f*actor of *a*ctivated *T* cells) (FERRARI et al. 1999b). NFAT plays a central role in the regulation of T cell immune gene expression following activation of the T cell receptor (TCR). As is true for NFAT regulation in T lymphocytes, the activation of microglial cell NFAT by $P2X_7$ receptors is absolutely dependent on Ca^{2+} influx and stimulation of calcineurin, a calmodulin-regulated phosphatase, which dephosphorylates cytosolic NFAT to facilitate its translocation to the nucleus. Significantly, stimulation of the microglial $P2Y_2$ receptor with UTP was unable to mimic this effect of $P2X_7$ activation. Direct stimulation of microglial $P2X_7$ receptors also triggers activation of NFκB which plays a key roles in many types of immune and inflammatory gene regulation in monocytes, macrophages, and many other immune effector cells (FERRARI et al. 1997c). The mechanism by which $P2X_7$ receptors are coupled to NFκB induction is complex and, based on the effects of various pharmacological inhibitors, may involve generation of reactive oxygen radicals, stimulation of proteosome activity, and accumulation of the activated caspase-1 protease.

Recent studies have also implicated P2Y receptors and $P2X_7$ receptors in the modulation of NFκB activation by cytokines (TNFα) or microbial factors (LPS/endotoxin) that are well-characterized inducers of NFκB. Although low concentrations (<300 μmol/l) of extracellular ATP or UTP by themselves do not stimulate the nuclear translocation of NFκB or the activation of NFκB-dependent gene expression in murine macrophages, both of these nucleotides can potentiate (by ca. twofold) the activation of NFκB-based signaling that is stimulated by LPS (CHEN et al. 1998; TONNETI et al. 1995). The potentiating effect of UTP on LPS-dependent NFκB activation requires phospholipase C activity, Ca^{2+} mobilization, and CaMKII (calmodulin-dependent protein kinase II) activity; these findings suggest mediation by either $P2Y_2$, $P2Y_4$, or $P2Y_6$ subtypes (CHEN et al. 1998). The mechanism involves potentiation at the level of increased phosphorylation and subsequent degradation of the IκB inhibitory subunit of the cytosolic NFκB complex.

In contrast to the potentiation of LPS-dependent NFκB activation produced by costimulation with UTP or ATP, inclusion of 2-methylthio-ATP or oxidized ATP (oATP) represses (by ca. twofold) the activation of NFκB-based gene expression by LPS in murine macrophages (DENLINGER et al. 1996; HU et al. 1998; SIKORA et al. 1999). Additionally, oATP attenuates (by ca. twofold) the acute activation of Erk-family MAP kinases and degradation of cytosolic IκB triggered by LPS (HU et al. 1998). Inhibition of a biological response by oATP inhibition usually suggests an involvement of $P2X_7$ receptors given the complete antagonism of $P2X_7$ function observed in oATP-treated cells (MURGIA et al. 1992). Thus, the ability of oATP to attenuate the immediate and sustained responses to LPS in the absence of added nucleotide agonists has led to the hypothesis that LPS triggers ATP release from macrophages leading to autocrine activation of P2X7 receptors and subsequent potentiation of LPS induction of NFκB signaling. Despite the appeal of this model, a recent study has reported that oATP can repress the activation of NFκB-based gene expression in macrophages from mice ($P2X_7$ knockout) in which the $P2X_7$ gene has been deleted by targeted disruption (Sikora et al. 1999). This suggests that oATP can additionally attenuate LPS-induced activation of NFκB and other signaling reactions by $P2X_7$-independent pathways. This may involve effects of oATP on: (i) P2X subtypes other than P2X7, (ii) P2Y receptors, or (iii) purinergic signaling elements other than P2 receptors (e.g., ecto-nucleotidases). Although oATP does not antagonize the $P2Y_2$ receptors that are expressed in macrophages (Murgia et al. 1993), possible effects of this reagent on other P2Y subtypes have not been reported. As noted above, activation of P2Y receptor subtypes can also potentiate LPS signaling. Finally, when mice are injected with 2-methythio-ATP, an agonist for multiple P2Y and P2X receptors, they are protected from the endotoxic shock and death that can be induced by uncontrolled activation of LPS signaling (Proctor et al. 1994). One plausible explanation for this striking observation is that 2-methylthio-ATP may induce a chronic, in vivo desensitization of P2 receptors that would otherwise potentiate LPS-induced expression of multiple proinflammatory genes.

3. Release of Reactive Oxygen and Nitrogen Species

The release of cytotoxic oxygen or nitrogen radicals is a major mechanism by monocytes/macrophages kill invading microbial pathogens and amplify inflammatory responses. The enzymes and regulatory proteins required for generating reactive oxygen intermediates (ROI), such as superoxide, are constitutively expressed in phagocytic leukocytes but require acutely triggered signals, such as increased protein kinase C and G protein activity for assembly and catalytic activity. Extracellular ATP, acting via a pertussis toxin-sensitive G protein, can trigger these signals and so induce superoxide release in guinea pig macrophages (Nakanishi et al. 1991). Other studies have indicated UTP and ATP, acting via a $P2Y_2$-like receptor, can rapidly stimulate ROI release from human monocyte-derived macrophages, but not human blood monocytes (Schmid-Antomarci et al. 1997). This acquisition of P2Y-stimulated ROI release during monocyte-to-macrophage differentiation was strongly correlated with the expression of charybdotoxin-sensitive K^+ channels. In contrast to the constitutively expressed cellular machinery for generating oxygen radicals, the generation of reactive nitrogen species such as nitric oxide requires activation of the inducible nitric oxide synthase (iNOS or NOS-II) gene via the binding of NFκB and other transcription factors to the iNOS promoter. As noted above, ATP, acting via either P2Y or $P2X_7$ receptors, and UTP, acting via $P2Y_2$ or $P2Y_{4/6}$ receptors, can potentiate LPS-induced nuclear accumulation of NFκB in macrophages and this is correlated with increased expression of iNOS mRNA, protein, and activity (Chen et al. 1998; Tonetti et al. 1995).

The ability of oATP to attenuate NFκB activation by LPS and TNFα is similarly complemented by an oATP-induced reduction of the cytokine-dependent increases in iNOS mRNA, protein, and activity (Hu et al. 1998; Sikora et al. 1999). However, as also noted above, the inhibitory effect of oATP on LPS-stimulated iNOS activity remains following targeted disruption of the murine $P2X_7$ gene (Sikora er al. 1999).

4. Release of Cytokines and Other Immune Mediators

Activated monocyte/macrophages also release a variety of cytokines, such as IL-1 and TNF, arachidonic acid-derived mediators, such as prostaglandins, and components of the blood clotting cascade, such as plasminogen and tissue factor. These released factors amplify the natural immune response or potentiate activation of the lymphocytes responsible for acquired immunity. ATP and UTP activation of P2Y-family receptor can rapidly trigger arachidonic acid release from murine macrophages (Lin 1997). Extracellular ATP (but not UTP), acting through the $P2X_7$ receptor, is a very effective stimulant of IL-1β release from human monocytes (Buell et al. 1998; Grahames et al. 1999), human monocyte-derived macrophages (Ferrari et al. 1997a), mouse macrophages (Perregaux and Gabel 1994, 1998), mouse microglial cells (Ferrari et al. 1997b), mouse dendritic cells (Mutini et al. 1999), and intact mice (Griffiths et al. 1995). The IL-1β gene is transcriptionally silent in non-activated monocyte/macrophages, but is rapidly activated by primary proinflammatory stimuli such as endotoxin/LPS. Release of the biologically active 17kDa IL-1β requires proteolytic processing via the caspase-1/interleukin converting enzyme (ICE) and facilitated efflux via an as yet uncharacterized plasma membrane transporter. Activation of the $P2X_7$ receptor can provide the signals required for both proteolytic processing of IL-1β by caspase-1/ICE and the efflux of the processed cytokine. One of these signals appears to the rapid and sustained loss of cytosolic K^+ that is induced via the $P2X_7$ channel/pore (Ferrari et al. 1997a; Perregaux and Gabel 1994). $P2X_7$ activation of mouse macrophages also triggers the proteolytic processing (via calcium-activated calpain-family proteases) and release of IL-1α (Perregaux and Gabel 1998). The large increase in cytosolic Ca^{2+} that characterizes $P2X_7$ receptor activation can also rapidly drive release of plasminogen, a key component of the clotting cascade, from mouse microglial cells (Inoue et al. 1998).

5. Growth, Death, and Differentiation of Monocyte/Macrophages

Given the strong expression of the pore-forming $P2X_7$ receptor in most monocytes and macrophages, it is not surprising that extracellular ATP can be an effective inducer of monocyte/macrophage death. Depending on the strength and duration of $P2X_7$ receptor activation, ATP will trigger either necrotic or apoptotic death of human THP1 monocytic leukemia cells (Humphreys and Dubyak 1996; Spranzi et al. 1993) human macrophages (Blanchard et al.

1991), murine macrophages (Chiozzi et al. 1996; Perregaux and Gabel 1994), murine microglial cells (Ferrari et al. 1996, 1999a), human and murine dendritic cells (Coutinho-Silva et al. 1999). Consistent with the ability of $P2X_7$ receptors to stimulate caspase-1 signaling during IL-1β release, $P2X_7$ receptors can also trigger other caspase-family proteases that play key roles in apoptotic progression (Coutinho-Silva et al. 1999; Ferrari et al 1999a). Thus, the $P2X_7$ receptor may have a bifunctional role in inflammatory macrophages: an initial induction of IL-1β release followed by death of the cytokine-releasing cell. This latter function may provide a mechanism for reducing the load of activated macrophages at inflammatory loci, thereby permitting resolution of the inflammatory process.

The activation of P2Y receptors in various cell types can modulate signaling cascades involved in cell growth, survival, or differentiation (Neary and Abbracchio, Chap. 11, first volume). While tissue monocytes and macrophages generally reenter the cell cycle only rarely, the myeloid progenitor cells of monocytes within bone marrow are actively cycling. Extracellular ATP has been reported to inhibit proliferation of human myeloid progenitors (Cowen et al. 1991; Jiang et al. 1997), murine myelomonocytic progenitors (Yamaguchi et al. 1994), and guinea pig alveolar macrophages which proliferate spontaneously when cultured in vitro (Nakanishi et al. 1994). ATP-induced withdrawal of the myeloid progenitor cells from the cell cycle is associated with differentiation into more mature monocytic cells (Cowen et al. 1991; Jiang et al. 1997; Nakanishi et al. 1994). This appears to involve either the G protein-coupled $P2Y_2$ receptor (Cowen et al. 1991; Yamaguchi et al. 1994) or a novel cyclic AMP-coupled $P2Y_{11}$ subtype (Conigrave et al. 1998; Jiang et al. 1997). These in vitro observations raise the interesting possibility that nucleotides locally released within the bone marrow, can modulate the rate at which myeloid progenitor cells mature into granulocytes or monocytes prior to release into the blood.

6. Microbial Killing

The ability of P2 receptors to trigger release of various reactive oxygen/nitrogen species and proinflammatory cytokines undoubtedly contributes to the killing of invading pathogens. Recent studies have also implicated the $P2X_7$ receptor in the regulated fusion of phagosomes that contain internalized bacteria with the lysosomes that facilitate lytic degradation of the phagocytosed microbes (Lammas et al. 1997; Molloy et al 1994; Sikora et al. 1999). Mycobacteria, which repress the constitutive fusion of phagosomes with lysosomes, can proliferate within macrophages and subsequently lyse the host macrophage, leading to persistent infection. However, when mycobacteria-infected human (Lammas et al. 1997; Molloy et al 1994) or murine (Sikora et al. 1999) macrophages are treated with ATP, a rapidly induced fusion of the phagosomes with lysosomes triggers killing of the microbes before the macrophages themselves die.

C. Granulocytes: Neutrophils, Eosinophils, and Basophils/Mast Cells

Granulocytes, like the mononuclear phagocytes, descend from a common set of myeloid progenitor cells in the bone marrow (Fig. 1). The term "granulocyte" is used to denote all myeloid-derived leukocytes (neutrophils, eosinophils, and basophils) that contain conspicuous intracellular granules or vesicles containing antimicrobial factors. The very short-lived neutrophils play a central role in the earliest phases of natural immunity. Microbial invasion or tissue damage stimulates endothelial cells to release acutely inflammatory agents, such as histamine, platelet activating factor, and interleukin-8, that rapidly (within an hour or so) stimulate neutrophil movement across the endothelium and so result in a massive local recruitment of activated bacteriocidal neutrophils. Like monocyte/macrophages, neutrophils both phagocytose bacteria and release toxic factors such as reactive oxygen intermediates and various digestive enzymes. In contrast to the predominant role of neutrophils in native immunity, eosinophils and basophils are primarily involved in specialized phases of the acquired immune response that involve antigens recognized by IgE-type immunoglobulins (Fig. 1). Antigen-specific IgE antibodies are targeted to the extracellular surface of basophils or mast cells residing in various tissue spaces, including skin, that are also in close contact with the outside environment.. Antigen activation of mast cells or basophils precoated with IgE can trigger powerful immune reactions (immediate hypersensitivity) minutes after the antigen enters a host organism. When exposed to the antigen recognized by surface-bound IgE, these leukocytes rapidly degranulate to release histamine and other paracrine signaling molecules. Eosinophils represent other important effector cells for IgE-mediated immunity. Eosinophils are particularly effective in killing IgE-coated helminthic parasites due to their ability to secrete a specific protein (major basic protein) that is toxic to helminths.

I. Expression of P2 Receptor Subtypes in Granulocytes

1. P2Y Receptors

Ca^{2+}-mobilizing P2Y receptors activated by micromolar concentrations of ATP have been characterized in human neutrophils (Cowen et al. 1989; Verghese et al. 1996; Walker et al. 1991), human HL-60 leukemic granulocytes (Cowen et al. 1989), rat mast cells (Osipchuk and Cahalan 1992), human eosinophils (Burgers et al. 1993), and rat basophilic leukemia cells (Osipchuk and Cahalan 1992). Pharmacological studies in human neutrophils (Verghese et al. 1996; Walker et al. 1991) and HL-60 granulocytes (Cowen et al. 1989) have indicated that UTP is equipotent to ATP as a Ca^{2+}-mobilizing agonist, that ATP and UTP are cross-desensitizing, and that the effects of both nucleotides are strongly inhibited by pertussis toxin. These pharmacological data are consistent with expression of a $P2Y_2$ subtype receptor that can couple to G_i-family G proteins in neutrophils. $P2Y_2$ mRNA levels are high in HL-60 cells before (Clifford et al. 1997; Jin et al. 1998; Martin et al. 1997) and after granulocytic differentiation (Martin et al. 1997). Although $P2Y_2$ mRNA is also abundant in normal human bone marrow progenitor cells (Clifford et al. 1997), these transcripts are not detected in human blood neutrophils (Jin et al. 1998). This suggests that the $P2Y_2$ gene is poorly transcribed in circulating, postmitotic neutrophils, but that $P2Y_2$ receptors synthesized during development of neutrophil progenitor cells are retained in significant numbers following release of mature neutrophils (which are very short-lived) into the blood. Although human neutrophils do not express detectable levels of $P2Y_1$ mRNA, transcripts encoding $P2Y_4$ and $P2Y_6$ receptors are abundant in both circulating neutrophils and various types of myeloid leukemic progenitor cell lines (Jin et al. 1998). However, as noted above, ATP and UTP efficiently cross-desensitize each other in neutrophils and this implies low expression of functional uridine nucleotide-selective receptors. It is interesting to speculate whether the functional expression of $P2Y_4$ and $P2Y_6$ receptors may be upregulated following neutrophil emigration to inflammatory tissues as occurs with T lymphocytes (Somers et al. 1998). Although the molecular identity of the Ca^{2+}-mobilizing P2Y receptors in eosinophils or mast cell/basophils has not been explicitly defined, functional studies indicate that ADP is a poor agonist in eosinophils (Burgers et al. 1993) but a potent agonist in mast cell/basophils (Osipchuk and Cahalan 1992). This suggests expression of P2Y1 receptors in the latter but not former granulocytes. In addition to Ca^{2+} mobilizing P2Y receptors, human HL-60 granulocytic leukocytes do express the ATP-selective $P2Y_{11}$ receptor that can activate both phospholipase C and adenylate cyclase (Communi et al. 1997; Conigrave et al. 1998; Jiang et al. 1997). However, the expression of functional $P2Y_{11}$ receptors in circulating human neutrophils has not been evaluated.

2. P2X Receptors

Despite the marked expression of $P2X_1$ receptor mRNA and function in various human myeloid progenitor cells (Buell et al. 1996; Clifford et al. 1998), analysis of purified human neutrophils revealed no expression of $P2X_1$ receptor protein or function (Clifford et al. 1998). The expression of this receptor in eosinophils and basophils has not been studied. As noted previously, $P2X_4$ receptor mRNA is abundant in the total pool of human blood leukocytes (Soto et al 1996). Whether neutrophils per se or other granulocytes express this ionotropic ATP receptor is unknown. Rat mast cells were among the first cell types in which expression of the pore-forming $P_{2Z}/P2X_7$ receptor was extensively characterized (Cockcroft and Gomperts 1979). At present, there are no reported analyses of $P2X_7$ receptor expression or function in neutrophils and eosinophils.

II. Ecto-Nucleotidase Expression and Function in Granulocytes

Although expression of immunoreactive CD39 protein is very low in circulating human neutrophils (KANSAS et al. 1991), these cells do exhibit a modest level of CD39-like ecto-apyrase activity (CLIFFORD et al. 1997). In contrast, no significant expression of CD73-like ecto-5′-nucleotidase activity is measurable in neutrophils (CLIFFORD et al. 1997; MADARA et al. 1993).

III. Sources of Extracellular Nucleotides for Activation of Granulocytes

As for circulating monocytes, granulocyte P2 receptors are most likely activated by nucleotides released by activated platelets or endothelial cells. Indeed, platelets (at physiological cytocrits) have been shown to release ATP in amounts sufficient for activating P2Y-mediated chemotaxis in human eosinophils (BURGERS et al 1993) and $P2Y_2$-mediated Ca^{2+} mobilization in human neutrophils (WARD et al. 1988). Granulocytes per se can be significant sources of released nucleotides in certain biological settings. Significantly, ATP is rapidly released during the degranulation of rat mast cells or rat basophilic leukemia cells stimulated via paracrine activation of P2Y or antigen-engagement of presensitized IgE/Fce receptor complexes (OSIPCHUK and CAHALAN 1992). Whether neutrophils or eosinophils also release ATP via regulated exocytosis has not been systematically evaluated. However, inflammatory neutrophils that migrate into crypts within intestinal epithelial monolayers do release AMP in amounts sufficient to support the generation of adenosine via CD73 ecto-5′-nucleotidase localized on the apical surface of the epithelial cells (MADARA et al. 1993). This results in paracrine stimulation of epithelial A2 adenosine receptors that trigger both accumulation of cyclic AMP and activation of Cl^- secretion. It remains to be established whether the neutrophils directly release AMP or whether they release ATP that is then locally metabolized to AMP by CD39 ecto-apyrases. Since neutrophils also express adenosine receptors (MONTESINOS and CRONSTEIN, Chap. 24, this volume), this neutrophil-derived AMP/ adenosine may also function in the autocrine modulation of neutrophil function.

IV. Specific Roles for P2 Receptors in Granulocyte Function

1. Cell Adhesion and Chemotaxis

As with circulating monocytes, treatment of neutrophils with extracellular ATP or UTP increases their adherence to endothelial cells (DAWICKI et al. 1995) and to each other (SIEFERT et al. 1989). This effect, which is presumably mediated by $P2Y_2$ receptors, is correlated with very rapid activation of the CD11b/CD18 integrin (FREYER et al. 1988). In addition to stimulating the adhesive properties of granulocytes, ATP acts a chemoattractant for eosinophils (BURGER et al 1993). Both ATP and UTP are chemoattractants for human neutrophils and HL-60 granulocytes (VERGHESE et al. 1996) Thus, elevated levels of extracellular ATP or UTP at the surface of blood vessels and within injured tissues may serve to stimulate extravasation of circulating granulocytes.

2. Degranulation and Release of Immune/Inflammatory Mediators

ATP and UTP induce a rapid exocytosis of the primary granules contained with human neutrophils or HL-60 granulocytes (SIEFERT et al. 1989). These granules contain myeloperoxidase, elastase, and other degradative enzymes. This ATP- and UTP-induced degranulation is substantially repressed by pertussis toxin. As noted above, stimulation of rat mast cells or basophils with ATP or ADP induces rapid degranulation of vesicles containing adenine nucleotides (OSIPCHUK and CAHALAN 1992). Previous studies (COCKROFT and GOMPERTS 1979) demonstrated that ATP could also trigger rapid histamine release from mast cells.

3. Release of Reactive Oxygen Species

In contrast to macrophages which generate superoxide in response to extracellular ATP, human neutrophils show no significant activation of a respiratory burst or superoxide release when acutely stimulated with either ATP or UTP (SIEFERT et al. 1989; WARD et al. 1988; WALKER et al. 1991). However, both ATP and UTP act to prime or sensitize neutrophils such that larger amounts of superoxide are generated when the cells are subsequently stimulated with proinflammatory agonists such as bacterial formyl peptides or phorbol esters. Since this priming action of ATP and UTP is inhibited by pertussis toxin, it is likely mediated by $P2Y_2$ receptors (SIEFERT et al. 1989; WALKER et al. 1991).

4. Growth, Death, and Differentiation

Although neutrophils live for only brief periods following their release into the blood and peripheral tissue spaces, other granulocytes, such as mast cells or basophils, can survive in tissue for prolonged times. ATP, acting through $P2X_7$ receptors, can induce rapid necrotic death of tissue mast cells (COCKCROFT and GOMPERTS 1979). Nucleotides may also modulate growth and differentiation of granulocyte progenitor cells, such as promyelocytes, within the bone marrow. Thus, treatment of HL-60 human leukemic promyelocytes with ATP induces both growth arrest and induction of gene expression programs that typify commitment to granulocytic differentiation (COWEN et al. 1991; JIANG et al. 1997). These effects may be mediated by a the adenylate cyclase-coupled $P2Y_{11}$ receptor known to be expressed in these cells (COMMUNI et al. 1997; CONIGREAVE et al. 1998).

D. Lymphocytes: B Cells, T Cells, and Natural Killer Cells

All lymphocytes are derived from common lymphoid progenitor cells of the fetus; these include B cells, T cells, and natural killer (NK) cells (Fig. 1).

Antigen-activated B-lymphocytes synthesize and secrete large quantities of soluble immunoglobulin (Ig) antibodies that selectively bind and sequester the particular antigen. Since these antibodies can circulate freely through the blood, B cell responses underlie so-called humoral immunity. Antigen-binding receptors are expressed on both of the two major classes of T lymphocytes: the $CD4^+$ helper T cells and $CD8^+$ cytolytic T cells. In contrast to B cells, T cells recognize antigens only when antigens are bound to major histocompatibility complex (MHC) proteins on the surface of so-called antigen-presenting cells (APC). Antigen activation of helper T lymphocytes induces these cells to release cytokines. Some of these cytokines stimulate growth and activation of B lymphocytes to provide feed-forward amplification of the acquired immune system while other cytokines activate nearby macrophages. In contrast, antigen activation of the $CD8^+$ cytolytic T cells (CTL) causes CTL to bind to a target cell that displays the antigenic molecule on its cell surface. This binding triggers the exocytotic release of CTL granules that contain both pore-forming proteins (perforin and granulysin) and other cytotoxic proteins. The release of these proteins is highly localized to the synapse-like contact between the CTL and its target cell. Like CTL, NK cells also bind tightly to various target cells, such as those infected with pathogenic viruses, and directly release pore-forming proteins and other cytotoxic enzymes into the intercellular space between the NK cell and its target. Unlike CTL, NK cells bind to target cells by a poorly characterized mechanism that does not involve prior antigenic exposure. Thus, NK cells function as a component of native, rather than acquired, immunity.

Lymphocytes undergo selection and initial maturation in the primary lymphoid tissues: the thymus for T cells and the bone marrow for B cells (Fig. 1). Although the lymphocytes released into the circulation from these primary tissues are mature (rather than progenitor) cells, they are termed naïve lymphocytes because they have not yet been exposed to specific antigens. Naïve B and T cells continuously migrate from the blood into the secondary lymphoid tissues such as peripheral lymph nodes, the spleen, the tonsils, and intestinal Peyer's patches. If a naïve lymphocyte encounters its specific antigen within these secondary tissues, this will trigger the complex sequence of altered gene expression (termed activation) that results in proliferation of that particular naïve lymphocyte and acquisition of differentiated effector functions such as secretion of antibodies by activated B cell or the secretion of various cytokines by helper T cells. If a naïve lymphocyte does not encounter its specific antigen within several hours following its passage into a secondary lymphoid tissue, it will eventually be released back into the blood for another cycle of recirculation and immune surveillance. The appropriate trafficking of lymphocytes out of the blood, through the secondary lymphoid tissues, and back into the blood is dependent on the dynamic but highly regulated expression of many cell surface adhesion molecules and receptors. As evident in the discussion below, several P2 receptors and ecto-nucleotidases can be included among these dynamically regulated lymphocyte surface proteins.

I. Expression of P2 Receptor Subtypes in Lymphocytes

1. P2Y Receptors

In contrast to findings with circulating monocytes and granulocytes, multiple studies have reported that freshly isolated human blood lymphocytes (which comprise ~15% B cells, 50% helper T cells, 25% CTLs, and 10% NK cells) do not mobilize IP_3-sensitive Ca^{2+} stores in response to extracellular ATP, ADP, or UTP (Cowen et al. 1989). Likewise, Ca^{2+}-mobilization in response to P2Y agonists is not apparent in lymphocytes isolated from the blood of patients with either B cell or T cell leukemias (Wiley and Dubyak 1989) or freshly purified human blood T cells (Baricordi et al. 1996). These various data indicate that $P2Y_1$, $P2Y_2$, $P2Y_4$, or $P2Y_{11}$ receptors are not expressed in functionally significant numbers in either circulating B cells or T cells. In contrast to this lack of P2Y receptor function, circulating human lymphocytes (unfractionated) do express significant steady-state levels of mRNA transcripts for $P2Y_1$, $P2Y_2$, $P2Y_4$, and $P2Y_6$ receptors (Jin et al. 1998). Thus, while the $P2Y_1$, $P2Y_2$, $P2Y_4$, and PY_6 genes are transcriptionally active in at least certain subpopulations of circulating lymphocytes, the corresponding receptors are either not translated or are not trafficked to the plasma membrane in functionally significant numbers. This suggests that lymphocytes may rapidly upregulate their complement of functional P2Y receptors by posttranscriptional mechanisms. Moreover, this upregulation may occur only when lymphocytes exit the circulation and take up residence in the peripheral tissue sites of lymphocyte activation. Consistent with this possibility, Padeh et al. (1991) described the expression of a Ca^{2+}-mobilizing $P2Y_1$-like receptor in the B lymphocytes, but not T lymphocytes, isolated from human tonsillar lymph nodes. More recently, Somers et al. (1998) reported that $P2Y_6$ receptor mRNA and UDP-dependent Ca^{2+} mobilization is upregulated when human blood-derived T lymphocytes are cultured in vitro with phytohemagglutinin, a T cell mitogen that also induces expression of genes required for immune responses. In situ hybridization analysis revealed that $P2Y_6$ mRNA was strongly expressed in the activated T cells infiltrating the colonic epithelia of patients within various inflammatory bowel diseases; but was absent in the resident and presumably quiescent T cells of nondiseased colonic segments (Somers et al. 1998). Thus, expression of functional P2Y receptors in lymphocytes appears to require close physical proximity to other cell types that express or release molecules used for the modulation of lymphocyte activation. This is also supported by recent observations indicating $P2Y_2$ receptor mRNA expression in the progenitors of both helper and cytotoxic T cells that develop within the confines of the mouse thymus (Koshiba et al. 1997). These latter investigators also reported that activation of the TCR (T cell receptor) in tissue cultured thymic lymphocytes induced a strong but transient upregulation of $P2Y_2$ mRNA levels. However, it should also be stressed that several functional studies have failed to reveal P2Y receptor-dependent Ca^{2+} mobilization in any numerically significant subpopulation of murine thymocytes (El-Moatassim et al. 1987;

PIZZO et al. 1991; ROSS et al. 1997). Thus, lymphocytes exhibit complex and multiphasic patterns of P2Y receptor expression with very transient expression occurring during development, suppression during circulation in the blood, and strong upregulation in selected lymphocyte subsets following their trafficking to secondary lymphoid tissues.

2. P2X Receptors

The expression of ionotropic P2X receptors, as assayed by ATP-gated inward currents, ATP-induced depolarization, or ATP-induced Ca^{2+} influx, has been characterized in many types of lymphocytes including human blood T cells (BARICORDI et al. 1996), normal and virus-transformed human tonsillar B cells (BRETSCHNEIDER et al. 1995; MARKWARDT et al. 1997), leukemic human lymphocytes (WILEY and DUBYAK 1989; WILEY et al. 1992, 1998), and murine thymic lymphocytes (CHUSED et al. 1996; EL-MOATASSIM et al. 1990; FREEDMAN et al. 1999; PIZZO et al. 1991; ROSS et al. 1997). The low agonistic potency of ATP as well as the sensitivity to inhibition by oATP strongly suggests that these lymphocyte responses reported are mediated by the $P2X_7/P_{2Z}$ receptor. Recent studies have verified expression of mRNA for this latter receptor subtype in human lymphocytes (WILEY et al. 1998), rat bone marrow B cells (COLLO et al. 1997), murine thymocytes (FREEDMAN et al. 1999), and the BW5147 mouse thymocyte cell line (HUMPHREYS et al. 1998). Although the $P2X_7$ receptor appears to be present in many subsets of B and T lymphocytes, significant differences characterize the functional expression of these receptors in various lymphoid cell types. At the qualitative level, some lymphocyte preparations (WILEY et al. 1998; CHUSED et al. 1996) but not others (PIZZO et al. 1991; HUMPHREYS et al. 1998) respond to $P2X_7$ activation with the increased permeability to large organic dyes that typifies $P2X_7$ function in macrophages. At the quantitative level, $P2X_7$-dependent Ca^{2+} influx varies in different murine thymocytes with large, actively dividing cells showing stronger responses than small, terminally differentiated cells (ROSS et al. 1997). Significantly, when murine thymocytes are selected by maturation-specific markers, $P2X_7$ function is observed in both the CD4+ progenitors of helper T cells and the CD8+ progenitors of CTL (CHUSED et al. 1996; ROSS et al. 1997).

The ability of ATP to trigger depolarization and Ca^{2+} influx but not nonselective pore formation in some classes of lymphocytes (PIZZO et al. 1991; HUMPHREYS et al. 1998) has raised the question of whether such lymphocytes may express P2X receptor subtypes other than $P2X_7$. Moreover, the initial characterization of the $P2X_1$ cDNA by VALERA et al. (1994) was attended by the surprising observation that the $P2X_1$ mRNA sequence was highly homologous to that of RP-2, a previously cloned gene product upregulated during apoptotic induction of rat thymocytes (OWENS et al. 1991). Subsequent studies have verified the expression of $P2X_1$ receptor mRNA or protein in developing murine thymocytes (CHVATCHKO et al. 1996; KOSHIBA et al. 1997; FREEDMAN et al. 1999) and rat thymocytes (CHVATCHKO et al. 1996; JIANG et al. 1996), but not in circulating human lymphocytes (CLIFFORD et al. 1998) or mature T lymphocytes in peripheral mouse lymph nodes (CHVATCHKO et al. 1996). RT-PCR analysis has revealed the additional expression of $P2X_2$ receptor mRNA in subpopulations of murine thymocytes (FREEDMAN et al. 1999). FREEDMAN et al. (1999) further confirmed the expression of functional $P2X_1$ receptors in developing murine T cells by demonstrating that α,β-methylene ATP, a $P2X_1$-selective agonist, can activate inward currents in patch-clamped murine thymocytes. In contrast, electrophysiological analysis of human tonsil-derived B cells failed to identify ATP-gated channels with the pharmacological characteristics of $P2X_1$-type receptors (MARKWARDT et al. 1997). Thus, as with P2Y receptors, the expression of $P2X_1$ receptors in lymphocytes is highly dependent on the developmental status and /or tissue location of the particular lymphocyte population.

II. Ecto-Nucleotidase Expression and Function in Lymphocytes

Prior to the identification of CD39-family proteins and CD73 as major ecto-nucleotidases, BARANKIEWICZ et al. (1988) determined that ecto-ATPase, ecto-ADPase, and ecto-AMPase activities were strongly expressed by resident B cells of human peripheral lymph nodes and by various transformed human B cell progenitor cells. In contrast, none of the ecto-nucleotidase activities were significantly expressed in normal human T lymphocytes or in various immortalized T cell lines. These striking differences in ecto-nucleotidase expression between B and T lymphocytes suggested that B cells, which lack adenosine receptors, could be significant producers of extracellular adenosine used for the paracrine modulation of T lymphocytes and other immune/inflammatory cells that express adenosine receptors (APASOV et al. 1997; MONTESINOS and CRONSTEIN, Chap. 24, this volume).

Subsequent analysis of CD39 expression (prior to its identification as an ecto-apyrase) revealed that immunoreactive CD39 protein expression was low in circulating human B and T lymphocytes but was greatly upregulated in B cells and subsets of T cell following in vitro activation with growth-stimulatory cytokines (KANSAS et al. 1991). These latter studies also showed that CD39 expression was particularly high in Epstein-Barr virus transformed B lymphocytes. Analysis of ecto-ATPase function also indicated low enzyme activity in circulating B cells, T cells, and NK cells, but significant activity in all of these lymphoid cell types following in vitro activation or transformation (DOMBROWSKI et al. 1998). Localization studies of CD39 immunoreactivity and ecto-apyrase activity in various pig lymphoid tissues showed highest activity in the spleen and bone marrow and lower activities in thymus and peripheral lymph nodes (BENREZZAK et al. 1999). Thus, the pattern of CD39/ecto-apyrase expression in lymphoid cells and tissues reiterates the pattern of P2 receptor expression: low or no expression in circulating cells but strong upregulation following recruitment to peripheral tissue sites of further development or immune activation. Similar analyses of CD73 immunoreactivity and ecto-5′-

nucleotidase/ecto-AMPase activities revealed more complicated patterns of expression with positive signals in 70% of circulating human B cells and 30% circulating T cells (GUTENSOHN et al. 1995). However, like CD39, CD73 expression can be significantly upregulated in B, T, and NK cells following activation by various immune stimuli (PEOLA et al. 1996; RESTA et al. 1998).

The ability of lymphoid cells to regulate the expression of CD39, CD73, and perhaps other ecto-nucleotidases during development or immune activation suggests that these cells can strongly modulate the interstitial concentrations of nucleotides or adenosine within the tightly packed tissue sites of lymphocyte development (the thymus and bone marrow) or immune activation (the spleen and peripheral lymph nodes). In turn, the highly localized regulation of extracellular nucleotide and adenosine levels may positively or negatively modulate the survival, development, and activation of T lymphocytes (APASOV et al. 1997; DOMBROWSKI et al. 1995b; RESTA et al. 1998), B lymphocytes (DOMBROWSKI et al. 1998), and NK cells (DOMBROWKI et al. 1995a; SCHMIDT et al. 1984; WILLIAMS et al. 1997) within these tissues. It should also be stressed that the various ecto-nucleotidase proteins can modulate immune cell activation by mechanisms unrelated to the catabolism of extracellular nucleotides. These include functions as adhesion proteins and direct signaling proteins. Thus, antibodies against CD39 induce homotypic adhesion of lymphocytes (KANSAS et al. 1991) while CD73 has been implicated as an adhesion molecule to mediate the binding of the lymphocytes to endothelial cells (ARVILOMMI et al. 1997). Moreover, treatment of T cells with antibodies against CD73 generate accessory signals that strongly potentiate activation of the cells by antigen-occupied T cell receptors (RESTA et al.). Significantly, mutations of CD73 that eliminate its ecto-5′-nucleotidase activity do not repress its ability to function as a positive modulator of T cell activation (GUTENSOHN et al. 1995).

III. Sources of Extracellular Nucleotides for the Activation of Lymphocytes

Because various P2 receptors are predominantly expressed in lymphocytes localized within tissues specialized for lymphoid development (e.g., the thymus) or activation (e.g., peripheral lymph nodes or inflammatory loci), it is likely that the nucleotides locally released within these tissues are used for the activation of lymphocyte P2 receptors. As noted in the discussions of mononuclear phagocytes and granulocytes, these nucleotides may be derived from invading pathogenic microbes, from injured host cells, or from cells, such as endothelial cells and epithelial cells, that selectively release nucleotides via facilitated transport mechanisms. Moreover, cytotoxic T lymphocytes may also release nucleotides by nonlytic pathways stimulated during as a consequence of antigen presentation (BLANCHARD et al. 1995) or ligation of TCR by activating antibodies (FILIPPINI et al. 1990). If a bound target cell of a CTL or NK cell expresses pore-forming $P2X_7$-type receptors, the released ATP may provide an alternative to perforin-based or Fas-based pathways of cell killing (MACINO et al. 1996; REDEGELD et al. 1991).

IV. Specific Roles for P2 Receptors in Lymphocytes

1. Remodeling of Lymphocyte Cell Surface Composition

Lymphocytes may recirculate many times between the blood and secondary lymphoid tissues during their surveillance for foreign antigens. Certain adhesion molecules, such as L-selectin, facilitate the binding of circulating lymphocytes to the endothelial cells (EC) of the blood vessels that supply these lymphoid tissues. Following binding to the EC, lymphocyte surface adhesion molecules are rapidly shed via the actions of various matrix metalloproteases (MMP). Wiley and colleagues have demonstrated that treatment of human leukemic B cells with extracellular ATP can trigger rapid remodeling of the lymphocyte cell surface. These cell surface changes, which are specifically mediated by the $P2X_7$ receptor, include rapid loss of L-selectin (JAMIESON et al. 1996) and the shedding of CD23, the low affinity IgE receptor (GU et al. 1998) via the activation of at least two MMP pathways. In addition to this modulation of surface protein composition, stimulation of lymphocyte $P2X_7$ receptors induces a strong phospholipase D activity which can lead to significant changes in plasma membrane phospholipid composition (GARGETT et al. 1996). Thus, $P2X_7$ and possibly other P2 receptors may play significant roles in modulating the trafficking of lymphocytes to, and within, the secondary lymphoid tissue sites of immune activation.

2. Modulation of NK Cell Function

Incubation of NK cells with either ATP (SCHMIDT et al. 1984) or adenosine (WILLIAMS et al. 1997) can significantly reduce the secretion of cytotoxic factors used by these lymphocytes to induce death of bound target cells. The ability of 5′-*p*-(fluorosulfonyl)benzoyl adenosine (which can covalently modify diverse cell surface nucleotide binding sites) to irreversibly inhibit the cytotoxic capacity of NK cells provides additional indirect evidence for the involvement of purinergic signaling elements in NK function.

3. Modulation of Transcription Factors Involved in Immune Gene Regulation

Activation of the Ca^{2+}-mobilizing $P2Y_1$-like receptors in human tonsillar B lymphocytes induces a large and rapid increase in c-fos mRNA (PADEH et al. 1991). This is significant because c-fos protein is a component in the AP-1 transcription and AP-1 binding sites are common elements in the *cis*-regulatory regions of many immune and inflammatory genes. Given $P2X_7$ receptors can drive the strong activation of the NFAT (nuclear factor of activated T cells) transcription factor expressed in microglial cells (FERRARI et al. 1999b), it is likely that ATP may also activate NFAT in those subsets of T cells that also express $P2X_7$ receptors. NFAT plays a critical role in the regulated expression of IL-2, an important T cell growth factor. As noted below, $P2X_7$ receptor activation has been associated with mitogenic stimulation of human T lymphocytes (Baricordi et al. 1996).

4. Growth, Death, and Differentiation

Many studies have described the effects of extracellular ATP and other nucleotides on the survival or differentiation of lymphocytes. Given the diversity of lymphocytes and the complex cellular interactions and multiple tissue sites that characterize lymphocyte development and activation, it is not surprising that these studies have yielded a wide range of sometimes contradictory findings. However, a number of general observations may be noted.

Consistent with the expression of $P2X_7$ receptors in many subpopulations of T and B cells, ATP or BzATP has been shown to induce either necrotic or apoptotic death of lymphocytes. $P2X_7$-dependent apoptosis or lysis has been described for tissue cultured murine thymocytes (Apasov et al. 1997; Chused et al. 1996; Pizzo et al. 1991) and a variety of murine lymphoma cell lines (Zanovello et al. 1990; Zheng et al. 1991). Apasov et al. (1997) have emphasized that the ability of ATP and $P2X_7$ receptors to induce nonselective lysis vs protein synthesis-dependent apoptosis of murine thymocytes may vary depending on the lymphocyte subset, the local generation of adenosine, and the presence of other modulators of T cell survival such as TCR cross-linking or glucocorticoids. In contrast to the usual emphasis on $P2X_7$-dependent cell death, Ross et al. (1997) have proposed that the major role of $P2X_7$ receptors in murine thymocytes may be to provide Ca^{2+}-dependent signals for the fine-tuning of thymocyte positive selection and differentiation. To make matters even more complex, activation of $P2X_7$-like receptors in some subpopulations of T cells, such as medullary thymocytes of the mouse (El-Moatassim et al. 1987) and tissue-cultured human T lymphocytes (Baricordi et al. 1996) can drive signaling cascades that result in mitogenesis and cell proliferation.

As noted previously, increased expression of the $P2X_1$ receptor has been associated with apoptotic induction of thymocytes (Owens et al. 1991; Chvatchko et al. 1996). Significantly, in situ hybridization and immunocytochemical analyses demonstrated that $P2X_1$ receptor expression is specifically upregulated in apoptotic T cells within the thymus, but not in apoptotic T cells within peripheral lymph nodes, of mice injected with staphylococcal enterotoxin B (Chvatchko et al. 1996). These latter investigators also reported that dexamethasone induction of apoptosis in tissue cultured rat thymocytes was potentiated by ATP but attenuated by apyrase (added as an ATP scavenger) and various P2X antagonists (suramin and PPADS). These functional effects were correlated with elevated levels of $P2X_1$ mRNA transcripts in the dexamethasone-induced thymocytes. In contrast, Jiang et al. (1996) observed no significant effects of ATP, suramin, or PPADS on glucocorticoid-induced apoptosis of cultured rat thymocytes. While Jiang et al. (1996) noted the presence of $P2X_1$ mRNA in rat thymocytes, the steady-state level of these transcripts did not change with glucocorticoid treatment. Further complications regarding the role of $P2X_1$ receptors in thymocyte apoptosis were presented by Koshiba et al. (1997) who reported that in vitro glucocorticoid treatment upregulated $P2X_1$ mRNA levels in rat but not mouse thymocytes. It should be noted that Chvatchko et al. (1996) took care to remove apoptotic thymocytes which are naturally present in thymic isolates before treating the cultured cells with dexamethasone; this initial removal of apoptotic thymocytes may considerably improve the signal/noise ratio of such assays and thereby facilitate measurements of dexamethasone-dependent increases in $P2X_1$ expression. Finally, Freedman et al. (1999) recently reported that the number of CD4+/CD8+ thymocytes is significantly elevated in mouse thymi following in situ treatment with PPADS; this further supports a role for $P2X_1$ or other PPADS-sensitive P2X receptors in modulating the survival of thymic lymphocytes.

E. Conclusions

Two recurring themes may be gleaned from these studies of P2 receptor function in immune and inflammatory leukocytes. First, various P2 receptor subtypes are often functionally expressed only at particular stages of leukocyte development or activation state. Given that leukocytes develop and function within multiple tissue compartments (bone marrow, blood, thymus, spleen, secondary lymph nodes, and peripheral tissues), the ability of these cells to adjust rapidly or modulate their repertoire of expressed P2 subtypes suggests that particular receptors are used for leukocyte signaling functions that occur only within specific tissue contexts. Second, although the genes encoding various P2 receptor subtypes can be transcriptionally active throughout multiple stages of leukocyte development or activation, there is often a poor correlation between the presence of particular P2 receptor mRNA transcripts and the expression of the encoded receptor in functionally significant amounts. Such disparities between transcriptional activity and functional expression indicate that posttranscriptional mechanisms may significantly limit the expression of functional cell surface P2 receptors until leukocytes leave the initial sites of early development (marrow, thymus) and enter either the circulating blood or the peripheral tissue sites of leukocyte activation. These mechanisms could include low stability of receptor mRNA transcripts, poor translational efficiency of the transcripts, or fundamental differences in plasma membrane targeting/retention of the expressed P2 receptor protein. These potential complications emphasize the need for appropriate assays of receptor protein levels, receptor function, and receptor pharmacology when positing significant roles for any particular P2 subtype at different stages of leukocyte development or activation.

Given the combined diversity of leukocyte subtypes and P2 receptor subtypes, the continued or anticipated development of subtype-selective P2 receptor antibodies and antagonists will be critical for defining the roles for these receptors at various stages of the immune and inflammatory responses. Of course, the other group of key reagents for such studies will be mice with targeted deletions of particular P2 receptor genes or ecto-nucleotidases. These analyses should be facilitated by the extensive use of the mouse as the experimental organism of choice for analysis of immune and inflammatory responses at the systemic level. At present, immune or inflammatory phenotypes have not been described for the various P2 receptor knockout mice reported thus far. These include mice lacking the $P2Y_1$ (FABRE et al. 1999), $P2Y_2$ (HOMOLYA et al. 1999), $P2X_1$ (MULRYAN et al. 2000), and $P2X_7$ (SIKORA et al. 1999) receptor genes. As noted previously, macrophages isolated from the $P2X_7^{-/-}$ mice exhibit no obvious differences in the activation of NFκB-dependent gene expression when stimulated in vitro with LPS, TNFα, or *Mycobacteria* (SIKORA et al., 1999). Significantly, an altered immune/inflammatory phenotype is apparent in mice which lack the CD39 ecto-apyrase gene (ENJYOJI et al. 1999) and exhibit intrinsic prothrombotic disorders. Moreover, when $CD39^{-/-}$ organs are transplanted into rats, the xenografted organs are more rapidly rejected by antibody-dependent mechanisms. These early studies provide an initial, fascinating glimpse of how knockout mice will allow us to define expected or unexpected roles for P2 nucleotide receptors in immune and inflammatory signaling.

References

Alonso-Torre SR, Trautmann A (1993) Calcium responses elicited by nucleotides in macrophages. Interaction between two receptor subtypes. J Biol Chem 268: 18640–18647

Altieri DC, Wiltse WL, Edgington TS (1990) Signal transduction initiated by extracellular nucleotides regulates the high affinity ligand recognition of the adhesive receptor CD11b/CD18. J Immunol 145:662–670

Apasov SG, Koshiba M, Chused TM, Sitkovsky MV (1997) Effects of extracellular ATP and adenosine on different thymocyte subsets: possible role of ATP-gated channels and G protein-coupled purinergic receptor. J Immunol 158:5095–5105

Arvilommi AM, Salmi M, Airas L, Kalimo K, Jalkanen S (1997) CD73 mediates lymphocyte binding to vascular endothelium in inflamed human skin. Eur J Immunol 27:248–254

Baricordi OR, Ferrari D, Melchiorri L, Chiozzi P, Hanau S, Chiari E, Rubini M, Di Virgilio F (1996) An ATP-activated channel is involved in mitogenic stimulation of human T lymphocytes. Blood 87:682–690

Barankiewicz J, Dosch H, Cohen A (1988) Extracellular nucleotide catabolism in human B and T lymphocytes. J Biol Chem 263:7094–7098

Benrezzak O, Grondin G, Sevigny J, Gendron FP, Rousseau E, D'Orleans-Juste P, Beaudoin AR (1999) Identification and immunolocalization of two isoforms of ATP-diphosphohydrolase (ATPDase) in the pig immune system. Arch Biochem Biophys 370:314–22

Blanchard DK, McMillen S, Djeu JY (1991) IFN-γ enhances sensitivity of human macrophages to extracellular ATP-mediated lysis. J Immunol 147:2579–2585

Blanchard DK, Wei S, Duan CN, Pericle F, Diaz JI, Djeu JY (1995) Role of extracellular adenosine triphosphate in the cytotoxic T-lymphocyte-mediated lysis of antigen presenting cells. Blood 85:3173–3182
Bretschneider F, Klapperstuck M, Lohn M, Markwardt F (1995) Nonselective cationic currents elicited by extracellular ATP in human B-lymphocytes. Pflugers Arch 429:691–698
Buell G, Chessell IP, Michel AD, Collo G, Salazzo M, Herren S, Gretener D, Grahames C, Kaur R, Kosco-Vilbois MH, Humphrey PP (1998) Blockade of human $P2X_7$ receptor function with a monoclonal antibody. Blood 92:3521–3528
Buell, G, Michel AD, Lewis C, Collo G, Humphrey PPA, Surprenant A (1996) $P2X_1$ receptor activation in HL-60 cells. Blood 87:2659–2663
Burgers JA, Schweizer RC, Koenderman L, Bruijnzeel PL, Akkerman JW (1993) Human platelets secrete chemotactic activity for eosinophils. Blood 181:49–55
Chen BC, Chou CF, Lin WW (1998) Pyrimidinoceptor-mediated potentiation of inducible nitric oxide synthase induction in J774 macrophages. Role of intracellular calcium. J Biol Chem 273:29754–29763
Chessell IP, Michel AD, Humphrey PP (1997) Properties of the pore-forming $P2X_7$ purinoceptor in mouse NTW8 microglial cells. Br J Pharmacol 121:1429–1437
Chiozzi P, Sanz JM, Ferrari D, Falzoni S, Aleotti A, Buell GN, Collo G, Di Virgilio F (1997) Spontaneous cell fusion in macrophage cultures expressing high levels of the $P_{2Z}/P2X_7$ receptor. J Cell Biol 138:697–706
Chiozzi P, Murgia M, Falzoni S, Ferrari D, Di Virgilio F (1996) Role of the purinergic P_{2Z} receptor in spontaneous cell death in J774 macrophage cultures. Biochem Biophys Res Commun 218:176–181
Chused TM, Apasov S, Sitkovsky M (1996) Murine T lymphocytes modulate activity of an ATP-activated P2Z-type purinoceptor during differentiation. J Immunol 157:1371–1380
Chvatchko Y, Valera S, Aubry JP, Renno ST, Buell G, Bonnefoy JY (1996) The involvement of an ATP-gated ion channel, $P2X_1$ in thymocyte apoptosis. Immunity 5:275–283
Clifford EE, Parker K, Humphreys BD, Kertesy SB, Dubyak GR (1998) The $P2X_1$ receptor, an adenosine triphosphate-gated cation channel, is expressed in human platelets but not in human blood leukocytes. Blood 91:3172–3181
Clifford EE, Martin KA, Dalal P, Thomas R, Dubyak GR (1997) Stage-specific expression of P2Y receptors, ecto-apyrase, and ecto-5′-nucleotidase in myeloid leukocytes. Am J Physiol 273:C973-C987
Cockcroft S, Gomperts BD (1979) ATP induces nucleotide permeability in rat mast cells. Nature (Lond). 279:541–542
Collo G, Neidhart S, Kawashima E, Kosco-Vilbois M, North RA, Buell G (1997) Tissue distribution of the $P2X_7$ receptor. Neuropharmacol 36:1277–1283
Communi D, Govaerts C, Parmentier M, Boeynaems JM (1997) Cloning of a human purinergic P2Y receptor coupled to phospholipase C and adenylyl cyclase. J Biol Chem 272:31969–31973
Conigrave AD, Lee JY, van der Weyden L, Jiang L, Ward P, Tasevski V, Luttrell BM, Morris JB (1998) Pharmacological profile of a novel cyclic AMP-linked P2 receptor on undifferentiated HL-60 leukemia cells. Br J Pharmacol 124:1580–1585
Coutinho-Silva R, Persechini PM (1997) P2Z purinoceptor-associated pores induced by extracellular ATP in macrophages and J774 cells. Am J Physiol 273: C1793–C1800
Coutinho-Silva R, Persechini PM, Bisaggio RC, Perfettini J-L, Sá-Neto ACT, Kanellopoulos JM, Motta-Ly I, Dautry-Varsat A, Ojcius DM (1999) $P_{2Z}/P2X_7$ receptor-dependent apoptosis of dendritic cells. Am J Physiol 276:C1139–1147
Cowen DS, Lazarus HM, Shurin SB, Stoll SE, Dubyak G (1989) Extracellular adenosine triphosphate activates calcium mobilization in human phagocytic leukocytes and neutrophil/monocyte progenitor cells. J Clin Invest 83:1651–1660

Cowen DS, Berger M, Nuttle L, Dubyak GR (1991) Chronic treatment with P_2-purinergic receptor agonists induces phenotypic modulation of the HL-60 and U937 human myelogenous leukemia cell lines. J Leukocyte Biol 50:109–122
Dawicki DD, McGowan-Jordan J, Bullard S, Pond S, Rounds S (1995) Extracellular nucleotides stimulate leukocyte adherence to cultured pulmonary artery endothelial cells. Am J Physiol 12:L666–L673
Denlinger LC, Fisette PL, Garis KA, Kwon G, Vazqueztorres A, Simon AD, Nguyen B, Proctor RA, Bertics PJ, Corbett JA (1996) Regulation of inducible nitric oxide synthase expression by macrophage purinoceptors and calcium. J Biol Chem 271:337–342
Di Virgilio F, Bronte V, Collavo D, Zanovello P (1989) Responses of mouse lymphocytes to extracellular adenosine 5′-triphosphate (ATP): lymphocytes with cytotoxic activity are resistant to the permeabilizing effects of ATP. J Immunol 143: 1955–1960
Dombrowski KE, Cone JC, Bjorndahl JM, Phillips CA (1995a) Irreversible inhibition of human natural killer cell natural cytotoxicity by modification of the extracellular membrane by the adenine nucleotide analog 5′-*p*-(fluorosulfonyl)benzoyl adenosine. Cell Immunol 160:199–204
Dombrowski KE, Ke Y, Brewer KA, Kapp JA (1998) Ecto-ATPase: an activation marker necessary for effector cell function. Immunol Rev 161:111–118
Dombrowski KE, Ke Y, Thompson LF, Kapp JA (1995b) Antigen recognition by CTL is dependent upon ecto-ATPase activity. J Immunol 154:6227–6237
El-Moatassim C, Dornand J, Mani J-C (1987) Extracellular ATP increases cytosolic free calcium in thymocytes and initiates the blastogenesis of the phorbol 12-myristate 13-acetate-treated medullary population. Biochim Biophys Acta 927:437–444
El-Moatassim C, Mani J-C, Dornand J (1990) Extracellular ATP^{4-} permeabilizes thymocytes not only to cations but also to low-molecular-weight solutes. Eur J Pharmacol 181:111–118
El-Moatassim C, Dubyak GR (1992) A novel pathway for the activation of phospholipase-D by P_{2Z} purinergic receptors in BAC1.2F5 macrophages. J Biol Chem 267:23664–23673
Enjyoji K, Sevigny J, Lin Y, Frenette PS, Christie PD, Esch JS 2nd, Imai M, Edelberg JM, Rayburn H, Lech M, Beeler DL, Csizmadia E, Wagner DD, Robson SC, Rosenberg RD (1999) Targeted disruption of cd39/ATP diphosphohydrolase results in disordered hemostasis and thromboregulation. Nature Med 5:1010–1017
Fabre JE, Nguyen M, Latour A, Keifer JA, Audoly LP, Coffman TM, Koller BH (1999) Decreased platelet aggregation, increased bleeding time and resistance to thromboembolism in $P2Y_1$-deficient mice. Nature Med 5:1199–1202
Falzoni S, Munerati M, Ferrari D, Spisani S, Moretti S, Di Virgilio F (1995) The purinergic P_{2Z} receptor of human macrophage cells – characterization and possible physiological role. J Clin Invest 95:1207–1216
Ferrari D, Chiozzi P, Falzoni S, Dal Susino M, Melchiorri L, Baricordi OR, Di Virgilio F (1997a) Extracellular ATP triggers IL-1β release by activating the purinergic P_{2Z} receptor of human macrophages. J Immunol 159:1451–1458
Ferrari D, Chiozzi P, Falzoni S, Hanau S, Di Virgilio F (1997b) Purinergic modulation of interleukin-1β release from microglial cells stimulated with bacterial endotoxin. J Exp Med 185:579–582
Ferrari D, Los M, Bauer MK, Vandenabeele P, Wesselborg S, Schulze-Osthoff K (1999a) P_{2Z} purinoreceptor ligation induces activation of caspases with distinct roles in apoptotic and necrotic alterations of cell death. FEBS Lett 447:71–75
Ferrari D, Stroh C, Schulze-Osthoff K (1999b) $P2X_7/P_{2Z}$ purinoreceptor-mediated activation of transcription factor NFAT in microglial cells. J Biol Chem 274: 13205–13210
Ferrari D, Villalba M, Chiozzi P, Falzoni S, Ricciardi-Castagnoli P, Di Virgilio F (1996) Mouse microglial cells express a plasma membrane pore gated by extracellular ATP. J Immunol 156:1531–1539

Ferrari D, Wesselborg S, Bauer MKA, Schulze-Osthoff K (1997c) Extracellular ATP activates transcription factor NF-κB through the P_{2Z} purinoreceptor by selectively targeting NF-κB p65. J Cell Biol 139:1635–1643

Filippini A, Taffs RE, Sitkovsky MV (1990) Extracellular ATP in T-lymphocyte activation: possible role in effector functions. Proc Natl Acad Sci USA 87: 8267–8271

Freedman BD, Liu QH, Gaulton G, Koltikoff MI, Heschler J, Fleischmann BK (1999) ATP-evoked Ca^{2+} transients and currents in murine thymocytes: possible role for P2X receptors in death by neglect. Eur J Immunol 29:1635–1646

Freyer DR, Boxer LA, Axtell RA, Todd 3d RF (1988) Stimulation of human neutrophil adhesive properties by adenine nucleotides. J Immunol 141:580–586

Gargett CE, Cornish EJ, Wiley JS (1996) Phospholipase D activation by P_{2Z}-purinoceptor agonists in human lymphocytes is dependent on bivalent cation influx. Biochem J 313:529–535

Grahames, CBA, Michel AD, Chessel IP, Humphrey PPA (1999) Pharmacological characterization of ATP- and LPS-induced IL-1β release in human monocytes. Br J Pharmacol 127; 1915–1921

Greenberg S, Di Virgilio F, Steinberg TH, Silverstein SC (1988) Extracellular nucleotides mediate Ca^{2+} fluxes in J774 macrophages by two distinct mechanisms. J Biol Chem 263:10337–10343

Griffiths RJ, Stam EJ, Downs JT, Otterness IG (1995) ATP induces the release of IL-1 from LPS-primed cells in vivo. J Immunol 154:2821–2828

Gu B, Bendall LJ, Wiley JS (1998) Adenosine triphosphate-induced shedding of CD23 and L-selectin (CD62L) from lymphocytes is mediated by the same receptor but different metalloproteases. Blood 92:946–951

Gutensohn W, Resta R, Misumi Y, Ikehara Y, Thompson LF (1995) Ecto-5′-nucleotidase activity is not required for T cell activation through CD73. Cell Immunol 161:213–217

Hickman SE, Elkhoury J, Greenberg S, Schieren I, Silverstein SC (1994) P_{2Z} adenosine triphosphate receptor activity in cultured human monocyte-derived macrophages. Blood 84:2452–2456

Homolya L, Watt WC, Lazarowski ER, Koller BH, Boucher RC (1999) Nucleotide-regulated calcium signaling in lung fibroblasts and epithelial cells from normal and $P2Y_2$ receptor (–/–) mice. J Biol Chem 274:26454–26460

Hu Y, Fisette PL, Denlinger LC, Guadarrama AG, Sommer JA, Proctor RA, Bertics PJ (1998) Purinergic receptor modulation of lipopolysaccharide. Signaling and inducible nitric-oxide synthase expression in RAW 264.7 macrophages. J Biol Chem 273:27170–27175

Humphreys BD, Dubyak GR (1996) Induction of the $P_{2Z}/P2X_7$ nucleotide receptor and associated phospholipase D activity by lipopolysaccharide and IFN-γin the human THP-1 monocytic cell line. J Immunol 157:5627–5637

Humphreys BD, Dubyak GR (1998) Modulation of $P2X_7$ nucleotide receptor expression by pro- and anti-inflammatory stimuli in THP-1 monocytes. J Leukocyte Biol 64:265–273

Humphreys BD, Virginio C, Surprenant A, Rice J, Dubyak GR (1998) Isoquinolines as antagonists of the $P2X_7$ nucleotide receptor: high selectivity for the human versus rat receptor homologues. Mol Pharmacol 54:22–32

Inoue K, Nakajima K, Morimoto T, Kikuchi Y, Koizumi S, Illes P, Kohsaka S (1998) ATP stimulation of Ca^{2+}-dependent plasminogen release from cultured microglia. Br J Pharmacol. 123:1304–1310

Jamieson GP, Snook MB, Thurlow PJ, Wiley JS (1996) Extracellular ATP causes loss of L-selectin from human lymphocytes via occupancy of P_{2Z} purinoceptors. J Cell Physiol 166:637–642

Jiang, L, Foster FM, Ward P, Tasevski V, Luttrel BM, Conigrave AD (1997) Extracellular ATP triggers cyclic AMP-dependent differentiation of HL-60 cells. Biochem Biophys Res Commun 232:626–630

Jiang S, Kull B, Fredholm BB, Orrenius S (1996) P2X purinoceptor is not important in thymocyte apoptosis. Immunol Lett 49:197–201
Jin JG, Dasari VR, Sistare FD, Kunapuli SP (1998) Distribution of P2Y receptor subtypes on haematopoietic cells. Br J Pharmacol 123:789–794
Kansas GS, Wood GS, Tedder TF (1991) Expression, distribution, and biochemistry of human CD39. J Immunol 146:2235–2244
Koshiba M, Apasov S, Sverdlov V, Chen P, Erb L, Turner JT, Weisman GA, Sitkovsky MV (1997) Transient up-regulation of $P2Y_2$ nucleotide receptor mRNA expression is an immediate early gene response in activated thymocytes. Proc Natl Acad Sci USA 94:831–836
Koshlukova SE, Lloyd TL, Araujo MW, Edgerton M (1999) Salivary histatin 5 induces non-lytic release of ATP from *Candida albicans* leading to cell death. J Biol Chem 274:18872–18879
Lammas DA, Stober C, Harvey CJ, Kendrick N, Panchalingam S, Kumararatne DS (1997) ATP-induced killing of mycobacteria by human macrophages is mediated by purinergic P_{2Z}($P2X_7$) receptors. Immunity 7:433–444
Langosch JM, Gebicke-Haerter PJ, Norenberg W, Illes P (1994) Characterization and transduction mechanisms of purinoceptors in activated rat microglia. Br J Pharmacol 113:29–34
Lin WW (1997) Priming effects of lipopolysaccharide on UTP-induced arachidonic acid release in RAW 264.7 macrophages. Eur J Pharmacol 321:121–127
Macino B, Zambon A, Milan G, Cabrelle A, Ruzzene M, Rosato A, Mandruzzato S, Quintieri L, Zanovello P, Collavo D (1996) CD45 regulates apoptosis induced by extracellular adenosine triphosphate and cytotoxic T lymphocytes. Biochem Biophys Res Commun 226:769–776
Madara JL, Patapoff TW, Gillece-Castro B, Colgan SP, Parkos CA, Delp C, Mrsny RJ (1993) 5′-Adenosine monophosphate is the neutrophil-derived paracrine factor that elicits chloride secretion from T84 intestinal epithelial cell monolayers. J Clin Invest 91:2320–2325
Markwardt F, Lohn M, Bohm T, Klapperstuck M (1997) Purinoceptor-operated cationic channels in human B lymphocytes. J Physiol(London) 498:143–151
Martin KA, Kertesy SB, Dubyak GR (1997) Down-regulation of P_{2U}-purinergic nucleotide receptor messenger RNA expression during in vitro differentiation of human myeloid leukocytes by phorbol esters or inflammatory activators. Mol Pharmacol 51:97–108
Molloy A, Laochumroonvorapong P, Kaplan G. (1994) Apoptosis, but not necrosis, of infected monocytes is coupled with killing of intracellular bacillus Calmette-Guerin. J Exp Med 180:1499–1509
Mulero JJ, Yeung G, Nelken ST, Ford JE (1999) CD39-L4 is a secreted human apyrase, specific for the hydrolysis of nucleoside diphosphates. J Biol Chem 274: 20064–20067
Mulryan K, Gitterman DP, Lewis CJ, Vial C, Leckie BJ, Cobb AL, Brown JE, Conley EC, Buell G, Pritchard CA, Evans RJ (2000) Reduced vas deferens contraction and male infertility in mice lacking $P2X_1$ receptors. Nature 403:86–89
Murgia M, Hanau S, Pizzo P, Rippa M, Di Virgilio F (1993) Oxidized ATP an irreversible inhibitor of macrophage purinergic P_{2Z} receptor. J Biol Chem 268:8199–8203
Mutini C, Falzoni S, Ferrari D, Chiozzi P, Morelli A, Baricordi OR, Collo G, Ricciardi-Castagnoli P, Di Virgilio F (1999) Mouse dendritic cells express the $P2X_7$ purinergic receptor: Characterization and possible participation in antigen presentation. J Immunol 163:1958–1965
Nakanishi M, Takihara H, Minoru Y, Yagawa K (1991) Extracellular ATP itself elicits superoxide generation in guinea pig peritoneal macrophages. FEBS Lett 282:91–94
Nakanishi M, Kawasaki M, Ogino H, Yoshida M, Yagawa K (1994) Extracellular ATP regulates the proliferation of alveolar macrophages. Am J Respir Cell Mol Biol 10:560–564

Naumov AP, Kiselyov KI, Mamin AG, Kaznacheyeva EV, Kuryshev YA, Mozhayeva GN (1995) ATP-operated calcium-permeable channels activated via a guanine nucleotide-dependent mechanism in rat macrophages. J Physiol (London) 486: 339–347
Nuttle LC, Dubyak GR (1994) Differential activation of cation channels and non-selective pores by macrophage P_{2Z} purinergic receptors expressed in *Xenopus* oocytes. J Biol Chem 269:13988–13996
Osipchuk Y, Cahalan M (1992) Cell-to-cell spread of calcium signals mediated by ATP receptors in mast cells. Nature 359:241–244
Owens GP, Hahn WA, Cohen JJ (1991) Identification of mRNAs associated with programmed cell death in immature thymocytes. Mol Cell Biol 11:4177–4188
Padeh S, Cohen A, Roifman CM (1991) ATP-induced activation of human B lymphocytes via P2-purinoceptors. J Immunol 146:1626–1632
Peola S, Borrione P, Matera L, Malavasi F, Pileri A, Massaia M (1996) Selective induction of CD73 expression in human lymphocytes by CD38 ligation: a novel pathway linking signal transducers with ecto-enzyme activities. J Immunol 157:4354–4362
Perregaux DG, Gabel CA (1994) Interleukin-1β maturation and release in response to ATP and nigericin. J Biol Chem 269:15195–15203
Perregaux DG, Gabel CA (1998) Post-translational processing of murine IL-1: evidence that ATP-induced release of IL-1α and IL-1β occurs via a similar mechanism. J Immunol 160:2469–2477
Pizzo P, Zanovello P, Bronte V, Di Virgilio F (1991) Extracellular ATP causes lysis of mouse thymocytes and activates a plasma membrane ion channel. Biochem J 274:139–144
Proctor RA, Denlinger LC, Leventhal PS, Daugherty SK, van de Loo J-W, Tanke T, Firestein GS, Bertics PJ (1994) Protection of mice from endotoxic death by 2-methylthio-ATP. Proc Natl Acad Sci USA 91:6017–6020
Rassendren F, Buell GN, Virginio C, Collo G, North RA, Surprenant A (1997) The permeabilizing ATP receptor, $P2X_7$. Cloning and expression of a human cDNA. J Biol Chem 272:5482–5486
Resta R, Yamashita Y, Thompson LF (1998) Ecto-enzyme and signaling functions of lymphocyte CD73. Immunol Rev 161:95–109
Redegeld F, Filippini A, Sitkovsky M (1991) Comparative studies of the cytotoxic T lymphocyte-mediated cytotoxicity and of extracellular ATP-induced cell lysis: Different requirements in extracellular Mg^{2+} and pH. J Immunol 147:3638–3645
Robson SC, Kaczmarek E, Siegel JB, Candinas D, Koziak K, Millan M, Hancock WW, Bach FH (1997) Loss of ATP diphosphohydrolase activity with endothelial cell activation. J Exp Med 185:153–63
Ross PE, Ehring GR, Cahalan MD (1997) Dynamics of ATP-induced calcium signaling in single mouse thymocytes. J Cell Biol 138:987–998
Schmid-Antomarchi H, Schmid-Alliana A, Romey G, Ventura MA, Breittmayer V, Millet MA, Husson H, Moghrabi B, Lazdunski M, Rossi B (1997) Extracellular ATP and UTP control the generation of reactive oxygen intermediates in human macrophages through the opening of a charybdotoxin-sensitive Ca^{2+}-dependent K^+ channel. J Immunol 159:6209–6215
Schmidt A, Ortaldo JR, Herberman RB (1984) Inhibition of human natural killer cell reactivity by exogenous adenosine 5′-triphosphate. J Immunol 132:146–150
Schwiebert EM (1999) ABC transporter-facilitated ATP conductive transport. Am J Physiol (Cell) 276:C1–8
Seifert R, Wenzel K, Eckstein F, Schultz G (1989) Purine and pyrimidine nucleotides potentiate activation of NADPH oxidase and degranulation by chemotactic peptides and induce aggregation of human neutrophils via G proteins. Eur J Biochem 181:277–285
Sikora A, Liu J, Brosnan C, Buell G, Chessel I, Bloom BR (1999) Purinergic signaling regulates radical-mediated bacterial killing in macrophages through a $P2X_7$-independent mechanism. J Immunol 163:558–561

Somers GR, Hammet FMA, Trute L, Southey MC, Venter DJ (1998) Expression of the $P2Y_6$ purinergic receptor in human T cells infiltrating inflammatory bowel disease. Lab Invest 78:1375–1383
Soto F, Garcia-Guzman M, Gomez-Hernandez JM, Hollmann M, Karschin C, Stuhmer W (1996) $P2X_4$: an ATP-activated ionotropic receptor cloned from rat brain. Proc Natl Acad Sci USA 93:3684–3688
Sperlagh B, Hasks G, Nemeth Z, Vizi ES (1998) ATP released by LPS increases nitric oxide production in raw 264.7 macrophage cell line via $P_{2Z}/P2X_7$ receptors. Neurochem Int 33:209–215
Spranzi E, Djeu JY, Hoffman SL, Epling-Burnette PK, Blanchard DK (1993) Lysis of human monocytic leukemia cells by extracellular adenosine triphosphate – mechanism and characterization of the adenosine triphosphate receptor. Blood 82: 1578–1585
Steinberg TH, Newman AS, Swanson JA, Silverstein SC (1987) ATP^{4-} permeabilizes the plasma membrane of mouse macrophages to fluorescent dyes. J Biol Cheml 262:8884–8888
Sud'ina GF, Mirzoeva OK, Galkina SI, Pushkareva MA, Ullrich V (1998) Involvement of ecto-ATPase and extracellular ATP in polymorphonuclear granulocyte-endothelial interactions. FEBS Lett 423:243–248
Sung S-SJ, Young JDE, Origlio AM, Heiple JM, Kaback HR, Silverstein SC (1985) Extracellular ATP perturbs transmembrane ion fluxes, elevates cytosolic $[Ca^{2+}]$, and inhibits phagocytosis in mouse macrophages. J Biol Chem 260:13442–13449
Surprenant A, Rassendren F, Kawashima E, North RA, Buell G (1996) The cytolytic P_{2Z} receptor for extracellular ATP identified as a P_{2X} receptor ($P2X_7$). Science 272:735–738
Tonetti M, Sturla L, Giovine M, Benatti U, Deflora A (1995) Extracellular ATP enhances mRNA levels of nitric oxide synthase and TNFα in lipopolysaccharide-treated RAW 264.7 murine macrophages. Biochem Biophys Res Commun 214:125–130
Valera S, Hussy N, Evans RJ, Adami N, North RA, Surprenant A, Buell G (1994) A new class of ligand-gated ion channel defined by P2X receptor for extracellular ATP. Nature 371:516–519
Ventura MA, Thomopoulos P (1995) ADP and ATP activate distinct signaling pathways in human promonocytic U-937 cells differentiated with 1,25-dihydroxy-vitamin D3. Mol Pharmacol 47:104–114
Verghese MW, Kneisler TB, Bouchron JA (1996) P_{2U} agonists induce chemotaxis and actin polymerization in human neutrophils and differentiated HL60 cells. J Biol Chem 271:15597–15601
Walker BAM, Hagenlocker BE, Douglas VK, Tarapchak SJ, Ward PA (1991) Nucleotide responses of human neutrophils. Lab Invest 64:105–112
Ward PA, Cunningham TW, McCulloch KK, Phan SH, Powel J, Johnson KJ (1988) Platelet enhancement of O2-. responses in stimulated human neutrophils. Identification of platelet factor as adenine nucleotide. Lab Invest 58:37–47
Wiley JS, Dubyak GR (1989) Extracellular adenosine triphosphate increases cation permeability of chronic lymphocytic leukemic lymphocytes. Blood 73:1316–1323
Wiley JS, Gargett CE, Zhang W, Snook MB, Jamieson GP (1998) Partial agonists and antagonists reveal a second permeability state of human lymphocyte $P_{2Z}/P2X_7$ channel. Am J Physiol (Cell) 275:1224–1231
Wiley JS, Chen R, Wiley MJ, Jamieson GP (1992) The ATP^{4-} receptor-operated ion channel of human lymphocytes: inhibition of ion fluxes by amiloride analogs and by extracellular sodium ions. Arch Biochem Biophys 292:411–418
Williams BA, Manzer A Blay J, Hoskin DW (1997) Adenosine acts through a novel extracellular receptor to inhibit granule exocytosis by natural killer cells. Biochem Biophys Res Commun 231:264–269
Yamaguchi M, Hirayoshi K, Okuma M, Nagata K (1994) Enhancement of differentiation induction of mouse myelomonocytic leukemic cells by extracellular ATP. J Cell Physiol 159:441–449

Yamashita Y, Hooker SW, Jiang H, Laurent AB, Resta R, Khare K, Coe A, Kincade PW, Thompson LF (1998) CD73 expression and fyn-dependent signaling on murine lymphocytes. Eur J Immunol 28:2981–2990
Zanovello P, Bronte V, Rosato A, Pizzo P, Di Virgilio F (1990) Responses of mouse lymphocytes to extracellular ATP. II. Extracellular ATP causes cell type-dependent lysis and DNA fragmentation. J Immunol 145:1545–1550
Zheng LM, Zychlinsky A, Liu C-C, Ojcius DM, Young JDE (1991) Extracellular ATP as a trigger for apoptosis or programmed cell death. J Cell Biol 112:279–288

CHAPTER 26

On the Role of the P2X$_7$ Receptor in the Immune System

F. Di Virgilio, V. Vishwanath, and D. Ferrari

A. Introduction

Twenty years ago Gomperts and Cockcroft were the first to describe ATP-dependent permeability increases in a cell type involved in the immune response (mast cells), hypothesizing that "a possible mechanism would involve channel formation by the aggregation of transmembrane monomeric units..." (Cockcroft and Gomperts 1979). Ever since, similar responses to ATP have been described in many other immune and inflammatory cells (Table 1), leading many authors to propose that expression of the "ATP permeabilizing receptor" is a feature of cells involved in host defense. In mast cells, as well as in other immune cells, the active form was found to be ATP in its fully dissociated form. Accordingly, this hypothetical receptor was named the ATP^4-receptor, later to become P$_{2Z}$ when Gordon (1986) carefully defined the properties of the P2 receptor of mast cells and lymphocytes, and found that this receptor did not entirely fit the P2X/P2Y subclassification originally proposed by Burnstock and Kennedy (1985). The functional "signature" of the P$_{2Z}$ receptor (reversible permeabilization of the plasma membrane) was so peculiar that many doubted the actual "receptor" nature of P2Z, reckoning that ATP might cause in immune cell types a nonspecific perturbation of the plasma membrane that in turn led to permeability transition. However, even before the cloning of the actual molecule responsible for ATP-dependent permeabilization, evidence supporting the receptor nature of P$_{2Z}$ was compelling as:

1. Other nucleosides or nucleotides could not mimic this effect, even at high concentrations (Cockcroft and Gomperts 1980; Steinberg et al. 1987).
2. It was possible to select from ATP-sensitive lines cell clones fully resistant to ATP (Steinberg and Silverstein 1987; Murgia et al. 1992).
3. Periodate-oxidized ATP (oATP) was shown to completely block permeabilization, and another ATP analog, 2-methylthio-9-β-L-ribofuranosyladenine 5′-triphosphate (2-MeS-L-ATP) was 50% inhibitory (Murgia et al. 1993; Tatham et al. 1988).

The conclusive proof that P$_{2Z}$ was a real entity came with the cloning and the demonstration that this receptor was related to the P2X subfamily and was therefore named P2X$_7$ (Surprenant et al. 1996; Rassendren et al. 1997).

Table 1. Cells expressing the $P2X_7$/P2Z receptor

Cell type	Method	References
Rat mast cells	Functional responses	Cockcroft and Gomperts (1979)
Mouse T and B lymphocytes	Functional responses (?)	Di Virgilio et al. (1989); Filippini et al. (1990a,b); Chused et al. (1996)
Human B lymphocytes	Functional responses (?), Abs, RT-PCR	Wiley and Dubyak (1989); Ferrari et al. (1994); Bretschneider et al. (1995); O.R. Baricordi et al. (in preparation)
Human T lymphocytes	Functional responses (?); RT-PCR	Baricordi et al. (1996); O.R. Baricordi et al. (in preparation)
Mouse macrophages	Functional responses, cloning, Abs, in-situ hybridization, RT-PCR	Steinberg et al. (1987); Surprenant et al. (1996); Chiozzi et al. (1997)
Human macrophages	Functional responses, cloning, Abs, RT-PCR	Hickman et al. (1994); Falzoni et al. (1995); Dubyak et al. (1996); Rassendren et al. (1997)
Mouse/rat microglial cells	Functional responses, Abs, in situ hybridization	Ferrari et al. (1996, 1997a); Collo et al. (1997)
Human Langerhans cells	Functional responses	Girolomoni et al. (1993)
Human dendritic cells	Abs	Buell et al. (1998)
Mouse dendritic cells	Functional responses, Abs	Mutini et al. (1999)
Human fibroblasts	Functional responses, Abs, RT-PCR	Solini et al. (1999)
Porcine and bovine endothelial cells	Functional responses, RT-PCR	von Albertini et al. (1998)
Rat retina neurons	Functional responses, Abs, RT-PCR	Brandle et al. (1998)
Rat mesangial cells	Functional responses, Northern	Schulze-Lohoff et al. (1998)
CHO cells	Functional responses, RT-PCR	Michel et al. (1998)
Rat supraoptic neurons	RT-PCR	Shibuya et al. (1999)
Rat salivary gland	RT-PCR	Turner et al. (1998)
Rat submandibular glands	Functional responses, RT-PCR	Alzola et al. (1998)
Rat parotid acinar cells	Functional responses, RT-PCR	Tenneti et al. (1998)
Rat hepatic arteries	RT-PCR	Phillips et al. (1998)
Human saphenous vein smooth muscle	Functional responses, RT-PCR	Cario-Toumaniantz et al. (1998)

B. Are P2Z and P2X$_7$ One and the Same Molecule?

There is no doubt that P2X$_7$ is the receptor responsible for the membrane-permeabilizing effects of extracellular ATP (Surprenant et al. 1996; Rassendren et al. 1997); there is less certainty, however, whether P2X$_7$ is the sole constitutive subunit of the immune cell receptor phenotypically known as P$_{2Z}$. On the one hand, transfection of P2X$_7$ confers sensitivity to ATP-dependent permeabilization, but on the other permeability properties of the lymphocyte P$_{2Z}$ receptor are clearly different as compared to those of the mast-cell and macrophage receptor. Despite extensive investigation, neither T nor B lymphocytes have ever been shown to take up solutes larger than ethidium bromide following activation of the ATP-gated pore (El Moatassim et al. 1989; Di Virgilio et al. 1989; Wiley and Dubyak 1989; Wiley et al. 1993; Ferrari et al. 1994; Markwardt et al. 1997), although the pharmacological profile, supported by preliminary RT-PCR data, is strongly suggestive of P2X$_7$ receptor expression by these cell types (Ferrari et al. 1994; Bretschneider et al. 1995; Gargett et al. 1997). A possible interpretation, as yet unsubstantiated by experimental findings, of these conflicting data is that the various lymphocyte lineages express heteromeric P2X$_7$/P2X$_7$ rather than homomeric P2X$_7$ receptors, while macrophages and other mononuclear phagocytes express a homomeric P2X$_7$ receptor.

C. Distribution of P2X$_7$ Receptor Subunits

P2X$_7$ was originally cloned from a rat brain library; however, and quite surprisingly, central neurons were found not to express P2X$_7$, thus raising the possibility that this receptor was indeed cloned from non-neuronal brain constituents such as microglia (Surprenant et al. 1997; Collo et al. 1997). Screening of cells and tissue sections with molecular probes recently become available has basically confirmed previous studies on P$_{2Z}$ receptor localization assessed by necessity by functional assays, thus providing an a posteriori demonstration of the reliability of the "P$_{2Z}$ signature" (ATP-dependent plasma membrane permeabilization) as proof of P2X$_7$ expression. Nevertheless, tissue and cell distribution of P2X$_7$ is still somewhat ill defined as, in contrast to the common belief that this receptor is almost exclusively expressed by immune cells, it is now clear that it can also be found outside the immune system, albeit to a much lower level (Table 1). P2X$_7$ has been detected by RT-PCR or Northern blotting in cells as different as retinal neurons (Brandle et al. 1998), endothelial cells (von Albertini et al. 1998), fibroblasts (Solini et al. 1999), and epithelia (Christoffersen et al. 1998; Tenneti et al. 1998). Nevertheless, immune cells, with the relevant exception of polymorphonuclear granulocytes, express by far P2X$_7$ to the highest level and in these cells this receptor has been more thoroughly characterized and its functions better defined than in any other cell type.

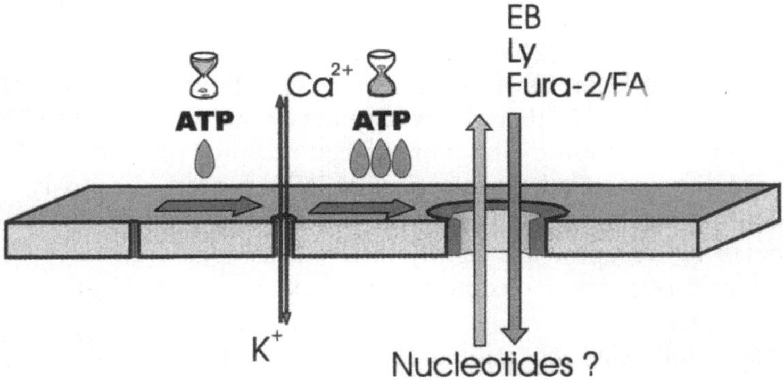

Fig. 1. ATP induces formation of plasma membrane channels/pores in cells expressing $P2X_7$ receptor. A short exposure to ATP induces formation of plasma membrane channels permeable to monovalent and divalent cations (*left*), while a more prolonged incubation or repeated ATP pulses cause formation of pores permeable to low molecular weight extracellular markers and small intracellular substrates (*right*)

D. $P2X_7$ as a Cytotoxic Receptor

The most intriguing feature of $P2X_7$ is its ability to undergo a channel/pore transition upon sustained stimulation or repetitive ATP applications (Fig. 1), as if this receptor exhibited memory for previous stimulations (see Cook et al. 1998 for an example of a Ca^{2+}-based memory in other P2X subtypes). It is clear that this behavior may have deleterious effects as no cell can survive with an open pore on the plasma membrane.

I. Early Observations

Interest in the possible role of $P2X_7$ in immunomediated cytotoxic reactions was initially raised by the serendipitous observation made by our laboratory and that of Sitkovsky that, while freshly isolated mouse lymphocytes and lymphoid cell lines were sensitive to ATP-mediated cytotoxicity, T lymphocytes with cytotoxic activity (CTL), whether derived from established CTL clones or raised by allogeneic in vivo stimulation, were completely resistant (Di Virgilio et al. 1989; Filippini et al. 1990a). Furthermore, lymphokine activated killer (LAK) cells were also resistant to ATP and development of refractoriness to ATP paralleled acquisition of the LAK phenotype (Di Virgilio et al. 1989). At least two mechanisms were invoked to explain the insensitivity of cytotoxic lymphocytes to ATP:

1. A protective effect due to expression of very active plasma membrane ecto-ATPases (Filippini et al. 1990a)
2. A lack of the "cytotoxic ATP receptor" (Di Virgilio et al. 1989, 1990)

There was no follow-up to these early observations, but now that molecular reagents for $P2X_7$ have become available it would be interesting to re-investigate the issue. Susceptibility to ATP-mediated cytotoxicity seems to be a distinctive feature not just of lymphocytes but of any cell expressing $P2X_7$ (macrophages, microglia, mesangial cells, dendritic cells). Whether it then occurs by necrosis or apoptosis is dependent on the ATP dose, length of stimulation, and the given cell type (ZANOVELLO et al. 1990). Furthermore, cell clones selected for refractoriness to ATP-mediated cytotoxicity can be shown to lack the $P2X_7$ protein, and $P2X_7$ transfection into resistant cells confers susceptibility to ATP-mediated cytotoxicity (FERRARI et al. 1996, 1997a; SURPRENANT et al. 1996).

II. What Role for ATP-Mediated Cytotoxicity in the Immune System?

An obvious candidate function would be CTL-mediate killing, as initially proposed by our group and that of Sitkovsky. This hypothesis has been put to the test in a few laboratories, and evidence obtained suggests that it is likely incorrect, at least in its original, somewhat naive, form. Were ATP a primary mediator of CTL-mediated killing, one would expect that soluble ATP-hydrolyzing enzymes (e.g., apyrase or hexokinase) should, at least partially, inhibit CTL-mediated target cell lysis. However, experiments aimed at testing this prediction gave conflicting results (FILIPPINI et al. 1990b; REDEGELD et al. 1993). Furthermore, cells lacking $P2X_7$, or cells in which $P2X_7$ was irreversibly blocked with oATP, should be a less susceptible target. Again, this prediction was not fulfilled (ZAMBON et al. 1993). Nevertheless, it is well known that CTL-mediated cytotoxicity depends on the activation of multiple possibly redundant pathways; thus the inability of $P2X_7$ inhibitors to block fully this complex response is not, after all, totally unexpected.

An interesting proposal is that $P2X_7$-mediated cytotoxicity might be related to antigen-presenting activity. This suggestion originates from studies by BLANCHARD et al. (1995) reporting that lysis of antigen-presenting macrophages by specific CTLs was prevented by the addition of hexokinase to the reaction medium. In the same study, these authors provided the first evidence that KN-62, supposedly a calmodulin inhibitor, blocked a reportedly "ATP receptor"-mediated response. This compound decreased by 50% CTL-dependent lysis of a macrophage line expressing the $P2X_7$ receptor, while it had no effect on another line lacking this receptor. Blanchard and coworkers interpreted the inhibitory activity of KN-62 as suggestive of an involvement of calmodulin in ATP-mediated lysis; however with hindsight and after the thorough characterization of KN-62 (GARGETT and WILEY 1997) in human B lymphocytes, it is likely that calmodulin played no role in the experiments of Blanchard et al. and the inhibition of lysis was due to $P2X_7$ receptor blockade. The hypothesis that $P2X_7$ might be involved in cellular interactions during antigen presentation is further supported by the recent demonstration by

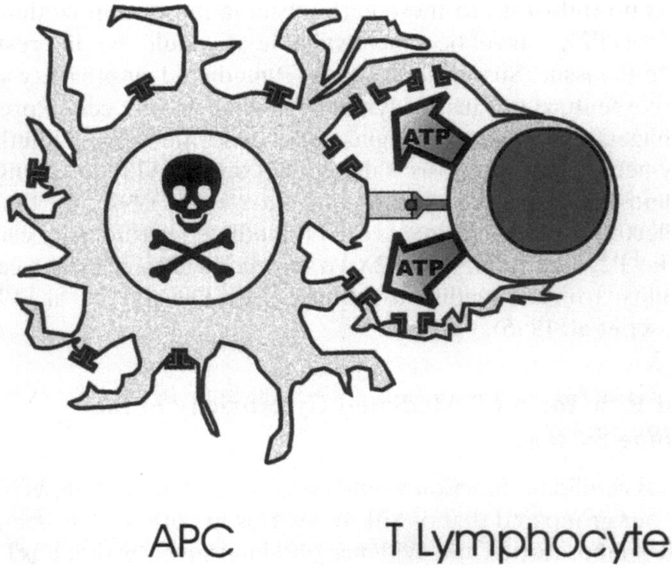

Fig. 2. Hypothetical mechanism for ATP-mediated killing of antigen presenting cells (APC) by T lymphocytes. It is proposed that ATP is released by the T cell in a protected compartment at the site of T lymphocyte-APC interaction. $P2X_7$ receptor simulation by ATP then induces death of the APC

COUTINHO-SILVA et al. (1996) and BUELL et al. (1998) that mouse and human follicular dendritic cells express $P2X_7$. It might be possible therefore that $P2X_7$-dependent cytotoxicity is one of the mechanisms used by CTLs to eliminate APCs once these cells have exhausted their task as stimulatory cells, and thus to down-modulate the immune response (Fig. 2). For example, Langerhans cells are known to migrate to lymph nodes once they are antigen stimulated, where they activate antigen-specific lymphocytes; however, there are no dendritic cells in the efferent lymph, a finding indicating that activated Langerhans cells must die after lymphocytes have been stimulated (BANCHERAU and STEINMAN 1998).

III. $P2X_7$ and Disease

An intriguing hint to the importance of $P2X_7$-dependent cytotoxicity in disease control by the host comes from experiments by MOLLOY et al. (1994) and LAMMAS et al. (1997). During an investigation of the effect of extracellular ATP on macrophage cultures infected with *Mycobacterium bovis* (*bacillus* Calmette Guerin, BCG), both groups reported that $P2X_7$ activation caused killing of the phagocyte as well as of the intracellular pathogen. Lammas and coworkers performed a thorough pharmacological characterization of ATP-mediated killing of intracellular BCG, concluding that the receptor involved was $P2X_7$. Of particular relevance is the observation that oATP, the most widely used $P2X_7$ blocker, could completely prevent killing of the phagocyte and of the ingested pathogen in parallel with the inhibition of plasma membrane permeabilization. It is of interest that of the several cytotoxic molecules tested by Lammas and co-workers, i.e., complement, anti-CD69, and anti-CD95 antibodies, only ATP exhibited the potent cytotoxic effect on intracellular BCG. The mechanism responsible for ATP-mediated killing of the intracellular pathogen is mysterious. LAMMAS et al. 1997 also elegantly demonstrated that nitric oxide (NO) and reactive oxygen species can be ruled out, as inhibitors of the inducible NO synthase (iNOS) did not prevent ATP-mediated killing of BCG, and the killing also occurred in macrophages from patients with X-linked autosomal recessive chronic granulomatous disease (CGD). MOLLOY et al. (1994) reported that, upon ATP challenge, striking morphological changes could be observed in the vacuolar apparatus of the macrophages: BCG-containing vesicles, that in the absence of ATP were closely juxtaposed to the bacterial cells and originally contained only one or few bacilli, now were swollen, contained many bacilli, and appeared to originate from the fusion of hitherto discrete cytoplasmic vacuoles. In our laboratory we obtained very similar data, suggesting that one of the mechanism by which ATP enhances BCG killing is by increasing the rate of cytoplasmic vesicle (phagosome-lysosome?) fusion (Fig. 3).

Mycobacterial infection is the only model in which the cytotoxic effect of ATP has been investigated; however, it is not unlikely that other intracellular pathogens may also be sensitive to killing by ATP, especially those that are harbored within phagocytic vacuoles (e.g., *Lysteria, Chlamidia*, etc.).

Very often macrophages can be induced to kill the intracellular parasite upon activation with IFN-γ; thus it may not be merely coincidental that this cytokine and other pro-inflammatory factors also up-regulate $P2X_7$ expression in macrophages (BLANCHARD et al. 1991; FALZONI et al. 1995; HUMPHREY and DUBYAK 1998). In this light, the cytotoxic activity mediated by $P2X_7$ could be an extreme measure to which the immune system resorts in those cases in which there is no other alternative means to get rid of an intracellular pathogen ("Samson shall die with all of the Philistines").

Over the last few years cytotoxic effects mediated by $P2X_7$ have also attracted the attention of nephrologists concerned with the damage suffered by mesangial cells during kidney diseases. Mesangial cells are smooth muscle-like elements that in vitro undergo apoptosis upon stimulation with ATP (in a concentration range from 300 μmol/l to 5 mmol/l) and at the same time up-regulate the tumor suppressor protein p53 in an oATP-inhibitable fashion (SCHULZE-LOHOFF et al. 1998). In addition, they are very active producers of IL-1β, a cytokine that not only has a central role as a modulator of the immune system, but also acts as an autocrine growth factor stimulating proliferation of the intrinsic mesangial cells. Given the compelling evidence implicating $P2X_7$ in IL-1β release (see below), it is clear that involvement of this receptor in kidney diseases might be profound.

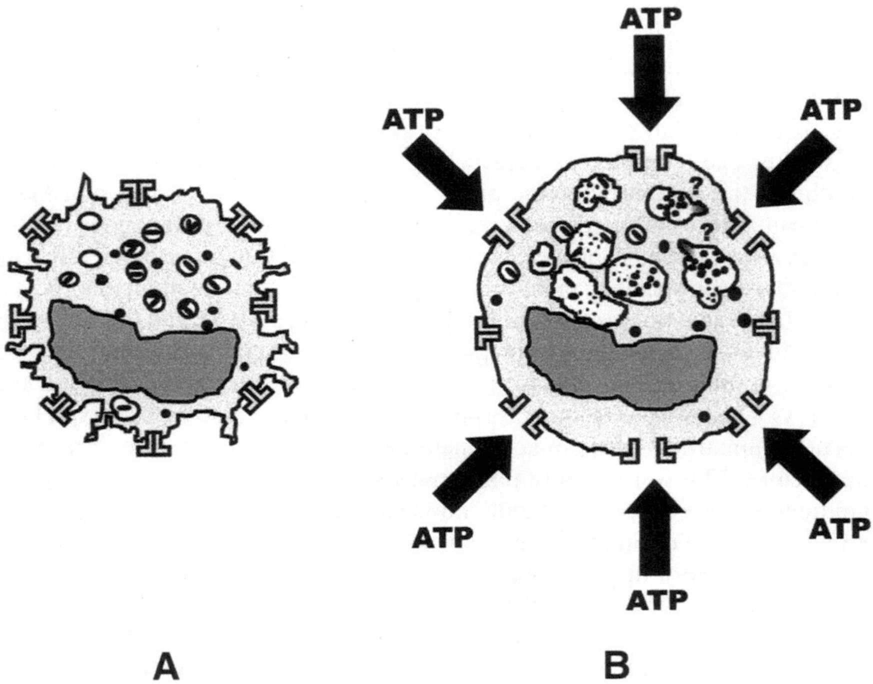

Fig. 3A,B. Hypothetical model for ATP-mediated killing of intracellular parasites (*Bacillus* Calmette Guerin, BCG and *Mycobacterium tuberculosis*, MTB). **A** BCG or MTB are phagocytosed but survive within the phagosomes as a consequence of inhibition of phagosome/lysosome fusion. **B** Upon stimulation of $P2X_7$ receptor, phagosomes swell and undergo an enhanced rate of phagosome/phagosome or phagosome/lysosome fusion, thus causing death of the intracellular parasites

IV. $P2X_7$ in Spontaneous Cell Death

It is always appealing to link a biological process to a disease or a pathological alteration; however we should not neglect the possibility that $P2X_7$ also plays a role in spontaneous cell death, especially given the widespread evidence that ATP release is a common event. Chiozzi et al. (1996) observed that J774 macrophages that hyperexpress $P2X_7$ ($P2X_7$hyper) are extremely fragile and very difficult to propagate in vitro as they undergo a high rate of spontaneous death. This is in patent contrast with the parental J774 cell line that is easily grown in culture, and in even sharper contrast with J774 clones selected for low $P2X_7$ expression ($P2X_7$hypo) that show a rate of death at least ten times lower than the $P2X_7$hyper cells. Furthermore, cell death was drastically reduced by incubation of the J774 cultures in the presence of oATP, or the ATP-hydrolyzing enzymes apyrase or hexokinase, suggesting that it was caused by an accumulation of ATP in the culture medium that led to opening of the $P2X_7$ pore. We did not further characterize the mechanism of cell death due to this chronic exposure to presumably low but sustained extracellular ATP concentrations; however, cell morphology was strongly indicative of an apoptotic process (CHIOZZI et al. 1996). Such an autocrine/paracrine mechanism could also trigger an amplifying loop by causing ATP release from dying cells and therefore further ATP accumulation in the pericellular environment. We think that in an intact tissue the cytotoxic effect of released ATP may be even more prominent as it is likely that in a condition under which physiological cell-to-cell interactions are conserved, ATP will be diluted into the bulk phase much more slowly, and thus it will have the chance to reach higher concentrations at the level of the plasma membrane. Low ATP affinity for $P2X_7$ has been felt by many as a substantial obstacle for the acceptance of major role for ATP as a cytotoxic molecule in vivo. However, we think that dose-dependency curves obtained *in vitro* should be taken with a pinch of salt as many other factors acting *in vivo* may modify receptor number or affinity, thus enhancing the cytotoxic activity of ATP, as clearly documented for IFN-γ, TNFα and bacterial endotoxin (LPS) (BLANCHARD et al. 1991; FALZONI et al. 1995; HUMPHREY and DUBYAK 1998). Furthermore, it should not be a surprise that a receptor with a cytotoxic potential, such as $P2X_7$, has a low affinity for ATP, as otherwise it would be activated far more often, given the widespread occurrence of this extracellular nucleotide, with deleterious effects on cell viability.

An intriguing hypothesis is that adverse side effects of some drugs may also be mediated by $P2X_7$. One such compounds is the anti-inflammatory drug tenidap that, although by itself harmless to J774 macrophages at concentrations as high as 500 μmol/l, in the presence of 2 mmol/l ATP, a threshold lethal concentration in this cell type, causes 60% lysis even at a dose as low as 50 μmol/l (SANZ et al. 1998). Accordingly, low tenidap concentrations sensitize macrophages to non-cytotoxic (high micromolar) ATP doses. Interestingly, macrophage clones lacking $P2X_7$ are resistant to the synergistic cytotoxic effect of tenidap plus ATP. Potentiation of a response mediated by another P2X subtype (contraction of rat vas deferens) has also been demonstrated for a chemical poison, bis(2-chloroethyl)sulfide, thus pointing to a more widespread role for P2X receptors in the mediation of cell responses to injury (LUNDY et al. 1998).

Given the well accepted role of ATP as a neurotransmitter, microglial cells are an obvious model for the investigation of $P2X_7$ functions. Illes and colleagues have extensively investigated ionotropic P2 receptors in this cell type (NORENBERG et al. 1994; ILLES et al. 1996) and have recently provided evidence that ATP is involved via $P2X_7$ in the release of plasminogen from activated rat microglia (INOUE et al. 1998). Plasminogen is a factor that promotes development of dopaminergic neurons and neurite outgrowth, and therefore partakes in post-injury healing and regeneration.

E. $P2X_7$ receptors and IL-1β Release

LPS-dependent release of the pro-inflammatory cytokine IL-1β from macrophages and microglial cells, in contrast to peripheral blood monocytes, is a slow and inefficient process that leads to extracellular accumulation of minute amounts of this cytokine and mainly of the high MW (31–34 kDa), uncleaved, biologically-inactive, pro-cytokine form. This finding led many investigators to postulate that a second stimulus is needed to trigger processing and secretion of the cytokine in its low MW (17 kDa) biologically active form, but the identity of this second stimulus has remained elusive. Hogquist et al. (1991) observed that extracellular ATP triggered IL-1β processing and release, and in 1992, Gabel and co-workers (Perregaux et al. 1992) reported that mature IL-1β formation could be induced by the K^+ ionophore nigericin. What is common to nigericin and ATP? Perregaux et al. (1992) reasoned that both nigericin and ATP decrease intracellular K^+ levels (see Steinberg and Silverstein 1987) and that perhaps this ionic perturbation was needed to activate the enzyme that cleaves pro-IL-1β into mature IL-1β, i.e., interleukin-1β-converting enzyme (ICE), also known as caspase-1. Later studies fulfilled this prediction as ATP was shown to trigger IL-1β release via a non-lytic mechanism in many different mononuclear phagocytic cells, and release was inhibited by procedures that prevented K^+ efflux (Perregaux and Gabel 1994; Ferrari et al. 1996; Ferrari et al. 1997b). In support of a key role for K^+, Cheneval et al. (1998) showed that a reduction in the K^+ concentration also leads to proteolytic activation of isolated recombinant ICE. Interestingly, while proteolytic activation of the isolated enzyme could be induced by a reduction in the concentration of other cations beside K^+, autoprocessing of cytoplasmic ICE showed an absolute requirement for K^+ depletion. That ATP acts via ICE is also demonstrated by the ability of a specific ICE inhibitor, the tetrapeptide YVAD (Tyr-Val-Ala-Asp), to block ATP-dependent IL-1β release (Perregaux and Gabel 1998a). Furthermore, macrophages isolated from mice deficient in ICE were unable to process pro-IL-1β upon challenge with LPS plus ATP (Li et al. 1995). Finally, involvement of $P2X_7$ in ATP-mediated ICE activation is supported by agonist and antagonist profile of cytokine release (Ferrari et al. 1997b), blockade by a specific anti $P2X_7$ monoclonal Ab (Buell et al. 1998) and detection of ICE proteolytic fragments (p20 and p10) in ATP-stimulated microglial cells (Ferrari et al. 1997c). There are no clues as to how a decrease in K^+ concentration may activate ICE autoprocessing; nevertheless, K^+ provides a straightforward link between $P2X_7$ and ICE as opening of the $P2X_7$ channel/pore provides a very fast pathway for K^+ efflux (Fig. 4) Perregaux and Gabel, 1998b. It would be interesting to test whether the same K^+-based mechanism of activation also applies to other caspases and how this may be involved in apoptosis.

It is obviously too early to draw from these experiments any conclusions as to the role of ATP and $P2X_7$ in IL-1β processing and release during the inflammatory reaction *in vivo*; however, a recent report from our laboratory

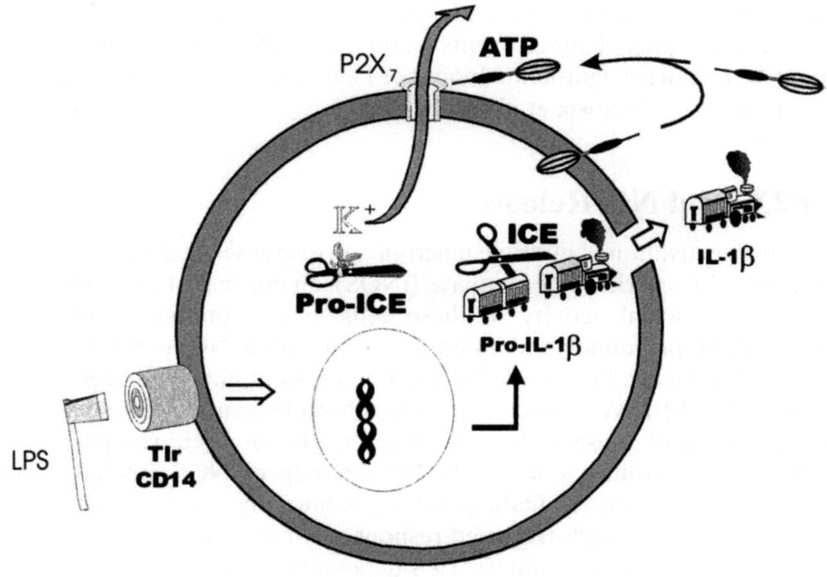

Fig. 4. $P2X_7$ receptor stimulation by ATP of LPS-primed cells induces interleukin-1β secretion. Binding of bacterial endotoxin (*LPS*) to membrane receptors induces synthesis of the 34 KDa immature form of IL-1β (*pro-IL-1β*). Activation of $P2X_7$ receptor by paracrine and/or autocrine ATP release, causes a decrease in the intracellular potassium concentration thus activating the interleukin-1β-converting enzyme (*ICE*) which in turn processes pro-IL-1β to its biologically active form (*IL-1β*)

(FERRARI et al. 1997d) suggests that this purinergic receptor has a more central role than previously thought, even in the absence of exogenously added ATP. We have observed that LPS causes non-lytic release of ATP from mouse microglia, and that pre-treatment of these cells with oATP or apyrase blocks LPS-dependent IL-1β release, suggesting that the LPS signal for IL-1β release consists, at least in part, of an autocrine/paracrine stimulation mediated by ATP. Besides microglia, macrophages also release ATP in response to LPS (FERRARI et al. 1997d; SPERLAGH et al. 1998); thus this mechanism could be widespread.

Participation of $P2X_7$ in LPS-dependent activation of immune cells might have very interesting and far reaching practical applications in the treatment of sepsis caused by Gram-negative bacteria. PROCTOR et al. (1994) showed that the ATP analogue, 2-methylthio-ATP (2-MeS-ATP) inhibited endotoxin-stimulated release of toxic mediators such as TNFα and IL-1β and protected mice from endotoxin-induced death. Interpretation of this early experiment is not straightforward, as 2-MeS-ATP is an agonist at P2Y as well as P2X purinoceptors (ABBRACCHIO and BURNSTOCK 1994); however, at $P2X_7$, 2-MeS-ATP acts as a partial agonist, and it is thus conceivable that it behaves as an effective ATP antagonist. Altogether, these observations suggest that $P2X_7$

(and maybe other P2 receptors) take part in some crucial but as yet unknown step in the complex chain of events leading to septic shock, either as a component of a paracrine/autocrine loop (see Ferrari et al. 1997d) or as a binding site for LPS (see Proctor et al. 1994).

F. P2X$_7$ and NO Release

LPS causes activation of multiple macrophage responses, including stimulation of the inducible nitric-oxide synthase (iNOS), an enzyme of great importance for the bactericidal activity of these cells. It was previously shown that 2-meS-ATPH profoundly inhibited LPS-stimulated NO generation in the murine macrophage line RAW 264.7 (Denlinger et al. 1996). More recently, Hu et al. (1998) have further extended these observations by showing that pre-incubation of these cells with either oATP or pyridoxal-phosphate-6-azophenyl-2′,4′-disulfonic acid (PPADS) also reduces iNOS stimulation due to LPS, confirming previous findings by our laboratory on the ability of P2X$_7$ antagonists to block LPS-triggered responses (Ferrari et al. 1997d). Furthermore, oATP treatment reduces LPS-dependent iNOS expression without affecting macrophage viability. The signal transduction chain set in motion by LPS plausibly involves several pathways, some of which are poorly known. Hu and coworkers discovered that both LPS-mediated NFκB and MAP kinase activation are strongly blocked by oATP. MAP kinase in RAW 264.7 macrophages can also be stimulated by P2Y (presumably P2Y$_2$) receptor; thus to stress the specificity of action of oATP these authors showed that MAP kinase activation in response to UTP was not inhibited by oATP, elegantly confirming the original data on the receptor selectivity of this dialdehyde compound (Murgia et al. 1993). The finding that a P2X$_7$ receptor antagonist is able to block NFκB activation mediated by LPS is particularly intriguing in the light of the recent demonstration by Ferrari et al. (1997c) that direct stimulation of P2X$_7$ by ATP causes NFκB activation.

G. P2X$_7$ and Transcription Factor Activation

A few reports in recent years have provided evidence for transcription factor activation by purinergic agonists (Neary et al. 1996; Ferrari et al. 1997c; von Albertini et al. 1998). Ferrari and co-workers (1997c) dissected the pathway leading to P2X$_7$-dependent NFκB activation in microglial cells and showed that it was inhibited by the ICE/caspase-1 blocker YVAD, clearly suggesting that ICE was involved in this process. To support this suggestion, Ferrari and coworkers also showed that P2X$_7$ stimulation triggered the formation of proteolytic caspase-1 fragments, a response that was blocked by oATP and was absent in microglial cell clones selected for lack of P2X$_7$. NFκB is an omo/hetero-dimer generated from at least five different subunits (p65, c-Rel, RelB, p50, and p52) that confer to the complex distinct biological activities

(BALDWIN 1996). A surprising finding of $P2X_7$-mediated NFκB activation is that it only causes formation of the p65 homodimer. The p65 homodimer, in contrast to the more usual p65/p50 complex, binds to a κB motif present in the promoter regions of several genes encoding pro-inflammatory factors such as IL-6 and IL-8, or adhesion molecules such as ICAM-1 (KUNSCH and ROSEN 1993; TAKEUCHI and BAICHWAL 1995; LOREDO and BENTON 1998). This finding led FERRARI et al. (1997c) to suggest that $P2X_7$ presumably triggers expression of NFκB target genes different from those induced by pro-inflammatory stimuli that use the more classical p50/p65 heterodimers. In this regard it may be relevant that human fibroblasts also express $P2X_7$ (SOLINI et al. 1999) and that in these cells stimulation of this receptor drives secretion of IL-6, a cytokine that plays a central role in the degenerative changes observed in inflammatory joint diseases. Is K^+ depletion also needed for NFκB activation? This is presently unknown, but the hypothesis is clearly worthy of scrutiny.

H. $P2X_7$ and Multinucleated Giant Cell Formation

Multinucleation is a common event in many processes involving mononuclear phagocytes: osteoclasts normally fuse to generate multinucleated giant cells (MGCs) during bone resorption; furthermore, during chronic inflammatory reactions, very often macrophages and other inflammatory cells generate typical formations called granulomas. Within these structures, macrophages differentiate into epithelioid cells that eventually fuse into large polykaryons also named MGCs. This is a very important aspect of widespread infectious diseases such as tuberculosis, but little is known about the functions of MGCs and nothing about the mechanism by which fusion occurs. In 1995, we started an extensive characterization of the $P2X_7$ receptor of monocyte-derived human macrophages and observed that its blockade by oATP completely prevented MGCs formation in vitro (FALZONI et al. 1995). However, other responses, such as concanavalin A-dependent cytoplasmic Ca^{2+} changes or chemotaxis, and expression of plasma membrane molecules thought to take part in cell fusion (e.g., CD11a, CD18 and CD54), were unaltered. We have further extended these studies by investigating cell fusion in J774 macrophage clones selected for high ($P2X_7$hyper) or low ($P2X_7$hypo) expression of the $P2X_7$ receptor, respectively (CHIOZZI et al. 1997). $P2X_7$hyper cells are extremely fragile and easily die when they reach confluence, as previously described (CHIOZZI et al. 1996). However, in carefully handled cultures, $P2X_7$hyper monolayers survive long enough to form dense aggregates within which they spontaneously fuse. Spontaneous fusion generates MGCs of different size and shape, containing from 3 up to 20 nuclei and sometimes more. We have never found MGCs in $P2X_7$hypo monolayers, and only very seldom have we detected small MGCs containing 2–3 nuclei in J774-wt cultures. Incidentally, fusion greatly accelerates the death rate of $P2X_7$hyper cells. Within 24–36 h after fusion, vacuoles appear in the cytoplasm of MGCs, nuclei start to condense, and finally extensive nuclear and cytoplasmic condensation occurs. Interestingly, a MGC fate does not depend on the age of the culture but rather on that of the individual MGC. Occasional spontaneous MGC formation also occurs in monolayers of HEK293 cells stably transfected with $P2X_7$ (Chiozzi et al. 1997).

I. $P2X_7$ and Plasma Membrane Antigen Shedding

It is well known that many molecules found in the plasma membrane of circulating leukocytes are normally shed via poorly characterized mechanisms and accumulate in the serum in a variety of physiological and pathological conditions. Sometimes plasma levels of shed molecules increase during disease states and can be taken as a prognostic indicator. Activation of $P2X_7$ has been recently associated to release of soluble forms of two molecules expressed on the plasma membrane of lymphocytes from patients with B-cell-chronic lymphocytic leukemia (B-CLL), the cell adhesion molecule L-selectin (CD62L), and the low affinity receptor for IgE (CD23) (Jamieson et al. 1996; Gu et al. 1998). CD23 and L-selectin are normally found in high amounts in sera from patients with B-CLL, while lower levels are found in B-cell immunocytomas and low grade non-Hodgkin's lymphomas. Sera of normal subjects usually contain only very little amounts of them. These findings suggest that soluble CD23 levels may reflect disease activity and be a prognostic indicator in B-CLL (Sarfati et al. 1996). Lymphocytes isolated from patients with B-CLL express to a very high level an ATP receptor that shares many of the features of $P2X_7$ (although the size of the pore appears to be smaller with respect to the prototypical macrophage ATP receptor) and exhibit a strong cytotoxic response to ATP (Wiley and Dubyak 1989; Wiley et al. 1993). ATP doses needed to trigger L-selectin and CD23 shedding were very similar (ATP EC_{50} of $31.3\,\mu$mol/l and $29.7\,\mu$mol/l, respectively), but were at least one order of magnitude lower than those needed to elicit cytotoxicity, indicating that B-CLL can undergo a graded response to varying extracellular ATP concentrations.

J. Conclusions

It has been suggested that purine derivatives might have been primordial mediators of intercellular communication (Burnstock 1996), and effects of ATP on ameboid cells have been recognized for many years (Zimmermann et al. 1958). It is therefore somewhat ironic that little attention was paid until recently to the effects of ATP on phagocytic cells, known for over a century to be the most primitive form of immunity present in lower and higher vertebrates. The flurry of data produced over the last five years is clearly showing that, despite our negligence, the immune system has learnt to take perfect advantage of this primeval mediator and of purinergic receptors for tuning its

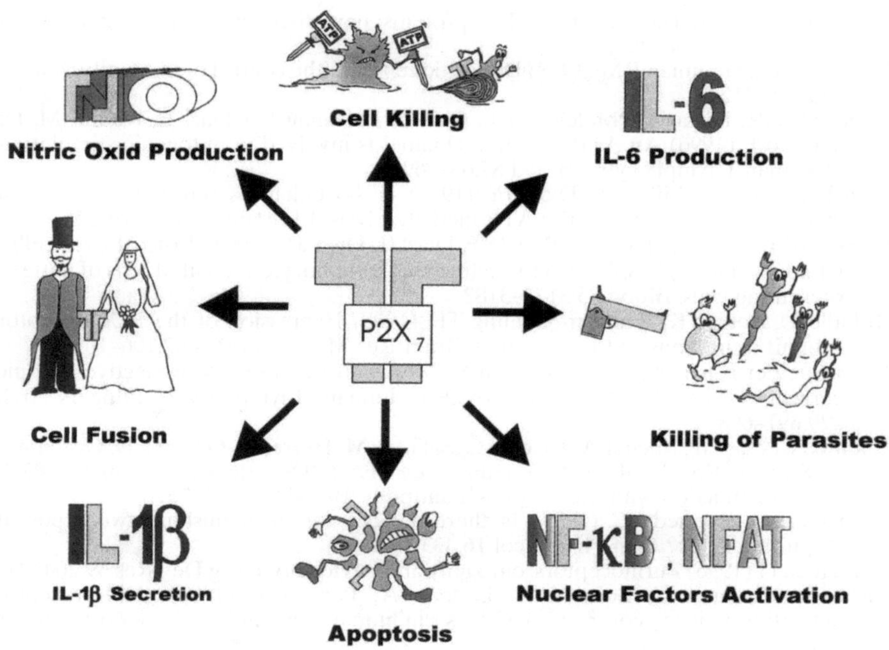

Fig. 5. Responses modulated by P2X$_7$ in immune cell

responses and fighting pathogenic microorganisms (Fig. 5). Our realization of the importance of ATP and purinergic receptors in leukocyte biology and pathology may pave the way to the discovery of innovative drugs for the treatment of many different chronic inflammatory illnesses.

Acknowledgements. This work was supported by grants from the National Research Council of Italy (target project on Biotechnology), the Ministry for Scientific Research (MURST), the Italian Association for Cancer Research (AIRC), the II Tuberculosis Project, and Telethon of Italy. V.V. was supported by a fellowship awarded by UNESCO-CNR (United Nations Educational and Cultural Organization-National Research Council).

References

Abbracchio MP, Burnstock G (1994) Purinoceptors: are there families of P2X and P2Y purinoceptors? Pharmac Ther 64:445–475

Alzola E, Perez-Extebarria A, Kabre E, Fogarty DJ, Metioui M, Chaib N, Macarulla JM, Matute C, Dehaye J-P, Marino A (1998) Activation by P2X$_7$ agonists of two phospholipases A_2 (PLA2) in ductal cells of rat submandibular gland. J Biol Chem 273:30208–30217

Baldwin AS (1996) The NFκB and IkB proteins: new discoveries and insights. Annu Rev Immunol 14:649–683
Banchereau J, Steinman RM (1998) Dendritic cells and the control of immunity. Nature 392:245–252
Baricordi OR, Ferrari D, Melchiorri L, Chiozzi P, Hanau S, Chiari E, Rubini M, Di Virgilio F (1996) An ATP-activated channel is involved in mitogenic stimulation of human T lymphocytes. Blood 87:682–690
Blanchard DK, McMillen S, Djeu JY (1991) IFN-γ enhances sensitivity of human macrophages to extracellular ATP-mediated lysis. J Immunol 147:2579–2585
Blanchard DK, Wei S, Duan C, Pericle F, Diaz JI, Djeu JY (1995) Role of extracellular adenosine triphosphate in the cytotoxic T lymphocyte-mediated lysis of antigen presenting cells. Blood 85:3173–3182
Brandle U, Kohler K, Wheeler-Schilling TH (1998) Expression of the $P2X_7$-receptor subunit in neurons of the rat retina. Brain Res Mol Brain Res 62:106–109
Bretschneider F, Klapperstuck M, Lohn M, Markwardt F (1995) Nonselective cationic currents elicited by extracellular ATP in human B-lymphocytes. Pflugers Arch 429:691–698
Buell G, Chessel IP, Michel AD, Collo G, Salazzo M, Herren S, Gretner D, Grahames C, Kaur R, Kosco-Vilbois MH, Humphrey PPA (1998) Blockade of human $P2X_7$ receptor function with a monoclonal antibody. Blood 92:3521–3528
Burnstock G, Kennedy C (1985) Is there a basis for distinguishing two types of P_2-purinoceptor? Gen Pharmacol 16:433–440
Burnstock G (1996) Purinoceptors: ontogeny and phylogeny. Drug Dev Res 39:204–242
Cario-Toumaniantz C, Loirand G, Ladoux A, Pacaud P (1998) $P2X_7$ receptor activation-induced contraction and lysis in human saphenous vein smooth muscle. Circ Res 83:196–203
Cheneval D, Ramage P, Kastelic T, Szelestenyi T, Niggli H, Hemmig R, Bachmann M, MacKenzie A (1998) Increased mature interleukin-1β (IL-1β) secretion from THP-1 cells induced by nigericin is a result of activation of p45 IL-1b-converting enzyme processing. J Biol Chem 273:17846–17851
Chiozzi P, Murgia M, Falzoni S, Ferrari D, Di Virgilio F (1996) Role of purinergic P2Z receptor in spontaneous cell death in J774 macrophage cultures. Biochem Biophys Res Commun 218:176–181
Chiozzi P, Sanz JM, Ferrari D, Falzoni S, Aleotti A, Buell GN, Collo G, Di Virgilio F (1997) Spontaneous cell fusion in macrophage cultures expressing high levels of the $P2Z/P2X_7$ receptor. J Cell Biol 138:697–706
Christoffersen BC, Hug MJ, Novak I (1998) Different purinergic receptors lead to intracellular calcium increases in pancreatic ducts. Pflugers Arch 436:33–39
Chused TM, Apasov S, Sitkovsky M (1996) Murine T lymphocytes modulate activity of an ATP-activated P2Z-type purinoceptor during differentiation. J Immunol 157:1371–1380
Cockcroft S, Gomperts BD (1979) ATP-induced permeability in rat mast-cells. Nature 279:541–542
Cockcroft S, Gomperts BD (1980) The ATP^4-receptor of rat mast cells. Biochem J 188:789–798
Collo G, Neidhart S, Kawashima E, Kosco-Vilbois M, North RA, Buell G (1997) Tissue distribution of the $P2X_7$ receptor. Neuropharmacology 36:1277–1283
Cook SP, Rodland KD, McCleskey EW (1998) A memory for extracellular Ca^{2+} by speeding recovery of P2X receptors from desensitization. J Neurosci 18:9238–9244
Coutinho-Silva R, Alves LA, Campos de Carvalho AC, Savino W, Persechini PM (1996) Characterization of P2Z purinergic receptors on phagocytic cells of the thymic reticulum in culture. Biochim Biophys Acta 1280:217–222
Denlinger LC, Fisette PL, Garis KA, Kwon G, Vazquez-Torres A, Simon AD, Nguyen B, Proctor RA, Bertics PJ, Corbett JA (1996) Regulation of inducible nitric oxide synthase expression by macrophages purinoceptors and calcium. J Biol Chem 271:337–342

Di Virgilio F, Bronte V, Collavo D, Zanovello P (1989) Responses of mouse lymphocytes to extracellular adenosine 5′-triphosphate (ATP). J Immunol 143:1955–1960
Di Virgilio F, Pizzo P, Zanovello P, Bronte V, Collavo D (1990) Extracellular ATP as a possible mediator of cell-mediated cytotoxicity. Immunol Today 11:274–277
Dubyak GR, Clifford EE, Humphreys BD, Kertesy SB, Martin KA (1996) Expression of multiple ATP receptor subtypes during the differentiation and inflammatory activation of myeloid leukocytes. Drug Dev Res 39:269–278
El-Moatassim C, Bernad N, Mani J-C, Dornand J (1989) Extracellular ATP induces a nonspecific permeability of thymocyte plasma membrane. Biochem Cell Biol 67:495–502
Falzoni S, Munerati M, Ferrari D, Spisani S, Moretti S, Di Virgilio F (1995) The purinergic P2Z receptor of human macrophage cells. Characterization and possible physiological role. J Clin Invest 95:1207–1216
Ferrari D, Munerati M, Melchiorri L, Hanau S, Di Virgilio F, Baricordi OR (1994) Responses to extracellular ATP of lymphoblastoid cell lines from Duchenne muscular dystrophy patients. Am J Physiol 267:C886–C892
Ferrari D, Villalba M, Chiozzi P, Falzoni S, Ricciardi-Castagnoli P, Di Virgilio F (1996) Mouse microglial cells express a plasma membrane pore gated by extracellular ATP. J Immunol 156:1531–1539
Ferrari D, Chiozzi P, Falzoni S, Dal Susino M, Collo G, Buell GN, Di Virgilio F (1997a) ATP-mediated cytotoxicity in microglial cells. Neuropharmacology 36:1295–1301
Ferrari D, Chiozzi P, Falzoni S, Dal Susino M, Melchiorri L, Baricordi OR, Di Virgilio F (1997b) Extracellular ATP triggers IL-1β release by activating the purinergic P2Z receptor of human macrophages. J Immunol 159:1451–1458
Ferrari D, Wesselborg S, Bauer MKA, Schulze-Osthoff K (1997c) Extracellular ATP activates transcription factor NFκB through the P2Z purinoceptor by selectively targeting NFκB p65 (RelA). J Cell Biol 139:1635–1643
Ferrari D, Chiozzi P, Falzoni S, Hanau S, Di Virgilio F (1997d) Purinergic modulation of interleukin-1β release from microglial cells stimulated with bacterial endotoxin. J Exp Med 185:579–582
Filippini A, Taffs RE, Agui T, Sitkovsky MV (1990a) EctoATPase activity in cytolytic T lymphocytes: Protection from the cytolytic effects of extracellular ATP. J Biol Chem 265:334–340
Filippini A, Taffs RE, Sitkovsky M (1990b) Extracellular ATP in T-lymphocyte activation: Possible role in effector functions. Proc. Natl Acad Sci USA 87:8267–8271
Gargett CE, Cornish JE, Wiley JS (1997) ATP, a partial agonist for the P2Z receptor of human lymphocytes. Br J Pharmacol 122:911–917
Gargett CE, Wiley JS (1997) The isoquinoline derivative KN-62: A potent antagonist of the P2Z receptor of human lymphocytes. Br J Pharmacol 120:1483–1490
Girolomoni G, Santantonio ML, Pastore S, Bergstresser PR, Giannetti A, Cruz PD Jr (1993) Epidermal Langerhans cells are resistant to the permeabilizing effects of extracellular ATP: in vitro evidence supporting a protective role of membrane ATPase. J Invest Dermatol 100:282–287
Gordon JL (1986) Extracellular ATP: effects, sources and fate. Biochem J 233:309–319
Gu B, Bendall LG, Wiley JS (1998) Adenosine triphosphate-induced shedding of CD23 and L-selectin (CD62L) from lymphocytes is mediated by the same receptor but different metalloproteases. Blood 92:946–951
Hickman SE, el Khoury J, Greenberg S, Schieren I, Silverstein SC (1994) P2Z adenosine triphosphate receptor activity in cultured human monocyte-derived macrophages. Blood 84:2452–2456
Hogquist KA, Nett MA, Unanue ER, Chaplin DD (1991) Interleukin 1 is processed and released during apoptosis. Proc Natl Acad Sci USA 88:8485–8450
Hu Y, Fisette PL, Denlinger LC, Guadarrama AG, Sommer JA, Proctor RA, Bertics PJ (1998) Purinergic receptor modulation of lipopolysaccharide signalling and inducible nitric-oxide synthase expression in RAW 264.7 macrophages. J Biol Chem 273:27170–27175

Humphreys BD, Dubyak GR (1998) Modulation of $P2X_7$ nucleotide receptor expression by pro- and anti-inflammatory stimuli in THP-1 monocytes. J Leuk Biol 64:265–273
Jamieson GP, Snook MB, Thurlow PJ, Wiley JS (1996) Extracellular ATP causes loss of L-selectin from human lymphocytes via occupancy of P2Z purinoceptors. J Cell Physiol 166:637–642
Illes P, Nieber K, Frohlich R, Norenberg W (1996) P2 purinoceptors and pyrimidinoceptors of catecholamine-producing cells and immunocytes. Ciba F Symp 198:110–125
Inoue K, Nakajima K, Morimoto T, Kikuchi Y, Koizumi S, Illes P, Kohsaka S (1998) ATP stimulation of Ca^{2+}-dependent plasminogen release from cultured microglia. Br J Pharmacol 123:1304–1310
Kunsch C, Rosen CA (1993) NFκB subunit-specific regulation of the interleukin-8 promoter. Moll Cell Biol 13:6137–6146
Lammas DA, Stober C, Harvey CJ, Kendrick N, Panchalingam S, Kumararatne DS (1997) ATP-induced killing of Mycobacteria by human macrophages is mediated by purinergic P2Z/$P2X_7$) receptors. Immunity 7:433–444
Li P, Allen H, Banerjee S, Franklin S, Herzog L, Johnston C, McDowell J, Paskind M, Rodman L, Salfeld J, Towne E, Tracey D, Wardwell S, Wei F-Y, Wong W, Kamen R, Seshadri T (1995) Mice deficient in IL-1b-converting enzyme are defective in production of mature IL-1β and resistant to endotoxic shock. Cell 80:401–411
Loredo GA, Benton HP (1998) ATP and UTP activate calcium-mobilizing P2U-like receptors and act synergistically with interleukin-1 to stimulate prostaglandin E2 release from human rheumatoid synovial cells. Arthritis Rheum 41:246–255
Lundy PM, Sawyer TW, Hand BT, Frew R (1998) Effects of Bis(2-chloroethyl)sulfide on ATP receptor-mediated responses of the rat vas deferens: possible relationship to cytotoxicity. J Pharmacol Exp Ther 285:299–306
Markwardt F, Lohn M, Bohm T, Klapperstuck M (1997) Purinoceptor-operated cationic channels in human B lymphocytes. J Physiol 498:143–151
Michel AD, Chessel IP, Hibell AD, Simon J, Humphrey PPA (1998) Identification and characterization of an endogenous $P2X_7$ (P2Z) receptor in CHO-K1 cells. Br J Pharmacol. 125:1194–1201
Molloy A, Laochumroonvorapong P, Kaplan G (1994) Apoptosis, but not necrosis, of infected monocytes is coupled with killing of intracellular Bacillus Calmette-Guerin. J Exp Med 180:1499–1509
Murgia M, Pizzo P, Steinberg TH, Di Virgilio F (1992) Characterization of the cytotoxic effect of extracellular ATP in J774 mouse macrophages. Biochem J 288:897–901
Murgia M, Hanau S, Pizzo P, Rippa M, Di Virgilio F (1993) Oxidized ATP. An irreversible inhibitor of the macrophage purinergic P2Z receptor. J Biol Chem 268:8199–8203
Mutini C, Falzoni S, Ferrari D, Chiozzi P, Morelli A, Baricordi OR, Collo G, Ricciardi-Castagnoli P, Di Virgilio F (1999) Mouse dendritic cells express the $P2X_7$ purinergic receptor: characterization and possible participation in antigen presentation. J Immunol 163:1958–1965
Neary JT, Zhu Q, Kang Y, Dash PK (1996) Extracellular ATP induces formation of AP-1 complexes in astrocytes via P2 purinoceptors. Neuroreport 25:2893–2896
Norenberg W, Langosch JM, Gebicke-Haerter PJ, Illes P (1994) Characterization and possible function of adenosine 5′-triphosphate receptors in activated rat microglia. Br J Pharmacol 111:942–950
Perregaux D, Barberia J, Lanzetti AJ, Geoghegan KF, Carty TJ, Gabel CA (1992) IL-1β maturation: evidence that mature cytokine formation can be induced specifically by nigericin. J Immunol 149:1294–1303
Perregaux D, Gabel CA (1994) Interleukin-1β maturation and release in response to ATP and nigericin. J Biol Chem 269:15195–15203
Perregaux DG, Gabel CA (1998a) Post-translational processing of murine IL-1: evidence that ATP-induced release of IL-1α and IL-1β occurs via a similar mechanism. J Immunol 160:2469–2477

Perregaux DG, Gabel CA (1998b) Human monocyte stimulus-coupled IL-1β post-translational processing: modulation via monovalent cations. Am J Physiol 275:C1538–C1547

Phillips JK, McLean AJ, Hill CE (1998) Receptors involved in nerve-mediated vasoconstriction in small arteries of the rat hepatic mesentery. Br J Pharmacol 124:1403–1412

Proctor RA, Denlinger LC, Leventhal PS, Daugherty SK, van de Loo J-W, Tanke T, Firenstein GS, Bertics PJ (1994) Protection of mice from endotoxic death by 2-methylthio-ATP. Proc Natl Acad Sci USA 91:6017–6020

Rassendren F, Buell GN, Virginio C, Collo G, North RA, Surprenant A (1997) The permeabilizing ATP receptor, P2X7. Cloning and expression of a human cDNA. J Biol Chem 272:5482–5486

Redegeld F, Filippini A, Trenn G, Sitkovsky MV (1993) Possible role of extracellular ATP in cell-cell interactions leading to CTL-mediated cytotoxicity. In: Sitkovsky MV, Henkart PA (eds) Cytotoxic cells. Birkhauser, Boston, pp 307–313

Sanz JM, Chiozzi P, Di Virgilio F (1998) Tenidap enhances $P2Z/P2X_7$ receptor signalling in macrophages. Eur J Pharmacol 355:235–244

Sarfati M, Chevret S, Chastang C, Biron G, Stryckmans P, Delespesse G, Binet JL, Merle-Beral H, Bron D (1996) Prognostic importance of serum soluble CD23 level in chronic lymphocytic leukemia. Blood 88:4259–4264

Schulze-Lohoff E, Hugo C, Rost S, Arnold S, Gruber A, Br ne B, Sterzel RB (1998) Extracellular ATP causes apoptosis and necrosis of cultured mesangial cells via $P2Z/P2X_7$ receptors. Am J Physiol 275:F962–F971

Shibuya I, Tanaka K, Hattori Y, Uezono Y, Harayama N, Noguchi J, Ueta Y, Izumi F, Yamashita H (1999) Evidence that multiple P2X purinoceptors are functionally expressed in rat supraoptic neurones. J Physiol 514:351–367

Solini A, Chiozzi P, Morelli A, Fellin R, Di Virgilio F (1999) Human primary fibroblasts in vitro express a purinergic $P2X_7$ receptor coupled to ion fluxes, microvesicle formation and IL-6 release. J Cell Sci, 112:297–305

Sperlagh B, Hasko G, Nemeth Z, Vizi ES (1998) ATP released by LPS increases nitric oxide production in raw 264.7 macrophage cell line via $P2Z/P2X_7$ receptors. Neurochem Int 33:209–215

Steinberg TH, Newman AS, Swanson JA, Silverstein SC (1987) ATP^{4-} permeabilizes the plasma membrane of mouse macrophages to fluorescent dyes. J Biol Chem 262:8884–8888

Steinberg TH, Silverstein SC (1987) Extracellular ATP^{4-} promotes cation fluxes in the J774 mouse macrophage cell line. J Biol Chem 262:3118–3122

Surprenant A, Rassendren F, Kawashima E, North RA, Buell G (1996) The cytolytic P2Z receptor for extracellular ATP identified as a P2X receptor ($P2X_7$). Science 272:735–738

Takeuchi M, Baichwal VR (1995) Induction of the gene encoding mucosal vascular addressin cell adhesion molecule 1 by tumor necrosis factor α is mediated by NFκB proteins. Proc Natl Acad Sci USA 92:3561–3565

Tatham PER, Cusack NJ, Gomperts BD (1988) Characterisation of the ATP^{4}-receptor that mediates permeabilisation of rat mast cells. Eur J Pharmacol 147:13–21

Tenneti L, Gibbons SJ, Talamo BR (1998) Expression and trans-synaptic regulation of $P2X_4$ and P2Z receptors for extracellular ATP in parotid acinar cells. J Biol Chem 273:26799–26808

Turner JT, Weisman GA, Landon LA, Park M, Camden JM (1998) Salivary gland nucleotide receptors: evidence for functional expression of both P2X and P2Y subtypes. Eur J Morphol 36:170–175

von Albertini M, Palmetshofer A, Kaczmarek E, Koziak K, Stroka D, Grey ST, Stuhlmeier KM, Robson SC (1998) Extracellular ATP and ADP activate transcription factor NFκB and induce endothelial cell apoptosis. Biochem Biophys Res Commun 248:822–829

Wiley JS, Chen R, Jamieson GP (1993) The ATP^{4-} receptor-operated channel (P2Z class) of human lymphocytes allows Ba^{2+} and $ethidium^+$ uptake: inhibition of fluxes by suramin. Arch Biochem Biophys 305:54–60

Wiley JS, Dubyak GR (1989) Extracellular adenosine triphosphate increases cation permeability of chronic lymphocytic leukemic lymphocytes. Blood 73:1316–1323

Zambon A, Bronte V, Di Virgilio F, Hanau S, Steinberg TH, Collavo D, Zanovello P (1994) Role of extracellular ATP in cell-mediated cytotoxicity: a study with ATP-sensitive and ATP-resistant macrophages. Cell Immunol 156:458–467

Zanovello P, Bronte V, Rosato A, Pizzo P, Di Virgilio F (1990) Responses of mouse lymphocytes to extracellular ATP. II. Extracellular ATP causes cell type-dependent lysis and DNA fragmentation. J Immunol 145:1545–1550

Zimmermann AM, Landau JV, Marsland D (1958) The effects of adenosine triphosphate and dinitro-*o*-cresol upon the form and movement of Amoeba proteus. Exp Cell Res 15:484–495

Section V
Endocrine System

CHAPTER 27

Purinergic Receptors and the Pharmacology of Type 2 Diabetes

P. Petit, D. Hillaire-Buys, M.M. Loubatières-Mariani, and J. Chapal

A. Introduction

Diabetes mellitus is the phenotypic expression of a disorder resulting from various environmental factors interacting with a genetic component. It is an heterogeneous metabolic syndrome consisting of a dysregulation of glucose homeostasis, with additional abnormalities in lipid and protein metabolism. This disorder induces both acute and long-term complications, particularly cardiovascular, with an increased rate of morbidity and mortality as compared to the general population. The incidence and associated complications of the disease are in continuous progression; indeed, by the year 2010, the total number of people with diabetes is projected to reach 221 million, as compared to an estimated 124 million in 1997 (Amos et al. 1997). The prevalence of diabetes in adults worldwide was estimated to be 4.0% in 1995 and to rise to 5.4% by the year 2025 (King et al. 1998).

Diabetes mellitus is classified into two different types (Expert Committee on the Diagnosis and Classification of Diabetes Mellitus 1997): type 1 (insulin dependent), characterized by a β cell loss usually of autoimmune origin, and type 2 (non-insulin dependent; NIDDM), characterized by an insulin secretory defect associated with various degrees of insulin resistance, leading to chronic hyperglycemia with fasting plasma glucose above 7.0 mmol/l. Type 2 diabetes is by far the most common form of diabetes, accounting for approximately 90% of cases in Western Countries; its prevalence, up to 5% of the population in European Countries, is increasing mainly due to increased life expectancy, increased prevalence of obesity, and decreased physical activity.

The results of the United Kingdom Prospective Diabetes Study (UK Prospective Diabetes Study (UKPDS) Group, 1998) point out that lowering blood glucose affects beneficially diabetes-related events: the intensive blood-glucose control reduces the incidence of microvascular complications in type 2 diabetes, as it does in type 1 diabetes (Diabetes Control and Complications Trial Research Group 1993). Hence, decreasing hyperglycemia to (near) normoglycemia is a major therapeutic goal in the treatment of diabetes.

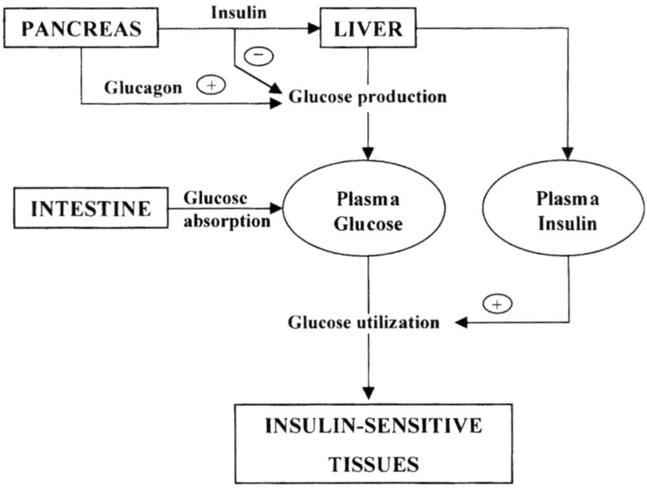

Fig. 1. Glucose homeostasis

The concentration of glucose in plasma is a strictly regulated parameter, which depends on the balance between glucose entry into the circulation on the one hand, involving absorption by the gastrointestinal tract in the prandial and post-prandial period and production by the liver in the postabsorptive state, and glucose utilization by insulin-sensitive tissues such as skeletal muscles and adipose tissue on the other hand (Fig. 1). Glucose concentration is a signal for the endocrine pancreas to release insulin or glucagon. Insulin lowers the concentration of glucose by inhibiting the hepatic glucose production and stimulating glucose utilization by skeletal muscles and adipocytes; in these latter cells, insulin also inhibits lipolysis, which subsequently contributes to decrease gluconeogenesis. Glucagon opposes the effects of insulin, particularly in the liver, where it stimulates glycogenolysis and gluconeogenesis.

Lowering of blood glucose concentration in type 2 diabetes may be achieved by different therapeutic measures such as diet, physical activity, and pharmacological interventions; these latter include drugs which stimulate insulin secretion, such as sulphonylureas, drugs which reduce insulin resistance such as metformine (biguanide), and drugs which delay or attenuate glucose absorption, such as α-glucosidase inhibitors.

A variety of P2 and adenosine/P1 receptors are expressed in the different tissues involved in the regulation of metabolic function, and purinergic ligands may be useful in stimulating insulin secretion or reducing insulin resistance. Hence, purinergic receptors may constitute new potential targets for pharmacological intervention in the treatment of type 2 diabetes.

B. Role of P2 Receptors in the Stimulation and Amplification of Insulin Secretion: Novel Pancreatic Target for Pharmacological Intervention

The progressive β-cell dysfunction in type 2 diabetes, with altered insulin production and secretion (particularly the impairment of the first phase insulin response to glucose), is an important factor of the evolution of the disease and a major cause of hyperglycemia. Hence, drugs which stimulate or amplify insulin release will reduce plasma glucose concentration. Chronic hyperglycemia can generate secondary insulin resistance (YKI-JÄRVINEN 1990), and a treatment which reduces blood glucose may therefore also partially improve insulin sensitivity.

I. Experimental Evidence In Vivo or in Diabetic Animal Models for a Potential Antidiabetic Effect of P2Y Receptor Agonists

Adenine nucleotides like ATP stimulate insulin release in vivo in the rat (CANDELA and GARCIA-FERNANDEZ 1963) and in primates (LEVINE et al. 1970). P2Y receptor agonists are effective in vivo in modulating insulin secretion and glycemia (Table 1). 2-MethylthioATP (2MeSATP) stimulated insulin release and slightly decreased glycemia in the dog; however, due to its rapid breakdown, this nucleotide had to be injected directly into the pancreatic-duodenal artery (RIBES et al. 1988). Further studies with ADPβS [adenosine-5′-*O*-(2-thiodiphosphate)], a stable ATP analogue, showed that this compound increased insulin secretion in dogs and improved glucose tolerance, not only after i.v. but also after oral administration (HILLAIRE-BUYS et al. 1993). In anesthetized rats, ADPβS also increased insulin secretion, an effect that was clearly dependent on the nutritional state of the animals and the plasma level of glucose. The compound also improved glucose tolerance (HILLAIRE-BUYS et al. 1993). ADPβS is an orally active insulin secretagogue in vivo improving glucose tolerance.

In experiments performed in the perfused pancreas from rats with streptozotocin-induced diabetes, in which the majority of β cells are destroyed, the biphasic insulin response to a stimulating glucose concentration that was observed in normal rats was completely suppressed while the insulin response of the remaining β cells to ADPβS was preserved (HILLAIRE-BUYS et al. 1992). In the perfused pancreas of Zucker diabetic fatty rats, the insulin secretory responses to P2 receptor activation were also preserved (TANG et al. 1996).

Activation of P2Y receptors leads to enhanced insulin secretion and can reduce hyperglycemia in vivo, while limiting the risk of hypoglycemia. This response is also effective in the pancreas from diabetic animals, suggesting a potential interest of P2Y receptor agonists in the treatment of type 2 diabetes. More tissue specific P2Y agonists are required in order to avoid side effects

Table 1. Effectiveness of P2Y receptor agonists on plasma insulin and glucose concentrations in vivo and on insulin secretion in diabetic animal models

Agonist	Experimental model	Dose and route of administration	Insulin	Plasma glucose	Reference
2-methylthio ATP	Anaesthetized dog, pancreatic-duodenal bypass	Infusion in the pancreatic-duodenal artery, 15 μmol/l for 15 min	Stimulated	Slight ↓	Ribes et al. (1988)
ADPβS	Anaesthetized rat	0.2 mg/kg i.v.			Hillaire-Buys et al. (1993)
	Fasted		Transient ↑	NS	
	Fed		Sustained ↑	↓	
	IVGTT		Marked ↑	↓ Hyperglycemia	
	Conscious dog	0.1 mg/kg p.o.			
	OGTT		Marked ↑	↓ Hyperglycemia	
ADPβS	Isolated pancreas, streptozotocin diabetic rat	Infusion, 15 μmol/l	Stimulated		Hillaire-Buys et al. (1992)
ADPβS	Isolated pancreas, Zucker diabetic rat	Infusion, 15 μmol/l	Stimulated		Tang et al. (1996)

↑, increase; ↓, decrease; NS, not significant; IVGTT, intravenous glucose tolerance test; OGTT, oral glucose tolerance test.

in vivo, such as platelet aggregation and cardiovascular effects, that would be deleterious in diabetic states.

II. Cellular Target and Mechanisms of Action

Adenine nucleotides (ATP and ADP) potentiate glucose-induced insulin secretion, acting as extracellular signals on pancreatic β-cell surface P2 receptors (Loubatières-Mariani et al. 1979; Chapal and Loubatières-Mariani 1981a,b). The relative amounts of ATP and ADP in the β cell determine the energy charge which plays a major role in the regulation of cell metabolism (Atkinson 1978). On the other hand, these nucleotides are also key signal molecules implicated in the coupling of intracellular metabolic changes to membrane electrical activity, modulating the sulfonylurea receptor 1 (SUR1) which, together with the inwardly rectifying potassium channel 6.2 (Kir6.2) form the β-cell type ATP-sensitive potassium channel (Aguilar-Bryan et al. 1998). Hence, adenine nucleotides (ATP and ADP) can affect insulin secretion in different ways, acting both as intracellular signals on the K_{ATP} channels, and as extracellular ligands of the P2 β-cell surface receptors. For the purposes of this review, only P2 receptor-mediated effects will be discussed.

P2 receptors of the pancreatic β cell have been pharmacologically characterized as being of the P2Y type (Bertrand et al. 1987). Thus, the P2Y agonist ADPβS, was approximately 100-fold more potent than ATP (Bertrand et al. 1991). More recently, a cDNA clone encoding a rat P2Y receptor was isolated from an insulinoma cDNA library (Tokuyama et al. 1995). A novel putative P2 receptor was also cloned, that closely resembles P2Y and P2U receptors; it is encoded by an intronless single copy gene that is exclusively expressed in pancreas, in contrast to the P2U and the P2Y purinergic receptors that are widely distributed throughout the periphery (Stam et al. 1996). The pharmacological properties and physiological relevance of purinergic receptors of the pancreatic β cell have been reviewed elsewhere (Loubatières-Mariani et al. 1995, 1997; Petit et al. 1996). Activation of P2Y receptors strongly potentiates the insulin secretion induced by a slightly stimulating glucose concentration.

A P2X receptor subtype may also exist on the surface on pancreatic β cells. Functionally, this receptor can be characterized at low, non-stimulating glucose concentrations; it is rapidly desensitized, thus limiting the insulin-secretory response in the presence of a low glucose concentration (Petit et al. 1998). A cDNA clone encoding a $P2X_4$ receptor has been isolated from a rat pancreatic islet cDNA library (Wang et al. 1996); the $P2X_4$ mRNA was shown to be expressed in various hormone-secreting cell lines, including the insulin-secreting cell lines, RINm5F and HIT-T15.

The stimulus-secretion coupling of β-cell P2 receptors has not yet been clearly established. An increase in cytoplasmic free calcium concentration has been consistently observed (Arkhammar et al. 1990; Theler et al. 1992) that may result from either an increased influx of extracellular calcium (Petit et al. 1987; GESCHWIND et al. 1989) or intracellular calcium mobilization (GYLFE and HELLMAN 1987; BLACHIER and MALAISSE 1988). The involvement of polyphosphoinositide hydrolysis in the potentiating effect of adenine nucleotides on insulin release is controversial (BLACHIER and MALAISSE 1988; PETIT et al. 1988) In addition, alterations in membrane potassium conductance by extracellular ATP or structural analogues have also been observed in mouse islets (PETIT et al. 1989) and RINm5F insulinoma cells (LI et al. 1991). A slight decrease in the membrane potassium conductance has been recently reported with a P2X but not a P2Y agonist (PETIT et al. 1998).

In human pancreatic islets and isolated human insulin-secreting cells, extracellular ATP leads to elevation of the cytoplasmic free Ca^{2+} concentration (KINDMARK et al. 1991; SQUIRES et al. 1994); preliminary data indicated that different P2 receptor agonists are effective in stimulating insulin release in human pancreatic islets; the effect of a P2Y agonist is essentially glucose-dependent (FERNANDEZ-ALVAREZ et al. 1998). The major role of glucose in the functional response brought about by P2Y receptor activation may be clinically relevant in the context of the risk of hypoglycemia, which is a common drawback of the currently available insulin secretagogues.

Most of the available P2 agonists examined as insulin secretagogues are non-selective in their tissue effects and can also induce vascular effects (HILLAIRE-BUYS et al. 1991, 1998; BERTRAND et al. 1991; FISCHER et al. 1998). The structure-activity relationships indicate that substitutions at the C2 position on the adenine ring of ATP and definite modifications of the polyphosphate chain enhance the insulin secreting activity of the ligands (CHAPAL et al. 1997). The further characterization of the different receptor subtypes and their coupling mechanism may allow the design of novel ligands with increased tissue specificity for the pancreatic β cell.

C. Role of A_1 Adenosine Receptors in the Enhancement of Insulin Action: Novel Extrapancreatic Target for Pharmacological Intervention

Resistance to insulin action is also a characteristic feature of type 2 diabetes and drugs which improve insulin sensitivity may be advantageous because of reduction in glycemia; also the decreased demand on β-cell secretion may help to retard the progressive β-cell dysfunction.

I. Experimental Evidence In Vivo for a Potential Antidiabetic Effect of Adenosine/A_1 Ligands

The metabolic effects of a number of adenosine A_1 agonists or antagonists as potential antidiabetic drugs have been investigated in vivo (Table 2).

The adenosine A_1 agonists, GR79236{N^6-[(1*S*, *trans*)-2-hydroxycyclopentyl] adenosine} and CPA (N^6-cyclopentyladenosine) have glucose-

Table 2. Effects of P1/Adenosine receptor ligands on plasma insulin and glucose concentrations in vivo

Ligand	Animal model	Dosage and route of administration	Plasma insulin	Plasma glucose	Reference
GR 79236 (A_1 *agonist*)	Conscious normal rat	0.2 μg/kg/min, 2 h, i.v.	ND	Decreased	MERKEL et al. (1995)
GR 79236 (A_1 *agonist*)	Fructose-induced insulin resistance; *IVGTT*	1 mg/kg/day, 8 days	ND	Improved glucose tolerance	QU et al. (1997)
RG 14202 (A_1 *agonist*)	Streptozotocin diabetic rat	0.1–1.0 mg/kg, p.o.	ND	Unchanged	COX et al. (1997)
SDZ WAG 994 (A_1 *agonist*)	SHR rat with hyperglycemia induced by streptozotocin	0.1 mg/kg, p.o.	Unchanged	Decreased	ISHIKAWA et al. (1998)
SPA (A_1 *agonist*)	Zucker diabetic rat	100 μg i.v. in 15 min	Decreased	Increased	VAN SCHAICK et al. (1998)
BW 1433 (A_1 *antagonist*)	Zucker diabetic rat, *IPGTT*	3–12 mg/kg, p.o., b.i.d., 6 weeks	Decreased peak	Improved glucose tolerance	XU et al. (1998)

ND, not determined; IVGTT, intravenous glucose tolerance test; IPGTT, intraperitoneal glucose tolerance test.

lowering effects in conscious fasted non-diabetic rats (MERKEL et al. 1995): a 2-h i.v. infusion of these compounds resulted in significant decreases in plasma glucose, glycerol, triglycerides, and free fatty acids. In a fructose-induced rodent model of insulin resistance, dyslipidemia, and hypertension, GR 79236 ameliorated the effects of fructose feeding on fatty acid and triglyceride levels as well as on blood pressure, and improved glucose tolerance (QU et al. 1997). RG 14202 [*N*-5′-ethyl-N^6(cyclopentyl) adenosine-5′-uronamide] reduced plasma triglycerides but not glucose, free fatty acids, or glycerol levels in streptozotocin diabetic rats (COX et al. 1997). In this study, CPA and SDZ WAG 994 (6-cyclohexyl-2′-*O*-methyl-adenosine) were also ineffective in reducing plasma glucose. In contrast, in spontaneously hypertensive rats (SHRs) with hyperglycemia induced by streptozotocin injection to neonates, SDZ WAG 994 induced a modest decrease in the serum glucose concentration (ISHIKAWA et al. 1998); this effect was associated with a decrease in blood pressure and a marked and long-lasting decrease in free fatty acids and triglycerides, but there was no significant change in insulin concentration. SDZ WAG 994 did not affect the serum glucose concentration when administered to normotensive diabetic controls. Since SDZ WAG 994 decreases plasma glucose in hypertensive but not in normotensive streptozotocin-induced diabetic animals, it is possible that a link could exist between the antihypertensive action of the drug and the lowering of plasma glucose, in accordance with the hemodynamic hypothesis of insulin resistance in hypertension (JULIUS et al. 1991). On the other hand, the A_1 receptor agonist N^6-(*p*-sulfophenyl)adenosine (SPA), administered acutely, increased blood glucose concentrations and decreased plasma insulin concentrations in diabetic Zucker rats (VAN SCHAICK et al. 1998), possibly reflecting an inhibition of pancreatic insulin secretion (CAMPBELL and TAYLOR 1982; HILLAIRE-BUYS et al. 1987).

The A_1 antagonist, 1,3-dipropyl-8-(*p*-acrylic)phenylxanthine (BW-1433), can improve glucose tolerance in Zucker diabetic rats, at lower levels of insulin (XU et al. 1998). This compound increased glucose disposal during an euglycemic hyperinsulinemic clamp in obese animals and increased glucose uptake in skeletal muscle, while decreasing glucose uptake in adipose tissue, suggesting that signaling from adenosine receptors may be a factor contributing to tissue-specific insulin resistance (CRIST et al. 1998).

II. Cellular Targets and Mechanisms of Action

1. Adipose Tissue

Impaired insulin suppression of free fatty acid release from adipose tissue is a key characteristic of the insulin resistance syndrome in type 2 diabetes (CHEN et al. 1987). The increased level of free fatty acids may contribute to hyperglycemia by impairing peripheral glucose utilization, as suggested by the glucose-fatty acid cycle of RANDLE et al. (1963), and by promoting hepatic glucose overproduction (BODEN 1997). Agents that reduce free fatty acid concentrations have been advocated for improving glycemic control in type 2 diabetes.

Adenosine plays an important role in the regulation of adipose tissue lipolysis (EBERT and SCHWABE 1973; SCHWABE et al. 1975; KATHER et al. 1985), and lipolysis in vivo was suggested to be under tonic inhibition by adenosine (KATHER et al. 1985). The effects of adenosine are mediated by A_1 receptors (LONDOS et al. 1980). The antilipolytic effects of adenosine agonists that result in decreased free fatty acids (energy substrate for hepatic gluconeogenesis) and glycerol (gluconeogenic precursor) also decrease plasma glucose concentration in vivo in some studies (Table 2), in agreement with the hypothesis that lowering of glucose concentration may result from inhibition of lipolysis (REAVEN et al. 1988).

Adenosine receptor agonists could improve insulin resistance by decreasing free fatty acid concentration from adipose tissue. In addition, adenosine can enhance insulin-stimulated glucose uptake in adipocytes, and its antagonism either by receptor antagonists or removal by adenosine deaminase resulted in a decreased glucose uptake (TAYLOR et al. 1979; JOOST et al. 1982; KURODA et al. 1987; VANNUCCI et al. 1992). The potentiation of insulin-stimulated glucose transport is cAMP-independent (KURODA et al. 1987), and it was proposed that adenosine promotes a conformation of the GLUT4 glucose transporter in the plasma membrane that is accessible to extracellular substrate (VANNUCCI et al. 1992).

Some of the studies reported in Table 2 are inconclusive or even show an increase in plasma glucose concentration, that could be the result of a decreased insulin secretion from pancreatic β cells. A_1 receptors that inhibit insulin release have been characterized in pancreatic β cells (HILLAIRE-BUYS et al. 1987) and are linked to a pertussis toxin-sensitive G protein signal transduction system (HILLAIRE-BUYS et al. 1989). Decreased electrical activity in β cells on activation of A_1 receptors (BERTRAND et al. 1989b) and inhibition of adenylate cyclase, rather than direct changes in membrane K^+ and Ca^{2+} permeabilities, underlie the action of A_1 agonists on insulin release (BERTRAND et al. 1989a).

Thus, the glucose lowering effect of adenosine receptor agonists due to improved insulin resistance may be hampered by their inhibition of insulin release from the pancreas.

On the other hand, a major drawback for the use of A_1 receptor agonists in type 2 diabetes is the occurrence of severe hemodynamic actions after peripheral administration. Most of the compounds that have been investigated so far (Table 2) induce their beneficial metabolic effects at doses that also produce significant bradycardia and hypotension. This point is all the more relevant since a decreased sensitivity to the antilipolytic action of adenosine A_1 receptor agonists occurs in Zucker diabetic animals (VAN SCHAICK et al. 1998). Adipocytes from streptozotocin-induced diabetic rats had a decreased sensitivity to adenosine, resulting from decreased coupling of the receptor to its G protein, apparently due to a functional abnormality in G_i (GREEN and JOHNSON 1991).

2. Skeletal Muscle

Skeletal muscle is a major site of body glucose disposal (BARON et al. 1988), and the rate of glucose uptake is increased by contraction and/or insulin. The literature on the role of adenosine in regulating glucose transport in skeletal muscle is confusing.

In the heart, adenosine potentiates insulin-stimulated myocardial glucose uptake (LAW et al. 1988; LAW and RAYMOND 1988; WYATT et al. 1989; ANGELLO et al. 1993). Adenosine receptor stimulation enhanced insulin-mediated glucose uptake in contracting, but not in resting, skeletal muscle in rat hindquarters; it was also shown to be involved in the synergistic stimulation of muscle glucose transport by insulin and by contractions (VERGAUWEN et al. 1994). Adenosine removal decreases insulin- and contraction-stimulated glucose transport in skeletal muscle (rat epitrochlearis and soleus muscle) by reducing the number of GLUT4 transporters at the cell surface (HAN et al. 1998), raising the possibility that decreased adenosine production or reduced action could play a role in the development of insulin resistance.

However, adenosine inhibits insulin-stimulated glucose uptake and metabolism in rat soleus muscle (ESPINAL et al. 1983; BUDOHOSKI et al. 1984; CHALLIS et al. 1984; LEIGHTON et al. 1988). It was proposed that adenosine exerts a post-receptor insulin modulatory action in skeletal muscle and that this action is mediated by A_1 receptors (CHALLIS et al. 1992). However, no A_1 or A_3 adenosine receptors are expressed in muscles, whereas low levels of A_2 receptor mRNA were detected using PCR (DIXON et al. 1996). The effects of the A_1 antagonist, BW-1433, indicate that the major result of A_1 receptor antagonism in Zucker rats is an increase in whole body glucose utilization (XU et al. 1998). Adenosine antagonism may also induce cardiovascular side effects including elevation of systolic blood pressure (KAROON et al. 1995). Blockade of A_1 receptors by antagonists like FK-838 or KW-3902 reduces salt-sensitive hypertension (UEHARA et al. 1995; NOMURA et al. 1995).

While adenosine A_1 receptors are potentially attractive targets for pharmacological intervention in diabetes, the complexity of the responses to agonists and antagonists in vivo underlines the complexity at the integrative level in vivo in contrast to the in vitro pharmacology in a single cell type.

D. Conclusions

Both purine nucleotides and nucleosides are involved in the regulation of metabolic functions, at different levels, that are impaired in the heterogeneous expressions of diabetes. Purines also affect the cardiovascular system, which may contribute to some pathophysiological aspects of the disease, particularly diabetes complications. The further characterization of the different purinergic receptors involved in the pathophysiology of diabetes will allow the generation and testing of more tissue (e.g., receptor) specific compounds to better target drugs to ameliorate dysfunctional purinergic-mediated mechanisms that participate to this metabolic disorder that has such a major effect on human health.

Acknowledgement. The authors thank V. Montesinos for her assistance in manuscript preparation.

References

Aguilar-Bryan L, Clement IV JP, Gonzalez G, Kunjilwar K, Babenko A, Bryan J (1998) Toward understanding the assembly and structure of K_{ATP} channels. Physiol Rev 78:227–245

Amos AF, McCarty DJ, Zimmet P (1997) The rising global burden of diabetes and its complications: estimates and projections to the year 2010. Diabetic Med 14 (Suppl 5):S1–85

Angello DA, Berne RM, Coddington NM (1993) Adenosine and insulin mediate glucose uptake in normoxic rat hearts by different mechanisms. Am J Physiol 265:H880–H885

Arkhammar P, Hallberg A, Kindmark H, Nilsson T, Rorsman P, Berggren PO (1990) Extracellular ATP increases cytoplasmic free Ca^{2+} concentration in clonal insulin-producing RINm5F cells. Biochem J 265:203–211

Atkinson DE (1978) Cellular energy metabolism and its regulation. Academic Press, San Diego

Baron AD, Brechtel G, Wallace P, Edelman SV (1988) Rates and tissue sites of non-insulin- and insulin-mediated glucose uptake in humans. Am J Physiol 255:E769–E774

Bertrand G, Chapal J, Loubatières-Mariani MM, Roye M (1987) Evidence for two different P_2-purinoceptors on β cell and pancreatic vascular bed. Br J Pharmacol 91:783–787

Bertrand G, Chapal J, Puech R, Loubatières-Mariani MM (1991) Adenosine-5′-*O*-(2-thiodiphosphate) is a potent agonist at P_2 purinoceptors mediating insulin secretion from perfused rat pancreas. Br J Pharmacol 102:627–630

Bertrand G, Nenquin M, Henquin JC (1989a) Comparison of the inhibition of insulin release by activation of adenosine and α_2-adrenergic receptors in rat β cells. Biochem J 259:223–228

Bertrand G, Petit P, Bozem M, Henquin JC (1989b) Membrane and intracellular effects of adenosine in mouse pancreatic β cells. Am J Physiol 257:E473–E478

Blachier F, Malaisse WJ (1988) Effect of exogenous ATP upon inositol phosphate production, cationic fluxes and insulin release in pancreatic islet cells. Biochem Biophys Acta 970:222–229

Boden G (1997) Role of fatty acids in the pathogenesis of insulin resistance and NIDDM. Diabetes 46:3–10

Budohoski L, Challis RAJ, McManus B, Newsholme EA (1984) Effects of analogues of adenosine and methyl xanthines on insulin sensitivity in soleus muscle of the rat. FEBS Lett 167:1–4

Campbell IL, Taylor KW (1982). Effects of adenosine, 2-deoxyadenosine and N^6-phenylisopropyladenosine on rat islet function and metabolism. Biochem J 204:689–696

Candela JLR, Garcia-Fernandez MC (1963) Stimulation of secretion of insulin by adenosine-triphosphate. Nature 197:1210 (Abstract)

Challis RAJ, Budohoski L, McManus B, Newsholme EA (1984) Effects of an adenosine-receptor antagonist on insulin-resistance in soleus muscle from obese Zucker rats. Biochem J 221:915–917

Challis RAJ, Richards SJ, Budohoski L (1992) Characterization of the adenosine receptor modulating insulin action in rat skeletal muscle. Eur J Pharmacol 226:121–128

Chapal J, Hillaire-Buys D, Bertrand G, Pujalte D, Petit P, Loubatières-Mariani MM (1997) Comparative effects of adenosine-5′-triphosphate and related analogues on insulin secretion from the rat pancreas. Fundam Clin Pharmacol 11:537–545

Chapal J, Loubatières-Mariani MM (1981a) Effects of phosphate-modified adenine nucleotide analogues on insulin secretion from perfused rat pancreas. Br J Pharmacol 73:105–110

Chapal J, Loubatières-Mariani MM (1981b) Attempt to antagonize the stimulatory effect of ATP on insulin secretion. Eur J Pharmacol 74:127–134

Chen YD, Golay A, Swislocki AL, Reaven GM (1987) Resistance to insulin suppression of plasma free fatty acid concentrations and insulin stimulation of glucose uptake in non-insulin-dependent diabetes mellitus. J Clin Endocrinol Metab 64:17–21

Cox BF, Clark KL, Perrone MH, Welzel GE, Greenland BD, Colussi DJ, Merkel LA (1997) Cardiovascular and metabolic effects of adenosine A1-receptor agonists in streptozotocin-treated rats. J Cardiovasc Pharmacol 29:417–426

Crist GH, Xu B, Lanoue KF, Lang CH (1998) Tissue-specific effects of in vivo adenosine receptor blockade on glucose uptake in Zucker rats. FASEB J 12:1301–1308

Diabetes Control and Complications Trial Research Group (1993) The effect of intensive treatment of diabetes on the development and progression of long-term complications in insulin-dependent diabetes mellitus. N Engl J Med 329:977–986

Dixon AK, Gubitz AK, Sirinathsinghui DJ, Richardson PJ, Freeman TC (1996) Tissue distribution of adenosine receptors mRNAs in the rat. Br J Pharmacol 118:1461–1468

Ebert R, Schwabe U (1973) Studies on the antilipolytic effect of adenosine and related compounds in isolated fat cells. Naunyn-Schmiedeberg's Arch Pharmacol 278:247–259

Espinal J, Challiss RAJ, Newsholme EA (1983) Effect of adenosine deaminase and adenosine analogue on insulin sensitivity in soleus muscle of the rat. FEBS Lett 158:103–106

Expert Committee on the Diagnosis and Classification of Diabetes Mellitus (1997) Report. Diabetes Care 20:1183–1197

Fernandez-Alvarez J, Hillaire-Buys D, Manteghetti M, Chapal J, Loubatières-Mariani MM, Gomis R, Petit P (1998) Comparative effects of P2 receptor agonists on insulin secretion from human and rat pancreatic islets. Drug Develop Res 43:62, A 239

Fischer B, Shulkin A, Sheihet L, Boyer JL, Harden KT, Gendron FP, Baudoin AR, Chapal J, Hillaire-Buys D, Petit P (1998) Adenosine 5′-phosphorothioates-2-thioether and 5′-boranophosphate-2-thioether derivatives as potential antidiabetic drugs. Drug Develop Res 43:28, A108

Geschwind JF, Hiriart M, Glennon MC, Najafi H, Corkey BE, Matschinsky FM, Prentki M (1989) Selective activation of Ca^{2+} influx by extracellular ATP in a pancreatic β-cell line (HIT). Bioch Biophys Acta 1012:107–115

Green A, Johnson JL (1991) Evidence for impaired coupling of receptors to G_i protein in adipocytes from streptozotocin-induced diabetic rats. Diabetes 40:88–94

Gylfe E, Hellman B (1987) External ATP mimics carbachol in initiating calcium mobilization from pancreatic β cells conditioned by previous exposure to glucose. Br J Pharmacol 92:281–289

Han DH, Hansen PA, Nolte LA, Holloszy JO (1998) Removal of adenosine decreases the responsiveness of muscle glucose transport to insulin and contractions. Diabetes 47:1671–1675

Hillaire-Buys D, Bertrand G, Chapal J, Puech R, Ribes G, Loubatières-Mariani MM (1993) Stimulation of insulin secretion and improvement of glucose tolerance in rat and dog by the P2Y purinoceptor agonist, adenosine-5′-*O*-(2-thiodiphosphate). Br J Pharmacol 109:183–187

Hillaire-Buys D, Bertrand G, Gross R, Loubatières-Mariani MM (1987) Evidence for an inhibitory A_1 subtype adenosine receptor on pancreatic insulin-secreting cells. Eur J Pharmacol 136:109–112

Hillaire-Buys D, Chapal J, Linck N, Blayac JP, Petit P, Loubatières-Mariani MM (1998) Involvement of K^+ channel permeability changes in the L-NAME and indomethacin resistant part of adenosine-5′-*O*-(2-thiodiphosphate)-induced relaxation of pancreatic vascular bed. Br J Pharmacol 124:149–156

Hillaire-Buys D, Chapal J, Petit P, Loubatières-Mariani MM (1991) Dual regulation of pancreatic vascular tone by P2X and P2Y purinoceptor subtypes. Eur J Pharmacol 199:309–314

Hillaire-Buys D, Gross R, Chapal J, Ribes G, Loubatières-Mariani MM (1992) P_{2Y} purinoceptor responses of β cells and vascular bed are preserved in diabetic rat pancreas. Br J Pharmacol 106:610–615

Hillaire-Buys D, Gross R, Loubatières-Mariani MM, Ribes G (1989) Effect of pertussis toxin on A_1 receptor-mediated inhibition of insulin secretion. Br J Pharmacol 96:3–4

Ishikawa J, Mitani H, Bandoh T, Kimura M, Totsuka T, Hayashi S (1998) Hypoglycemic and hypotensive effects of 6-cyclohexyl-2′-*O*-methyl-adenosine, an adenosine A1 receptor agonist, in spontaneously hypertensive rat complicated with hyperglycemia. Diabetes Res Clin Pract 39:3–9

Joost HG, Steinfelder HJ (1982) Modulation of insulin sensitivity by adenosine. Effects on glucose transport, lipid synthesis, and insulin receptors of the adipocyte. Mol Pharmacol 22:614–618

Julius S, Gudbrandsson T, Jamerson K, Shahab ST, Andersson O (1991) The hemodynamic link between insulin resistance and hypertension. J Hypertens 9:983–986

Karoon P, Rubino A, Burnstock G (1995) Enhanced sympathetic neurotransmission in the tail artery of 1,3-dipropyl-8-sulphophenylxanthine (DPSPX)-treated rats. Br J Pharmacol 116:1918–1922

Kather H, Bieger W, Michel G, Aktories K, Jakobs KH (1985) Human fat cell lipolysis is primarily regulated by inhibitory modulators acting through distinct mechanisms. J Clin Invest 76:1559–1565

Kindmark H, Köhler M, Nilsson T, Arkhammar P, Wiechel KL, Rorsman P, Efendic S, Berggren PO (1991) Measurements of cytoplasmic free Ca^{2+} concentration in human pancreatic islets and insulinoma cells. FEBS Lett 291:310–314

King H, Aubert RE, Herman WH (1998) Global burden of diabetes, 1995–2025: prevalence, numerical estimates, and projections. Diabetes Care 21:1414–1431

Kuroda M, Honnor RC, Cushman SW, Londos C, Simpson IA (1987) Regulation of insulin-stimulated glucose transport in the isolated rat adipocyte. J Biol Chem 262:245–253

Law WR, McLane MP, Raymond RM (1988) Adenosine is required for myocardial insulin responsiveness in vivo. Diabetes 37:842–845

Law WR, Raymond RM (1988) Adenosine potentiates insulin-stimulated myocardial glucose uptake in vivo. Am J Physiol 254:H970–H975

Leighton B, Lozeman FJ, Vlachonikolis IG, Challis RAJ, Pitcher JA, Newsholme EA (1988) Effects of adenosine deaminase on the sensitivity of glucose transport, glycolysis and glycogen synthesis to insulin in muscles of the rat. Int J Biochem 20:23–27

Levine RA, Oyama S, Kagan A, Glick SM (1970) Stimulation of insulin and growth hormone secretion by adenine nucleotides in primates. J Lab Clin Med 75:30–36

Li G, Milani D, Dunne MJ, Pralong WF, Theler JM, Petersen OH, Wollheim CB (1991) Extracellular ATP causes Ca^{2+}-dependent and -independent insulin secretion in RINm5F cells. J Biol Chem 266:3449–3457

Londos C, Cooper DMF, Wolff J (1980) Subclasses of external adenosine receptors. Proc Natl Acad Sci USA 77:2551–2554

Loubatières-Mariani MM, Chapal J, Lignon F, Valette G (1979) Structural specificity of nucleotides for insulin secretory action from the isolated perfused rat pancreas. Eur J Pharmacol 59:277–286

Loubatières-Mariani MM, Hillaire-Buys D, Chapal J, Bertrand G, Petit P (1997) P_2 purinoceptor agonists: new insulin secretagogues potentially useful in the treatment of non-insulin-dependent diabetes mellitus. In: Jacobson KA, Jarvis MF (eds) Purinergic approaches in experimental therapeutics. Wiley-Liss, pp 253–260

Loubatières-Mariani MM, Petit P, Chapal J, Hillaire-Buys D, Bertrand G, Ribes G (1995) Effects of purinoceptor agonists on insulin secretion. In: Belardinelli L, Pelleg A (eds) Adenosine and adenine nucleotides: From molecular biology to integrative physiology. Kluwer, Boston, pp 337–345

Merkel LA, Hawkins ED, Colussi DJ, Greenland BD, Smits GJ, Perrone MH, Cox BF (1995) Cardiovascular and antilipolytic effects of the adenosine agonist GR79236. Pharmacology 51:224–236

Nomura H, Nagashima K, Kusaka H, Karasawa A (1995) Antihypertensive effects of KW-3902, an adenosine A1-receptor antagonist, in Dahl salt-sensitive rats. Jpn J Pharmacol 68:389–396

Petit P, Bertrand G, Schmeer W, Henquin JC (1989) Effects of extracellular nucleotides on the electrical, ionic and secretory events in mouse pancreatic β cells. Br J Pharmacol 98:875–882

Petit P, Hillaire-Buys D, Manteghetti M, Debrus S, Chapal J, Loubatières-Mariani MM (1998) Evidence for two different types of P2-receptors stimulating insulin secretion from pancreatic B cell. Br J Pharmacol 125:1368–1374

Petit P, Loubatières-Mariani MM, Keppens S, Sheehan MJ (1996) Purinergic receptors and metabolic function. Drug Develop Res 39:413–425

Petit P, Manteghetti M, Loubatières-Mariani MM (1988) Differential effects of purinergic and cholinergic activation on the hydrolysis of membrane polyphosphoinositides in rat pancreatic islets. Biochem Pharmacol 37:1213–1217

Petit P, Manteghetti M, Puech R, Loubatières-Mariani MM (1987) ATP and phosphate-modified adenine nucleotide analogues: effects on insulin secretion and calcium uptake. Biochem Pharmacol 36:377–380

Qu X, Cooney G, Donnelly R (1997) Short-term metabolic and haemodynamic effects of GR 79236 in normal and fructose-fed rats. Eur J Pharmacol 338:269–276

Randle PJ, Garland PB, Hales CN, Newsholme EA (1963) The glucose-fatty acid cycle: its role in insulin sensitivity and the metabolic disturbances of diabetes mellitus. Lancet i:785–789

Reaven G, Chang H, Ho H, Jeng CY, Hoffman B (1988) Lowering of plasma glucose in diabetic rats by antilipolytic agents. Am J Physiol 254:E23–E30

Ribes G, Bertrand G, Petit P, Loubatières-Mariani MM (1988) Effects of 2-methylthio ATP on insulin secretion in the dog in vivo. Eur J Pharmacol 155:171–174

Schwabe U, Ebert R, Erbler HC (1975) Adenosine release from fat cells: effect on cyclic AMP levels and hormone actions. Adv Cyclic Nucleotide Res 5:569–583

Squires PE, James RFL, London NJM, Dunne MJ (1994) ATP-induced intracellular Ca^{2+} signals in isolated human insulin-secreting cells. Pflügers Arch 427:181–183

Stam NJ, Klomp J, Van de Heuvel N, Olijve W (1996) Molecular cloning and characterization of a novel orphan receptor (P2P) expressed in human pancreas that shows high structural homology to the P2U purinoceptor. FEBS Lett 384:260–264

Tang J, Pugh W, Polonsky KS, Zhang H (1996) Preservation of insulin secretory responses to P2 purinoceptor agonists in Zucker diabetic fatty rats. Am J Physiol 270:E504–E512

Taylor WM, Halperin ML (1979) Stimulation of glucose transport in rat adipocytes by insulin, adenosine, nicotinic acid and hydrogen peroxide. Biochem J 178:381–389

Theler JM, Mollard P, Guérineau N, Vacher P, Pralong WF, Schlegel W, Wollheim CB (1992) Video imaging of cytosolic Ca^{2+} in pancreatic β cells stimulated by glucose, carbachol, and ATP. J Biol Chem 267:18110–18117

Tokuyama Y, Hara M, Jones EMC, Fan Z, Bell GI (1995) Cloning of rat and mouse P_{2y} purinoceptors. Biochem Biophys Res Commun 211:211–218

Uehara Y, Numabe A, Hirawa N, Kawabata Y, Nagoshi H, Kaneko H, Gomi T, Goto A, Toyo-oka T, Omata M (1995) A new adenosine subtype-1 receptor antagonist, FK-838, attenuates salt-induced hypertension in Dahl salt-sensitive rats. Am J Hypertens 8:1189–1199

UK Prospective Diabetes Study (UKPDS) Group (1998) Intensive blood-glucose control with sulphonylureas or insulin compared with conventional treatment and risk of complications in patients with type 2 diabetes (UKPDS 33). Lancet 352:837–853

Vannucci SJ, Nishimura H, Satoh S, Cushman SW, Holman GD, Simpson IA (1992) Cell surface accessibility of GLUT4 glucose transporters in insulin-stimulated rat adipose cells. Biochem J 288:325–330

Van Schaick EA, Zuideveld KP, Tukker HE, Langemeijer MW, Ijzerman AP, Danhof M (1998) Metabolic and cardiovascular effects of the adenosine A1 receptor agonist N^6-(*p*-sulfophenyl)adenosine in diabetic Zucker rats: influence of the disease on the selectivity of action. J Pharmacol Exp Ther 287:21–30

Vergauwen L, Hespel P, Richter EA (1994) Adenosine receptors mediate synergistic stimulation of glucose uptake and transport by insulin and by contractions in rat skeletal muscle. J Clin Invest 93:974–981

Wang CZ, Namba N, Gonoi T, Inagaki N, Seino S (1996) Cloning and pharmacological characterization of a fourth P2X receptor subtype widely expressed in brain and peripheral tissues including various endocrine tissues. Biochem Biophys Res Commun 220:196–202

Wyatt DA, Edmunds MC, Rubio R, Berne RM, Lasley RD, Mentzer RM Jr (1989) Adenosine stimulates glycolytic flux in isolated perfused rat hearts by A1-adenosine receptors. Am J Physiol 257:H1952–H1957

Xu B, Berkich DA, Crist GH, Lanoue KF (1998) A1 adenosine receptor antagonism improves glucose tolerance in Zucker rats. Am J Physiol 274:E271–E279

Yki-Järvinen H (1990) Acute and chronic effects of hyperglycaemia on glucose metabolism. Diabetologia 33:579–585

CHAPTER 28

Purinergic and Pyrimidinergic Receptor Signaling in Bone Cells

J. FERRIER

A. Introduction to Bone Cells

Bone-forming osteoblasts and bone-resorbing osteoclasts work together to remodel bone. During its lifetime each cell type expresses a variety of signaling receptors, including those for parathyroid hormone (PTH), calcitonin, various cytokines, steroids, prostaglandins, and nucleotides. These receptors and their linked intracellular signaling pathways regulate proliferation, differentiation, and bone remodeling activity. There are extracellular signaling molecules that convey information between osteoblasts and osteoclasts; one is known (interleukin-6), but the identities of others are unknown.

Work to date shows a number of distinct signaling pathways linked to a number of P2 receptors in osteoclasts. In osteoblasts and osteoblastic cells there is evidence for a number of P2 receptors, and for synergy between their signaling pathways and those linked to some of the other receptor types. There is also some evidence for P1 receptors in osteoblastic cells, but not in osteoclasts.

The bone-regulating molecules with the greatest spatial reach within an organism are PTH, calcitonin, the steroids, and some of the cytokines, while others of the cytokines and the prostaglandins are more localized. The nucleotides would seem by their nature to represent a highly localized signal, which could also occur over a very short duration. This presents a picture of various systemic signals and others of a more localized nature being modulated by nucleotides over possibly very small spatial domains (perhaps in some cases comprising only a few cells), and possibly very short durations (perhaps in some cases only a few minutes). However, even short duration modulations could have long-term consequences.

B. Osteoblasts

I. Purinergic and Pyrimidinergic Receptors in Osteoblastic Cells and Osteoblasts

The osteoblastic rat osteosarcoma line UMR-106 has at least two types of functional P2 receptor. The evidence consists of measurements of intracellular

[Ca^{2+}] responses to extracellular nucleotides. KUMAGAI et al. (1991) showed calcium pulses resulting from P2 receptor stimulation. Further work provided evidence for $P2Y_2$ and $P2Y_3$ receptors (REIMER and DIXON 1992; SISTARE et al. 1994), while other reports supported the presence of $P2Y_1$ and $P2Y_2$ receptors (YU and FERRIER 1993a; GALLINARO et al. 1995). Recent work, outlined below, shows that no simple combination of known receptor types can account for all of the calcium response data in UMR-106 cells (LUO 1995). Work based on molecular cloning has shown that the osteoblastic lines SaOS2 and Te-85, as well as isolated human osteoblasts, express transcripts for $P2Y_2$ receptors (BOWLER et al. 1995), and that the osteoblastic lines OHS-4 and MG-63 express mRNA for $P2Y_1$, $P2Y_2$, $P2Y_4$, $P2Y_6$, and $P2Y_7$ receptors (MAIER et al. 1997).

There is evidence that P2Y mediated signaling is heterogeneous in osteoblastic cells. Using the fluorescent intracellular calcium indicator Fluo-3, YU and FERRIER (1993a) showed fluorescence images suggesting that some UMR-106 cells respond more to ATP than to UTP, while other cells in the same field of view (comprising about 60 cells) have the opposite response. These cells could have differences in expression or functional modulation of two or more P2Y receptors or of some component of the signaling pathways leading to a calcium pulse. These differences could conceivably be linked to an intracellular rhythm or to phenotype divergence. There is similar variability for PTH receptor- and calcitonin receptor-mediated calcium responses in rat bone nodule osteoblasts (FERRIER et al. 1992). Recent work shows that differences in PTH receptor expression in these cells is based on phenotype heterogeneity (LIU et al. 1997).

There is also some evidence that P1 receptors occur in osteoblasts and osteoblastic cells. LERNER et al. (1987) reported that P1 (A_2-like) receptors are linked to cyclic AMP stimulation in rat calvarial bone and in isolated osteoblastic cells from the calvaria, but with no clear relation to cell function. SHIMEGI (1998) showed that P1 receptor stimulation by adenosine induces proliferation of MC3T3-E1 osteoblastic cells, but the signaling pathways could not be determined.

II. Signaling Involving Intracellular Calcium in Osteoblastic Cells

1. Internal Calcium Release Induced by P2 Receptor Activation

P2 receptor activation in UMR-106 cells by extracellular ATP, UTP, and some other nucleotides, leads to an intracellular calcium pulse. Measured calcium indicator fluorescence associated with the calcium pulse returns to baseline within one minute. Experiments based on depleting extracellular calcium or intracellular calcium stores, and G protein inhibitors, show that there is no calcium influx across the cell membrane: the calcium signal is entirely based on a G protein/phospholipase C/inositol trisphosphate pathway leading to internal calcium release. Measurements have been done with these cells in suspension, using a spectrophotometer and fluorescent calcium indicators

(KUMAGAI et al. 1991; REIMER and DIXON 1992; SISTERE et al. 1994, 1995; GALLINARO et al. 1995; KAPLAN et al. 1995), and also with cells attached to the bottom of culture dishes, using a laser scanning microscope (YU and FERRIER 1993a; LUO 1995; LUO et al. 1997). Similar results have been obtained for osteoblastic MC3T3-E1 cells, based on spectrophotometer measurements using a fluorescent indicator (SHIMEGI 1996). Other work that reported an ATP-stimulated calcium influx in MC3T3-E1 cells was based on an assay using radiolabeled calcium (SUZUKI et al. 1993). In this type of measurement the increased calcium exchange across the cell membrane accompanying and following an internal calcium release could incorrectly suggest a net calcium influx across the cell membrane.

There appears to be no P1 receptor-linked pathway involving calcium in osteoblastic cells. Application of extracellular adenosine does not produce an intracellular calcium response in UMR-106 cells (YU and FERRIER 1993a), nor in MC3T3-E1 cells (SHIMEGI 1998).

2. Desensitization by and Lack of Adaptation to ATP in Osteoblastic Cells

In the presence of 50 μmol/l ATP, UMR-106 cells have no calcium response to further application of ATP (REIMER and DIXON 1992; YU and FERRIER 1993a; LUO 1995; LUO et al. 1997). When ATP is removed from the extracellular medium, there is only partial recovery of the calcium response to ATP within 12 min, showing that some desensitization has occurred (LUO 1995; LUO et al. 1997). This may result from the activation of protein kinase C (PKC) during the calcium response. Since the calcium response derives from phospholipase C activation there should be a substantial activation of PKC via the synergistic effect of Ca^{2+} and diacylglycerol. It has been reported that PKC activation inhibits the calcium response produced via $P2Y_1$ receptors in these cells (GALLINARO et al. 1995), suggesting that PKC can produce desensitization via phosphorylation of $P2Y_1$ receptors.

3. Competition and Desensitization Studies with ATP, ADP, UTP, and 2-MeSATP

LUO (1995) and LUO et al. (1997) reported a series of experiments with UMR-106 cells intended to distinguish between competition between different nucleotides for the same receptors (which can be eliminated immediately by washing away prior application of a nucleotide) and desensitization (which cannot be eliminated quickly by washing away the first nucleotide). There is a strong calcium response at 50 μmol/l and 100 μmol/l to ADP, to UTP, and to 2-methylthio-ATP (2-MeSATP), which is fairly specific among the P2Y receptors for the $P2Y_1$ receptor (BURNSTOCK 1997). The presence of 100 μmol/l extracellular 2-MeSATP or ADP largely inhibits the calcium response to 100 μmol/l ATP. The inhibition is only partially removed by removal of the 2-MeSATP or ADP. This indicates that there is competition between ATP and 2-MeSATP and between ATP and ADP for at least some of the receptors that are producing a response to ATP, as well as some desensitization by 2-MeSATP and ADP of the response to ATP. This is consistent with the presence of a $P2Y_1$ receptor.

There is no effect of UTP on the response to 2-MeSATP at 100 μmol/l, and only a relatively small inhibition by UTP of the response to ADP, which is alleviated by washing away the UTP. These results are consistent with a response to 2-MeSATP and to ADP mediated by the $P2Y_1$ receptor.

On the other hand, with UTP applied first there is no inhibition of the response to ATP at 50 μmol/l for each nucleotide, but there is a large inhibition at 100 μmol/l. Washing away the UTP almost completely restores the response to ATP. These results show that UTP can compete with ATP for all of the receptors involved in the calcium response to ATP, with no desensitization produced. This does not seem consistent with having a substantial part of the calcium response to ATP mediated by a $P2Y_1$ receptor.

Prior application of 50 μmol/l ATP substantially inhibits the calcium response to 50 μmol/l UTP; washing away the ATP partially restores the response. This shows that ATP competes strongly for the same receptor that UTP uses to induce a calcium response, and that ATP partially desensitizes that receptor or the linked signaling pathway. These data are consistent with the response to UTP being produced by a $P2Y_2$ receptor, but would also agree with the reported properties of the $P2Y_7$ and $P2Y_8$ receptors (BURNSTOCK 1997).

However, prior application of 2-MeSATP at 100 μmol/l, but not at 50 μmol/l, produces a substantial inhibition of the response to UTP, which is completely alleviated by removal of the 2-MeSATP. This shows that 2-MeSATP can compete strongly for the receptors that provide the response to UTP. There is a similar effect with prior application of ADP at 100 μmol/l. These results are not consistent with the known properties of the $P2Y_2$ receptor, nor with those of the $P2Y_7$ or $P2Y_8$. These results could fit with the reported properties of the $P2Y_6$ receptor, but the $P2Y_6$ is not consistent with the strong competition by ATP (BURNSTOCK 1997).

4. Lack of Spatial Localization of the Calcium Response to P2 Activation

There is no evidence for spatial localization of the P2 receptor-mediated calcium signal in the UMR-106 cells as far as can be seen in the published laser scanning images based on a fluorescent indicator for calcium (YU and FERRIER 1993a). Recent work done at higher resolution shows that the nucleus has a higher fluorescence intensity than the cytoplasm in the initial steady state and at the peak of the fluorescence pulse following P2 receptor stimulation (E.N. Bazhenova and J. Ferrier, unpublished). A typical result is shown in Fig. 1. However, a detailed analysis of more than 50 cells from 9 separate experiments shows no significant difference between the normalized fluorescence intensity (the percent intensity change) in the nucleus and that in the cytoplasm. The most likely explanation for this is that the free calcium

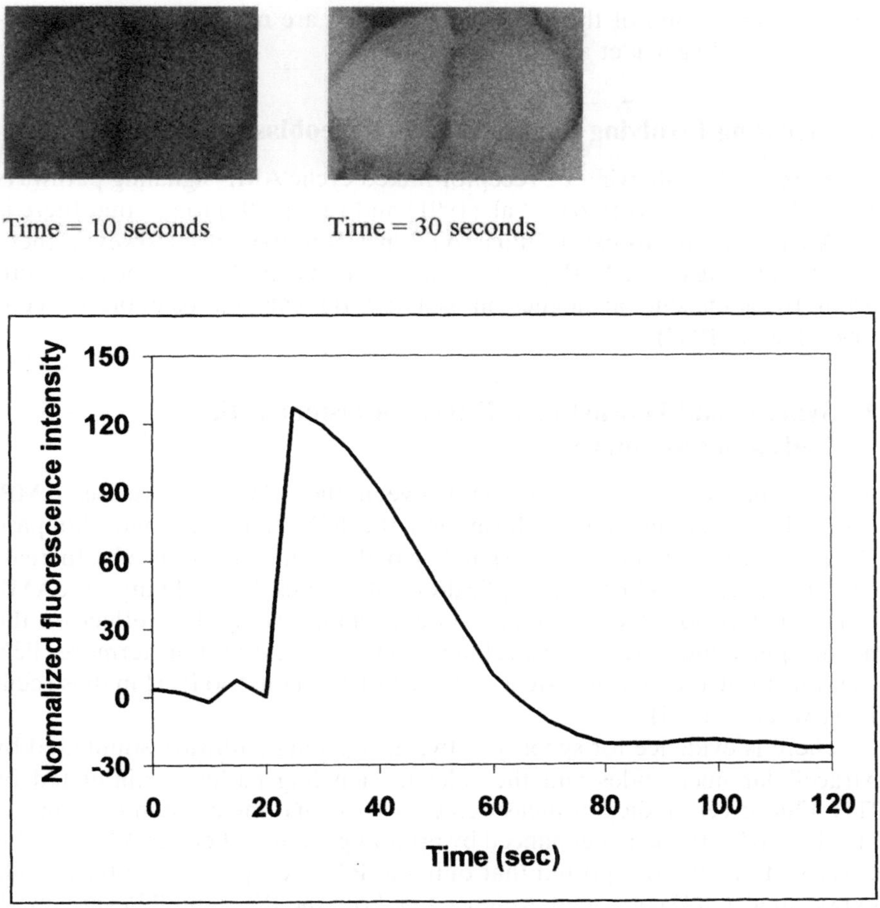

Fig. 1. Laser scanning images of two osteoblastic UMR 106 cells loaded with the fluorescent calcium indicator Fluo-3, and the time course of normalized fluorescence intensity (= percent change in intensity). The fluorescence pulse shows that cytosolic and nuclear free calcium has increased in response to an application of ATP (100 μmol/l) at about 10 s

concentration is the same in the nucleus and cytosol, with the higher fluorescence intensity in the nucleus resulting from a higher effective concentration of fluorescent indicator (probably because the cytoplasm includes many organelles from which the indicator is excluded).

However, since the resolution in these images is usually not less than a few μm, it is not possible to rule out free calcium concentration differences within the cytosol, or within the nucleus. Comparison of membrane potential spiking data to fluorescence-based calcium measurements in osteoblastic ROS-17/2.8 cells suggests that there can be calcium changes confined to a region within 200 nm of the cell membrane that are not seen in the fluorescence images (FERRIER et al. 1987, 1991).

III. Signaling Involving Cyclic AMP in Osteoblastic Cells

There are no data showing P2 receptor-linked cyclic AMP signaling pathways in osteoblastic cells. KUMAGAI et al. (1991) and LUO (1995) found that there is no cAMP response to extracellular ATP in UMR-106 cells. However, there are data showing cyclic AMP production in response to P1 (A2) receptor activation by adenosine analogues in isolated rat calvaria osteoblastic cells (LERNER et al. 1987).

IV. Synergy and Downstream Effects in Osteoblastic Cells and Osteoblasts

There seems to be no interaction between the adenylate cyclase/cAMP/protein kinase A signaling pathway and the P2Y-stimulated phospholipase C/inositol trisphosphate/calcium signaling pathway in UMR-106 cells. Increasing intracellular cAMP via application of extracellular dibutyryl cAMP (5 mmol/l) has no effect on steady state calcium level and no effect on the calcium pulse induced by extracellular ATP (LUO 1995). Furthermore, P2Y receptor stimulation has no effect on the cAMP response to PTH in these cells (KAPLAN et al. 1995).

There is evidence for synergy between signaling pathways stimulated by extracellular nucleotides and the calcium signaling pathway stimulated by PTH. Fluorescent indicator measurements show that the calcium response to PTH in UMR-106 cells is enhanced by prior application of either ATP or UTP. KAPLAN et al. (1995) reported that only the $P2Y_2$ receptor is involved, while SISTARE et al. (1995) provided evidence that both the $P2Y_1$ and $P2Y_2$ are implicated, but that PKC is not involved in producing this synergistic effect.

There is some evidence of a P2 signal producing a downstream effect on its own in osteoblasts. JONES et al. (1997) found, using in vitro rat calvarial cultures, that 2-MeSATP and ATP reduce bone formation, but to a lesser extent than PTH, while UTP at a low concentration (2 μmol/l) stimulates bone formation. These effects could result from changes in proliferation, differentiation, or bone-forming activity. On the other hand, SHIMEGI (1996) found that ATP enhances proliferation in osteoblastic MC3T3-E1 cells, via a pathway involving a calcium response but not involving PKC; there is also a synergistic enhancement of the effect of platelet-derived growth factor (PDGF) on proliferation. However, ATP has no measurable effect on proliferation or protein synthesis in UMR-106 cells (LUO 1995). Adenosine has also been found to stimulate proliferation of the MC3T3-E1 cells, and to synergistically enhance the effect on proliferation of PDGF, at least partly via a P1 receptor-linked pathway that includes a G protein but does not involve either calcium or cAMP (SHIMEGI 1996, 1998).

C. Osteoclasts

I. Purinergic and Pyrimidinergic Receptors in Osteoclasts

Studies on osteoclasts are more difficult than those on osteoblasts or osteoblastic cell lines, because true osteoclasts can be obtained only in small numbers, and kept in culture only for a few days. Work done with rabbit osteoclasts has provided evidence that osteoclasts have a number of P2 receptors (YU and FERRIER 1993b, 1994, 1995). Their identity is uncertain, but they probably include both P2Y and P2X receptors. Recently it has been reported that many cells in the bone marrow of rats, including cells of the monocyte-macrophage series, some of which are thought to be precursors of osteoclasts, express mRNA for the $P2X_7$ receptor (COLLO et al. 1997). Other workers have shown that human osteoclast-like tumor cells express $P2Y_2$ receptor transcripts, but that these transcripts do not lead to receptors that produce calcium signals (BOWLER et al. 1998). Work done with rodent osteoclasts has indicated that P2 receptor activation leads to stimulation of osteoclast formation and stimulation of osteoclast bone resorption activity (MORRISON et al. 1998).

There is no evidence for P1 receptors in osteoclasts. MORRISON et al. (1998) found no effect of adenosine on osteoclast activation or formation. YU and FERRIER (1993b) saw no calcium response to adenosine in osteoclasts.

II. Signaling Involving Intracellular Calcium in Osteoclasts

1. Internal Calcium Release and Influx Induced by P2 Receptor Activation

P2 receptor activation produces a large calcium pulse in rabbit osteoclasts, with a duration of about 1 min (YU and FERRIER 1993b, 1994; YU 1996). There are at least two separate pathways. One involves internal calcium release via a G protein/phospholipase C/inositol trisphosphate cascade, and is thus presumably linked to a P2Y receptor. Another involves influx of calcium across the cell membrane, and probably does not involve a G protein. This latter could utilize one of the P2X receptors, which include an ion channel. Only ATP and ADP have substantial calcium responses at 50 μmol/l and 100 μmol/l, while 2-MeSATP has a small response, and UTP, AMP, adenosine, and β,γ-methylene-ATP produce no calcium response. This fits with the known potency order of the $P2Y_5$ receptor (BURNSTOCK 1997), which has been reported in the T lymphocyte, like the osteoclast a cell of hematopoietic descent. The P2X receptor present in rabbit osteoclasts should have a greater response to ATP than to 2-MeSATP, which would fit with the $P2X_4$, $P2X_5$, $P2X_6$, and $P2X_7$ (BURNSTOCK 1997). The $P2X_7$ has been observed in bone marrow monocyte/macrophages and other cells of hematopoietic origin (BURNSTOCK 1997; COLLO et al. 1997). Since the $P2X_7$ has been implicated in macrophage fusion to form multinucleated cells (CHIOZZI et al. 1997), it is tempting to speculate that it is involved in formation of the multinucleated osteoclasts. On the other hand, multinucleated osteoclasts are stable and active for a few days in culture, much longer than the few hours reported for the macrophage-derived multinucleated cells. Furthermore, there is no sign in rabbit osteoclasts at 100 μmol/l ATP of the cytotoxicity reported with $P2X_7$ activation.

2. Adaptation to ATP in Osteoclasts

The P2 receptor/calcium signaling system in rabbit osteoclasts shows adaptation to the ambient ATP level, allowing a response to an increase in extracellular [ATP] (YU and FERRIER 1993b, 1994). This adaptation occurs up to at least 500 μmol/l ATP, allowing repeated calcium responses to 100 μmol/l ATP increases. PKC is not involved in the adaptation mechanism. Also, adaptation does not depend on the calcium influx pathway, since it occurs in calcium-free extracellular medium. LUO et al. (1997) found that the calcium response to ATP is partially inhibited by activation of protein kinase A (PKA). This suggests that the adaptation mechanism may be similar to one reported for A_{2A} receptors in cardiac cells (PALMER et al. 1994), in which the affinity of the receptor is reduced as ligand concentration increases, as a result of phosphorylation of the receptor by PKA. Thus, the adaptation mechanism could involve a P2Y-linked pathway leading to increased PKA activity.

3. Desensitization by UTP in Osteoclasts

Application of 100 μmol/l UTP to the extracellular medium produces no measurable calcium response in rabbit osteoclasts. However, there is a significant reduction in the calcium response to subsequent application of ATP, even if the UTP is removed from the bathing medium before application of the ATP (YU 1996; LUO et al. 1997). The mechanism of this desensitization is unknown, but it must involve a non-calcium signaling pathway linked to a P2Y receptor.

4. Spatial Localization of the Calcium Response to P2 Activation in Osteoclasts

Within the rabbit osteoclast, apparently equivalent regions of the cell body can show different calcium signals (XIA and FERRIER 1996). This is seen in response to mechanical perturbation and calcitonin application. However, calcium responses to nucleotide application appear to reach all equivalent regions of the multinucleated osteoclast cell body uniformly (YU and FERRIER 1993b, 1994; FERRIER and YU 1996). This suggests that the signaling effect of a hormone such as calcitonin, which may be more specific than that of the nucleotides, may be achieved by spatial segregation of the elements of its signaling pathway.

On the other hand, P2 receptor activation via ATP produces calcium release from the nuclear envelope into the nucleus, producing a greater free calcium concentration increase in the nucleus than in the cytosol (FERRIER and YU 1996). This is different from the calcium response to calcitonin, in which the nuclear free calcium increase appears to be the same as that in the neighboring cytosol (XIA and FERRIER 1996).

III. Signaling Involving Intracellular Protons in Osteoclasts

Another pathway in rabbit osteoclasts links P2 receptor activation to a large transient increase in the intracellular proton concentration (about a fivefold increase), lasting 1–2 min (YU and FERRIER 1995; YU 1996). The mechanism involves stimulation of the chloride/bicarbonate exchanger via an unknown pathway. The agonist potencies are similar to those of the calcium response. At 100 μmol/l, ADP produces about the same response as ATP, while 2-MeSATP produces a much smaller response, and UTP produces very little response. This suggests that the receptors are the same as those producing the calcium response. However, the pathway producing the pH response is separate from those that produce the calcium response: i.e., the pH response does not result from the calcium response. Clamping intracellular calcium with BAPTA/AM gives an increase in the effect of ATP on pH, and inducing an intracellular calcium pulse by adding calcium to the extracellular medium produces a large decrease in $[H^+]$. Thus, the pathway producing an increase in $[H^+]$ is opposing and overcoming a tendency of the calcium response to decrease $[H^+]$. The images obtained in these measurements, using the fluorescent indicator SNAFL-calcein, show a pH response that is uniform across the cell body (YU 1996).

The pH response measurements show the same kind of adaptation to increases in extracellular ATP as seen in the calcium responses (YU and FERRIER 1995). This is further evidence that the adaptation mechanism reflects a change in receptor affinity, rather than some change in the linked intracellular pathways (LUO et al. 1997), and that the receptors producing the pH response are the same as at least some of those producing the calcium response. These must be P2Y receptors.

IV. Downstream Effects in Osteoclastic Cells and Osteoclasts

There is some evidence that P2 signaling can stimulate bone resorption. BOWLER et al. (1998) report that human osteoclastic tumor cells show increased resorption activity when exposed to 10 μmol/l ATPγS during a 24-h assay. This is apparently via a pathway that does not involve an intracellular calcium signal. Furthermore, MORRISON et al. (1998) find increased rat osteoclast resorption activity in 0.2–2 μmol/l ATP during a 26-h assay; this effect is inhibited by the P2 receptor antagonist suramin. On the other hand, application of 100 μmol/l ATP to rabbit osteoclasts induces cellular contraction within 10–20 min (YU 1996), indicating that P2 stimulation could produce a short-term decrease in resorption activity, possibly via the calcium response pathway. MORRISON et al. (1998) also report that 0.2–2 μmol/l ATP increases the formation rate of mouse osteoclasts during 10 days in culture.

D. Differential P2 Receptor-Mediated Signaling and Possible Role in Bone Remodeling

The data discussed above provide the possibility that P2 receptor-mediated signaling could play a role in the differential regulation of bone cells. The mechanism in osteoclasts for adaptation to extracellular ATP, combined with the lack of such a mechanism in osteoblasts, and the partial homologous desensitization produced by ATP in osteoblastic cells, provide a means to have different signaling effects on these two cell types. That is, exposure to repeated pulses of ATP should have a greater calcium signaling effect on the osteoclasts than the osteoblasts, and continual exposure to 50 μmol/l ATP would make the osteoblasts but not the osteoclasts completely insensitive to a further increase in ATP.

Moreover, if we envision UTP as a signaling molecule in vivo (as reviewed by ANDERSON and PARKINSON 1997; CONNOLLY, Chap. 14, first volume), then the different effects of UTP on the osteoblasts and osteoclasts, with the osteoclasts showing almost complete desensitization by UTP of the response to ATP, while there is no such desensitization in osteoblasts, provide a further means of differential signaling. Thus, exposure of these cells in vivo to a 50 μmol/l pulse of UTP would render the osteoclasts but not the osteoblasts insensitive to subsequent pulses of ATP. These data provide the possibility of differential regulation of osteoblasts and osteoclasts.

There is a complex interplay between osteoblasts and osteoclasts. They respond differently to many of the extracellular signaling molecules that affect bone remodeling (FLUHMANN et al. 1998; PARTRIDGE et al. 1994; SUDA et al. 1997; WIMALAWANSA 1997), and there is evidence that they send signals back and forth (HILL et al. 1995; OWENS et al. 1996). For both cell types there are signaling molecules that elicit both a large calcium pulse and a large cAMP pulse and that produce inhibition of activity in that cell type and a further signal (perhaps a cytokine) that is transmitted to the other cell type, which serves to increase activity in the other cell type. Well-studied examples are PTH, which inhibits osteoblast activity (at least over the short-term) and enhances osteoclast activity, and calcitonin, which inhibits osteoclast activity and enhances osteoblast activity. It has recently been shown that at least one of the osteoblast to osteoclast signals induced by PTH is interleukin-6 (IL-6), and that the cAMP/PKA pathway is necessary and sufficient for this induction (GREENFIELD et al. 1996; ONYIA et al. 1997).

A hypothesis can be developed for the role of ATP and UTP in bone cell regulation, using the data concerning P2 receptor signaling presented above, and some conjecture. The evidence that long-term exposure to increased extracellular ATP inhibits bone formation in vitro (JONES et al. 1997), combined with the synergistic enhancement by ATP of the calcium response to PTH (KAPLAN et al. 1995; SISTARE et al. 1995), suggests that ATP by itself may inhibit osteoblast activity in the short-term. However, since ATP produces no cAMP response in osteoblastic cells (KUMAGAI et al. 1991; LUO 1995), and does not increase the cAMP response to PTH (KAPLAN et al. 1995), it would probably not produce or enhance osteoblast to osteoclast signaling via IL-6. In addition, since long-term exposure to a low concentration of UTP can enhance bone formation in vitro (JONES et al. 1997), it can be hypothesized that UTP may also stimulate osteoblast activity in the short-term.

The data regarding the effect of ATP on osteoclast activity indicate that long-term exposure to a fairly low concentration stimulates resorption activity (BOWLER et al. 1998; MORRISON et al. 1998), while short-term exposure to a higher concentration probably inhibits osteoclast activity (YU 1996), as discussed above. It can be speculated that the effect of UTP on resorption activity is small, since there is no osteoclast calcium response to UTP (YU and FERRIER 1994), and that the UTP-induced desensitization of the osteoclast calcium response to ATP (LUO et al. 1997) could block an effect of ATP on resorption.

This speculative model provides the possibility that ATP could be used as a message from osteoblasts to osteoclasts, to inhibit resorption in the short-term. This model also allows UTP to be used as a message from osteoclasts to osteoblasts, to stimulate osteoblast activity, possibly as a result of the effect of calcitonin on osteoclasts. Other neighboring cells, such as fibroblasts, could also conceivably use ATP and UTP as signals to osteoblasts and osteoclasts. Repeated pulses of ATP would have a larger effect on osteoclasts than osteoblasts, because of the partial homologous desensitization in osteoblasts. Thus, according to this speculative model, repeated pulses of ATP (such as may be produced in a controlled fashion), coming either from osteoblasts or neighboring cells such as fibroblasts, would result in a slight inhibition of bone formation activity by the osteoblasts, and a greater inhibition of resorption activity by the osteoclasts, yielding a net increase in bone formation. On the other hand, the data presented above show that a long-term continual (not pulsed) increase in ATP (such as may be produced by large scale cell death in the immediate neighborhood) would inhibit bone formation by osteoblasts (JONES et al. 1997) and stimulate bone resorption by osteoclasts (BOWLER et al. 1998; MORRISON et al. 1998), producing a large net increase in bone resorption.

However, multiple pulses of ATP following a pulse of UTP could, according to the model developed above, have a greater inhibiting effect on osteoblasts than osteoclasts, yielding a net increase in bone resorption. Also, it is conceivable that the UTP-induced desensitization of the calcium response in osteoclasts is not specific to the ATP-stimulated pathways, but could involve a more general inhibition of calcium pathways including those used in producing the IL-6 stimulation of resorption activity. This would mean that both the resorption-stimulating effect of IL-6 and the possible resorption-inhibiting effect of pulsed ATP could be blocked by a prior pulse of UTP. Furthermore, the resorption-stimulating effect of long-term exposure to elevated ATP could conceivably be blocked by the concurrent long-term presence of elevated UTP.

The data and the model presented here suggest that P2 receptor-linked signaling could play a role in some pathological conditions. For example, in the simplest case, the data suggest that a long-term increase in extracellular ATP, as might be produced by chronic abnormal apoptosis near bone surfaces, could lead to inhibition of bone formation and stimulation of resorption. This could result in osteoporosis, and lack of healing of bone fractures. The data and model suggest that application of UTP could then be a therapeutic treatment. The data showing synergistic interactions between P2-receptor linked signaling and signaling linked to other receptors (i.e., PTH and PDGF) suggest that abnormal levels of extracellular nucleotides could lead to abnormal bone remodeling. Furthermore, the model developed above indicates that abnormal extracellular nucleotide levels could interfere with normal signaling between osteoblasts and osteoclasts, and between these cells and other types of neighboring cells, that might be based on pulses of ATP and UTP. This could result in a further contribution to abnormal bone remodeling.

A speculative model such as developed here is useful to point out the complexities of the bone remodeling system and the possible role of P2 receptors. Such a model can be used to design further experimental studies, which are clearly needed before these complexities can be understood.

References

Anderson CM, Parkinson FE (1997) Potential signaling roles for UTP and UDP: sources, regulation and release of uracil nucleotides. Trends Pharmacol Sci 18:387–392

Bowler WB, Birch MA, Gallagher JA, Bilbe G (1995) Identification and cloning of human P2u purinoceptor present in osteoclastoma, bone, and osteoblasts. J Bone Miner Res 10:1137–1145

Bowler WB, Littlewood-Evans A, Bilbe G, Gallagher JA, Dixon CJ (1998) $P2Y_2$ receptors are expressed by human osteoclasts of giant tumor but do not mediate ATP-induced bone resorption. Bone 22:195–200

Burnstock G (1997) The past, present and future of purine nucleotides as signaling molecules, Neuropharmacol 36:1127–1139

Chiozzi P, Sanz JM, Ferrari D, Falzoni S, Aleotti A, Buell GN, Collo G, Di Virgilio F (1997) Spontaneous cell fusion in macrophage cultures expressing high levels of the $P2Z/P2X_7$ receptor. J Cell Biol 138:697–706

Collo G, Neidhart S, Kawashima E, Kosco-Vilbois M, North RA, Buell G (1997) Tissue distribution of the $P2X_7$ receptor. Neuropharmacol 36:1277–1283

Ferrier J, Ward-Kesthely A, Homble F, Ross S (1987) Further analysis of spontaneous membrane potential activity and the hyper-polarizing response to parathyroid hormone in osteoblast-like cells. J Cell Physiol 130:344–351

Ferrier J, Kesthely A, Lagan E, Richter C (1991) An experimental test of a model for repeated Ca^{2+} spikes in osteoblastic cells. Biochem Cell Biol 69:433–441

Ferrier J, Kesthely A, Xia S-L (1992) Hormone responses of in vitro bone nodule cells: studies on changes of intracellular calcium and membrane potential in response to parathyroid hormone and calcitonin. Bone Miner 19:103–116

Ferrier J, Yu H (1996) Nuclear versus perinuclear and cytoplasmic calcium in osteoclasts. Cell Calcium 20:381–388

Fluhmann B, Zimmermann U, Muff R, Bilbe G, Fischer JA, Born W (1998) Parathyroid hormone responses of cyclic AMP-, serum- and phorbol ester- responsive

reporter genes in osteoblast-like UMR-106 cells. Mol Cell Endocrinol 139:89–98
Gallinaro BJ, Reimer WJ, Dixon SJ (1995) Activation of protein kinase C inhibits ATP-induced $[Ca^{2+}]_i$ elevation in rat osteoblastic cells: selective effects on P2y and P2u signaling pathways. J Cell Physiol 162:305–314
Greenfield EM, Horowitz MC, Lavish SA (1996) Stimulation by parathyroid hormone of interleukin-6 and leukemia inhibitory factor expression in osteoblasts is an immediate-early gene response induced by cAMP signal transduction. J Biol Chem 271:10984–10989
Hill PA, Reynolds JJ, Meikle MC (1995) Osteoblasts mediate insulin-like growth factor-I and -II stimulation of osteoclast formation and function. Endocrinol 136:124–131
Jones SJ, Gray C, Boyde A, Burnstock G (1997) Purinergic transmitters inhibit bone formation by cultured osteoblasts. Bone 21:393–399
Kaplan AD, Reimer WJ, Feldman RD, Dixon SJ (1995) Extracellular nucleotides potentiate the cytosolic $[Ca^{2+}]_i$ but not cyclic adenosine 3′,5′-monophosphate response to parathyroid hormone in rat osteoblastic cells. Endocrinol 136:1674–1685
Kumagai H, Sacktor B, Filburn CR (1991) Purinergic regulation of cytosolic calcium and phosphoinositide metabolism in rat osteoblast-like osteosarcoma cells. J Bone Miner Res 6:697–707
Lerner UH, Sahlberg K, Fredholm BB (1987) Characterization of adenosine receptors in bone. Studies on the effect of adenosine analogues on cyclic AMP formation and bone resorption in cultured mouse calvaria. Acta Physiol Scand 131:287–296
Liu F, Malaval L, Aubin JE (1997) The mature osteoblast phenotype is characterized by extensive plasticity. Exp Cell Res 232:97–105
Luo LC (1995) Receptors for extracellular ATP and other nucleotides in osteoblastic cells. MSc Thesis, University of Toronto
Luo LC, Yu H, Ferrier J (1997) Differential purinergic receptor signaling in osteoclasts and osteoblastic cells. Cell Signal 9:603–607
Maier R, Glatz A, Mosbacher J, Bilbe G (1997) Cloning of $P2Y_6$ cDNAs and identification of a pseudogene: comparison of P2Y receptor subtype expression in bone and brain tissues. Biochem Biophys Res Commun 240:298–302
Morrison MS, Turin L, King BF, Burnstock G, Arnett TR (1998) ATP is a potent stimulator of the activation and formation of rodent osteoclasts. J Physiol 511:495–500
Onyia JE, Libermann TA, Bidwell J, Arnold D, Tu Y, McClelland P, Hock JM (1997) Parathyroid (1–34)-mediated interleukin-6 induction. J Cell Biochem 67:265–274
Owens JM, Gallagher AC, Chambers TJ (1996) Bone cells required for osteoclastic resorption but not for osteoclastic differentiation. Biochem Biophys Res Commun 222:225–229
Palmer TM, Gettys TW, Jacobson KA, Stiles GL (1994) Desensitization of the canine A2a adenosine receptor: delineation of multiple processes. Mol J Pharmacol 45:1082–1094
Partridge NC, Bloch SR, Pearman AT (1994) Signal transduction pathways mediating parathyroid hormone regulation of osteoblastic gene expression. J Cell Biochem 55:321–327
Reimer WJ, Dixon SJ (1992) Extracellular nucleotides elevate $[Ca^{2+}]i$ in rat osteoblastic cells by interaction with two receptor subtypes. Am J Physiol 263:C1040–C1048
Shimegi S (1996) ATP and adenosine act as a mitogen for osteoblast-like cells (MC3T3-E1). Calcif Tiss Int 58:109–113
Shimegi S (1998) Mitogenic action of adenosine on osteoblast-like cells, MC3T3-E1. Calcif Tiss Int 62:418–425
Sistare FD, Rosenzweig BA, Contrera JG, Jordan B (1994) Separate P2t and P2u purinergic receptors with similar second messenger signaling pathways in UMR-106 osteoblasts. J Pharmacol Exp Ther 269:1049–1061
Sistare FD, Rosenzweig BA, Contrera JG (1995) P2 purinergic receptors potentiate parathyroid hormone receptor-mediated increases in intracellular calcium and inositol trisphosphate in UMR-106 osteoblasts. Endocrinol 136:4489–4497

Suda T, Nakamura I, Jimi E, Takahashi N (1997) Regulation of osteoclast function. J Bone Miner Res 12:869–879
Suzuki A, Kotoyori J, Oiso Y, Kozawa O (1993) Prostaglandin E2 is a potential mediator of extracellular ATP action in osteoblast-like cells. Cell Adhesion Commun 1:113–118
Wimalawansa SJ (1997) Amylin, calcitonin gene-related peptide, calcitonin, and adrenomedullin: a peptide superfamily. Crit Rev Neurobiol 11:167–239
Xia S-L, Ferrier J (1996) Localized calcium signaling in multinucleated osteoclasts. J Cell Physiol 167:148–155
Yu H, Ferrier J (1993a) Osteoblast-like cells have a variable mixed population of purino/nucleotide receptors. FEBS Lett 328:209–214
Yu H, Ferrier J (1993b) ATP induces an intracellular calcium pulse in osteoclasts. Biochem Biophys Res Commun 191:357–363
Yu H, Ferrier J (1994) Mechanisms of ATP-induced Ca^{2+} signaling in osteoclasts. Cell Signal 6:905–914
Yu H, Ferrier J (1995) Osteoclast ATP receptor activation leads to a transient decrease in intracellular pH. J Cell Sci 108:3051–3058
Yu H (1996) P2 purinoceptor-linked Ca^{2+} signaling and pH changes in osteoclasts. PhD Thesis, University of Toronto

CHAPTER 29

Clinical Opportunities in Purinergic Neuromodulation

M. WILLIAMS

A. Introduction

With the characterization of the functional role of the purines, ATP, ADP, AMP, and adenosine, and the pyrimidine nucleotides, UTP and UDP, as extracellular messengers in both normal and diseased tissues (RALEVIC and BURNSTOCK 1998), an implicit challenge has become the use of the knowledge derived from this research over the past 25 years to gain a better understanding of the role of the purinergic receptor family in disease pathophysiology. This can then be used to discover novel therapeutic agents to redress or alter the cellular dysfunction associated with various disease states.

The process of discovering and developing drugs for human use is a lengthy, complex, and costly one (WILLIAMS 1996; LUTZ and KENAKIN 1999; VRETTOS and STEINER 1999) combining scientific feasibility and commercial attractiveness with a high degree of serendipity. Based on historical data, only 1 or 2 molecules in a cohort of 10,000 that are synthesized actually become drugs and only 1 or 2 of 10 that are advanced to human testing actually reach the market place. Thus greater than 80% of the effort focused on the discovery and development of new drugs does not reach fruition for a variety of scientific reasons that include: lack of human efficacy, unacceptable side effects, and poor pharmacokinetic characteristics, as well as commercial considerations that can change markedly in the 8–12 years it typically takes to bring a compound from the bench to market.

The process of drug discovery, especially in the current environment of enabling technologies like high throughput screening (HTS), combinatorial chemistry, library diversity, and bioinformatics (KENNY et al. 1998; TRIGGLE 1998; LUTZ and KENAKIN 1999), is focused on discrete molecular targets – receptors, ion channels or enzymes – which, because of a previously established association with therapeutically used agents, a known association with a disease state, novelty and localization in a target tissue, and/or genomic targeting are thought to be involved in some aspect of the disease process.

Inevitably, while a target of this type is of tremendous interest at the basic research level, the clinical viability of compounds interacting with the selected target is dependent on the degree of intrinsic activity/efficacy in activating (agonist) or inhibiting/blocking (antagonist) of the transduction processes associated with the target and also the associated side effect liabilities. Thus while many compounds have attractive efficacy profiles, their utility as potential drugs is limited by side effect liabilities that can extend from transitory skin rashes to life threatening prolongation of the cardiac QTc wave. Other factors impacting the viability of a target are tolerance with repeated administration, overt mechanism-based toxicities, the (in)ability to identify lead molecules with appropriate pharmacokinetic properties, and the absolute relevance of the identified molecular target in the disease pathway. Inevitably, while the scientist is intrigued by a novel mechanism, the interest of the patient is limited to whether the compound is effective in treating their disease/disorder with a minimum of side effect liabilities.

Drug discovery activities in the purine area initially focused on adenosine itself as a potential antihypertensive agent (HONEY et al 1930; JEZER et al. 1930), an approach that was precluded by the short half-life of the purine nucleoside. Medicinal chemistry efforts in the late 1970s facilitated by the availability of functional biochemical assays focused on elaborating on the structural diversity of adenosine agonists (JACOBSON and KNUTSEN, Chap. 6, first volume). While NECA was claimed as having utility as a rodent poison in initial patents (a use where a therapeutic index was probably not an issue), much of the early research in the pharmaceutical industry focused on the potential use of P1 receptor agonists as antihypertensive agents based on the by then extensive literature, extending over half a century, on the cardiovascular actions of the purine nucleoside (OLSSON and PEARSON 1990). The characterization of caffeine, the most widely consumed, legally available psychoactive drug (NEHLIG 1998; FREDHOLM et al. 1999), as an adenosine antagonist (SATTIN and RALL 1970) led to an additional focus for P1 receptor agonists and antagonists as CNS active agents.

Over the past two decades, many studies have provided evidence for an involvement of purines in the actions of a variety of CNS active drugs that include antipsychotics, antidepressants, anxiolytics, and cognition enhancers. The effects of known CNS drugs, representative of these various therapeutic classes, were examined for their ability to modulate adenosine-mediated responses in the CNS or, alternatively, focused on the study of the effects of various P1 ligands, agonists, or antagonists, on the effects of such prototypic CNS agents. In many instances only single, somewhat high, concentrations of a single compound, or a limited number of compounds, were used to generalize to a complete class of psychotherapeutic agents, frequently with no negative controls, thus limiting the value of the data (WILLIAMS and BURNSTOCK 1997).

The availability of transgenic mice with knockouts of the various P1 and P2 receptors will however provide additional information on the phenotype defined by a particular receptor (NYCE 1999). For instance, the A_{2A} receptor mouse knockout is more aggressive and has higher anxiety levels than controls (LEDENT et al. 1997).

To a major extent, given the lack of robust tools akin to the methylxanthine antagonists in the P1 area and also to the somewhat 'soft' definition of receptors based on the rank order potency of a limited series of agonists related to ATP, P2 receptors have been treated with low priority as drug discovery targets within the pharmaceutical industry. With the exception of the elegant strategy in identifying selective platelet $\mathbf{P_{2T}}$/$P2Y_{12}$ receptor antagonists for use as novel antithrombotic agents (HOURANI, Chap. 20, this volume), that led to the identification of ARL 67085 (HUMPHRIES et al. 1996) and AR-C 69331-MX (INGALL et al. 1999), it was not until the P2 receptor families (P2X and P2Y) were cloned and localized to discrete tissue systems, e.g., $P2Y_2$ receptors in lung tissue for the treatment of cystic fibrosis (WEGNER, Chap. 23, this volume), $P2X_3$ receptors in sensory pathways as novel analgesic targets (BURNSTOCK 1996; SALTER and SOLLEVI, Chap. 13, first volume), that there was renewed interest in their potential as drug discovery targets. Interestingly, much of this effort has originated from venture capital financed biotechnology companies working with leading academic researchers in the field of purinergic research (LINDEN and WILLIAMS 2001).

Drug discovery efforts for purines have, to date, been almost exclusively focused on developing P1 receptor ligands for the treatment of cardiovascular and CNS disease states. Strategically, given the plethora of highly effective treatments for hypertension (β-adrenoceptor blockers, angiotensin converting enzyme (ACE) inhibitors, angiotensin-II (A-II) antagonists, etc.), coupled with the inherent difficulty in developing novel psychotherapeutic agents, irrespective of the molecular target, it is not surprising, in retrospect, that few agents acting at P1 receptors – beyond adenosine itself (BELLARDINELLI et al. 1995) – have been approved for human use.

B. Purinergic Therapeutics

Compounds modulating P1 and P2 receptor function can be divided into three distinct classes:

1. Conventional agonist, partial agonist, and antagonist ligands
2. Allosteric modulators of receptor function
3. Modulators of the endogenous systems that regulate the extracellular availability of ATP, adenosine, UTP, and their respective nucleotides

This latter group includes the various ecto-ATPases (E-NTPases) that catalyze the degradation of purine nucleotides (ZIMMERMANN and BRAUN 1999; ZIMMERMANN, Chap. 8, first volume), adenosine deaminase (ADA), adenosine kinase (AK), and the bidirectional transporter systems that remove adenosine from the extracellular environment (GEIGER et al. 1997; BALDWIN et al. 1999).

Based on successful precedents with compounds like the benzodiazepines which are allosteric or indirect modulators of the $GABA_A$ receptor, partial agonists, allosteric modulators, and novel modulators of ATP and adenosine metabolism may prove to be clinically useful agents with improved therapeutic indices compared to the directly acting compounds evaluated to date (IJZERMAN and VAN DER WENDEN 1997).

Medicinal chemistry efforts in the area of P1 ligands have been based on modification of the purine nucleoside pharmacophore to develop potent, receptor selective agonists and on the theophylline/caffeine alkylxanthine nucleus (DALY 1982) and a number of empirically identified pharmacophores to develop antagonists (JACOBSON and VAN RHEE 1997).

Ligands for the P2 receptor family have been less facile in their development. Stable, nonhydrolyzable analogs of ATP have been the focus of P2 agonist ligands (CUSACK and HOURANI 1990) while antagonists have, again, been empirically identified in a variety of tissue systems with no equivalent of the xanthine pharmacophore to aid in defining receptor function and structure activity relationships. The nearest approximation has been the truncation of suramin, a weak antagonist of P2 receptors with a multitude of other activities (BHAGWAT and WILLIAMS 1997) that has led to NF023 and XAMR0721 (JACOBSON and VAN RHEE 1997). Allosteric modulators (KN 62) and potent, nanomolar noncompetitive antagonists (TNP-ATP; JACOBSON and KNUTSEN, Chap. 6, first volume) represent newer ligands that potently modulate P2 receptor function. Minimal efforts have been expended on developing synthetic probes for the P2 receptor family with the majority of selective compounds in recent years emerging from the continuing efforts at the Molecular Recognition Section at the NIH (JACOBSON and WILLIAMS 2001).

I. Cardiovascular Disorders

Preclinical data (for review see OLSSON and PEARSON 1990) and limited human studies (SOLLEVI 1986; OWALL et al. 1988) have provided considerable evidence for the therapeutic potential of purinergic ligands in modulating the cardiovascular system and in the treatment of cardiovascular disorders. By reducing peripheral resistance, intravenous adenosine can be used to produce hypotension to control intraoperative hemorrhage and avoid hypertensive crisis, to reduce pain (see below and SALTER and SOLLEVI, Chap. 13, first volume), especially in the postoperative setting and to reduce the amount of anesthetic used during surgical procedures (FUKUNAGA 1997). The development of adenosine itself as a cardiac imaging agent, a diagnostic tool in assessing sinoatrial node function, and as an effective treatment for supraventicular tachycardia reflect a highly pragmatic and successful approach to the transition of animal findings to the human situation (BELLARDINELLI et al. 1995). CVT 3146 is a low affinity (Ki = 1.3 μmol/l), short acting A_{2A} receptor agonist under study for the treatment of cardiovascular disorders that may have pharmacodynamic properties similar to adenosine (ELZEIN et al. 2000).

Based on the demonstrated effects of adenosine and its nucleotides, purinergic targets for drug therapy include antagonists, the treatment of arrythmias (sinus bradycardia, atrioventricular node), and agonists for the treatment of ventricular conduction disturbances. Adenosine can act as a protective agent in ischemia, cardiac and cerebral, and in reperfusion injury. During ischemia, adenosine is produced from ATP and in ischemic preconditioning where brief periods of ischemia followed by reperfusion can protect cardiac myocytes from subsequent ischemic and myocardial infarction (MI), it has been shown that adenosine is the effector agent acting via A_1 and/or A_3 receptors (LINDEN 1994; LIU et al. 1997; PARSONS et al. 2000). The adenosine uptake blocker, dipyridamole in a major stroke prevention study in Europe involving 6602 patients with prior stroke or transient ischemic episodes, was found to reduce stroke recurrence both alone and in combination with aspirin (PICANO and ABBRACCHIO 1998). Adenosine antagonists may also be useful in the treatment of angina pectoris (OLSSON and PEARSON 1990).

ADP and synthetic purine nucleotide analogs like AR-C69931MX (INGALL et al. 1999) are highly effective antithrombic agents acting by blockade of the recently cloned platelet $\mathbf{P_{2T}}$/$P2Y_{12}$ receptor (HOLLOPETER et al. 2000; HOURANI, Chap. 20, this volume) with minimal reported side effect liabilities and proven superiority to platelet glycoprotein IIb/IIIa antagonists. CS-747 is a newer antithrombotic agent that appears to act via antagonism of the platelet $\mathbf{P_{2T}}$/$P2Y_{12}$ receptor (SUGIDACHI et al. 2000).

In $P2Y_1$ knockout mice, platelets require higher than normal ADP concentrations to aggregate and, when they do aggregate, do not undergo any shape change (FABRE et al. 1999; LEON et al. 1999). ADP can also inhibit platelet adenylate cyclase activity in $P2Y_1$ knockouts, indicating the presence of a second ADP-responsive P2 receptor (?P_{2T} receptor; LEON et al. 1999). $P2Y_1$ knockouts show increased bleeding time and are resistant to thromboembolism.

The "site and event specific" adenosine regulating agent, acadesine (AICA riboside; TSUCHIDA et al. 1994), a compound with modest effects on a number of the enzymes involved in regulating the removal of adenosine from the extracellular space, was shown to prevent the consequences of cardiac ischemia in animal models and delay the decay of preconditioning. In some, but not all, human studies, acadesine was found to reduce early cardiac death, MI, and adverse cardiac outcomes associated with coronary artery bypass graft (CABG) surgery, and a meta-analysis of five randomized, placebo-controlled, double-blind trials from 81 centers in North America and Europe did show an overall benefit from the infusion of acadesine (MANGANO 1997). Despite these findings, additional trials are still required for registration of this compound for human use.

Early attempts at developing novel P1 agonists beyond adenosine and its nucleotides for the treatment of cardiovascular disorders led to the identification of compounds like the nonselective P1 agonist, metrifudil, which entered the clinic in the early 1970s as a treatment for coronary heart disease (SCHAUUMANN and KUTSCHA 1973) and CGS 21680C, the first potent and selective A_{2A} agonist (HUTCHISON et al. 1989). The latter was targeted as a peripherally selective antihypertensive agent but never entered the clinic. A_{2A}

knockout mice showed increased blood pressure, heart rate, and platelet aggregation (LEDENT et al. 1997). The selective A_1 receptor agonist, SDZ WAG 994 (BELLARDINELLI et al. 1994) has potential as a treatment for cardiac arrhythmias. AMP 579 is a nonselective P1 agonist in development for the treatment of cardiac ischemia, inhibiting adherence-dependent superoxide radical generation, PMN degranulation, and preserving vascular endothelial function (MCVEY et al. 1999; ZHAO et al. 2000).

Several novel coronary vasodilators that produce their effects via P1 receptors including WRC 0470, CVT 3033, and DWH146e are currently being examined as cardiac imaging/diagnostic agents. NNC 21-0136 is an unusual A_1 receptor agonist which has reduced unwanted cardiovascular pharmacodynamics as compared to other A_1 receptor agonists when used as a neuroprotective agent (KNUTSEN et al. 1998).

The xanthine-based P1 antagonists, BG 9719, FK 453, KW3905, and related structures are potent renal vasodilators (JACKSON, Chap. 17, this volume) currently being evaluated for their use in the treatment of renal dysfunction and congestive heart failure. However, the effects of P1 antagonists on renin release in chronic renal failure and their augmentation of angiotensin-II glomerular hypertension has raised concerns in regard to their therapeutic utility. Nonetheless, BG 9719 has shown positive clinical data in congestive heart failure (WOLFF et al. 1998) and a backup to this P1 antagonist is being advanced to the clinic.

II. Respiratory/Pulmonary Function

The use of theophylline as an antiasthmatic agent and action as an albeit weak antagonist of P1 receptors has focused attention on the development of novel P1 antagonists as asthmatic medications (BIAGGIONI and FEOKTISTOV, Chap. 22, this volume). Adenosine can produce bronchoconstriction in atopic and nonatopic asthmatic patients but in normals (CUSHLEY et al. 1983) potentiates mast cell mediator release (MARQUARDT et al. 1978). The effects of adenosine on bronchoconstriction appear to be indirect, involving histamine release from mast cells. In addition, adenosine levels are increased in bronchoalveolar lavage fluid (BAL) in asthmatics as compared to controls (DRIVER et al. 1993) while isolated bronchii from asthmatics were more sensitive to the contractile effects of adenosine than tissue from nonasthmatics (BJORCK et al. 1993). Finally, the adenosine uptake blocker, dipyridamole, can elicit severe bronchospasm in asthmatics (EAGLE and BOUCHER 1989).

Despite these data, the role of adenosine as a key mediator in the pathophysiology of asthma remains highly controversial (BARNES et al. 1998) with the antiasthmatic actions of theophylline being ascribed to its ability to inhibit phosphodiesterase activity. The development of novel, potent, and selective P1 receptor antagonists would provide a means to test the adenosine hypothesis. However, CGS 15943, a nonxanthine P1 antagonist targeted for use in asthma in the late 1980s, was found to be a highly potent skin irritant, fortunately during its formulation as an aerosol for delivery to patients rather than in the clinic. Enprofylline, a weak A_{2B} receptor antagonist, has been shown to be effective in the treatment of asthma (FEOKISTOV et al. 1998) and JW-VI-08 is a nanomolar xanthine-based A_{2B} antagonist that blocks *ex vivo* NECA-induced mast cell activation in BAL from asthmatics (BIAGGIONI et al. 2000).

Of additional interest regarding the role of P1 receptors in asthma is the fact that EPI 2010, a respirable antisense oligonucleotide to the initiation codon of the human A_1 receptor, attenuated sensitivity to adenosine, house mite dust, and histamine in a rabbit model of allergic asthma and was also effective in the *Ascaris*-sensitized primate model of asthma (NYCE 1999). Adenosine and related P1 agonists may also have potential in the treatment of adult respiratory distress syndrome (BIAGGIONI and FEOKTISTOV, Chap. 22, this volume).

Theophylline is used in the treatment of COPD (chronic obstructive pulmonary disease; WEGNER, Chap. 23, this volume) where obstruction of airway flow due to chronic bronchitis or emphysema is potentially life-threatening. In addition to theophylline and steroid administration, lung volume reduction surgery and lung transplantation are current approaches to the treatment of COPD.

Cystic fibrosis (CF) is an autosomal recessive disease resulting from a gene mutation on chromosome 7 involved in the production of CF transmembrane regulator (CFTR) that functions as an epithelial cyclic AMP-regulated chloride channel (STUTTS and BOUCHER 1999). Mutations in the CFTR affect epithelial function, not only in the lung but also in intestinal epithelia, sweat duct, and pancreas.

In lung epithelia, CFTR mutations result in defects in Na^+ and Cl^- transport leading to dehydrated secretions that are poorly cleared and are associated with chronic infections. Chronic sinusitis, nasal obstruction, and rhinorrhea reflect the upper respiratory tract consequences of CF while in the lower respiratory tract, cough, increased sputum volume, and impaired pulmonary function (decreased forced vital capacity (FVC) and forced expiratory volume; FEV_1) occur leading to chronic airway destruction and bronchiolitis resulting in respiratory failure.

In both CF and chronic bronchitis, reduced mucus clearance and hydration and reduced ciliary beat frequency reduce the function of the lung epithelium and the ability of the lung to defend itself from bacterial and viral infection (WEGNER, Chap. 23, this volume). In CF, inhaled nucleotides, including the $P2Y_2$ receptor agonist, UTP, activate chloride secretion in the airway epithelium (KNOWLES et al. 1991; BENNETT et al. 1996). INS 365 and other stable analogs of UTP (e.g., INS 37217; YERXA 2000) are currently in development for the treatment of COPD and, via their effects on ciliary beat function (BRAIMAN eta l. 2000), may also represent adjunctive therapy to enhance the effectiveness of antibiotics used in the treatment of respiratory infections (BOUCHER 2000). Further confirmation of the role of the $P2Y_2$ receptor in chloride transport has come from studies in $P2Y_2$ receptor knockout mice (CRESSMAN et al. 1999).

ATP can activate pulmonary vagal C-fiber terminals (PELLEG and HURT 1996) eliciting a neurogenic inflammatory response that results in transient bonchoconstriction (KATCHANOV et al. 1998) and increased lung resistance, the nucleotide being more potent than adenosine in producing this effect. ATP can also potentiate anti-IgE-induced mast cell histamine release (SCHULMAN et al. 1998). Antagonists selective for the P2 receptors present on human eosinophils and lung mast cells represent a novel approach to the treatment of asthma.

III. CNS Disorders

A large body of preclinical data exists studying the effects of adenosine, ATP, and UTP on central and peripheral nervous system function.

1. Stroke/Ischemia

Hypoxia and focal ischemia lead to marked increases in extracellular adenosine levels (RUDOLPHI et al. 1992) and in animal models, P1 receptor agonists like CHA reduce stroke-related cell death and hippocampal neurodegeneration while P1 antagonists exacerbate ischemic brain damage by enhancing glutamate release. The neuroprotective effects of adenosine are mediated by several P1 receptors: A_1 receptor activation stabilizes neuronal membrane potential, inhibits neuronal excitability, and reduces release of the excitatory amino acid, glutamate (VON LUBITZ 1997; KNUTSEN et al. 1998) thus preventing initiation of the stroke cascade (DIRNAGEL et al. 1999). P1 receptor agonists can hyperpolarize astrocyte membranes thus limiting extracellular glutamate and potassium accumulation and also affect local cerebral blood flow and local inflammatory responses, e.g., platelet aggregation, neutrophil recruitment, and adhesion acting via the A_{2A} receptor (FIRESTEIN 1996). Selective A_3 receptor agonists have biphasic effects on cell survival. At nanomolar concentrations, these agonists are neuroprotective and inhibit the process of apoptosis but at micromolar concentrations are neurotoxic (JACOBSON 1998). Selective A_1 and A_3 receptor agonists are neuroprotectants in animal models of stroke (KNUTSEN et al. 1998). A_3 receptor agonists like NNC 53-0055 can inhibit production of the cytotoxic cytokine, TNFα which is a key mediator in cell damage associated with cerebral ischemia (FIRESTEIN 1996; FEUERSTEIN et al. 1997). Paradoxically however, substituted adenosine analogs with weak P1 receptor activity, e.g., NNC 53-0017, are most effective in blocking TNFα production (KNUTSEN et al. 1998).

P2 receptors are also involved in events subsequent to stroke-related ischemia in brain tissue. mRNA for the $P2X_7$ receptor is upregulated on microglia in the ischemic penumbra 24h after middle cerebral artery occlusion in the rat (COLLO et al. 1997) indicating that cytolytic pore formation and inflammatory cytokine release (DI VIRGILIO et al. 1999; DI VIRGILIO et al., Chap. 26, this volume) may be key events in the neural trauma and neurodegeneration associated with stroke. $P2X_7$ receptor knockout mice have been developed but there is currently limited knowledge available on their phenotype (SIKORA et al. 1999).

2. Epilepsy

In epileptic patients with spontaneous onset seizures (DURING and SPENCER 1992), rapid and marked increases in CNS adenosine concentrations occur and similar results have been derived in animal models of epilepsy (KNUTSEN and MURRAY 1997). Chemically – and electrically – evoked seizure activity can be reduced by exogenous adenosine and other P1 agonists (KNUTSEN and MURRAY 1997) acting via A_1 receptors, effects that are blocked by methylxanthines. Adenosine has thus been described as an endogenous anticonvulsant (YOUNG and DRAGUNOW 1994). In terms of developing adenosine agonists as novel anticonvulsant therapies, the cardiovascular side effect liabilities and sedative effects of the nucleosides need to be put into the context of the degree of unmet medical need with existing anticonvulsant agents like phenytoin, carbamazepine, valproate, ethosuximide, phenobarbital, primidone, and gabatril, and the high barrier to introducing new therapies for this CNS disorder.

3. Neurodegeneration – Alzheimer's and Parkinson's Disease

The precise causes and etiology of neurodegenerative diseases like Alzheimer's (AD) and Parkinson's (PD) disease are complex with considerable evidence for multiple genetic association components and environmental factors. The CNS stimulant properties of the xanthine-based adenosine antagonists including caffeine, theophylline, propentofylline, and BIIP 20 (the active, S-enantiomer of KFM-19) and their ability to improve cognitive performance in animals, including primates, and humans (FREDHOLM et al. 1999) by blocking the actions of endogenous adenosine, has led to their clinical evaluation in cognitive disorders including AD, but none have shown sufficiently robust efficacy for their continuation.

Purinergic ligands can however modulate trophic factor production in nervous tissue increasing neuronal viability and regeneration (NEARY et al. 1996; RATHBONE et al. 1998, 2000; NEARY and ABBRACCHIO, Chap. 11, first volume [HEP 151/I]). Nerve growth factor (NGF) exerts a tonic cell death suppressing signal, and its withdrawal, or blockade of its actions, leads to neuronal death.

Polypeptide growth factors linked to receptor tyrosine kinase pathways, e.g., fibroblast growth factors, epidermal growth factor, and platelet-derived growth factor, are increased as the result of neural injury (NEARY et al. 1996) and ATP can act in combination with these factors to stimulate astrocyte proliferation resulting in reactive astrogliosis, a hypertrophic/hyperplastic response typically associated with brain trauma, stroke/ischemia, seizures, and various neurodegenerative disorders. ATP and GTP induce trophic factor (NGF, NT-3, FGF) synthesis in astrocytes and neurons although the effects of the latter nucleotide are not consistent with any defined P2 receptor profile. During reactive astrogliosis, astrocytes undergo process elongation expressing GFAP (glial fibrillary acidic protein), an astrocyte specific intermediate filament protein with an increase in astroglial cellular proliferation. ATP increases GFAP and AP-1 complex formation in astrocytes mimicking the effects of bFGF.

The hypoxanthine analog, neotrofin (AIT-082), has similar trophic actions to GTP, upregulating neurotrophin production and can enhance working memory and restore age-induced memory deficits in mice (RATHBONE et al. 2000). Neotrofin has shown positive effects in limited Phase II trials in AD patients and is advancing to further trials.

The role of purines in striatal function was first established over 30 years ago when FUXE and UNGERSTEDT (1976) showed that methylxanthines, like caffeine, stimulated rotational behavior and potentiated the effects of dopamine (DA) agonists in rats with unilateral striatal lesions, a model of PD. The anatomical links between dopamine and adenosine systems in the striatum, nucleus accumbens, and olfactory tubercule have been extensively documented (FERRE et al. 1997; SVENNINGSSON et al. 1999) as have their functional interactions in controlling movement via direct (cortical activating) and indirect (cortical inhibiting) striatal dopaminergic pathways that provide a tonic regulation of normal motor activity. From these studies, it appears that striatal adenosine A_{2A} receptors play a key role in neurological disorders that involve basal ganglia dysfunction, e.g., PD. Given intrastriatally, the selective A_{2A} agonist, CGS 21680, can attenuate rotational behavior produced by DA agonists in unilaterally lesioned rats. The interactions between striatal dopaminergic and purinergic systems in striatum has suggested that dopaminergic dysfunction can be indirectly ameliorated by adenosine receptor modulation. Selective adenosine A_{2A} receptor antagonists like KF 17837, KW 6002, and SCH 58261 have positive effects in MPTP-lesioned marmosets and cynomolgus monkeys, well characterized animal models of PD, enhancing the effects of the gold standard therapy for PD, the dopamine precursor, L-dopa (RICHARDSON et al. 1997; KANDA et al. 1998). KW-6002 is currently being advanced to Phase III clinical trials based on positive Phase II data in PD patients. Additional support for a role of P1 antagonists as a therapeutic approach to PD has come from a thirty-year longitudinal study of 8004 Japanese. American males in the Honolulu Heart Program (WEBSTER et al. 2000) of whom 102 developed PD. The age adjusted incidence of PD was inversely associated with caffeinated coffee consumption. Thus the incidence of PD was 10.4/10,000 person years in individuals who did not consume coffee and 1.9/10,000 person years in coffee drinkers.

Since P1 antagonists facilitate striatal dopaminergic function, agonists, based on the dopaminergic hypothesis of schizophrenia may be anticipated to have antipsychotic-like activity and preclinical studies would support this hypothesis (MARTIN et al. 1993; HAUBER and KOCH 1997). However, CI-936, a selective A_{2A} receptor agonist that entered clinical trials over a decade ago as a novel antipsychotic, was discontinued for unstated reasons. In a preliminary clinical study, the adenosine uptake blocker, dipyridamole, potentiated the effects of haloperidol in decreasing the positive and general symptoms of schizophrenia (AKHONDZADEH et al. 2000).

4. Sleep

The hypnotic and sedative effects of adenosine have been well established. Direct adenosine administration into the brain elicits an EEG profile similar to that seen in deep sleep, an increase in REM sleep with a reduction in REM sleep latency resulting in an increase in total sleep, and microdialysis studies show that extracellular adenosine concentrations increase in basal forebrain in direct proportion to periods of sustained wakefulness and decline during sleep, indicating that adenosine functions as an endogenous sleep regulator (PORKKA-HEISKANEN et al. 1997). In contrast, caffeine can suppress REM sleep and decrease total sleep time (FREDHOLM et al. 1999). Infusion of the A_{2A} agonist, CGS 21680 into the subarachnoid space associated with the ventral surface of the rostral basal forebrain, an area designated as the prostaglandin D_2-sensitive sleep-promoting zone, increased slow wave (SWS) and paradoxical (PS) sleep, effects that were blocked by the A_{2A} antagonist, KF 17837 (SATOH et al. 1996) while the A_1 selective agonist, CHA, suppressed SWS and PS prior to eliciting an increase in SWS.

The market potential for a new hypnotic agent acting via a P1 receptor mechanism given concerns regarding CV side effects is probably limited, especially given the success in the marketplace of newer benzodiazepines like zolpidem that are also used to avoid the consequences of jet lag. A slow release form of caffeine, in addition to its CNS stimulant activity, can reverse the effects of 36h sleep deprivation in humans (PATAT et al. 2000).

Purinergic receptor-mediated mechanisms are involved in the control of respiratory function (MCQUEEN and RIBERIO 1986; MCQUEEN et al. 1998) and site specific gene ablation using antisense to the rat A_1 receptor can attenuate baroreceptor reflex responses (NYCE 1999). This purinergic control of respiratory function led to the suggestion that adenosine agonists might be useful in the treatment of sleep apnea (CARLEY and RADULOVACKI 1997). Subsequent studies however failed to demonstrate a physiological role for adenosine in sleep apnea expression (CARLEY and RADULOVACKI 1999).

5. Pain and Anesthesia

The understanding of the dynamic nature of pain perception and its intensity has increased exponentially in the past decade (WILLIAMS et al. 1999; WOLFF and SALTER 2000). The role of purines in pain perception is well established (BURNSTOCK 1996, 2000; SAWYNOK 1999; SALTER and SOLLEVI, Chap. 13, first volume) with both P1 agonists and P2X antagonists representing novel approaches to pain control (BURNSTOCK 1996).

Adenosine, adenosine receptor agonists, and AK inhibitors are potent inhibitors of nociceptive processing in the brain and spinal cord. Given intrathecally, these agents have analgesic effects in a broad spectrum of animal models (e.g., mouse hot plate, mouse tail flick, rat formalin, mouse abdominal constriction, rat neuropathic pain models) that can blocked by systemic or intrathecal administration of adenosine antagonists (SAWYNOK 1999).

Adenosine A_1 receptor agonists modulate acutely-evoked and inflammation-evoked responses of spinal cord dorsal horn nociceptive neurons and inhibit pain responses evoked by spinal injection of substance P and the glutamate agonist, NMDA, the latter agent reflecting the role of glutamate as a key mediator of the abnormal hyperexcitability of spinal cord dorsal horn neurons (central sensitization or wind up) associated with clinical pain states. A_1 agonists inhibit the spinal cord release of glutamate and reduce CSF levels of substance P in rat. P1 receptor mechanisms in pain involve both pre- and post-synaptic effects on transmission from primary afferent fibers to neurons of the substantia gelatinosa of the spinal dorsal horn, reflecting both peripheral and supraspinal mechanisms (SAWYNOK 1999; SALTER and SOLLEVI, Chap. 13, first volume). CHA and NECA are 10- to 1000-fold more potent in inhibiting ACh-induced writhing in mice when administered i.c.v. than orally, indicating a supraspinal site of action The ability of adenosine to inhibit peripheral neurotransmitter release (MASINO and DUNWIDDIE, Chap. 9, first volume) and inflammatory processes (FIRESTEIN 1996) may block peripheral sensitization, a key feature of the pain resulting from tissue injury and inflammation.

Spinal administration of the A_1 agonist, R-PIA, to a neuropathic pain patient relieved allodynia without affecting normal sensory perception (KARSTEN and GORDH 1996) while adenosine infusion at doses without effect on the cardiovascular system improved pain symptoms, reducing spontaneous pain, ongoing hyperalgesia, and allodynia in patients with neuropathic pain. Low dose infusion of adenosine during surgery reduced volatile anesthetic requirement and that for postoperative opioid analgesia (FUKUNAGA 1997) while AK inhibitors like CP 3269 and ABT-702 are effective analgesic agents in animal pain models via effects that can be blocked by xanthine adenosine antagonists but not by opioid antagonists (JARVIS et al. 2000). Limitations to the AK approach have been discussed (ABBRACCHIO and WILLIAMS, Chap. 1, first volume).

ATP has opposite effects to adenosine in pain processing. When applied to sensory afferents, ATP produces neuronal hyperexcitability and perception of intense pain (BURNSTOCK 1996) effects are mediated via $P2X_3$ and $P2X_{2/3}$ receptors present on sensory afferents and in the spinal cord. The nucleotide also induces nociceptive responses at local sites of administration (BLAND-WARD and HUMPHREY 1998) and facilitates nociceptive responses to other noxious stimuli including substance P.

P2 receptor antagonists including suramin and PPADS, even while limited in their in vivo effects, reduce nociceptive responses in animal models of acute and persistent pain (BLAND-WARD and HUMPHREY 1998). ATP is released from a number of cell types (e.g., sympathetic nerves, endothelial cells, visceral smooth muscle) in response to trauma and $P2X_3$ receptor expression is upregulated in sensory afferents and spinal cord following damage to peripheral sensory fibers. $P2X_3$ receptor knockout mice have reduced nociceptive responses (COCKAYNE et al. 2000). Since the effects of ATP are opposite to those of adenosine, it is likely that the nociceptive effects of ATP can be autoregulated by adenosine production from the nucleotide.

6. Auditory and Ocular Function

P2 and P2 receptors are present in the vestibular system with novel $P2X_2$ receptor splice variants being present in rat and guinea pig cochlea (HOUSLEY, Chap. 12, first volume) and in the visual system where purines are involved in modulating neurotransmission processes as well as affecting retinal blood flow and intraocular pressure. $P2X_2$ receptors are present on the endolymphatic surface of the rat cochlear endothelium, an area associated with sound transduction (THORNE and HOUSLEY 1996). ATP can inhibit auditory nerve activity regulating cochlear mechanics and hair cell function.

P2Y receptors are found in the marginal cells of the *stria vascularis*, an area involved in generating the ionic and electrical gradients of the cochlea (HOUSLEY 1998). As in the lung, ATP regulates fluid homeostasis in the auditory apparatus and modulates hearing sensitivity and development. Perilymphatic ATP depresses the sound-evoked gross compound action potential of the auditory nerve and the distortion product otoacoustic emission, the latter a measure of the active transduction process of the outer hair cells (KUJAWA et al. 1984). P1 receptors are also present in the auditory system and cisplatin-induced ototoxicity is accompanied by upregulation of P1 receptors in the cochlea (FORD et al. 1997). The A_1 receptor agonist, R-PIA, can attenuate the deleterious effects of repeated exposure to high-intensity high frequency noise (HU et al. 1997), reducing hair cell damage via antioxidant effects.

The increased environmental pollution resulting from ambient noise, e.g., traffic, airplane engines, etc., in the environment and the ubiquitous use of headphones to listen to music, has led to an increased load on the auditory system. Hearing loss, a phenotype represented by more than 200 distinct syndromes, affects more than 70 million individuals with progressive age-associated hearing loss (presbycusis) being common. Hearing loss occurs as the result of genetic and environmental factors. The role of purinergic systems in auditory function suggests that P1 agonists may protect against hearing loss while P2 systems may have potential in the treatment of hearing disorders including Meniere's disease and tinnitus.

In the eye, adenosine can modulate GABAergic and cholinergic neurotransmission, has mitogenic properties, and can modulate intraocular pressure (HOUSLEY, Chap. 12, first volume). Retinal ischemia enhances adenosine release. A_1 receptors are localized to the inner retina with A_2-like receptors present in the retinal pigmented layer. ATP also functions as a neurotransmitter in the retina and both P2X and P2Y receptors are present in the eye. In the ocular mucosa, $P2Y_2$ receptor activation can increase salt, water, and mucus secretion and may represent a potential treatment for dry eye disease (YERXA 2000). In the retinal pigmented layer, $P2Y_2$ receptor activation promotes fluid absorption and experimentally may represent a novel treatment for retinal detachment. Glaucoma, diabetic retinopathy, and macular degeneration are additional disease targets that may be amenable to treatment with purinergic receptor agonists and antagonists (HOUSLEY, Chap. 12, first volume).

IV. Diabetes

P1 receptor agonists including GR 79236 and CPA lower glucose levels and antagonists improve glucose tolerance via effects on the pancreas and via effects on adipose tissue lipolysis (PETIT et al., Chap. 27, this volume) which may involve A_1 receptor-mediated effects on adipose tissue leptin production (OZECK et al. 2000). ATP can stimulate pancreatic insulin release via a glucose-dependent mechanism involving $P2Y_1$ receptors on pancreatic β-cells (LOUBATRIERES-MARIANI et al. 1997; PETIT et al., Chap. 27, this volume) reducing hyperglycemia. P2Y receptor agonists may thus have potential as antidiabetic agents. ATP, acting via a P2Y-like receptor, can redistribute the glucose transporters, GLUT1 and GLUT4, from the plasma membrane to microsomal membranes in cardiomyocytes (FISCHER et al. 1999) and can also modulate insulin secretion via intracellular interactions with ATP-sensitive potassium channels in islet β-cells (NICHOLS et al. 1996).

The potential role of P1 and P2 agonists as novel treatments for diabetes, especially Type II, has yet to be fully explored with bioavailable compounds that are selective for the receptors present in the pancreas and fat tissue.

V. Osteoporosis

P2X receptors are present on osteoclasts with P2Y receptors being present on both osteoblasts and osteoclasts (DIXON and SIMS 2000; FERRIER, Chap. 28, this volume). Extracellular ATP and adenosine can act as osteoblast mitogens potentiating the effects of growth factors on these bone cells (SHIMEGI 1996). ATP can be released in response to shear stress (BURNSTOCK 1999) and may function as mechanotransducer in skeletal tissue. Low concentrations of ATP, but not adenosine, stimulate osteoclast formation and their resorptive actions in vitro (MORRISON et al. 1998) and also inhibit osteoblast-dependent bone formation. The action of bisphosphonates like clodronate in the treatment of Paget's disease and tumor-induced osteolysis has been suggested to involve interaction with P2 receptors on osteoclasts (DIXON and SIMS 2000). P2 receptor modulators may thus have potential in the treatment of osteoporosis, rheumatoid arthritis, periodontitis, and osteopenia.

VI. Inflammation

Adenosine is released at sites of inflammation and exerts anti-inflammatory effects via multiple mechanisms involving all P1 receptor subtypes (FIRESTEIN 1996; MONTESINOS and CRONSTEIN, Chap. 24, this volume), inhibiting neutrophil rolling and adhesion to vascular endothelium, decreasing free radical production, and affecting endothelial cell permeability, reducing bradykinin- and histamine-induced vascular leakage. Adenosine inhibits production of the pro-inflammatory cytokine, TNFα, and suppresses TNFα mRNA expression and plasma levels in vivo (KNUTSEN et al. 1998). The purine nucleoside also inhibits the matrix metalloprotease, collagenase (MMP-1), but not TIMP-1 or stromelysin production and gene expression on synoviocytes (A_{2B} receptor). The AK inhibitor, GP 515 (FIRESTEIN et al. 1994) improved survival in a murine septic shock model and a rat model of bacterial peritonitis, suggesting that the anti-inflammatory actions of adenosine do not suppress normal immune responses to infection. AK inhibitors also decreased carrageenan-induced pleurisy and paw edema, and air pouch and dermal neutrophil accumulation after local injection of the inflammatory mediators. A_1 receptor agonists and AK inhibitors also inhibit pleural and peritoneal inflammation in rats (FIRESTEIN 1996).

Some of the anti-inflammatory actions of NSAIDs (e.g., aspirin, sodium salicylate, and sulfasalzine) appear to be independent of cyclooxygenase (COX) inhibition since their anti-inflammatory actions are maintained in COX-2 and NFκB knockout mice. Aspirin increases adenosine levels in mouse air pouch exudates 17-fold vs controls (224 nmol/l vs 13 nmol/l). The anti-inflammatory effects of the NSAID can be reversed by the A_{2A} receptor antagonist, DMPX, and adenosine deaminase, leading to the proposal that adenosine acts as an anti-inflammatory autacoid independent of COX-1, COX-2, or NFκB p105 (CRONSTEIN et al. 1999). However, in reactive gliosis, P2Y receptor activation results in the induction of COX-2, an effect reversed by the COX-2 inhibitor, NS-398 (BRAMBILLA et al. 1999).

VII. Wound Healing

Topical administration of adenosine and A_{2A} receptor agonists can promote wound healing in both normal and diabetic mice (MONTESINOS et al. 1997). This effect is dependent upon modulation of local inflammatory events (MONTESINOS and CRONSTEIN, Chap. 24, this volume) and tissue plasminogen activator (tPA) since application of the A_{2A} receptor agonist, CGS 21680, to mice with a tPA knockout as well as A_{2A} receptor knockouts showed a diminished effect of CGS 21680 on wound healing (DESAI et al. 2000). The presence of $P2X_5$ receptor immunostaining in stratified epithelium and a potential role for $P2X_7$ receptor -mediated apoptosis in skin sloughing (BURNSTOCK 1998) suggest that P2 receptor modulation may also have a benefit in wound healing.

Given the topical application of agents for wound healing, the side effects associated with the use of systemically administered P1 agonists can be avoided.

VIII. Infectious Diseases

ATP can induce cytolysis in macrophages infected with mycobacterium via $P2X_7$ receptor-mediated apoptotic and necrotic processes (LAMMAS et al. 1997). This novel antimicrobial activity of ATP suggested a potential utility for the purine nucleotide in the treatment of tuberculosis. However, additional studies in a $P2X_7$ receptor knockout mouse suggested that activation of the $P2X_7$ receptor is a contributory but not essential component of macrophage responses to ATP (SIKORA et al. 1999; DUBYAK, Chap. 25, this volume).

IX. Erectile and Reproductive Function

P1 receptors in canine penile cavernosal smooth muscle control blood flow, resulting in a dose-dependent increase in intercavernosal pressure and a full erection in response to adenosine (TAKAHASI et al. 1992). ATP has variable effects on cavernosal smooth muscle depending on the level of basal tone, relaxing corporal smooth muscle at high basal tension and contracting at low tension (WU et al. 1993). In diabetic males, adenosine and ATP show an enhanced sensitivity in penile tissue responses although, in diabetic rats, ATP-induced relaxation is decreased while the effects of adenosine, which are thought to be mediated via potassium channels, are enhanced (GÜR and ÖZTÜRK 2000). Given the tremendous interest in the treatment of erectile dysfunction as a life style disease in diabetics and in the elderly (MORELAND et al. 2000), purines may represent a novel approach, especially if they can be used topically.

Deletion of the $P2X_1$ receptor gene in male mice results in an approximately 90% reduction in fertility with no change in copulatory performance, the amount of sperm in the ejaculate being decreased as a result of a 60% reduction in the contraction sensitivity of the vas deferens to sympathetic nerve stimulation. (MULRYAN et al. 1999). While this suggests that selective $P2X_1$ receptor antagonists may represent novel, nonhormonal male contraceptives, the need to attain an absolute elimination of sperm rather than a 90% reduction may limit this as a reliable approach to birth control.

X. Neurourology

Detrusor malfunction due to age, childbirth, or exercise stress can lead to urge urinary incontinence (UUI), a major health problem in the aging female population. The urinary bladder is controlled by both sympathetic and parasympathetic nervous system input. ATP mimics the effects of parasympathetic stimulation resulting in bladder contraction (Burnstock et al. 1978; Dean and Downie 1978; O'Reilly et al. 1999; Burnstock, Chap. 13, first volume) via activation of P2X receptors present in the smooth muscle of the urinary bladder detrusor muscle involved in bladder emptying (Chancellor et al. 1992).

Micturition involves urethral relaxation with ATP functioning as a co-transmitter with nitric oxide (NO). NO mediates the first stage of relaxation (Pinna et al. 1998) and ATP the second, acting via P2 receptors present in the bladder urothelium and serosal layers (Burnstock, Chap. 15, first volume). ATP release occurs in rabbit bladder as a result of the hydrostatic pressure changes associated with bladder filling (Yoshimura and de Groat 1997). While the muscarinic receptors through which antagonists like oxybutynine and tolteridine produce their beneficial effects in UUI mediate 15% of rat urinary bladder neurogenic contraction, another 50% is mediated by P2X receptor mechanisms (Hashimoto and Kokubun 1995), suggesting that bladder selective P2X receptor antagonists may represent improvements over muscarinics in the treatment of UUI in both terms of efficacy and side effect liabilities (dry mouth etc.) provided they do not produce their own mechanism based side effects. Purinergic receptor dysfunction may also be involved in the etiology of interstitial cystitis and diabetic bladder dysfunction (Burnstock, Chap. 15, first volume).

XI. Cancer

ATP given i.p. doubled hepatic ATP pools and prevented weight loss in tumor-bearing mice (Rapaport and Fontaine 1989), an effect variously attributed to inhibition of gluconeogenesis from lactate via the Cori cycle, inhibition of the acute-phase response, and decreased production of the proinflammatory cytokines, IL-1 and IL-6 (Wang et al. 1992). The latter play a key role in the weight loss associated with cancer (Ottery et al. 1998). In a nonrandomized clinical trial, infusion of ATP for 96 h at 28-day intervals at doses of 50 μg/kg per min to patients with advanced non-small-cell lung cancer (NSCLC) increased ATP pools in red blood cells, an effect associated with an inhibition of weight loss, stabilization of the Karnofsky performance status, a measure of the impairment associated with cancer, reduced cachexia, and improved survival (Rapaport 1997). A subsequent randomized trial (Agteresch et al. 2000) in patients with advanced (stage IIIB or IV) NSCLC where ATP was given via intravenous infusion for 30 h at 2–4 week periods and outcome parameters assessed every month for 7 months showed that in controls not receiving ATP mean weight changes/month were –1.0 kg vs +0.2 kg in the ATP-treated group. Serum albumin (–1.3 g/l), elbow flexor muscle strength (–6%) were decreased in controls but were unchanged in the ATP-treated group. There was also less deterioration in quality of life measures with positive ATP effects on body weight, muscle strength, and albumin concentration being especially marked in cachectic patients.

In cystic fibrosis transmembrane conductance regulator homozygous and heterozygous nude mice, a decrease in breast tumor implantability was observed and ascribed to elevated blood ATP levels. Similarly, ATP could reduce humor breast tumor cell growth in vitro, supporting the concept of ATP as an antitumor agent (ABRAHAM et al. 1996).

Another P2X receptor designated as P2XM, which has high homology with the $P2X_6$ receptor subunit and can be induced by the tumor oncogene, p53, has been implicated in soft tissue tumor genesis (NAWA et al. 1999).

C. Challenges for Purine-Based Drug Discovery

The tremendous advances in the basic understanding of the receptors and signaling pathways involved in purinergic neurotransmission processes that have occurred over the past two decades have resulted in the identification of a plethora of potential disease states where agonists and antagonists of P1 and P2 receptors may have potential as novel therapeutic agents. Many of these have been reviewed in detail in the present volume of the *Handbook of Experimental Pharmacology* (151/I and II).

However, the key to producing a successful drug, beyond the fact that it needs to be efficacious in its targeted disease state, is that its use should provide a high margin of safety and, in addition, not have a major impact on the quality of life of the patient. The acceptable margin of safety, or therapeutic index, is in direct proportion to the severity of the disease and what other medications are available to treat the disease. The acceptable side effect profile for an agent that will be used acutely to treat cancer, a life-threatening disease, is very much different to a drug used on a chronic basis to treat a "lifestyle disease" like stress urinary incontinence or nonmorbid obesity.

As already noted, the majority of the research effort targeted toward compounds acting via purinergic receptors over the past 20 years has been focused on P1 receptor ligands. However, the transition of interesting P1 agonists and antagonists from basic in vitro studies and animal models approximating human disease states to the clinic has been very limited. Despite the considerable efforts of a number of established pharmaceutical companies (Abbott, Boehringer Ingelheim, Bristol Myers Squibb, Burroughs Wellcome, Byk Gulden, CIBA-Geigy, Glaxo, Kyowa Hakko, Marion/Marion Merrell Dow/ Hoechst Marion Roussel, Rhone Poulenc Rorer/Aventis, Merck, Nelson/ Whitby, Novo Nordisk, Parke-Davis, Pfizer, Schering-Plough, Takeda, Zeneca) in the P1 ligand area with the discovery of agonists like metrifudil, CV-1808, CI-936, GR 79236, NNC 21-0136, CGS 21680, and SDZ WAG 944, and antagonists including BWA-522, KW 3902, CP 66713, N-0861, CGS 15943, CGS 15943, FK 453, SCH 52861, KFM-19/BIIP 20, ZM 241385, PD 115199, MDL 102234, and L-268605, very few of these compounds entered clinical trials and nearly all those that did did not make it to the marketplace. Only adenosine has been approved for the treatment of supraventricular tachycardia (SVT) and as a cardiac imaging agent.

This lack of clinical success can be attributed to issues with side effect liabilities including sedation and hypotension as well as the choice of therapeutic target (e.g., hypertension, schizophrenia, stroke, asthma) where either superior competing compounds acting via nonpurinergic mechanisms exist or where the disease targeted, irrespective of the molecular approach, has an inherently high risk of failure, e.g., depression.

One success story from "big pharma," yet to be reflected in a marketed product but certainly showing promise in the clinic, is reflected in the ongoing efforts of Humphries, and colleagues at AstraZeneca in the UK in developing ADP antagonists like AR-C 69331-MX (INGALL et al. 1999) for use as novel antithrombotic agents. Taking basic research into platelet physiology (CUSACK and HOURANI 1990; HOURANI, Chap. 20, this volume) and building on Cusack's pioneering work on bioisosteres of adenine nucleotides (CUSACK et al. 1979), a series of modified nucleotide analogs were identified that showed an improved profile antithrombotic profile vs GPIIb/IIIa antagonists and with a safer side effect profile than aspirin (HUMPHRIES et al. 1996; JARVIS GE et al. 2000).

The development of adenosine as a human therapeutic agent is perhaps more of a case history for future activities in developing purinergic drugs (BELLARDINELLI et al. 1995). The basic enabling technology regarding the effects of adenosine on SVT were developed in an academic environment and subsequently reviewed in depth by large pharmaceutical companies who either decided that there was little in the way of unmet medical need where the purine nucleoside would fill a niche or that the fact that adenosine was not a proprietary molecule limited the return on the investment necessary to sustain the cost of the R & D effort. It was eventually left to a small pharmaceutical company that was in turn purchased by a Japanese major to commercialize these cardiovascular uses of adenosine effectively. A similar situation appears to exist for the compelling human data set on the use of adenosine as an anesthetic adjuvant and prophylatic analgesic (FUKUNAGA 1997).

The development of adenosine can be appropriately complemented by efforts associated with smaller, biotech companies like Medco Research King Pharmaceutints Scientific, Gensia/Metabasis, Adenosine Therapeutics, Discovery Therapeutics, Duska, Gilead, Inspire, Rine, CV Therapeutics, and SciClone. Much of the present effort at these companies is focused on developing P1 ligands for use in cardiac imaging or P1 antagonists for use in congestive heart failure. Among the more notable examples of compounds targeted at purinergic receptors are a compound related to BG9719, a xanthine P1 antagonist developed by CV Therapeutics that is being developed by Biogen for congestive heart failure and INS 365, a $P2Y_2$ receptor agonist, discovered and developed by Inspire in collaboration with the Boucher/Harden groups at UNC that is being developed in collaboration with Genentech and Kissei for the treatment of CF and COPD-related diseases.

The innovative, intellectual focus and ability of smaller biotech companies to work on disease targets that fall below the level of commercial interest of the larger pharmaceutical companies suggests that the optimal approach to drug discovery in the purinergic area may lie in the purview of smaller companies who can effectively bring academic concepts to proof of principle stage and may thus increasingly serve as the conduit between basic, academic research and big pharma (DREWS 1999); the latter can then focus its efforts and resources on the genomics/high throughput screening (HTS)/combinatorial chemistry approach to drug discovery.

However, the ability of smaller companies to identify novel pharmacophores even when part of a chemical library consortium, the paucity of potent, selective, bioavailable P2 receptor ligands, and the emerging complexity of the receptor targets also suggest that HTS approaches to identify novel pharmacophores may be one area in which big pharma can have a major impact on purine-based drug discovery research.

While much is known regarding the structure activity relationships for compounds interacting with P1 G-protein coupled receptors (GPCRs), recent evidence for the oligimerization of GPCRs (MILLIGAN 2000) and the emerging data on P2X heteromers and their interactions with other LGICs, e.g., neuronal nicotinic cholinergic receptors (SEARL et al. 1998; KHAKH et al. 2000) suggests that the dynamics and the actual composition of systems targeted by purinergic receptor ligands are potentially very complex. Unlike the interaction of acetylcholine with neuronal nicotinic cholinergic receptors (LLOYD and WILLIAMS 2000), very little is known regarding the nature of the ATP binding site on P2X receptors. Is it, like the neuronal nicotinic cholinergic receptor, formed on the interface between two subunits? What stoichiometry is required for receptor activation? – e.g., could the symmetrical structures of P2 receptor antagonists like suramin and the ability of dinucleotide polyphosphates to activate (Ap4A) and block (Ip^5I) P2X receptors indicate that interaction with two ATP receptors (not subunits) is required for function? The answers to these questions will require an increased focus on the biophysical aspects of receptor structure in conjunction with point mutation studies, the development of better P2 receptor ligands, and a better understanding of the allosteric properties of both P2X and P2Y receptors.

In addition to P1 and P2 receptors and enzymes involved in purine uptake and phosphorylation, the ecto-kinases that use ATP as a substrate may be drug discovery targets (REDEGOLD et al. 1999). For instance, platelets can be activated by inhibition of the phosphorylation state of CD36, a collagen/thrombospondin receptor (HATMI et al. 1999), while the PP2a phosphatase inhibitor, microcystin, which inhibits dephosphorylation inhibits ionophore-induced platelet aggregation (NAIK et al. 1991).

As with science in general, in the drug discovery process, as more is learnt regarding a molecular target, the more complex it becomes and open to side effect liabilities. Thus, in the purinergic receptor area, the initial disappointments in developing P1 receptor ligands has led to a pragmatic reconsideration of disease targets where the understanding of the basic physiology and pharmacology of the systems can be used to develop novel drugs. With the continuing polarization of the pharmaceutical industry into mega R & D companies at one end of the spectrum and biotechs at the other, it may be anticipated that the examples given above with adenosine, BG9719 and INS 365 will be repeated as more biotech companies focus on purinergic receptors as drug targets.

In the P2 area, the situation may be more complex given the need for a better understanding of the receptor targets, of the role of E-NTPases and ecto-ATPases in defining P2 ligand function, and the imperative to identify new ligands. Given a clearer understanding of the history of P1 receptor ligand drug discovery and a broader understanding of the challenges in drug discovery at the in vitro, in vivo, and human levels, it may be anticipated that the iterative generation of knowledge regarding P2 receptor function in normal and diseased tissues, using the receptor targets to define the properties of these ligands and vice versa, will result in the speedier prioritization of therapeutic targets that are amenable to modulation by P2 receptor ligands.

References

Abraham EH, Vos P, Kahn J, Grubmann SA, Jefferson DM, Ding I, Okunieff P (1996) Cystic fibrosis hetero- and homo-zygosity is associated with inhibition of breast cancer growth. Nature Med 2:593–596

Agteresch HJ, Dagnelie PC, van der Gaast A, Stijnen T, Wilson, JHP. Randomized clinical trial of adenosine 5′-triphosphate in patients with advanced non-small-cell lung cancer. J Natl Cancer Inst 92:321–328

Akhondzadeh S, Shasavand E, Jamilian HR, Shabestari O, Kamalipour A (2000) Dipyridamole in the treatment of schizophrenia: adenosine-dopamine receptor interactions. J Clin Pharm Ther 25:131–137

Baldwin SA, Mackey JR, Cass CE, Young JD (1999) Nucleoside transporters: molecular biology and implications for therapeutic development. Mol Med Today 5:216–224

Barnes PJ, Chung KF, Page CP (1998) Inflammatory mediators of asthma: an update. Pharmacol Rev 50:515–596

Belardinelli L, Lu J, Dennis D, Martens J, Shryock JC (1994) The cardiac effects of a novel A_1-adenosine receptor agonist in guinea pig isolated heart. J Pharmacol Exp Ther 271:1371–1382

Bellardinelli L, Shyrock JC, Song Y, Wang D, Shinivas M (1995) Ionic basis of the electrophysiological actions of adenosine on cardiomyocytes. FASEB J 9:359–365

Bennett WD, Olivier KN, Zeman KL, Hohneker KW, Boucher RC, Knowles MR (1996) Effect of uridine 5′-triphosphate plus amiloride on mucociliary clearance in adult cystic fibrosis. Am J Respir Crit Care Med 153:1796–1801

Bhagwat SS, Williams M (1997) P2 Purine and pyrimidine receptors: emerging superfamilies of G-protein coupled and ligand gated ion channel receptors. Eur J Med Chem 32:183–193

Biaggioni I, Feoktistov I, Sheller JR, Grant M, Belardinelli L, Wells JN (2000) Development of new selective A2B antagonists with greater potency than enprofylline. Drug Dev Res 50:104

Bjorck T, Gustafsson LE, Dahlem SE (1992) Isolated bronchi from asthmatics are hyperresponsive to adenosine which apparently acts indirectly by liberation of leukotrienes and histamine. Am Rev Respir Dis 145:1087–1091

Bland-Ward PA, Humphrey PPA (1997) Acute nociception mediated by hindpaw P2X receptor activation in the rat. Br J Pharmacol 122:365–371

Braiman A, Silberberg SD, Priel Z (2000) Purinergic stimulation of ciliary activity. Drug Dev Res 50:550–554
Brambilla R, Burnstock G, Bonazzi A, Ceruti S, Cattabeni F, Abbracchio MP (1999) Cyclo-oxygenase-2 mediates P2Y receptor-induced reactive astrogliosis. Br J Pharmacol 126:563–567
Boucher RC (2000) The physiologic and therapeutic roles of nucleotides in the lung in health and disease. Drug Dev Res 50:5
Burnstock G (1996) A unifying purinergic hypothesis for the initiation of pain. Lancet 347:1604 –1605
Burnstock G (1998) Purinergic signalling: therapeutic potential. Drug Dev Res 45:86–92
Burnstock G (1999) Release of vasoactive substances from endothelial cells by shear stress and purinergic mechanosensory transduction. J Anat 194:335–342
Burnstock G (2000) P2X receptors in sensory neurones. Br J Anaesth 84:476–488
Burnstock G, Cocks T, Crowe R, Kasakov I (1978) Purinergic innervation of the guinea-pig urinary bladder. Br J Pharmacol 63:125–138
Carley DW, Radulovacki M (1997) Adenosine effects in sleep apnea. In: Jacobson KA Jarvis MF (eds) Purinergic Approaches in Experimental Therapeutics. Wiley-Liss, New York, pp 515–525
Carley DW, Radulovacki M (1999) Role of peripheral adenosine A_1 receptors in the regulation of sleep apneas in rats. Exp Neurol 159:545–550
Chancellor MB, Kaplan SA, Blaivas JG (1992) The cholinergic and purinergic components of detrussor contractility in a whole rabbit bladder model. J Urol 148:906–909
Cockayne DA, Zhu Q-M, Hamilton S, Dunn PM, Zhong Y, Berson A, Kassotakis L, Bardini M, Muraski J, Novakovic S, Lachnit WG, Burnstock G, McMahon SB, Ford APDW (2000) $P2X_3$-deficient mice display urinary bladder hyporeflexia and reduced nocifensive behavior. Drug Dev Res 50:50
Collo G, Neidhart S, Kawashima E, Kosco-Vilbois M, North RA, Buell G (1997) Tissue distribution of the $P2X_7$ receptor. Neuropharmacol 36:1277–1283
Cressman VL, Lazorowski E, Homolya L, Boucher RC, Koller BH, Grubb BR (1999) Effect of loss of $P2Y_2$ receptor gene expression on nucleotide regulation of murine epithelial Cl^- transport. J Biol Chem 274:26461–26468
Cronstein BN, Montesinos MC, Weissman G (1999) Salicylates and sulfasalazine, but not glucocorticoids, inhibit leukocyte accumulation by an adenosine-dependent mechanism that is independent of prostaglandin synthesis and p105 of NFκB. Proc Natl Acad Sci USA 96:6377–6381
Cusack NJ, Hourani SMO (1990) Structure activity relationships for adenine nucleotide receptors on mast cells, human platelets, and smooth muscle. In: Jacobson KA, Daly JW, Manganiello V (eds) Purines in Cellular Signalling: Targets for New Drugs. Springer, New York, pp 254–259
Cusack NJ, Hickman ME, Born GVR (1979) Effects of D- and L-enantiomers of adenosine, AMP and ADP and their 2-choro and 2-azido-analogues on human platelets. Proc Roy Soc Lond B 206:139–144
Cushley MJ, Tattersfield AE, Holgate ST (1983) Inhaled adenosine and guanosine on airway resistance in normal and asthmatic subjects. Br J Clin Pharmacol 15:161–165
Daly JW (1982) Adenosine receptors. Targets for future drugs. J Med Chem 25: 197–207
Dean DM, Downie JW (1978) Contribution of adrenergic and "purinergic" neurotransmission to contraction in rabbit detrusor. J Pharmacol Exp Ther 207:430–437
Desai A, Montesinos MC, Cronstein BN (2000) CGS-21680. An adenosine A_{2A} receptor agonist, enhances wound healing in normal mice and urokinase plasminogen activator knockout (UPA-KO) mice but not in tissue plasminogen activator knockout (TPA-KO) mice. Drug Dev Res 50:71

Dirnagl U, Iadecola C, Moskowitz MA (1999) Pathobiology of ischaemic stroke: an integrated view. Trends Neurosci 22:391–397
Di Virgilio F, Sanz JM, Chiozzi P, Falzoni S (1999) The P2Z/P2X$_7$ receptor of microglial cells: a novel immunomodulatory receptor. Prog Brain Res 120:355–370
Dixon SJ, Sims SM (2000) P2 purinergic receptors on osteoblasts and osteoclasts: potential targets for drug development. Drg Dev Res 49:187–200
Drews J (1999) In Quest of Tomorrow's Medicines. Springer- Verlag, New York
Driver AG, Kukoly CA, Ali S, Mustafa SJ (1993) Adenosine in bronchoalveolar lavage fluid in asthma. Amer Rev Respir Dis 148:91–97
During MJ, Spencer DD (1992) Adenosine: a potential mediator of seizure arrest and postictal refractoriness. Ann Neurol 32:618–624
Eagle KA, Boucher RC (1989) Intravenous dipyridamole infusion causes severe bronchospasm in asthmatic patients. Chest 95:258–259
Elzein EO, Palle V, Gothe SA, Li Z, Gao Z, Meyer S, Blackburn B, Bellardinelli L, Zablocki J (2000) Design, synthesis and biological evaluation of 2-(4-substituted-N-pyrazolyl)-adenosine derivatives as novel short acting adenosine A_{2A} receptor agonists. Drug Dev Res 50:64
Fabre JE, Nguyen M, Latour A, Keifer JA, Audoly LP, Coffman TM, Koller BH (1999) Decreased platelet aggregation, increased bleeding time and resistance to thromboembolism in P2Y$_1$-deficient mice. Nature Med 5:1199–1202
Feoktistov I, Polosa R, Holgate ST, Biaggioni I (1998) Adenosine A_{2B} receptors: a novel therapeutic target in asthma? Trends Pharmacol Sci 19:148–153
Ferre S, Fredholm BB, Morrelli M, Popoli P, Fuxe K (1997) Adenosine-dopamine receptor-receptor interactions as an integrative mechanism in the basal ganglia. Trends Neurosci 20:482–487
Feuerstein GZ, Wang X, Barone FC (1997) Inflammatory gene expression in cerebral ischemia and trauma: potential new therapeutic targets. Ann NY Acad Sci 825:179–193
Firestein GS (1996) Anti-inflammatory effects of adenosine kinase inhibitors in acute and chronic inflammation. Drug Dev Res 39:371–376
Firestein GS, Paine MM, Boyle DL (1994) Mechanisms of methotrexate action in rheumatoid arthritis. Selective decrease in synovial collagenase gene expression. Arth Rheum 37:193–200
Fischer Y, Becker C, Loken C (1999) Purinergic inhibition of glucose transport in cardiomyocytes. J Biol Chem 274:755–761
Ford MS, Nie Z, Whitworth C, Rybak LP, Ramkumar V (1997) Up-regulation of adenosine receptors in the cochlea by cisplatin. Hearing Res 111:143–152
Fredholm BB, Bättig K, Holmén J, Nehlig A, Zvartau EE (1999) Actions of caffeine in the brain with special reference to factors that contribute to its widespread use. Pharmacol Rev 51:83–133
Fukunaga AF (1997) Purines in anesthesia. In: Jacobson KA, Jarvis MF (eds) Purinergic Approaches in Experimental Therapeutics. Wiley-Liss, New York, pp 471–494
Fuxe K, Ungerstedt U (1974) Action of caffeine and theophylline on supersensitive dopamine receptors: considerable enhancement of receptor responses to treatment with dopa and dopamine. Med Biol 52:48–54
Geiger JD, Parkinson FE, Kowaluk E (1997) Regulators of endogenous adenosine levels as therapeutic agents. In: Jacobson KA, Jarvis MF (eds) Purinergic Approaches in Experimental Therapeutics. Wiley-Liss, New York, pp 55–84
Gür S, Öztürk B (2000) Altered relaxant responses to adenosine and adenosine 5′-triphosphate in the corpus cavernosum from men and rats with diabetes. Pharmacology 60:105–112.
Hashimoto M, Kokubun S (1995) Contribution of P2-purinoceptors to neurogenic contraction of rat urinary bladder smooth muscle. Br J Pharmacol 115:636–640
Hatmi M, Gavaret JM, Elalamy I, Vargaftig BB, Jacquemin C (1996) Evidence for cAMP-dependent platelet ectoprotein kinase activity that phosphorylates platelet glycoprotein IV (CD36). J Biol Chem 271:24776–24780

Hauber W, Koch M (1997) Adenosine A_{2A} receptors in the nucleus accumbens modulate prepulse inhibition of the startle response. Neuroreport 8:1515–1518
Hollopeter G, Jantzen H-M, Vincent D, Li G, England L, Ramakrishnan V, Yang R-B, Nurden P, Nurden A, Julius D, Conley PB (2000) Identification of the Platelet ADP Receptor Targeted by Antithrombotic Drugs. Nature, in press
Hone RM, Ritchie WT, Thompson WAR (1930) The action of adenosine upon the human heart. Quart J Med 23:485–490
Housley GD (1998) Extracellular nucleotide signalling in the inner ear. Mol Neurobiol 16:21–48
Hu BH, Zheng XY, McFadden SL, Kopke RD, Henderson D (1997) R-phenylisopropyladenosine attenuates noise-induced hearing loss in the chinchilla. Hearing Res 113:198–206
Humphries RG, Leff P, Robertson MJ (1996) P_{2T}-purinoceptor antagonists: a novel class of anti-thrombotic agents. Drug Dev Res 37:175 Abstract
Hutchison AJ, Webb RL, Oei HH, Ghai GR, Zimmerman MB, Williams M (1989) CGS 21680 C, an A_2 selective adenosine receptor agonist with preferential hypotensive activity. J Pharmacol Exp Ther 251:47–55
IJzermann AP, van der Wenden NM (1997) Modulators of adenosine uptake, release, and inactivation. In: Jacobson KA, Jarvis MF (eds) Purinergic Approaches in Experimental Therapeutics. Wiley-Liss, New York, pp 129–148
Ingall AH, Dixon J, Bailey A, Coombs ME, Cox JD, McInally JI, Hunt SF, Kindon ND, Teobald BJ, Willis PA, Humphries RG, Leff P, Clegg JA, Smith JA, Tomlinson W (1999) Antagonists of the platelet P_{2T} receptor: A novel approach to antithrombotic therapy. J Med Chem 42:213–220
Jacobson KA (1998) Adenosine A_3 receptors: novel ligands and paradoxical effects. Trends Neurosci 19:184–191
Jacobson KA, van Rhee AM (1997) Development of selective purinoceptor agonist and antagonists. In: Jacobson KA, Jarvis MF (eds) Purinergic Approaches in Experimental Therapeutics. Wiley-Liss, New York, pp 101–128
Jacobson KA, Williams M (2001) Perspective: P2 receptors as drug targets. J Med Chem (In preparation)
Jarvis GE, Nassim MA, Humphries RG, Kirk IP, Tomlinson W, Cusworth EA, Midha A, Perrett JH, Mobbs EG, Hammersley MD, Watts IS (2000) Superior inhibition of ADP-induced human platelet aggregation by AR-C69931MX than clopidogrel. Drug Dev Res 50:90
Jarvis MF, Yu H, Kohlhaas K, Alexander K, Lee CH, Jiang M, Bhagwat SS, Williams M, Kowaluk EA (2000) ABT 702 (4-amino-5-(3-bromophenyl)-7-(6-morpholino-pyridin-3-yl)pyrido [2,3,-d] pyrimidine), a novel orally effective adenosine kinase inhibitor with analgesic and anti-inflammatory properties. I. In vitro characterization and acute antinociceptive effects in the mouse. J Pharmacol Exp Ther 295, in press
Jezer A, Oppenheimer BS, Schwartz SP (1933) The effect of adenosine on cardiac irregularities in man. Am Heart J 9:252–258
Kanda T, Jackson MJ, Smith LA, Pearce RK, Nakamura J, Kase H, Kuwana Y, Jenner P (1998) Adenosine A_{2A} antagonist: a novel antiparkinsonian agent that does not provoke dyskinesia in parkinsonian monkeys. Ann Neurol 43:507–513
Karsten R, Gordh TJ (1996) An A_1-selective adenosine agonist abolishes allodynia elicited by vibration and touch after intrathecal injection. Anesth Analg 80: 844–847
Katchanov G, Xu J, Schulman ES, Pelleg A (1998) ATP-triggered reflex bronchoconstriction. Drug Dev Res 45:342–349
Kenny BA, Bushfield M, Parry-Smith DJ, Fogarty S, Treherne JM (1998) The application of high throughput screening to novel lead discovery. Prog Drug Res 41:246–269
Khakh BS, Zhou X, Sydes J, Galligan JJ, Lester HA (2000) State-dependent cross inhibition between transmitter gated cation channels. Nature 406:405–410

Kujawa SG, Erosteguli C, Fallon M, Christ J, Bobin RP (1984) Effects of adenosine 5′-triphosphate and related agonists on cochlear function. Hearing Res 76:87–100
Knowles MR, Clarke LL, Boucher RC (1991) Activation by extracellular nucleotides of chloride secretion in the airway epithelia of patients with cystic fibrosis. N Engl J Med 8:533–538
Knutsen LJS, Murray TF (1997) Adenosine and ATP in epilepsy. In: Jacobson KA, Jarvis MF (eds) Purinergic Approaches in Experimental Therapeutics. Wiley-Liss, New York, pp 423–447
Knutsen LJS, Sheardown MJ, Roberts SM, Mogensen JP, Olesen UB, Thomsen C, Bowler AN (1998) Adenosine A_1 and A_3 selective *N*-alkoxypurines as novel cytokine modulators and neuroprotectants. Drug Dev Res 45:214–221
Lammas DA, Stober C, Harvey CJ, Kendrick N, Panchalingam S, Kumararatne DS (1997) ATP-induced killing of mycobacteria by human macrophages is mediated by purinergic P2Z ($P2X_7$) receptors. Immunity 7:433–444
Ledent C, Vaugeois JM, Schiffmann SN, Pedrazzini T, El Yacoubi M, Vanderhaeghen JJ, Costentin J, Heath JK, Vassart G, Parmentier M (1997) Aggressiveness, hypoalgesia and high blood pressure in mice lacking the adenosine A_{2a} receptor. Nature 388:674–678
Leon C, Hechler B, Freund M, Eckly A, Vial C, Ohlmann P, Dierich A, LeMur M, Cazenave J-P, Gachet C (1999) Defective platelet aggregation and increased resistance to thombosis in purinergic $P2Y_1$ receptor-null mice. J Clin Invest 104:1731–1737
Linden J (1994) Cloned adenosine A_3 receptors: pharmacological properties, species differences and receptor functions. Trends Pharmacol Sci 15:298–306
Linden J, Williams M (2001) The academic/biotechnology/pharmaceutical interface in drug discovery impact on purinergic receptor-based therapeutics. Drug Dev Res 52, in press
Liu G-S, Downey JM, Cohen MC (1997) Adenosine, ischemia and preconditioning. In: Jacobson KA, Jarvis MF (eds) Purinergic Approaches in Experimental Therapeutics. Wiley-Liss, New York, pp 153–172
Lloyd GK, Williams M (2000) Neuronal nicotinic receptors as novel drug targets. J Pharmacol Exp Ther 292:461–467
Loubatrieres-Mariani M-M, Hillaire-Buys D, Chapal J, Betrand G, Petit P (1997) P2 purinoceptor agonists: new insulin secretogogues potentially useful in the treatment of non-insulin-dependent diabetes mellitus. In: Jacobson KA, Jarvis MF (eds) Purinergic Approaches in Experimental Therapeutics. Wiley-Liss, New York, pp 253–260
Lutz M, Kenakin TP (1999) Quantitative Molecular Pharmacology and Informatics in Drug Discovery. Wiley, Chichester, UK, pp 1–31
Mangano DT (1997) Effects of acadesine on myocardial infarction, stroke, and death following surgery. A meta-analysis of the 5 international randomized trials. The Multicenter Study of Perioperative Ischemia (McSPI) Research Group. JAMA 277:325–332
Marquardt DL, Parker CW, Sullivan TJ (1978) Potentiation of mast cell release by adenosine. J Immunol 120:871–878
Martin GE, Rossi D, Jarvis MF (1993) Adenosine agonists reduce conditioned avoidance responding in the rat. Pharmacol Biochem Behav 45:951–958
McQueen DS, Riberio JA (1986) Pharmacological characterization of the receptor involved in chemoexcitation induced by adenosine. Br J Pharmacol 88:615–620
McQueen DS, Bond SM, Moores C, Chessell I, Humphrey PP, Dowd E (1998) Activation of P2X receptors for adenosine triphosphate evokes cardiorespiratory reflexes in anaesthetized rats. J Physiol 507:843–855
McVey M, Smits GJ, Cox BF, Kitzen J, Clark KL, Perrone MH (1999) Cardiovascular pharmacology of the adenosine A_1/A_2 receptor agonist AMP 579: coronary hemodynamic and cardioprotective effects in the canine myocardium. J Cardiovasc Pharmacol 33:703–710

Milligan G (2000) Receptors as kissing cousins. Science 288:65–67

Montesinos MC, Gadangi P, Longaker M, Sung J, Levine J, Nilsen D, Reibman J, Li M, Jiang CK, Hirschhorn R, Recht PA, Ostad E, Levin RI, Cronstein BN (1997) Wound healing is accelerated by agonists of adenosine A_2 (G αs- linked) receptors. J Exp Med 186:1615–1620

Moreland RB, Nakane M, Hsieh G, Brioni JD (2000) Prospectives for pharmacotherapy of male erectile dysfunction. Curr Opin CPNS Invest Drugs 2:283–202

Morrison MS, Turin L, King BF, Burnstock G, Arnett TR (1998) ATP is a potent stimulator of the activation and formation of rodent osteoclasts. J Physiol (Lond). 511:495–500

Mulryan K, Gitterman DP, Lewis CH, Vial C, Leckie BJ, Cobb AL, Brown JE, Conley EC, Buell G, Pritchard CA, Evans RJ (2000) Reduced vas deferens contraction and male infertility in mice lacking $P2X_1$ receptors. Nature 403:86–89

Naik UP, Kornecki E, Ehrlich YH (1991) Phosphorylation and dephosphorylation of human platelet surface proteins by an ecto-protein kinase/phosphatase system. Biochem Biophys Acta 1092:256–264

Nawa G, Miyoshi Y, Yoshikawa H, Ochi T, Nakamura Y (1999) Frequent loss of expression or aberrant alternative splicing of P2XM, a p53-inducible gene, in soft tissue tumors. Br J Cancer 80:1185–1189

Neary JT, Rathbone MP, Cattabeni F, Abbracchio MP, Burnstock G (1996) Trophic actions of extracellular nucleotides and nucleosides on glial and neuronal cells. Trends Neurosci 19:13–18

Nehlig A (1999) Are we dependent upon coffee and caffeine? A review on human and animal data. Neurosci Biobehav Rev 23:563–576

Nichols CG, Shyng S-L, Nestorowicz A, Glaser B, Clement IV JP, Gonzalez G. Aguilar-Bryan L, Permutt MA, Bryan J Adenosine diphosphate is an intracellular regulator of insulin secretion. Science 272:1785–1787

Nyce JW (1999) Insight into adenosine receptor function using antisense and gene-knockout approaches Trends Pharmacol. Sci 20:79–83

Olsson RA, Pearson JD (1990) Cardiovascular purinoceptors. Physiol Rev 70:761–845

O'Reilly BA, Knight G, Popert R, Rymer JM, Burnstock G, McMahon SB (1999) The role of P2X receptors in the overactive bladder. J Urol 163(4 Suppl):41

Ottery FD, Walsh D, Strawford A (1998) Pharmacologic management of anorexia/cachexia. Semin Oncol 25:35–44

Owall A, Lagenkramer M, Sollevi A (1988) Effects of adenosine induced hypotension on myocardial hemodynamics and metabolism during cerebral aneurysm surgery Anesth Analg 67:228–232

Ozeck MJ, Liang HX, Belardinelli L, Shyrock JC (2000) A_1 adenosine receptor activation increases leptin production by rat white adipose tissue. Drug Dev Res 50: 93

Parsons M, Young L, Lee JE, Jacobson KA, Liang BT (2000) Distinct cardioprotective effects of adenosine mediated by differential coupling of receptor subtypes to phospholipases C and D. FASEB J 14:1423–1431

Patat A, Rosenzweig P, Enslen M, Trocherie S, Miget N, Bozon M-C, Alalin H, Gandon JM (2000) Effects of a new slow release formulation of caffeine on EEG, psychomotor and cognitive functions in sleep-deprived subjects. Human Psychopharmacol. Clin Exp 15:153–170

Pelleg A, Hurt CM (1996) Mechanism of ATP's action on canine pulmonary vagal fibre nerve terminals. J Physiol (Lond.) 490:265–275

Picano E, Abbracchio MP (1998) European Stroke Prevention Study-2 results: serendipitous demonstration of neuroprotection induced by endogenous adenosine accumulation? Trends Pharmacol Sci 19:14–16

Pinna C, Puglis C, Burnstock G (1998) ATP and vasoactive intestinal polypeptide relaxant responses in hamster isolated proximal urethra. Br J Pharmacol 124:1069–1074

Porkka-Heiskanen T, Strecker RE, Thakkar M, Bjorkum AA, Greene RW, McCarley RW (1997) Adenosine: a mediator of the sleep-inducing effects of prolonged wakefulness. Science 276:1265–1268

Ralevic V, Burnstock G (1998) Receptors for purines and pyrimidines. Pharmacol Rev 60:413–492

Rapaport E (1997) ATP in the treatment of cancer. In: Jacobson KA, Jarvis MF (eds) Purinergic Approaches in Experimental Therapeutics. Wiley-Liss, New York, pp 545–553

Rapaport E, Fontaine J (1989) Anticancer activities of adenone nucleotides in mice are mediated through expansion of erythrocyte ATP pools. Proc Natl Acad Sci USA 86:1662–1666

Rathbone MP, Middlemiss PJ, Gysbers J, Diamond J, Holmes M, Pertens E, Juurlink BH, Glasky A, Ritzmann R, Glasky M, Crocker CE, Ramirez JJ, Lorenzen A, Fein T, Schulze E, Schwabe U, Ciccarelli R, Di Iorio P, Caciagli F (1998) Physiology and pharmacology of natural and synthetic nonadenine- based purines in the nervous system. Drug Dev Res 46:356–372

Rathbone MP, Middlemiss PJ, Gysbers JW, Andrew C, Herman MAR, Redd JK, Ciccarelli R, Di Iorio P, Caciagli R (2000) Trophic effects of purines in neurons and glial cells. Prog Neurobiol 59:663–690

Redegeld FA, Caldwell CC, Sitkovsky MV (1999) Ecto-protein kinases: ecto-domain phosphorylation as a novel target for pharmacological manipulation? Trends Pharmacol Sci 20:453–459

Richardson PJ, Kase H, Jenner PG (1997) Adenosine A_{2A} receptor antagonists as new agents for the treatment of Parkinson's disease. Trends Pharmacol Sci 18:338–344

Rudolphi KA, Schubert P, Parkinson FE, Fredholm BB (1992) Neuroprotective role of adenosine in cerebral ischaemia. Trends Pharmacol Sci 13:439–445

Satoh S, Matsumura H, Suzuki F, Hayasihi O (1996) Promotion of sleep mediated by the A_{2a}-adenosine receptor and possible involvement of this receptor in the sleep induced by prostaglandin D2 in rats. Proc Natl Acad Sci USA 93:5980–5984

Sattin A, Rall TW (1970) The effect of adenosine and adenine nucleotides on the cyclic adenosine 3′,5′-monphosphate content of guinea pig cerebral cortex slices. Mol Pharmacol 6:13–23

Sawynok J (1999) Purines in pain management. Curr Opin CPNS Drugs 1:37–38

Schaumann E, Kutscha W (1972) Clinical-pharmacological studies with a new orally active adenosine derivative. Drug Res 22:783–790

Schulman ES, Glaum MC, Post T, Wang Y, Raible DG, Mohanty J, Butterfield JH, Pelleg A (1999) ATP modulates anti-IgE-induced release of histamine from human lung mast cells. Am J Respir Cell Molecular Biol 20:530–537

Searl TJ, Redman RS, Silinsky EM (1998) Mutual occlusion of P2X ATP receptors and nicotinic receptors on sympathetic neurons of the guinea-pig. J Physiol (Lond) 510:783–791

Sikora A, Liu J, Brosnan C, Buell G, Chessel I, Bloom BR (1999) Cutting Edge: Purinergic signaling regulates radical-mediated bacterial killing mechanism in macrophages through a $P2X_7$-independent mechanism. J Immunol 163:558–561

Sollevi A (1986) Cardiovascular effects of adenosine in man: possible clinical implications. Prog Neurobiol 27:319–349

Shimegi S (1996) ATP and adenosine act as a mitogen for osteoblast-like cells (MC3T#-E1). Calcif Tissue Inter 58:109–113

Stutts MJ, Boucher RC (1999) Cystic fibrosis gene and functions of CFTR implications of dysfunctional ion transport for pulmonary pathogenesis. In: Yankaskas JR, Knowles MR (eds) Cystic Fibrosis in Adults. Lippincott-Raven, Philadelphia, pp 3–25

Sugidachi A, Asai F, Ogawa T, Inoue T, Koike H (2000) The in vivo pharmacological profile of CS-747, a novel antiplatelet agent with platelet ADP receptor antagonist properties. Br J Pharmacol 129:1439–1446

Svenningsson P, Le Moine C, Fisone G, Fredholm BB (1999) Distribution, biochemistry and function of striatal adenosine A_{2A} receptors. Prog Neurobiol 59:355–396

Takahashi Y, Ishii N, Lue TF, Tanagho EA (1993). Effects of adenosine on canine penile erection. J Urol 148:1323–1325

Thorne PR, Housley GD (1996) Purinergic signalling in sensory systems. Seminar Neurosci 8:233–246

Triggle DJ (1998) Chemical Diversity, Current Protocols Pharmacol. Wiley, New York, 9.0.1–9.0.18

Tsuchida A, Yang XM, Burckhartt B, Mullane KM, Cohen MV, Downey JM (1994) Acadesine extends the window of protection afforded by ischaemic preconditioning. Cardiovasc Res, 28:379–383

Von Lubitz DKJE (1997) Adenosine and acute treatment of cerebral ischemia and stroke – "put out more flags". In: Jacobson KA, Jarvis MF (eds) Purinergic Approaches in Experimental Therapeutics. Wiley-Liss, New York, pp 349–470

Vrettos N, Steiner M (1999) Unleashing managerial advantage in pharmaceutical research and development. Drugs Made In Germany 42:13–19

Wang P, Ba ZF, Morrison MH, Ayala A, Dean RE, Chaudry IH (1992) Mechanism of the beneficial effects of ATP-MgCl2 following trauma-hemorrhage and resuscitation: down regulation of inflammatory cytokine (TNF, IL-6) release. J Surg Res 52:364–371

Webster RG, Abbott RD, Petrovitch H, Morens DM, Grandinetti A, Tung K-H, Tanner CM, Maskai KH, Blanchette PL, Curb DJ, Popper JS, White LR (2000) Association of coffee and caffeine with the risk of Parkinson Disease. J Amer Med Assoc 283:2674–2679

Williams M (1996) Challenges in developing P_2 purinoceptor based therapeutics. In: Burnstock G (ed) P_2-Purinoceptors Localization, function and transduction mechanisms. CIBA Foundation Symposium 198, John Wiley, Chichester, UK, pp 309–321

Williams M, Burnstock G (1997) Purinergic neurotransmission and neuromodulation: a historical perspective. In: Jacobson KA, Jarvis MF (eds) Purinergic Approaches in Experimental Therapeutics Wiley-Liss, New York, pp 3–26.

Williams M, Jarvis MF (2000) Purinergic and pyrimidinergic receptors as potential drug targets. Biochem Pharmacol 59:1173–1185

Williams M, Kowaluk EA, Arneric SP (1999) Perspective: Emerging Molecular Approaches to Pain. Therapy J Med Chem 42:1481–1500

Woolf AA, Skettino S, Beckman E, Belardinelli L (1998) Renal effects of BG9719, a specific A_1 adenosine receptor antagonist, in congestive heart failure. Drug Dev Res 45:166–171

Woolf CJ, Salter MW (2000) Neuronal plasticity: increasing the gain in pain. Science 288:1765–1768

Wu HY, Broderick GA, Suh JK, Hypolite JA, Levin RM (1993) Effects of purines on rabbit corpus cavernosum contractile activity. Int J Impot Res 5:161–167

Yerxa BR (2000) Therapeutic use of nucleotides in respiratory and ophthalmic diseases. Drug Dev Res 50:22

Yoshimura N, de Groat WC (1997) Neural control of the lower urinary tract. Inter J Urol 4:111–125

Young D, Dragunow M (1994) Status epilepticus may be caused by loss of adenosine anticonvulsant mechanisms. Neurosci 58:245–261

Zhao Z-Q, Clark KL, Wang NP, Velez DA, Guyton RA, Vinten-Johansen J (2000) Comparison of AMP579 and adenosine in inhibition of cell-cell interaction between human neutrophil and vascular endothelial cell. Drug Dev Res 50:266–272

Zimmerman H, Braun N (1999) Ecto-nucleotidases – molecular structures, catalytic properties, and functional roles in the nervous system. Prog Brain Res 120:371–385

Subject Index

Printing: Mercedes-Druck, Berlin

Printing (Computer to Film): Saladruck, Berlin
Binding: H. Stürtz AG, Würzburg

MIX
Papier aus verantwortungsvollen Quellen
Paper from responsible sources
FSC® C105338

If you have any concerns about our products,
you can contact us on
ProductSafety@springernature.com

In case Publisher is established outside the EU,
the EU authorized representative is:
Springer Nature Customer Service Center GmbH
Europaplatz 3, 69115 Heidelberg, Germany

Printed by Libri Plureos GmbH
in Hamburg, Germany